Remote Sensing Techniques for Landslides Studies and Their Hazards Assessment

Remote Sensing Techniques for Landslides Studies and Their Hazards Assessment

Editor

Rachid El Hamdouni

Basel • Beijing • Wuhan • Barcelona • Belgrade • Novi Sad • Cluj • Manchester

Editor
Rachid El Hamdouni
University of Granada
Granada
Spain

Editorial Office
MDPI
St. Alban-Anlage 66
4052 Basel, Switzerland

This is a reprint of articles from the Special Issue published online in the open access journal *Remote Sensing* (ISSN 2072-4292) (available at: https://www.mdpi.com/journal/remotesensing/special_issues/landslides_rs_geo).

For citation purposes, cite each article independently as indicated on the article page online and as indicated below:

Lastname, A.A.; Lastname, B.B. Article Title. *Journal Name* **Year**, *Volume Number*, Page Range.

ISBN 978-3-7258-1045-1 (Hbk)
ISBN 978-3-7258-1046-8 (PDF)
doi.org/10.3390/books978-3-7258-1046-8

© 2024 by the authors. Articles in this book are Open Access and distributed under the Creative Commons Attribution (CC BY) license. The book as a whole is distributed by MDPI under the terms and conditions of the Creative Commons Attribution-NonCommercial-NoDerivs (CC BY-NC-ND) license.

Contents

Kyrillos M. P. Ebrahim, Sherif M. M. H. Gomaa, Tarek Zayed and Ghasan Alfalah
Recent Phenomenal and Investigational Subsurface Landslide Monitoring Techniques: A Mixed Review
Reprinted from: *Remote Sens.* **2024**, *16*, 385, doi:10.3390/rs16020385 1

Arthur Charléty, Mathieu Le Breton, Eric Larose and Laurent Baillet
2D Phase-Based RFID Localization for On-Site Landslide Monitoring
Reprinted from: *Remote Sens.* **2022**, *14*, 3577, doi:10.3390/rs14153577 58

Ioannis Farmakis, Efstratios Karantanellis, D. Jean Hutchinson, Nicholas Vlachopoulos and Vassilis Marinos
Superpixel and Supervoxel Segmentation Assessment of Landslides Using UAV-Derived Models
Reprinted from: *Remote Sens.* **2022**, *14*, 5668, doi:10.3390/rs14225668 76

Yifei He and Yaonan Zhang
Comparison of Three Mixed-Effects Models for Mass Movement Susceptibility Mapping Based on Incomplete Inventory in China
Reprinted from: *Remote Sens.* **2022**, *14*, 6068, doi:10.3390/rs14236068 95

Ali Bounab, Younes El Kharim and Rachid El Hamdouni
The Suitability of UAV-Derived DSMs and the Impact of DEM Resolutions on Rockfall Numerical Simulations: A Case Study of the Bouanane Active Scarp, Tétouan, Northern Morocco
Reprinted from: *Remote Sens.* **2022**, *14*, 6205, doi:10.3390/rs14246205 123

Yiting Gou, Lu Zhang, Yu Chen, Heng Zhou, Qi Zhu, Xuting Liu, et al.
Monitoring Seasonal Movement Characteristics of theLandslide Based on Time-Series InSAR Technology: The Cheyiping Landslide Case Study, China
Reprinted from: *Remote Sens.* **2023**, *15*, 51, doi:10.3390/rs15010051 144

Zongji Yang, Bo Pang, Wufan Dong and Dehua Li
Spatial Pattern and Intensity Mapping of Coseismic Landslides Triggered by the 2022 Luding Earthquake in China
Reprinted from: *Remote Sens.* **2023**, *15*, 1323, doi:10.3390/rs15051323 164

Belizario A. Zárate, Rachid El Hamdouni and Tomás Fernández del Castillo
Characterization and Analysis of Landslide Evolution in Intramountain Areas in Loja (Ecuador) Using RPAS Photogrammetric Products
Reprinted from: *Remote Sens.* **2023**, *15*, 3860, doi:10.3390/rs15153860 185

María Camila Herrera-Coy, Laura Paola Calderón, Iván Leonardo Herrera-Pérez, Paul Esteban Bravo-López, Christian Conoscenti, Jorge Delgado, et al.
Landslide Susceptibility Analysis on the Vicinity of Bogotá-Villavicencio Road (Eastern Cordillera of the Colombian Andes)
Reprinted from: *Remote Sens.* **2023**, *15*, 3870, doi:10.3390/rs15153870 224

Muhammad Afaq Hussain, Zhanlong Chen, Ying Zheng, Yulong Zhou and Hamza Daud
Deep Learning and Machine Learning Models for Landslide Susceptibility Mapping with Remote Sensing Data
Reprinted from: *Remote Sens.* **2023**, *15*, 4703, doi:10.3390/rs15194703 262

Sanshao Ren, Yongshuang Zhang, Jinqiu Li, Zhenkai Zhou, Xiaoyi Liu and Changxu Tao
Deformation Behavior and Reactivation Mechanism of the Dandu Ancient Landslide Triggered by Seasonal Rainfall: A Case Study from the East Tibetan Plateau, China
Reprinted from: *Remote Sens.* **2023**, *15*, 5538, doi:10.3390/rs15235538 . . . . **292**

Maria Francesca Ferrario and Franz Livio
Rapid Mapping of Landslides Induced by Heavy Rainfall in the Emilia-Romagna (Italy) Region in May 2023
Reprinted from: *Remote Sens.* **2024**, *16*, 122, doi:10.3390/rs16010122 . . . . **309**

Review

Recent Phenomenal and Investigational Subsurface Landslide Monitoring Techniques: A Mixed Review

Kyrillos M. P. Ebrahim [1,2], Sherif M. M. H. Gomaa [1,2,*], Tarek Zayed [1] and Ghasan Alfalah [3]

1 Department of Building and Real Estate, Faculty of Construction and Environment, The Hong Kong Polytechnic University, Hung Hom, Kowloon, Hong Kong SAR 999077, China
2 Structural Engineering Department, Faculty of Engineering, Mansoura University, Mansoura 35516, Egypt
3 Department of Architecture and Building Sciences, King Saud University, Riyadh 145111, Saudi Arabia
* Correspondence: sherif.gomaa@polyu.edu.hk

Citation: Ebrahim, K.M.P.; Gomaa, S.M.M.H.; Zayed, T.; Alfalah, G. Recent Phenomenal and Investigational Subsurface Landslide Monitoring Techniques: A Mixed Review. *Remote Sens.* **2024**, *16*, 385. https://doi.org/10.3390/rs16020385

Academic Editors: Alex Hay-Man Ng and Rachid El Hamdouni

Received: 24 November 2023
Revised: 3 January 2024
Accepted: 8 January 2024
Published: 18 January 2024

Copyright: © 2024 by the authors. Licensee MDPI, Basel, Switzerland. This article is an open access article distributed under the terms and conditions of the Creative Commons Attribution (CC BY) license (https:// creativecommons.org/licenses/by/ 4.0/).

Abstract: Landslides are a common and challenging geohazard that may be caused by earthquakes, rainfall, or manmade activity. Various monitoring strategies are used in order to safeguard populations at risk from landslides. This task frequently depends on the utilization of remote sensing methods, which include the observation of Earth from space, laser scanning, and ground-based interferometry. In recent years, there have been notable advancements in technologies utilized for monitoring landslides. The literature lacks a comprehensive study of subsurface monitoring systems using a mixed review approach that combines systematic and scientometric methods. In this study, scientometric and systematic analysis was used to perform a mixed review. An in-depth analysis of existing research on landslide-monitoring techniques was conducted. Surface-monitoring methods for large-scale landslides are given first. Next, local-scale landslide subsurface monitoring methods (movement, forces and stresses, water, temperature, and warning signs) were examined. Next, data-gathering techniques are shown. Finally, the physical modeling and prototype field systems are highlighted. Consequently, key findings about landslide monitoring are reviewed. While the monitoring technique selection is mainly controlled by the initial conditions of the case study, the superior monitoring technique is determined by the measurement accuracy, spatiotemporal resolution, measuring range, cost, durability, and applicability for field deployment. Finally, research suggestions are proposed, where developing a superior distributed subsurface monitoring system for wide-area monitoring is still challenging. Interpolating the complex nonlinear relationship between subsurface monitoring readings is a clear gap to overcome. Warning sign systems are still under development.

Keywords: landslide monitoring; subsurface monitoring; investigational monitoring; wireless monitoring; early warning monitoring; real-time monitoring

1. Introduction

The practice of landslide monitoring is the systematic observation and collection of data to enhance the understanding and analysis of this geological event. Any effective monitoring methodology should include the following goals: consistent and systematic data collection, the use of appropriate equipment, accurate timing of measurements, and the use of proper analytic techniques (i.e., how to interpret the collected data). These goals can respond to the following questions: (1) what has to be monitored (such as displacement, stress, and pore water pressure), (2) the number of devices to be utilized, and (3) the frequency and data collection methods. These goals and inquiries may be used to establish the budget, resources, planning, and monitoring system [1].

Geotechnical investigations have been conducted for years to discover the stability conditions of slopes under various geological and environmental circumstances [2,3]. To answer the first aforementioned question, landslide monitoring is used to track and measure slope stability parameters, such as ground movement (surface movement, subsurface movement, heights, and cracks), subsurface water conditions (depth of water table, pore water pressure, soil suction, and soil moisture), and climatic parameters (rainfall, snowfall, temperature, and humidity). These factors can subsequently be used in landslide prediction approaches, which were not the focus of this work [4,5]. The number, type, and location of sensors are determined by the local geology, subsurface conditions, and landslide area in answer to the second question [6]. Concerning the third question, comprehending the geographical and temporal distributions of these factors is critical for realizing landslide dynamics and controlling the associated risk [7].

Determining the most effective monitoring system requires a thorough understanding of the reasons that generate events (initial conditions). For instance, the use of tilt measurement may not be suitable for translational landslides or slow-moving landslides since the occurrence of tilting is improbable under such conditions. Similarly, when deep soil underneath an installation site becomes saturated, it might lead to landslides, which can damage topsoil moisture sensors [8]. Another example is if the effective rainfall value is the cumulative value of one day, then collecting data at 15 min intervals may not be necessary [1]. Landslides are classified as shallow or deep-seated based on the depth of the slip surface. Both of these types of landslides have distinct features and produce varying degrees of damage. As a result, defining the type of landslide and estimating the potential risk of a prospective landslide by measuring the depth of the sliding surface are both necessary [1]. The monitoring of landslides is divided into phenomena, investigation, and performance categories. The change in the slope over time in a particular geologic location is monitored using phenomena. To ascertain the temporal and physical parameters of an identified landslide, investigation monitoring is performed. A stabilizing system that is already in place can be evaluated for efficacy via performance monitoring [1].

Monitoring systems can be classified into surface and subsurface techniques [9]. The former cannot follow internal changes, but the latter can. Thus, this research focused mostly on subsurface monitoring approaches, where the optimal criteria for a monitoring system, according to the literature, should have the following features: provide real-time data; high sensitivity; high spatiotemporal resolution; cost-effectiveness; low power consumption; reliability; scalability; not affected by signal noise, such as temperature effects; limit the uncertainty caused by missing data; and be suitable for both shallow and deep landslides, as well as harsh environment conditions (i.e., the device should be coated) [10–12].

Both scientometric and systematic methodologies are covered in this paper. This paper is organized as follows: The research technique is presented in Section 2. The scientometric analysis is highlighted in Section 3. The systematic analysis is emphasized in Section 4, which is divided into four subsections: surface displacement, subsurface monitoring, wireless sensing networks, and physical and prototype systems. Section 5 lists the research gaps and future directions. Section 6 presents the conclusion and future recommendations.

According to the author's knowledge and available data, Table 1 shows various review studies that investigated landslide-monitoring techniques. Many of them focused on a specific methodology and approach. There is a lack of review publications on subsurface monitoring techniques. Scientometric analysis has rarely been used. As a result, this study's uniqueness may be summarized as follows:

(1) A mixed scientometric and systematic review is presented.
(2) All existing subsurface-monitoring technologies (movement, forces and stresses, groundwater, temperature, and warning signs) were comprehensively addressed in this study.
(3) A deep illustration of the data-transferring techniques is included (i.e., manual, wiring, wireless).
(4) A detailed demonstration of the adopted physical laboratory and field-monitoring systems is presented.
(5) This article presents the most recent research up until 2023.

Table 1. Available review articles for landslide-monitoring techniques.

Study	Year	Approach	Content
Angeli et al. [2]	2000	Systematic	Discussing the management, problems, and solutions of different systems.
Shamshi [13]	2004	Systematic	Landslide-monitoring instruments were reviewed briefly.
Eyo et al. [14]	2014	Systematic	Applications of low-cost GPS tools.
De Graff [1]	2011	Systematic	Illustrating how to obtain and build a better monitoring system.
Chae et al. [15]	2017	Systematic	Landslide prediction, monitoring (remote sensing and in situ based), and early warning.
So et al. [16]	2021	Systematic	LiDAR applications in Hong Kong.
Lapenna & Perrone [17]	2022	Systematic	Discussing time-lapse electrical resistivity tomography applications.
Breglio et al. [18]	2023	Systematic	The uses of photonic technology for monitoring deformation (slopes and tunnels), temperature, and soil humidity for agricultural soil.
Huang et al. [19]	2023	Systematic	Real-time monitoring using GNSS.
Auflič et al. [20]	2023	Bibliometric	Landslide-monitoring techniques based on questionnaire analysis.

This study examined the progression from one approach to another through a macroscopic view based on the technology utilized and the initial conditions, followed by a microscopic demonstration of the different system characteristics.

2. Research Methodology

A mixed review strategy was employed in this study, which consisted of scientometric and systematic techniques. The methodology is provided to help researchers improve systematic review reporting through the use of scientometric analysis. Furthermore, it highlights the complexity of conducting manual searches on database engines [21–24].

Identifying, screening, and qualifying are the three main steps of a systematic review, as shown in Figure 1. The steps involved in doing scientometric analysis are shown in Figure 2. These steps typically include collecting bibliometric data, exporting it to the suitable software, evaluating it, and finally, discussing the findings.

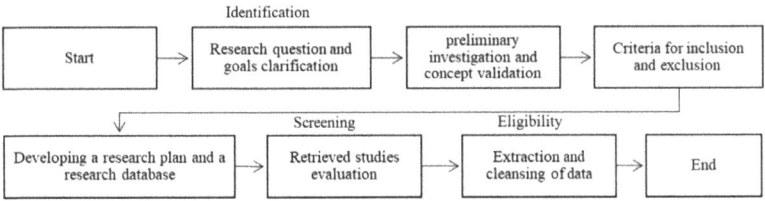

Figure 1. Systematic review process.

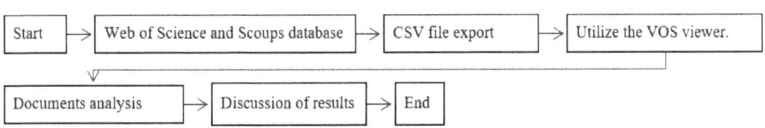

Figure 2. Scientometric analysis process, where CSV refers to comma-separated values text files and VOSviewer is an open-source software application (van Eck & Waltman, 2009) [25].

2.1. Identification Process

Geology, engineering, environmental science, ecology, meteorology, atmospheric science, geochemistry and geophysics, physical science, and water resources are some of the aspects used to study landslides [24]. Furthermore, as shown in Zou and Zheng's [24] keyword mapping, landslides have a large number of linked terms. As a consequence, the research method began by extracting important studies about landslides from the author's perspective. In this section, keywords, search databases, and inclusion and exclusion criteria were utilized to filter the papers acquired.

2.2. Selection of Database and Keywords

It is advisable to select numerous databases in a systematic review to obtain and review a thorough selection of relevant publications. Scopus, Web of Science, and Google Scholar are the three most often used databases in engineering research. Scopus and Web of Science are also compatible with modern scientific mapping programs, such as VOSviewer. In this study, we only used Scopus and Web of Science as preliminary search database sources for landslide monitoring, although Google Scholar was also employed in the snowballing approach. Following the selection of a search database, relevant keywords were chosen, namely, "landslide monitoring", to take into account all accessible datasets for monitoring approaches.

2.3. Inclusion and Exclusion Criteria

In any systematic review, inclusion and exclusion are crucial for limiting search results and focusing on the most relevant ones. This research used the following criteria: (1) research focusing on subsurface landslide monitoring, (2) studies published between 2000 and 2023, (3) articles published in peer-reviewed journals, and (4) studies published as articles and review submissions. The exclusion criteria were as follows: (1) papers published in a language other than English, (2) studies with no full text accessible, (3) manuscripts published in a subject area other than engineering, and (4) publications published in a source type other than a journal.

2.4. Screening and Evaluation of Collected Articles

As of May 2023, the Scopus and Web of Science databases revealed a total of 173 and 98 articles, respectively. The selected publications were then evaluated and assessed using the systematic reviews and meta-analyses (PRISMA) process (see Figure 3) [26]. Following this method, 143 papers were eliminated because they were duplicated, irrelevant, or did not have a complete text accessible. After reviewing the whole texts of each included article, 128 articles met the inclusion criteria. The backward and forward snowballing methods were then used to find more studies that were not found using Scopus or Web of Science searches [27]. In addition to the manual search, this search method yielded 26 more relevant articles, for a total of 154 articles suitable for inclusion.

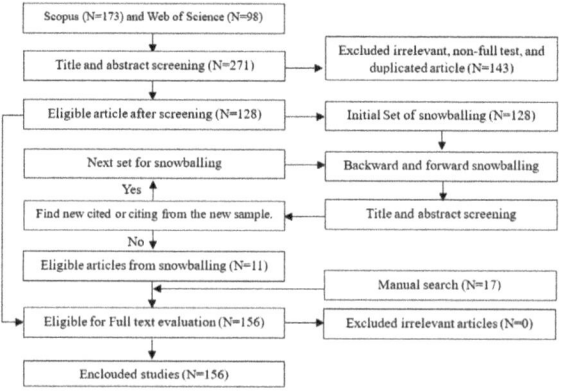

Figure 3. PRISMA screening and selection process diagram.

3. Scientometric Analysis

The scientometric study was conducted with the open-source VOSviewer software application version 1.6.20 [25]. This scientometric review was used to provide citation and co-authorship analyses of nations, organizations, authors, and keywords involved in the study topic, as shown in the following subsections. The resulting maps, networks, and analyses (i.e., the VOS output; please refer to Sections 3.1–3.5) highlight the link between these many aspects, whereas tables were mainly utilized to illustrate the statistics associated with these network maps. VOSviewer software was used to assess the 154 articles retrieved via snowballing and manual searching. The primary goal of scientometric analysis is to guarantee that the findings are relevant enough to be included in a systematic review.

3.1. Landslide Monitoring Annual Publications

Figure 4 depicts the overall number of landslide-monitoring-related papers published each year. From 2000 to 2016, the average yearly publishing rate was approximately two articles. Between 2017 and 2023 (until May 2023), the publishing rate increased significantly to record an average annual publishing rate of approximately 15 articles. The figure's second-degree polynomial trend line (refer to the trend equation presented in Figure 4) depicts how landslide monitoring has evolved. This trend is not surprising considering the world's growing concern over the loss of human life, property, and economic resources due to landslides.

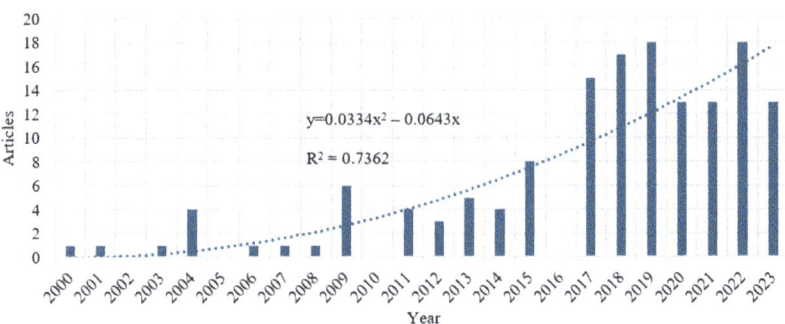

Figure 4. The total number of papers published each year about landslide monitoring.

3.2. Top Journals Contributing to Landslide Monitoring

The VOSviewer tool can now highlight the journals that frequently publish landslide-monitoring articles. As a consequence, this will help researchers choose a reputable journal in this field. When utilizing the VOSviewer program, the author employed two thresholds: a source has to include at least five documents and at least 10 citations. "Sources" were employed as the unit of analysis, and "bibliographic coupling" was the type of analysis. As a result, 7 journals out of 62 hit the threshold (Figure 5). In Figure 5, the node size illustrates the influence of journals as weighted by the number of citations. The overall link strength of a journal represents the number of links it has with other journals [25]. *Engineering Geology* was the most widely published and cited journal and had 449 citations and 12 publications.

3.3. Active Nations and Institutes in Landslide Monitoring

Understanding the scientific collaboration network makes it simpler to identify top laboratories, organizations, and nations. Furthermore, academic and industry practitioners seeking innovative landslide solutions should be aware of the cooperation network of nations investing more in this field. The aforementioned criteria were utilized, using "countries" as the unit of analysis and "bibliographic coupling" as a type of analysis. Only 10 of the 43 nations met the criterion. Figure 6 depicts the most frequently publishing nations, with China, the United States, and Italy having the most publications globally, with 50, 23, and 23 articles, respectively. Table 2 reports the top five institutions involved in landslide monitoring by using "bibliometric coupling" as the kind of analysis and "organization" as the unit of analysis. With five papers and 76 citations, the most frequently contributing institute was the School of Civil Engineering of Chongqing University, China.

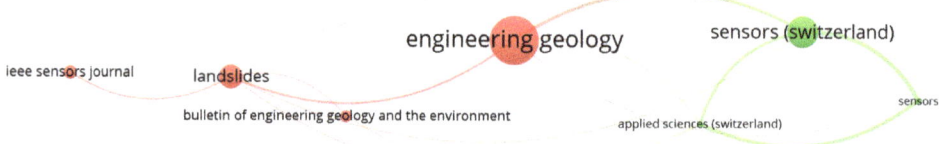

Figure 5. Top journals publishing in landslide monitoring.

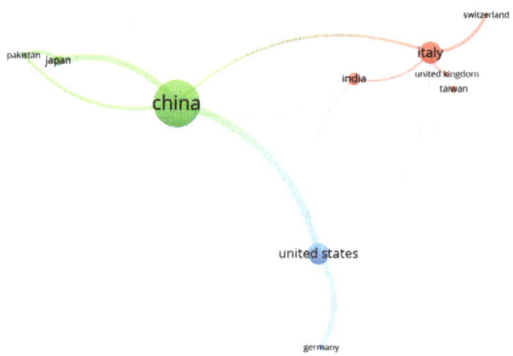

Figure 6. Top countries publishing in landslide prediction.

Table 2. Top five institutions publishing in the field of landslide monitoring.

Organization	Country	Articles	Citations
School of Civil Engineering of Chongqing University	China	5	76
Key Laboratory of New Technology for Construction of Cities in Mountain Areas, Chongqing University	China	4	67
Dept. of Civil Engineering/Research Center for Hazard Mitigation and Prevention, National Central University, Zhongda	Taiwan	3	28
Department of Civil Engineering, University of Tokyo	Japan	2	57
State Key Laboratory of Hydroscience and Engineering, Tsinghua University, Beijing	China	2	57

3.4. Active Scholars and Article Co-Citation Analysis in Landslide Monitoring

An author's total number of publications and citations on a certain topic can be used to calculate their influence on that topic. The top five authors based on the number of publications and citations were assembled using Excel Microsoft 365 software, as shown in Table 3. To solve the issue of older research obtaining more citations than more recent research, a normalized citation metric was also utilized in this study. The number of citations in an article was normalized by dividing the number of citations by the average number of citations in all publications published that year [25]. As a result, Table 4 lists the top five publications based on normalized citations.

Table 3. The top five authors on the subject of landslide monitoring.

Authors	Documents	Citations
Giri P.; Ng K.; Phillips W.	3 [8,28,29]	62
Seguí, C. and Veveakis, M.	2 [30,31]	8
Huisman, J. A., Hubbard, S. S., Redman, J. D. and Annan, A. P.	1 [32]	728
Iai Susumu	1 [33]	595
Babaeian, E., Sadeghi, M., Jones, S. B., Montzka, C., Vereecken, H. and Tuller, M.	1 [34]	230

Table 4. The top ten most cited publications based on normalized citations.

Study	Journal	Citation	Normalized Citations
Babaeian et al. [34]	Reviews of Geophysics	230	6.50
Zhang et al. [35]	Landslides	7	4.81
Chae et al. [15]	Geosciences Journal	199	4.60
Iverson [36]	Geomorphology	214	4.33
Buurman et al. [37]	IEEE Access	60	3.48

3.5. Co-Occurrence Mapping of Keywords in Landslide Prediction

By selecting "co-occurrence" as the kind of analysis and "all keywords" as the unit of analysis, VOSviewer software version 1.6.20 can identify the most frequently used keywords (i.e., the keywords used in literature). In this analysis, the author fixed the minimum number of occurrences at 10; only 17 keywords out of 1050 matched this requirement. The size of a keyword node correlates to its occurrence frequency. To illustrate, the most commonly used terms were "landslides" and "landslide monitoring," which had the largest node sizes of any keywords. Figure 7 highlights that there were three distinct clusters: green, blue, and red. The blue clusters show monitoring related to rainfall and soil moisture; the red clusters highlight keywords linked to monitoring sensors and wireless networks; and the green clusters highlight subsurface displacement monitoring systems, such as optical fiber techniques.

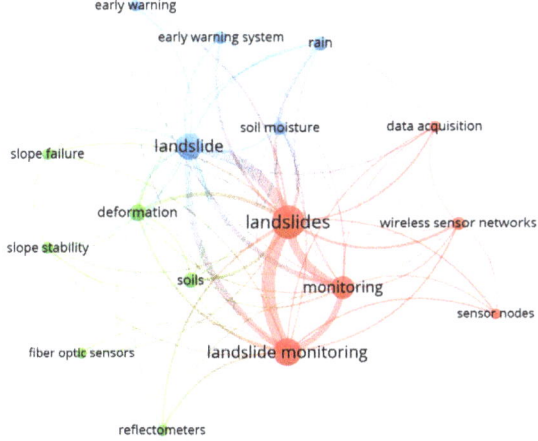

Figure 7. Keyword mapping in landslide monitoring weighted by occurrence.

4. Systematic Review of Monitoring Techniques

Landslide monitoring can be divided into two main categories: (1) surface- and (2) subsurface-monitoring techniques. In the following sections, both surface and subsurface monitoring are illustrated in detail. Figure 8 lists the surface-monitoring techniques, while Figure 9 presents the subsurface-monitoring process, including the testing procedure and data transfer mechanism. This paper first discusses the existing surface-monitoring approaches (refer to Figure 8 and Section 4.1), which have been well evaluated in the literature (refer to Table 1). Then, an in-depth investigation of subsurface procedures (such as movement, forces and stresses, water and temperature, and warning techniques) is presented. The warning approaches are subsurface monitoring devices that simply give a warning indication and no quantifiable data. The subsurface monitoring system is consistent with data-collection challenges, as well as prototype and physical modeling systems that are assessed to present a complete picture of such a topic (see Figure 9).

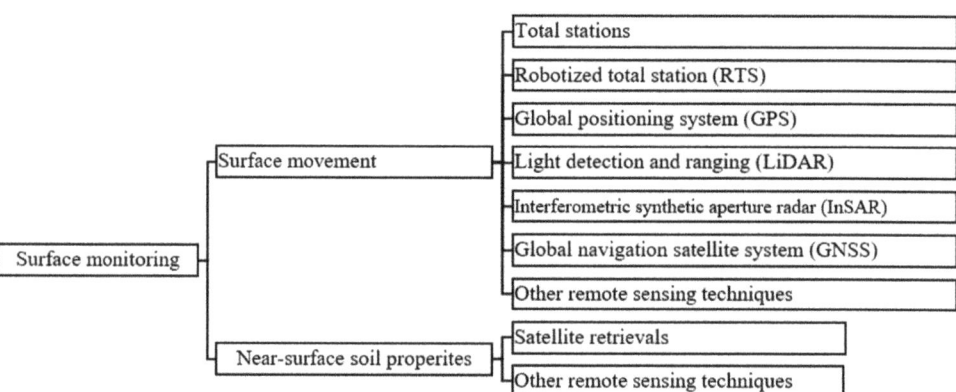

Figure 8. Surface-monitoring techniques.

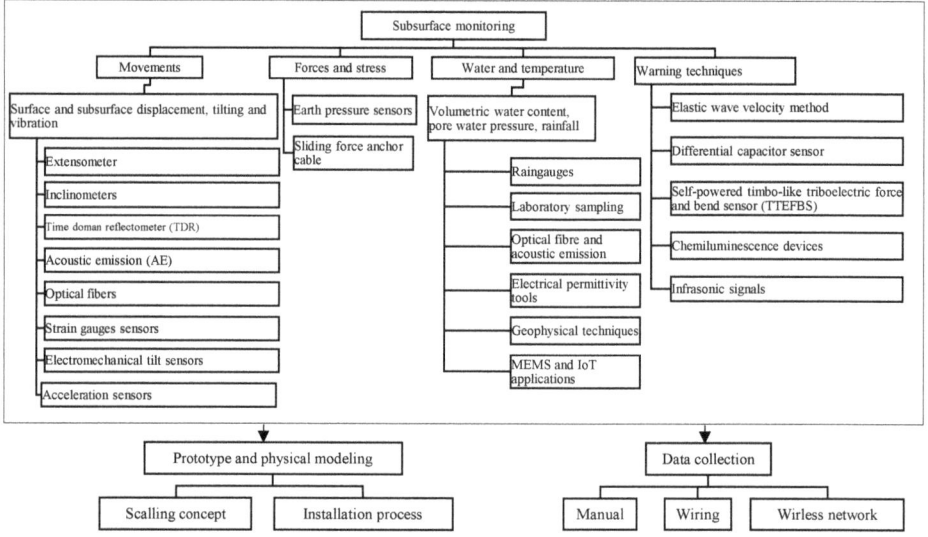

Figure 9. The subsurface-monitoring processes.

4.1. Surface Displacements

Surface displacement can be measured using various techniques, such as total stations, global positioning system (GPS) [38], robotized total station (RTS) [39], light detection and ranging (LiDAR) [16], synthetic aperture radar (SAR), interferometric synthetic aperture radar (InSAR) [40], persistent scatterer interferometry (PS-InSAR) [41], differential synthetic aperture radar (SAR) interferometry (DInSAR) [42], ground-based InSAR (GB-InSAR) [43], terrestrial laser scanning [44], global navigation satellite system (GNSS) [19], aerial photography [45], and satellite remote sensing techniques [46]. These techniques can only provide information about ground movements and are useful for wide-area surveillance. However, such instruments cannot determine the subsurface physical mechanism of landslides [14–16,19,20,47,48].

The GPS technique works on the basis that GPS satellites give navigation positioning signals for space resection measurement, hence calculating the 3D coordinate of the measuring point. However, high-power radio-transmitting stations and high-voltage transmission lines have a significant impact on GPS [9]. Furthermore, one GPS monitoring site costs approximately USD 6000 for a single device [49]. A robotized total station (RTS) is beneficial for distributing information about the present landslide condition and can give near real-time data, such as the ADVICE system [50]. However, false alarms owing to data inconsistencies caused by instrument faults, physical changes at the measurement location, and/or extremely local/shallow reactivations are always possible [50].

Remote sensing techniques (space-borne, aerial, and terrestrial surveys) can monitor broad regions without physical contact with the ground, though these technologies are expensive, have low resolution, and have difficulty collecting real-time data [7]. Although InSAR offers a better spatial resolution than GPS, it is hampered by atmospheric delay. Although PS-InSAR, which is an advanced radar interferometric measurement type that is representative of DInSAR, offers good accuracy, it is impacted by shadows and dense vegetation [9]. In satellite- and airborne-based SAR applications, the technique of differential synthetic aperture radar (SAR) interferometry (DInSAR) has been utilized to monitor vast regions of longer-distance landslides. DInSAR-based systems estimate displacements in millimeters by measuring phase changes between pairs of ground pictures acquired at various time intervals. The drawbacks are that the monitoring time intervals are excessive, ranging from hours (airborne) to weeks (satellite), and that daily or hourly monitoring is costly. Ground-based SAR (GB-InSAR), which is utilized over ranges ranging from a few hundred meters to a few kilometers, was created to alleviate the aforementioned difficulties. However, when a large bandwidth signal (for a high resolution in the range direction) is employed, a costly instrument is required [51,52]. The 3D laser scanning method has the added advantage of rapidly collecting (every 5 min) field deformation topography data with high accuracy and resolution [53]. However, the performance of a laser-light-based device is also influenced by weather conditions, such as severe fog or snow/rain [51]. By employing radio waves to scan a large area of the slope and provide temporal pictures, radar devices can track the movement of the slope. Nevertheless, there are drawbacks to using radar systems, such as the inability to monitor the slope in the event of snowfall or rain or when the line of sight (LoS) between the scanning device and the target slope is blocked. The technology is also useless for providing real-time warnings of sudden movements (i.e., seconds) since it takes several minutes to hours to scan the slope and interpret the photos to detect changes in the slope state [28].

A global navigation satellite system (GNSS) has been suggested to eliminate the requirement for line of sight (LOS) and to offer high-precision 3D monitoring. However, this technology has a significant maintenance cost and time requirements, as well as the presence of a single point of failure [54]. LiDAR-derived digital elevation models (DEMs) can quantify minor displacements across broad regions. Nevertheless, choosing an appropriate DEM resolution (i.e., pixel size, grid resolution, grid size) for constructing susceptibility maps is sometimes difficult since the scale of observation influences the evaluation, results, and interpretation [55].

The gradual degradation of slope stability generates landslides, and the sliding surface plays a vital role in landslide evolution. To illustrate, the landslide initiation is generated from the subsurface deep layers: only when the slope mass changes sufficiently can the slope surface deform macroscopically [56]. Surface monitoring systems may detect millimeter-scale deformation and can monitor wide regions with good spatial resolution and 3D capabilities, which is appropriate for landslide susceptibility, vulnerability, and risk maps for planning. The considerable time these systems need to spend returning to the same location, however, prevents these systems from offering real-time monitoring [57], and is not adequate for rapid landslides [29,58]. Therefore, there is a need for improvements in the small-scale subsurface monitoring of landslides [59,60]. To demonstrate this, Mucchi et al. [54] compared wireless sensor networks for ground instability monitoring (Wi-GIM) with RTSs and GB-InSAR to illuminate the fact that such sensor networks suffer from durability, precision, environmental impact, and maximum measuring range issues, and thus, further improvement is needed, as presented in Table 5. Surface displacement techniques are widely discussed in the literature (refer to Table 1), while this study mainly focused on subsurface monitoring systems.

Table 5. A comparison between Wi-GIM, RTS, and GB-In-SAR [54].

Case	Wi-GIM	RTS	GB-In-SAR
Cost (area = 100,000 m^2)	EUR 5220	EUR 18,150	EUR 58,100
Environmental impact	Good	Very good	Very good
Installation effort	Excellent	Good	Very good
Influence of rain/snow	Very good/very good	Good/good	Poor/very good
Completeness of measurement	Very good	Excellent	Excellent
Durability	Fair	Good	Excellent
Precision	Fair (2–3 cm)	Very good	Excellent
Maximum range	Fair	Excellent	Excellent

4.2. Subsurface Monitoring

Landslide deep displacement monitoring, where landslide initiation begins, is important for early warning forecasting and stability assessment [4,5,9]. In addition to displacement monitoring, subsurface monitoring techniques provide the added benefit of tracking internal forces, stress, moisture content, and temperature changes. Furthermore, such methods can provide early signs for emergencies.

4.2.1. Movement-Monitoring Devices

Extensometer Device

A conventional wire extensometer can provide a continual check of surface movement that may lead to a landslide. During emergencies, data can be obtained at regular intervals of 1–3 h, yet during routine situations, measurement intervals are 6 h. However, to obtain meaningful readings, the wire must be continually tensioned [2]. The quantity and rate of movement are measured and calculated manually within a centimeter range. However, key events might be missed if measurements are not obtained on time. To overcome the aforementioned issues, potentiometric extensometers detect displacement using a variable resistance mechanism, where a movable arm makes electrical contact along a fixed resistance strip. This type has the advantage that the wiring can be buried [61]. Crawford et al. [62] used a cable-extension transducer, which is a stainless-steel cable connected to a potentiometer housed in a protective casing, where the voltage output is then transformed to a linear absolute displacement. Fibreglass extensometers were initially placed (drilled horizontally in boreholes) in the S landslide to provide more precise data [2]. This type of extensometer is suited for rock slide applications since it can detect movement in the millimeter range [2]. Setiono et al. [63] created optical-based wire extensometers with an optical rotary encoder to count optical pulse signals and transform them into length units (refer to Figure 10). This approach offers a high resolution of 0.011 ± 0.0083 mm and a speed limit of approximately 36 mm/s.

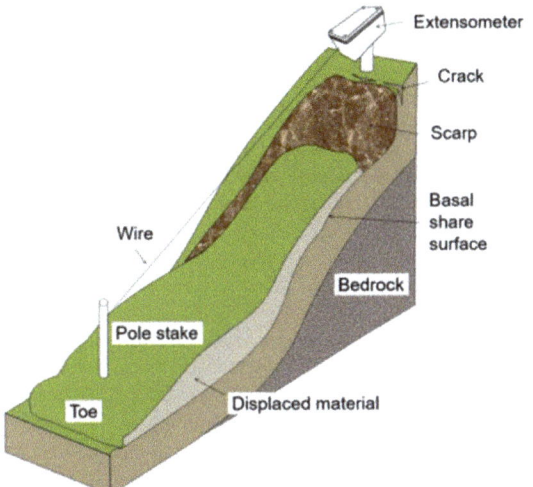

Figure 10. Schematic view of the extensometer system (From Setiono et al. [63]).

Nevertheless, the wire extensometer has the drawback of collecting data at the landslide surface, making it hard to analyze the deep displacement distribution, and being overly expensive, costing approximately EUR 1000 for a single monitoring site [64]. Moreover, this technique is a single-point measuring technique and cannot provide distributed monitoring [8,65]. With technological advancement, wire extensometers may now deliver real-time and high-resolution measurements. Wire extensometers can be linked to particular data logging units and can be combined with other sensors for landslide dynamic analysis. While this technique is more suited for translational landslides, it can additionally be used in roto-translational landslides and has been validated in field experiments with land shifts ranging from 12 mm to 150 mm [63].

Inclinometers

Compared with extensometers, inclinometers have the benefit of measuring deep displacement with a spatial vertical resolution of 0.5–1 m [66,67]. Measurements are collected regularly by installing a single inclinometer into grooved vertical pipes installed in deep boreholes to analyze their deformation. Later investigations employed numerous analog inclinometers or a series of digital in-place inclinometers positioned at different depths inside these pipes for continuous measurement. Inclinometers, however, are difficult to install, laborious, lack sensitivity, and are vulnerable to environmental dangers [10,61]. Using an inclinometer to determine the precise location of sliding surfaces is limited by the spatial vertical resolution [68,69], especially when the shear band thickness is small [35]. Automating inclinometers is impractical because the wiring restricts the number of inclinometers that may be installed in a region, resulting in limited area spatial resolution [70]. This approach is impracticable for measuring significant lateral deflections for two reasons: the limited displacement range [9] and the high expense of guide casing (approximately 30 USD/m) [47] (600 USD/inclinometer) [71]. Electric-powered inclinometers are the most often used equipment for measuring subsurface displacements. However, in real applications, this technique (i.e., electric-powered inclinometers) suffers from limited stability and durability, poor resistance to electromagnetic interference, high gravity dependency, and significant signal loss for long-distance transmission [72].

Thus, inclinometers are appropriate for landslides that move very slowly to slowly [73] and have a thick shear bandwidth (refer to Figure 11), for which a lengthy monitoring interval and low spatial resolution would be sufficient. Intelligent monitoring for landslides has been widely studied [74]. Recently, numerous research studies have been conducted to overcome the inclination drawbacks by improving the spatiotemporal resolution, lowering the cost, giving real-time data, and enabling wireless data transmission.

Time Domain Reflectometry (TDR)

TDR is a relatively new method that, similar to radar, employs a coaxial cable and a cable tester (refer to Figure 11) [70,71]. A TDR device is made up of a TDR step pulse generator, an oscilloscope (or receiver), and a transmission line coupled to a multiplexer for multipoint and multifunction usage through various types of sensing waveguides [69,75]. An electrical pulse is sent down a coaxial cable that has been grouted into a borehole by the cable tester. The pulse is reflected when it encounters a crack or distortion in the cable. The reflection is represented by a "spike" in the cable signature. The relative magnitude and rate of displacement, as well as the position of the deformation zone, can be measured instantly and precisely [69]. Lin et al. [70] examined TDR behavior using laboratory and numerical simulations in an attempt to quantify TDR displacement. The main assumptions and findings can be summarized as follows: (1) TDR looks useless for quantifying shear displacement unless the shear mode is fixed (for example, if a sliding surface exists between soft soil and the bedrock layer). To fix the shear mode in the sensor cable, a hard, brittle grout with low tensile strength can be utilized. (2) The relationship between soil and grout stiffness has no effect on the TDR response. (3) Achieving high-strength grout is preferable because the sliding force required to kink the sensing cable is greater than the grout strength. Ho et al. [68] quantified the relationship between the horizontal displacement and reflection coefficients with an R^2 of 0.93 through laboratory and field tests. Chung and Lin [69] used recent literature findings to construct a field prototype monitoring system with the following model characteristics: (1) water/cement ratio = 1 to improve the cable–grout–soil interface contact; (2) sand and gravel were suggested to be mixed into the grout cement when grout loss occurs; and (3) the spatial resolution was 5 cm, which was higher than that of conventional inclinometers and can determine the location of sliding surfaces at different depths. Chung and Lin [69] found that TDR can work with inclinometers (IN) to allow for more precise geological and mechanical modeling of a landslide, determining the amount and direction of shear deformation. In this context, placing TDR wires outside IN casings was considered for economic reasons [76]. Using a high-gravity centrifuge, Chung et al. [75,77] enhanced the applicability of TDR. A flexible coaxial wire was modified to increase its sensitivity for detecting small-scale shear displacement (0.5 mm).

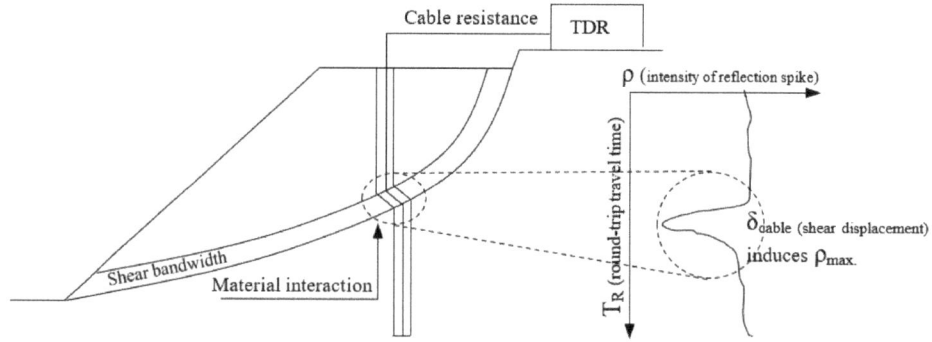

Figure 11. TDR deformation mechanism and affecting factors "Reprinted/adapted with permission from Chung & Lin [69], Elsevier".

This approach, however, faces challenges when quantifying the amount of displacement [68,78]. This is because numerous factors influence displacement, including (1) cable resistance, (2) soil–grout–cable contact, and (3) interaction and shear bandwidth. As a result, each cable has unique calibration measurement features. TDR is not suitable for multi-landslide failure zones [9]. TDR is not recommended for fast-flowing landslides or difficult-to-access steep slopes [29]. This method works best on rock slopes, and it is less effective on soft soils [70]. Reflection interference from closely spaced sliding surfaces requires additional investigation.

For the following reasons, this system is preferable over inclinometers: (1) low cost (in the United States, high-quality coaxial cable costs 13.5 USD/m, and the connection for installing each monitor hole costs USD 100.35) [47], (2) automated real-time data-collecting capability [9], (3) high spatial resolution to detect the exact location of the sliding surface (0.05 m), (4) TDR is capable of capturing the dynamics of shear deformation due to its unique characteristic of high temporal resolution (minute range) [69], and (5) the capability of measuring small displacements (0.5 mm) [77].

Acoustic Emission (AE)

The majority of AE monitoring studies are qualitative, determining the status of a slope based on the level of AE. A passive waveguide (i.e., grouted waveguide) is typically used for rock slope monitoring, whereas an active waveguide is used in soil slope monitoring by employing a steel pipe and granular backfill. The ringdown count (RDC), which is a frequent AE characteristic, is the number of times the AE signal amplitude surpasses the preset voltage threshold throughout a period. A certain frequency band of 20–30 kHz, which is the dominant frequency range produced from an active waveguide, is where AE signals are solely gathered to remove external noise [56,79,80]. Previous research employed metal tubes, which are problematic for large deformations because they are prone to failure from shear or bending. Deng et al. [56] created a unique flexible device to measure large movement (i.e., >500 mm) and quantified the deformation caused by AE using experimental shear testing in which a rubber tube was inserted into the borehole and passed through the sliding surface, as presented in Figure 12.

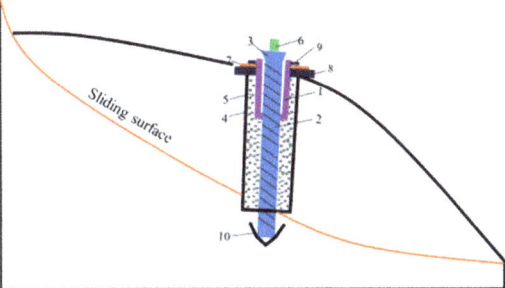

Figure 12. The AE flexible monitoring system: (1) sleeve with inner wall threaded; (2) anchor cable; (3) conical metal head; (4) rubber tube; (5) backfill material; (6) AE transducer; (7) ring dynamometer; (8) pedestal; (9) nut; (10) anchoring end (Modified from Deng et al. [56]).

AE technology is characterized by its dependability, low cost, great precision, and ability to be performed in real time. AE is sensitive to minor changes in displacement and velocity, allowing it to detect extremely slow-moving landslides with a high measuring range, outperforming both TDR and inclinometers. To illustrate, because of the hardness and brittleness of the inclinometer body, it can be bent excessively when the local shear displacement reaches approximately 50 mm, resulting in device failure. A comparison between GPS, extensometer, inclinometer, TDR, and acoustic emission systems is provided in Table 6.

Table 6. Comparison of various monitoring systems [56,71].

System	Range	Precision	Displacement	Cost (USD) [71]
GPS	-	3 mm	Surface	High (6000–10,000/station)
Extensometer [63]	Up to 1000 mm	0.011 ± 0.0083 mm	Surface	High (600–1500/station)
Inclinometer	<50 mm [79]	±0.01 mm per 500 mm	Deep	High (600/sensor)
TDR [68,81]	60 mm (210-mρ, reflection coefficient,)	0.5 mm [77]	Deep	Low (6–10/m)
AE	>500 mm [56]	0.0001 mm/h to 400 mm/h [80]	Deep	Low [56]

Optical Fiber System

Optical fiber technology has become more important, supplying a significant amount of the world's internet, television, and telephone networks. Because of the sensitivity of the propagating light signal to disturbances, such as strain and temperature change, optical fiber cables have been effectively employed as sensing devices that can transport high-quality data across large distances at remarkable speeds. Fibre-optic (FO) sensors can be inserted directly into the ground; linked to a stabilizing structure or reinforcement; or coupled to traditional monitoring equipment, such as an inclinometer [82].

First, Brillouin optical time domain reflectometry (BOTDR) was developed. However, BOTDR cannot detect strain and temperature at the same time [47]. A few years later, optical time domain reflectometry (OTDR) was developed as a distributed sensing technology [65] and considered a viable alternative to address the aforementioned limitation. Figure 13 depicts the essential components of the OTDR. A laser transmitter releases a short signal into the fiber, the timing of which is set by an electronic delay generator. The light is reflected to the source, and the delay generator measures the time delay relative to the start time of the pulse. Each time delay value is associated with a specific position along the fiber. Thus, in principle, backscattering and back reflections may be calculated in terms of their magnitude and pinpointed in terms of the distance along the connection [61]. Figure 14 is an example of a return signal obtained by using an OTDR system. Extrinsic and intrinsic sensors are the two types of OTDR displacement sensors. Extrinsic sensors that employ optical fiber as a transmission medium include reflexive, transmission, and interferometric sensors. Intrinsic sensors are commonly bend-loss-type sensors in which the optical fiber bends and creates macro bending loss, which is not favorable for long-distance optical data transmission. Fiber-optic displacement sensors based on the macro bending loss concept are intensity-based fiber-optical sensors, meaning that light transmission loss increases abruptly with large curvatures [83]. During light transmission, Rayleigh, Raman, and Brillouin scatterings occur and cause the light intensity to be attenuated. Rayleigh backscattering is the most powerful of the three [59,60], and OTDR can detect its light intensity as a function of time [47,58].

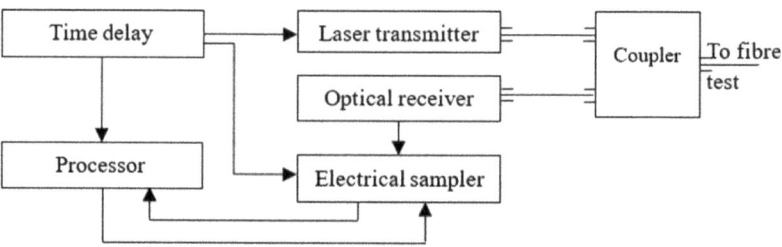

Figure 13. Basic elements of OTDR (modified after Aulakh et al. [61]).

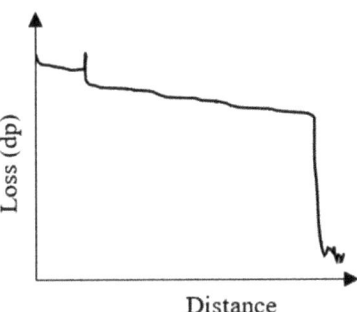

Figure 14. OTDR system sample return signal trace (modified after Aulakh et al. [61]).

The first single optical fiber can detect deformation with a high beginning accuracy of 0.3 mm; nevertheless, it has a limited sliding distance of 3.6 mm and a dynamic range of 3.3 mm [47]. The first generation of optical fiber has an unsatisfactory spatial distribution of twenty meters (one optical fiber was used to pass through a whole capillary steel pipe, with a suitable length of fiber left outside the pipe); it used a base material of PVC with no filler material within. To increase the spatial resolution, Aulakh et al. [61] developed a micro bend resolution-enhancer method that can improve the OTDR resolution up to 10 times. To increase the measuring range, Zhu et al. [47] created the second generation "combined optic fibre transducer" (COFT), with the base material being expansile polyester ethylene (EPS). Zheng et al. [84] used physical large-scale modeling to build an empirical formula for an innovative (COFT) that used OTDR based on the concepts of optical fiber micro bending loss. The COFT has a maximum sliding distance of 26.5 mm and an accuracy of 1 mm. The most effective material and the best cement-to-sand ratio in mortar were expandable polystyrene (EPS) and 1:5, respectively. The following are the capabilities of COFT fibers: (1) accurately predict the slide direction; (2) low cost; (3) remote, long-term, and real-time monitoring; (4) data collection takes seconds; and (5) distributed across numerous kilometers with great strain accuracy [47,58,84,85].

However, COFT finds it challenging to locate potential sliding surfaces and collect dispersed measurements of complex landslides, particularly the arrangement and interaction between multiple sliding surfaces. Therefore, a quasi-distributed measuring system and prospective sliding surfaces, especially on rock slopes, can be achieved using a parallel-series connected fiber-optic displacement sensor (PSCFODS) with bowknot bending modulation that makes it more bendable and sensitive [59,60,83]. The greatest value was 34 mm, and the initial measurement was 0.98 mm. Different lengths of capillary steel pipes were arranged to determine the sliding surface location with a spatial resolution of 250 mm. Zheng et al. [84] employed laboratory shear testing and field experiments in which many fiber-optic displacement sensors (FODSs) were linked in series. The starting measurement of the QDFODS was 0.98 mm, and the maximum displacement was 36 mm.

Interferometric "integral coherent measurements" are used in the coherent optical time-domain sensing principle (C-OTDR). The term comes from the sensing mechanism that produces an integral of the signal response throughout the whole length of the sensor. This sensing system demonstrates its suitability for providing an overall indication of the status of the monitored region, as well as the yielding strain and temperature change indicators with high temporal resolution [82]. Yu et al. [86] examined experimentally distributed coherent optical time-domain reflectometry (C-OTDR), which has a spatial resolution of one meter and a resolution of 0.1 m. The fiber was placed in a snake-like manner, as shown in Figure 15, to monitor the displacement in both directions at the same time.

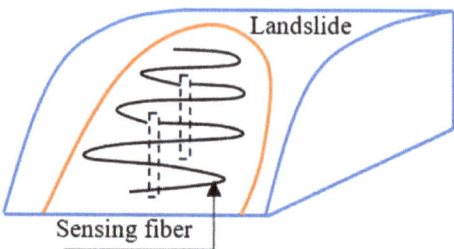

Figure 15. Schematic view of the snakelike distributed C-OTDR. "Reprinted/adapted with permission from Yu et al. [86], Elsevier".

OTDR, BOTDR, and C-OTDR have limited spatial distributions, which limit their usage. The spatial resolution has risen from 1 m for the Brillouin optical time domain reflectometry (BOTDR) technique to 0.1 m for the Brillouin optical time domain analysis (BOTDA) approach due to the rapid growth of fiber optic technology. However, their BOTDA installation is difficult, as BOTDA requires an incident laser from both ends of the optical fiber [78]. Schenato et al. [65] used experimental modeling to understand the evaluation of rainfall-induced landslides using a highly densely distributed optical fiber strain-sensing system with centimeter (10 mm) spatial resolution. Optical frequency domain reflectometry (OFDR) technology has recently been improved, allowing for strains to be measured with an exceptional spatial resolution (i.e., millimeters) [58]. Ivanov et al. [82] recommended a novel interrogation technique, namely, "Brillouin Optical Correlation Domain Analysis" (BOCDA), which will be used in a subsequent study to recover the whole strain profile along the deployed sensor fiber with centimeter spatial resolution.

Previous studies, however, were based on the micro bending theory or the beam theory, which does not consider mass movement kinematics. Zhang et al. [58] investigated the mechanism of distributed optical strain sensing (DFOSS) via a kinematic method through a parametric study on the sliding directions, shear zone width, and shear displacement. This approach simplifies the deformed sensing optical fiber (SOF) to be an arc or straight line, but the deformed shape might be rather complex since it is determined using the shearing angle, soil profile, grouting quality, etc. In contrast to simplistic techniques that assume the deformed shape to be rectangular or an arc, a more generic shear displacement calculation method (accumulative integral method (AIM)) is presented herein that does not presuppose the shape of the DSS cable [35]. In laboratory experiments, this suggested technique outperformed the triangle and arc models with a relative inaccuracy of approximately 6.5%.

To improve the stress transmission between the sensing cable and the surrounding soil, Ivanov et al. [82] concluded that the position of the sensors perpendicular to the sliding direction is preferable where better soil cable coupling is achieved. However, in such cases, these fibers are subjected to high shearing stresses, which limit their usage to shallow, slow-moving landslides. Minardo et al. [87] employed small anchors that were installed by placing pieces of geonet every 25 cm along the optical fiber. Zhang et al. [35] created a novel distributed-strain-sensing (DSS) cable based on the Brillouin frequency to improve the coupling behavior between the borehole-installed DSS cable and the surrounding soil. Anchors and deep confining pressures were used to enhance the coupling behavior, as shown in Figure 16.

A fiber Bragg grating (FBG)-based inclinometer can monitor quasi-distributed deformation at various depths (i.e., spatial vertical resolution) [72]. The FBG is a wavelength-selective filter. An FBG sensor will reflect light with a center wavelength matching the Bragg condition. Strain modifies the Bragg wavelength by causing the grating periodicity to expand or contract. Wang et al. [72] adopted a prototype monitoring system consisting of nine FBG-based inclinometers. This system has a spatial resolution of one meter and can detect horizontal displacement with high accuracy in the millimeter range. Zheng et al. [59,60] used the previous data to build a theoretical deflection relation using the Simpson integral model considering the cantilever beam where the displacement difference range was (−10% to 10%). Allil et al. [48] achieved a spatial resolution of 100 mm through laboratory tests. Despite FBG's numerous benefits, it is challenging to extract deformation directly from FBG strain sensors. Zheng et al. [78] and Zeng et al. [88] developed mathematical equations based on the conjugate beam approach that were validated using numerical analysis (ANSYS) and a large field shear test. The highest recorded value in laboratory experiments was approximately 50 mm, and the most significant absolute error between mathematical and field testing was approximately 10%. This system offers the benefits of low weight, small dimensions, corrosion resistance, high measurement precision, high instantaneity, anti-electromagnetic interference, and ease of installation [59,60]. Temperature, on the other hand, has an effect on FBG, C-OTDR, and BOTDR/A sensing technologies [72,83]. Thus, Zheng et al. [89] suggested a temperature compensation approach that can be used to reduce chirp change reflection peaks and offer temperature compensation.

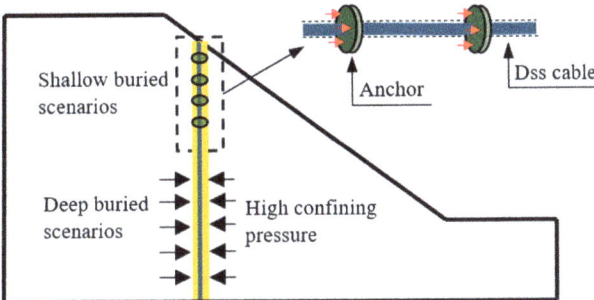

Figure 16. Soil cable coupling improvement. "Reprinted/adapted with permission from Zhang et al. [35], Springer Nature".

While FBG-based sensors offer discrete strain and temperature readings at predetermined places and are capable of providing dynamic measurements, this technology is unable to offer monitoring over a wide area. While BOTDR/A monitors strain and temperature change throughout the entire cable length, they are only capable of static monitoring, which can be over many kilometers (i.e., the distributed fiber length) [82]. Li et al. [90] created a novel system by merging BOTDA and XFG (fiber Bragg grating (FBG) and long-period fiber grating (LPFG)), resulting in a distributed system that can monitor a sizable region with discrete dynamic strain/temperature, as illustrated in Figure 17. However, laboratory studies were used to validate these findings. Table 7 lists the characteristics of FBG in comparison with some other techniques.

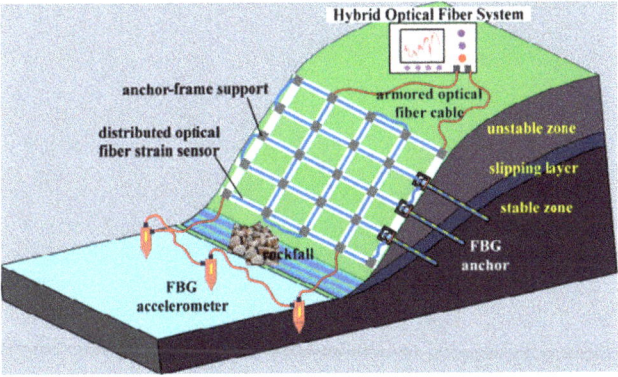

Figure 17. The developed hybrid system for multiparameter monitoring (From Li et al. [90]).

Table 7. A comparison between conventional inclinometers (IN), TDR, single optical fiber (SOF), combined optic fiber (COFT), parallel-series connected fiber-optic displacement sensor (PSCFODS), and FBG-based inclinometer.

Method	Initial Accuracy (mm)	Maximum Displacement (mm)	Spatial Resolution	Dynamic Range	Loading Direction	Sliding Location	Price (USD/m)
IN [66–68]	0.01	<50 [79]	500 mm	-	Yes	Yes	30 [47]
TDR [68,77,81]	0.5 mm	60 (210 mρ)	50 mm	0–20.4 [47]	Yes	Yes	13.5
SOF [91]	0.3	3.6	-	0–3.3	No	No	0.03
COFT [47,78,84,85]	0.98	36	-	0–34	Yes	Yes	0.45
PSCFODS [58,59,83]	0.98	36	250 mm	-	Yes	Yes	0.2
FBG [48,58,59,89]	0.02	50	100 mm	-	Yes	Yes	-

While numerous authors highlight the low cost and long lifespan of the sensor itself (i.e., optical fiber cables), the truth is quite contrary: costs may reach tens of thousands of euros and are often built to function in a controlled environment, such as a laboratory [82]. Optical fiber technology has not been used extensively for a long period in challenging outdoor environments. Additionally, even though the price of the optical sensor itself may be low, it is necessary to consider the price of additional optical data acquisition tools, such as fiber optic interrogators and optical fiber grating demodulators, as well as the need for highly skilled labor to manage and install these technologies. Additionally, the power consumption of the entire optical sensor system is not optimized for low-power field applications, where the entire piece of equipment must operate unattended for months on battery power [10].

Electromechanical Tilt Sensors

Fiber optics are widely used to improve performance, whereas electromechanical sensors appear to be a viable way to obtain both precision and a wide range of data [92]. However, because microelectromechanical systems (MEMS) are electronic devices, they must be charged, and their output signals must be transmitted outside by electric cables or wireless networks, which cannot be too lengthy, or the signals will be compromised by noise. Optical fiber sensors may be an alternative to electrically powered devices since they may be operated remotely and are powered by optical fiber cables, such as FBGs, without the requirement for electricity [48]. Nevertheless, many attempts were proposed to save sensor power consumption and to overcome wiring issues by developing wireless sensor networks (WSNs), which are further discussed. The low-power radio communication and modular architecture make installation and maintenance of the entire system easier than with cable-connected devices, and data transfer is more efficient [52]. Some types have excellent performance and are used to create sensors for structural monitoring [12,93,94].

Extensometers can only detect surface displacements, while inclinometers can only give subsurface displacement in one direction [8,65]. Both approaches require expert labor to install and maintain such instruments. Moreover, determining the landslide direction with both inclinometers and extensometers is challenging. Tilt measurements can indirectly detect two-dimensional shear deformation and determine the rotational direction in terms of tilt angle and sign convention [95]. When combined with MEMS and WSNS techniques, tilt sensors can provide the following benefits: (1) minimal cost, (2) simple installation (no deep boreholes required), and (3) real-time data [96]. Gian et al. [96] used a tiltmeter combined with other sensors to provide real-time data through WSNs. Chen et al. [97–99] adopted a MEMS sensor to measure tilt angle with a data frequency of 1 s. Abraham et al. [93,94] used a MEMS sensor with an accuracy of 0.017° and a resolution of 0.003° to measure the tilt angle in two directions (parallel and perpendicular to the slope movement).

Qiao et al. [52] investigated the relationship between the tilting direction and the depth of the tilt rod sensor using a MEMS tilt sensor (nominal resolution = 0.0025°) and temperature sensor, as shown in Figure 18. For diverse types of landslides, the depth of placement of tilt sensors with rods should be carefully determined. Both tilt sensors with short rods and tilt sensors with long rods can be employed for landslides with curved slip surfaces. Tilt sensors with short rods are ineffective for shallow translational landslides. To monitor these types of landslides, the tilt sensor rod must be placed in the stable layer. Artese et al. [12] created a novel sensor called the Position and Inclination Sensor (POIS), which is wireless, low cost, small, light, and consumes little power. This sensor costs approximately USD 400 and can measure the tilt in two directions. This sensor is suitable for both slow-moving and rapid landslides. Ruzza et al. [64] designed a multimodule system that consists of many biaxial tilt measuring units, as shown in Figure 19. When linked together, it may be deployed within a borehole supplied with a specialized inclinometer housing. Once mounted, the device continually collects tilt data at various depths and turns it into a displacement measurement. For landslides with depths ranging from 5 to 10 m, a multimodule in-place inclinometer costs EUR 700. The measurement accuracy is 0.37% of the inclinometer chain depth, the linear measuring range is ±20°, and it has good thermal efficiency.

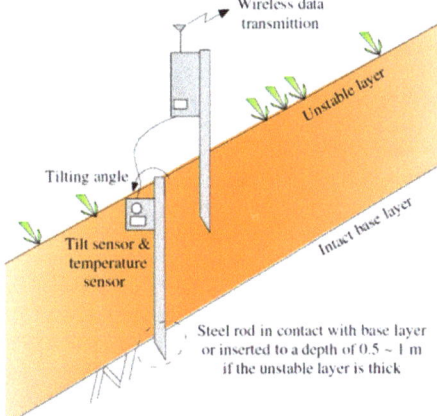

Figure 18. Wireless MEMS tilting monitoring system (From Qiao et al. [52]).

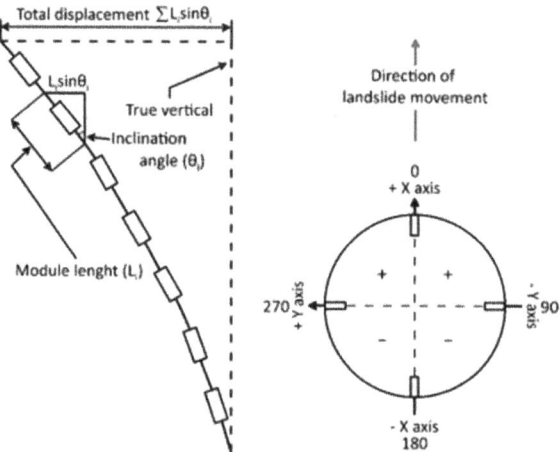

Figure 19. The multimodule inclinometers (From Ruzza et al. [64]).

Nevertheless, the inclination measurement accuracy is influenced by a variety of error causes, including noise, drift, and offset. Similar to Ruzza et al. [64] and to overcome the aforementioned accuracy limitation, Wielandt et al. [100] developed a low-cost, long-term wireless sensor that consists of three-axis accelerometers (MEMS) and a temperature sensor to monitor the change in sensor inclination, surrounding soil deformation, and subsurface temperature to reduce the draft error. The equipment achieved a resolution of 0.39 mm, a 95% confidence interval of ± 0.73 mm per meter of probe length, a depth spatial resolution of 100 mm, and an acceleration range of ± 2 g.

Tilt sensors, on the other hand, are point sensors and cannot extract deformations in regions where there is no inclination (i.e., translational landslides) [8,101]. This system has a high false alarm rate due to human or animal interventions. To show why there are so many false alarms from shallow sensors, consider the following: (1) erosion of the ground on rainy days may cause sensor tilting, although this does not affect landslides, and (2) external impacts from animals or human activity may cause tilt readings. Multimonitoring systems, therefore, have the benefits and ability to overcome these shortcomings [93,94].

Strain Gauge Sensors

Strain gauges can achieve cost-effective conditions compared with inclinometers. The strain gauge measures the strain experienced by the soil layer during slope instability and can be connected to a WSN to provide real-time data [102]. A strain gauge translates force, pressure, tension, weight, and other variables into a change in electrical resistance that can be measured. Before a landslide, strain gauges are used to quantify the micro movements within the unstable soil slope [103].

Ramesh and Vasudevan [6] used a casing up to 21 m long with strain gauges to assess deep subsurface movement. These strain gauges were attached to the inclinometer case's exterior diameter to detect displacement in the sliding direction and with 90, 120, and 240-degree angles to the sliding direction, respectively. Pipe strain gauges (i.e., strain gauges mounted on inclinometers) have a limited spatial resolution but can detect the depth of deformation in the soil surrounding the gauges [101]. If the casing has high bending stiffness in comparison with the surrounding soil, the small motions before the failure cannot be properly recorded. Additionally, it is difficult to utilize them to track changes in shallow strata above the bedrock. In comparison with PVC pipes and shapeAccelArray/Field (SAAF) devices, soil deformation sensors (SDSs) have been designed with bending stiffnesses that are 300 times and 50 times lower, respectively [104,105]. SDSs were created at the Institute for Geotechnical Engineering at ETH Zurich to track the subsurface motions of a silty sand slope. Askarinejad and Springman [105] investigated the behavior of SDS through experimental and numerical (PLAXIS) verification. Askarinejad et al. [73] developed fully automated novel slope deformation sensors (SDSs) that can measure fine movement (<1 mm) with a range of 0 to 25 mm and are suitable for rapid silty sand landslides. Kumar and Ramesh [10] created a unique Strain Gauge Deep Earth Probe (SG-DEP sensor) that consists of a basal body (grooved ABS pipe) that can flex/deform with the soil and strain gauges that can quantify the amount of flexion/deformation in this basal body. To obtain a full 360-degree directional measurement of the subsurface movement, 1000-ohm linear strain gauges (unaffected by temperature) are bonded to the midsection of the basal body in both orthogonal planes. The suggested SG-DEP sensor is also used in the system to monitor the change in curvature of the ABS pipe with a high sensitivity of 0.005799 m^{-1} (refer to Figure 16).

Acceleration Sensors

The majority of monitoring system components involve sensors for assessing soil tilting or displacement; however, acceleration sensors have yet to be commonly utilized. Independent of the trigger, acceleration sensors can be utilized to detect any movement (Giri et al., 2018). These sensors can be manufactured based on the technology of optical fibers [90], inertial measuring units (IMUs), and MEMS [8,28], in which sensor reading data can be transferred via wiring or wireless networks. By supplying a significant voltage differential V_{out}, the accelerometer reads a biaxial acceleration change.

Considering optical fiber technology, Li et al. [90] employed an FBG accelerometer to obtain rockfall vibrations using experimental tests. Inertial measuring units (IMUs) have the potential for real-time and distant applications in the monitoring and warning of landslides [28]. IMUs are used to combine different MEMS sensors, such as a three-axis accelerometer and three-axis gyroscope, to provide information about landslide movement and rock fall [28]. Based on the concept of a wireless sensor network, Kotta et al. [106] employed a vibration sensor (accelerometer) on Micaz devices to monitor vibrations brought on by landslides composed of montmorillonite (expansive clays) using prototype implementations. Similarly, Rosi et al. [7] adopted a prototype wireless sensor network to monitor acceleration using accelerometers. Ramesh and Rangan [102], Prabha et al. [103], and Gian et al. [96] used geophones to monitor vibrations caused by slope instability, which can provide real-time data when connected with WSNs.

The disadvantage of previous research is that it relied on a combination of linear and gravitational accelerations, or "raw acceleration data," to identify tilt or motion. As a result, a better system for keeping track of slides is needed. Giri et al. [28] used a MEMS wireless monitoring system to divide the acceleration into tilting and linear motions using a gyroscope, considering linear accelerations and gravity accelerations independently, as well as the angular velocities using experimental physical models. This method works well for translational landslides without tilting. The most obvious indication demonstrating the failure is the change in linear acceleration. Giri et al. [29] studied the behavior of shallow fast translational landslides in real time using the same system and scale model as Giri et al. [28]. According to the experimental study by Giri et al. [28], a translational slide is shown by a combination of low angular velocities within ± 10 deg/s, minor variations in gravity accelerations within ± 2 m/s^2, and linear accelerations of more than 1 m/s^2 in the longitudinal direction of the slope.

4.2.2. Force and Stress Monitoring

The majority of widely available monitoring and warning systems rely on displacement, which is affected by a variety of variables, such as rainfall, temperature, and soil moisture. Landslides, on the other hand, can be predicted in advance by monitoring the earth pressure and the sliding force in near real time, as the best metric for identifying the kinematic characteristics of landslides is the sliding force [9].

Earth pressure cells (EPCs) and seismic vibrators can be buried in the soil layer to measure the variation in earth pressure. Ma et al. [53] utilized an EPC to measure the earth pressure using experimental tests, where this device has a capacity of 500 kPa. Similarly, Askarinejad et al. [73] employed EPCs with an accuracy of 1 kPa, a range of 0–500 kPa, and a frequency of 100 Hz. Yunus et al. [107] developed a new smart wireless sensor to measure seismic vibrations. A set of weights placed on the cone transforms the loudspeaker (Visaton FR8 8-ohm) into a vibration sensor. When the loudspeaker detects seismic waves, the weights remain in place and apply stress on the cone, changing the distance between the coil and the base of the center pole. As a result, an output voltage is created at the loudspeaker's output terminal.

Using a constant resistance and large deformation (CRLD) anchor cable (refer to Figure 20), Tao et al. [108] created a monitoring system. The crucial warning threshold was set at 900 kN of cumulative sliding force, which allowed for an early forecast of a landslide 4 h prior to the event. Figure 20 depicts the monitoring system and monitoring curve stages, which are divided into three sections: (1) the steady stage, (2) the slowly rising stage, and (3) the stable stage. In the first stage, a few tensile cracks occur, whereas in the second stage, tensile cracks deeply penetrate the slope, and the shear plane inside the slope body extends. In the third stage, failure occurs, and the steady state occurs [108]. Chuan et al. [9] employed a prototype force sensor with a maximum capacity of 500 kPa and a precision of 1%. He et al. [109] established the "remote monitoring warning system of sliding force", which is a real-time and distant intelligence monitoring system based on the functional link between the sliding force and resistance force. Li et al. [110] developed a high-performance piezoelectric sensor that is able to adapt to both static and dynamic stresses through the self-structure pressure distribution method (SSPDM) and the capacitive circuit voltage distribution method (CCVDM). SSPDM was used to improve the compression capacity, and CCVDM was used to reduce the measuring error using the low-frequency method. This sensor can achieve a static range of 1500 kN and a dynamic range of 0–500 kN. However, this system was calibrated and verified using laboratory tests. It should be emphasized that the anchor cable must have the following characteristics: (1) strong strength, (2) low relaxation, and (3) high anticorrosion.

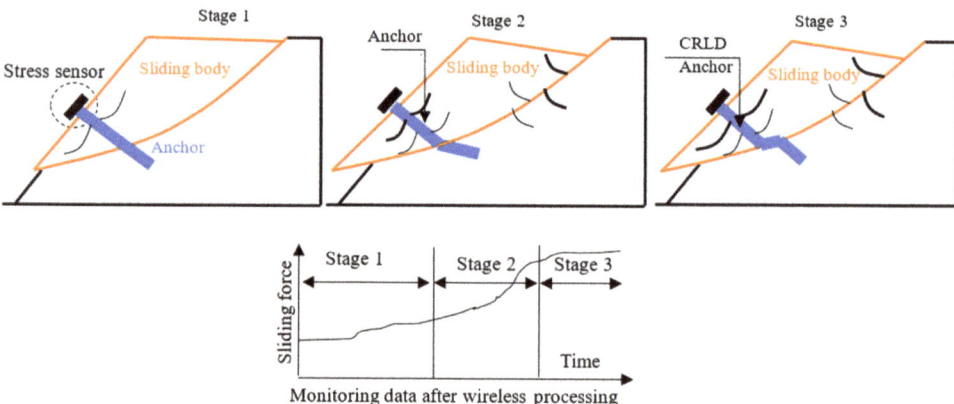

Figure 20. Sliding force monitoring system. "Reprinted/adapted with permission from Tao et al. [108], Springer Nature".

4.2.3. Water and Temperature Monitoring

There are three types of near-surface water monitoring: surface water monitoring, groundwater monitoring, and precipitation monitoring. Precipitation monitoring is primarily concerned with rainfall, whereas surface water monitoring covers near-surface soil moisture. Groundwater monitoring includes measures such as the groundwater level, pore water pressure, water temperature, water quality, and soil water content.

Precipitation Monitoring

Heavy rains are one of the most common causes of landslides. Table 8 shows, for example, the rainfall categories based on the Head of Meteorology, Climatology, and Geophysics Agency (BMKG) Regulation No. KEP.009 of 2010 [111].

Table 8. Rainfall intensity classifications.

Class	Per h (mm)	Per Day (mm)	Per Month		
			Rainy Days (Days)	Total Rainfall (mm)	Cumulative Rainfall (mm)
Very small	<1	<5	5–6	10-15	10-15
Small	1–5	5–20	6–7	60–70	70–85
Moderate	5–10	21–50	6–7	180–210	250–295
High	10–20	51–100	2–4	150–250	400–545
Very high	>20	>100	1–2	110–300	510–845

Rain gauges are classified into mechanical, optical, electrical, visual, and radar types, with the mechanical type, such as the traditional tipping bucket rain gauge (TBR), being the most extensively used and accurate. The mechanical type has the benefit of directly measuring the amount of rainfall, whereas the other methods adopt indirect measurements [112]. Ramesh and Vasudevan [6], Ramesh and Rangan [102], and Prabha et al. [103] adopted tipping bucket rain gauges to measure rainfall intensity. Latupapua et al. [111] developed a prototype wireless sensor network for measuring rainfall intensity using the Arduino Raindrop sensor. Crawford et al. [62] adopted a tipping bucket rain gauge (Rain Wise Inc) to measure rainfall with a data logger that has a 1 min resolution and is calibrated at 0.25 mm/tip.

However, TBRs suffer from limited measurement accuracy and significant abrasion under heavy rainfall conditions. Hu et al. [112] created a novel TBR based on multiple triboelectric nanogenerator (TENG) units capable of real-time rainfall monitoring via a freestanding TENG (F-TENG) unit and effective rainfall energy collection via a contact–separation mode TENG (CS-TENG) unit. The range of this system is 0 to 288 mm/d, and the resolution is 5.5 mm. It also features an excellent anti-humidity interference ability and a rainwater energy harvesting function, with a peak power generation capability of 7.63 mW under a rainfall intensity of 250 mm/d. As a result, this device is a self-powered wireless sensor with a high measuring range and resolution that can be used in hazardous conditions. Nevertheless, one disadvantage of utilizing rainfall records is that the rainfall criteria (i.e., empirical rainfall thresholds) do not account for the inner landslide mechanism [4,5]. Thus, understanding the subsurface changes in matric suction, moisture content, groundwater fluctuation, etc., is crucial for the better prediction of landslides.

Near-Surface Water Monitoring

Near-surface technologies, such as gamma-ray attenuation [113], soil heat flux [114], and ground penetration radar (GPR) [32], are costly, susceptible to noise, and incapable of providing deep moisture and temperature information [115]. Soil moisture may be monitored at the regional scale using remote sensing techniques, such as satellite retrievals, which are restricted to near-surface soil moisture. Although satellite-based soil moisture estimations [34] have been found to be beneficial for identifying landslide-prone situations, their application in landslide early warning systems is restricted by the coarse spatial resolution and the lower temporal resolution [116]. It should be highlighted that this study focused on subsurface monitoring techniques, which exclude the aforementioned investigations.

Subsurface Water Monitoring

Subsurface water monitoring approaches include site investigation and laboratory sampling, optical fiber and acoustic emission methods, electrical permittivity tools, geophysical techniques, and MEMS and IoT technology applications. Subsurface water monitoring includes soil moisture content (volumetric water content), pore water pressure (suction pressures), and groundwater level variation. It should be emphasized that soil moisture is a critical parameter for assessing and monitoring natural hazards, such as landslides. The volumetric water content response to rainfall events is more immediate than that of pore water pressure and retains its maximum value for some time before slope failure [115].

In the laboratory, the soil moisture content can be determined using the weight difference between the dry and wet states of the soil (soil drying technique) [117]. This technique has high local accuracy; however, it requires considerable time and is labor intensive. Thus, it is preferable only for small areas [118]. As for acoustic emission monitoring, the low-energy acoustic emission signals created in soils attenuate dramatically over short distances [119]. Nevertheless, it is challenging to link acoustic waves with soil moisture since it is impacted by the soil density, void ratio, effective stress, etc. [120]. The first generation of fiber Bragg gratings (FBGs) could monitor a volumetric water content (VMC) of just 5% when the humidity reached 90%. Consequently, Leone et al. [115] created a new generation of fiber-optic thermos hygrometer-based soil moisture sensors based on fiber Bragg gratings (FBGs) that can measure the VMC up to 37% in continuous real-time. This innovative system comprises a polyvinyl chloride (PVC) cylindrical structure with an upper section sealed by a hermetic stopper that interacts with the soil via a microporous hydrophobic membrane that covers its lower part (refer to Figure 21). Depending on the soil water content, a specified quantity of molecules of water in the vapor form can flow through and spread throughout the package volume when buried in the soil. A comparison between both systems is shown in Figure 22.

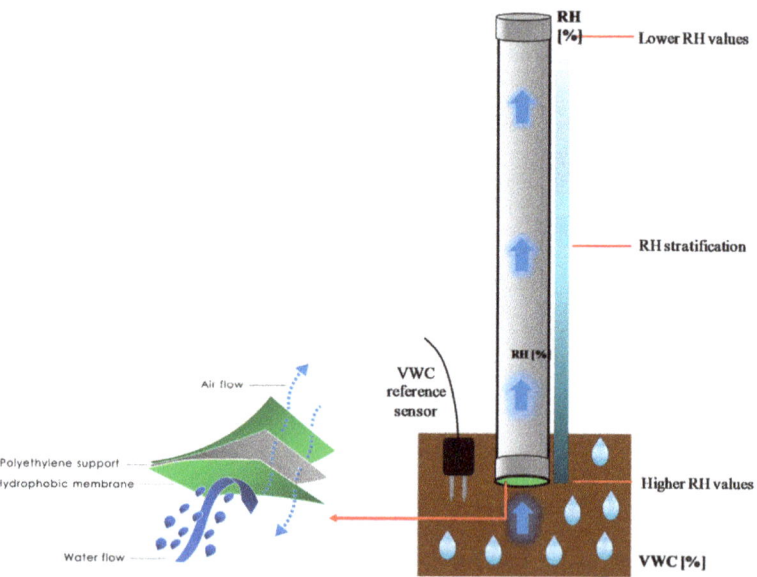

Figure 21. An optimized version of the optical soil moisture instrument (From Leone et al. [115]).

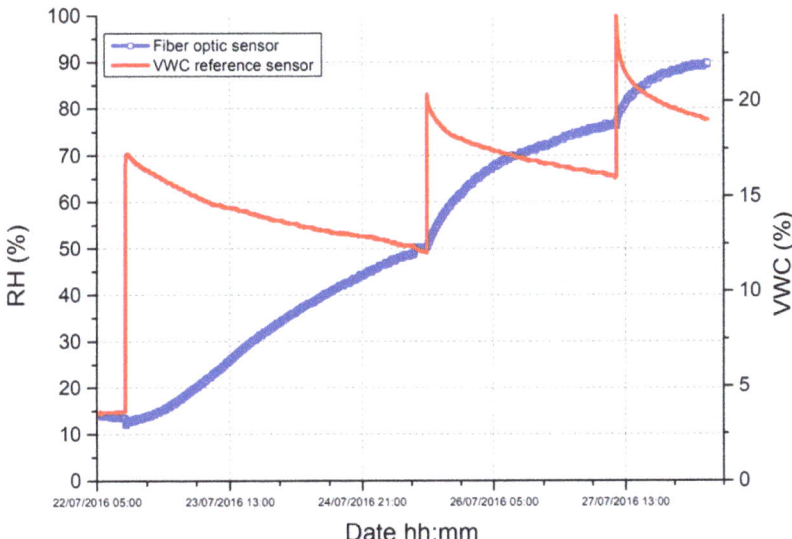

Figure 22. A comparison between the first generation (reference sensor) and the optimized sensor (optical fiber) (From Leone et al. [115]).

Dielectric permittivity technologies are used to estimate the volumetric water content (VWC), tensiometers are used to assess the soil water potential (SWP), and piezometers are used to monitor the water pressure. Ivanov et al. [82] employed a TDR probe for soil moisture monitoring. Minardo et al. [87] utilized tensiometers for soil suction stress monitoring. Ramesh and Vasudevan [6], Ramesh and Rangan [102], and Prabha et al. [103] adopted dielectric moisture sensors to quantify the volumetric water content and piezometers to measure pore water pressures. High-temporal-resolution measurements of soil moisture are possible. However, in situ sensors only monitor a small amount of material. Additionally, measurements may be impacted by local-scale phenomena (i.e., preferential flow, root development around the sensor) because of the small measurement volumes, which are in the range of several hundred to a few thousand cubic centimeters, thus making comparisons of data challenging [115,116].

Ultrahigh-frequency radio-frequency identification (UHF RFID) sensors are a promising option for soil moisture monitoring since the sensors are inexpensive, can be self-chargeable (no battery), can provide distance communications up to some meters, and can transfer real-time data [121]. In UHF RFID, the electrical characteristics of the tags change with the existence of water. Pichorim et al. [121] used two experimental methods to study UHF RFID tags for moisture content detection: one tag was buried into the ground as a sensor tag, and one tag was placed on the surface as a reference tag. This option, however, is expensive. The second method is affordable, which makes use of the SL900A chip and examines the relationship between soil moisture and sensor capacity. Sensor moisture readings vary from 6% in a dry condition to 16% in a saturated state. Both alternatives are long-term self-rechargeable sensors.

Geophysical techniques, such as electrical resistivity, are feasible approaches for correlating geotechnical observations since they are impacted by the soil profile, saturation degree, pore structure, effective stress, deformation, etc. As a result, these approaches can be employed as a monitoring system. They have the benefits of (1) being less expensive than typical geotechnical monitoring systems, (2) providing information across large regions rather than single points (i.e., plot-scale soil moisture fluctuation), (3) being nondestructive studies of ground parameters, (4) having a spatial resolution of meter-to-decameter scale, and (5) providing great temporal resolution [122]. A prime technique is electrical resistivity tomography (ERT), which determines the two- or three-dimensional distribution of electrical resistivity along one or more profile lines of electrodes installed on the soil surface or in boreholes. Electrical resistivity is calculated using pairs of electrodes that inject an electrical current into the ground and detect the potential difference [116,118,123,124]. ERT can offer information regarding the soil profile, moisture status, depth of the slide surface, and shape of a landslide. In general, single measurements are used for subsurface characterization, whereas repeated measurements at the same profile line (time-lapse tomography) are used to investigate time-variant processes in the subsurface. Two ERT profile lines, with one perpendicular to the slope direction (horizontal profile) and the other parallel to the slope direction (vertical profile), were placed on the plot, allowing for an assessment of the spatial variation in hydrological processes and lithological heterogeneity on the plot size [116]. The primary drawbacks of electrical resistivity measurement are the reduction in resolution with depth, the non-uniqueness of solutions for data inversion and interpretation, and the lack of direct information [99].

Crawford and Bryson [122] conducted research that correlated electrical resistivity measurements with shear strength within shallow landslides, in which prototype experiments were conducted to evaluate volumetric water content, soil water potential (suction), and electrical conductivity. In keeping with a prior study by Crawford and Bryson (2018), Crawford et al. [62] used electrical conductivity to estimate unsaturated soil properties (soil water characteristic curve (SWCC) and suction stress characteristic curve (SSCC)) based on the long-term field monitoring of movement, water content, water potential, and electric conductivity of rainfall-induced shallow landslides. A novel equation that uses electrical conductivity as a predictor of suction stress was developed. However, correctly measuring the water content of the landslide is extremely challenging [125]. When geophysical electrical monitoring (high-density resistivity) of soil moisture content is considered, it is shown that numerous factors impact the resistivity and moisture content, and the relationship is complicated and cannot be described using typical linear and nonlinear equations. As a result, Xiaochun et al. [118] used laboratory testing to train a hybrid artificial intelligence model, which was then tested using a large-scale model and used in field tests.

Some recent applications show that root zone soil moisture is often the most valuable hydrologic information for shallow landslide prediction; thus, its distributed monitoring should be considered by low-cost networks with easy installation and maintenance [126]. IoT technology applications have recently gained popularity. Marino et al. [126] explored the measurement of volumetric water content utilizing a network of low-cost capacitive sensors communicating through field testing within the space of Internet of things (IoT) technology. The correlation between the volumetric water content and the sensor output voltage (Vout^{-1}) reached an R^2 of 0.98. Similarly, MEMS sensors provide a viable way to provide real-time data at a low cost. Abraham et al. [93,94] used the MEMS volumetric water content, where the precision of the volumetric water content sensor was ±3%. Chuan et al. [9] measured the pore water pressure using a sensor with a capacity of 100 kPa and a precision of 0.3%. Chen et al. [97–99] used an EC–5 (by Decagon Devices, Inc., Washington, DC, USA) sensor to measure volumetric water content with a data frequency of 1 s. When these sensors are connected to a wireless network through ZigBee, Wi-Fi, or VSAT (satellite) networks, they can accomplish real-time monitoring [102]. Jeong et al. [92] used a wireless sensor to measure soil suction (tensiometer), groundwater content (soil moisture sensor), and rainfall (rain gauge). Using a combination of industry-tested sensors, Chu et al. [125] created SitkaNet, which is a cost-effective alternative. This sensor node can measure the soil moisture content at various depths (six sensors at various depths), water table, humidity, atmospheric pressure, temperature, and rainfall for a low cost of approximately 1000 USD/node. This device can send data in real time with a temporal resolution of 5 min and can operate for 6 months. However, these methods are point sensors with limited spatial resolution and suffer from high power consumption.

Temperature Monitoring

For deep-seated landslides, where thermal sensitivity plays a crucial role in the stability of the slide, Seguí and Veveakis [30,31] created a theoretical equation to quantify and decrease the uncertainty of the model parameters and use the temperature in the shear band. The feasibility of this study was confirmed using field tests, where a thermometer was employed to determine the potential thermal sensitivity of the material located in one of the most crucial regions of a landslide (the shear band). However, this system requires prior investigation to determine the location of the shear band. For shallow landslides, Ma et al. [53] experimentally showed that the surface temperature can provide early warning indicators, as the moving mass's surface temperature is much higher than the nonmoving mass's surface temperature. Prior to failure, the average change in surface temperature exhibits a significant increase, followed by a fall in the surface temperature.

4.2.4. Warning Techniques

Previous monitoring systems primarily focused on the accuracy of acquired data for improved prediction based on geological parameter monitoring; nevertheless, these approaches lack scene information and deal with emergency scenarios [127]. Thus, regardless of the precision and quantification of the monitored parameter, warning monitoring systems can offer an early warning indication. Sensors for moisture or slope deformation are point sensors that are exclusively sensitive to changes in physical characteristics in their immediate surroundings. As a result, several sensors are necessary to cover a large possible landslide region. This might drastically raise project costs, but limiting the number of sensors would reduce the landslide forecast efficiency, making the system itself doubtful. A promising technique where geological engineering uses damage-free studies of geotechnical parameters based on data delivered by elastic waves was developed [97]. Figure 23 illuminates the difference between the elastic wave velocity method and conventional methods [128].

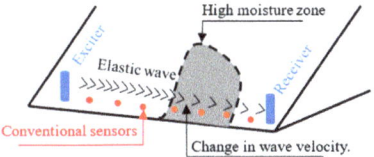

Figure 23. Comparison between elastic wave technique and conventional methods "Reprinted/adapted with permission from Irfan et al. [128], Elsevier".

Figure 24 shows how measuring the change in wave velocities may help to identify the time of failure start and post-failure strain rate. Such distinct differences in wave velocities during rainfall can help to construct a viable landslide prediction system. Since elastic waves are influenced by the internal structure of soil particles, shear modulus, void ratio, soil moisture content, soil deformation, and soil movement, they can be used to represent the internal mechanisms of soil [97,99]. Irfan et al. [128] proposed a unique method for monitoring slope deformation and soil moisture content by varying elastic wave velocities. The elastic wave characteristics were investigated through a series of triaxial tests. It was concluded that wave velocities decreased by nearly half when the soil saturation increased from 20% to ~80%: Figure 25 highlights the response of elastic wave velocities during rainfall-induced landslides (i.e., yielding).

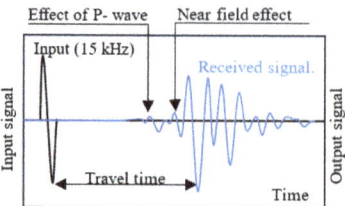

Figure 24. Received shear wave signal versus time "Reprinted/adapted with permission from Irfan et al. [128], Elsevier".

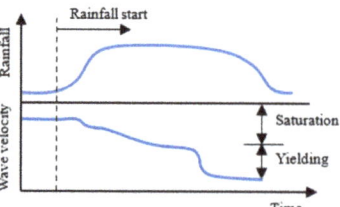

Figure 25. Elastic wave velocity during rainfall events "Reprinted/adapted with permission from Irfan et al. [128], Elsevier".

Chen et al. [97–99] constructed two physical models (small and large) to study the behavior of the elastic wave velocity in rainfall-induced landslides. The elastic wave velocity dropped continually in response to moisture content and deformation, and there was a clear increase in the rate of wave velocity decline when failure commenced. Chen et al. [98] proposed a threshold based on centrifuge experiments for predicting rainfall-induced landslides using a normalized shear velocity limit of 0.9. The scope of these investigations, however, was restricted to homogenous slopes and laboratory settings. Chen et al. [97,99] quantified the relationship between S-wave (V_S) and P-wave velocities (V_P) with the shear modulus (G_0) and constrained modulus (M_0). Furthermore, the shear wave velocity decreased with increasing deformation, which increased the water content, loss of matric suction, and effective stress in soil.

These studies (elastic wave), however, lacked a cost estimate for field deployment, and the location (i.e., near the toe, middle, or close to the crest) of the elastic wave transmitter/receiver in the field is unknown. According to experimental and laboratory studies by Chen et al. [98], the toe is best for monitoring waves. Exciting device selection is a complex problem (for example, powerful waves can harm slope stability, while weak waves may be influenced by noise) [97]. Furthermore, exciting devices require a constant high-power source to create excitation over a lengthy period. Because several receivers may be required to be deployed along the slope with a single transmitter, the receivers have to be cost-effective and energy-efficient [97,99]. The layered soil profile affects the wave velocity and direction, with each soil having unique characteristics that require future investigation of such complex behavior [98].

Previous research on acoustic emission (AE) was restricted to high-frequency signals in which AE is generated when a disordered material is subjected to stress, shear, or failure [129]. Low-frequency AE signals, such as infrasonic signals, have received less attention. In contrast to traditional monitoring systems (i.e., point systems, such as deformation systems or subsurface water systems), infrared signals can monitor several landslides within a local region with high penetration capacity and low attenuation. Zhang et al. [130] created a novel geophysical warning system based on experimental physical modeling in which infrasonic signals can reveal any microscopic variations in the subsoil caused by sliding forces (e.g., change in void ratio or porosity). When landslides begin, sensors can easily catch a high-energy infrasonic signal (refer to the pulse in Figure 26) as an indication of a macroscopic rearrangement of soil particles. The infrasonic signal can be converted to sound pressure using the short-time Fourier transform (STFT) and can be correlated with the sliding force, as presented in Figure 26. However, this method is influenced by external noise, such as wind, thunder, and motor vehicles, which must be filtered out.

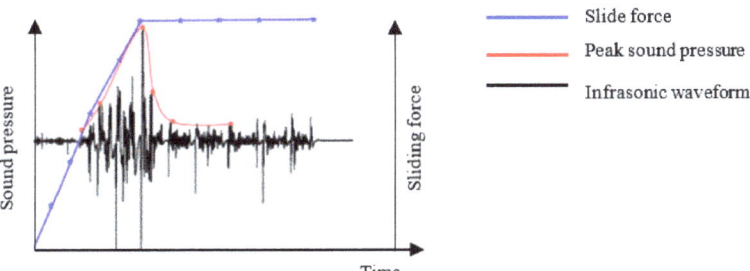

Figure 26. Time-series data for infrasonic signals "Reprinted/adapted with permission from Zhang et al. [130], Elsevier".

Motakabber and Ibrahimy [131] developed a wireless (almost 100 m sensor node distance) differential capacitor-type sensor using mathematical models and simulations. This sensor overcomes the limitations of capacitor-type sensors, which are noise and complex thermal adjustments, and has the advantages of being simple, robust, reliable, and cheaper. This system consists of an underground pretension cable with a capacitor gauge sensor attached at one end. When soil starts to deform, the formation of a force-on-force plate, as well as the pretension wire, results in a change in the differential capacitor.

Lin et al. [132] used a unique self-powered timbo-like triboelectric force and bend sensor (TTEFBS) to detect any rockfall movements or subsurface deformation as a voltage fluctuation. This system features a quick reaction time (<6 ms), long-term durability (>40,000 cycles), high compression and bending sensitivity (5.20 V/N and 1.61 V/rad, respectively), and distributed and wireless sensing capabilities. Similarly, Wang et al. [127] created a wireless sensor system using small-scale modeling that includes both an accelerometer and a camera sensor. The accelerometer was employed to provide early warning, and the camera sensor was used to perform visual analysis.

Through numerical (ANSYS) and experimental indoor experiments for soil deformation monitoring, Kuang [133] investigated a unique chemiluminescence-based approach. Chemiluminescence devices have reactants that are kept in distinct compartments and produce light instantly when distorted, making them easily detectable by inexpensive optoelectronics (i.e., light-dependent resistors (LDRs)). No power is needed for chemiluminescence to operate, as it is entirely passive. This device costs 1 USD/unit, where the dimensions of one unit are 400 mm in length and 15 mm in diameter. However, this system is sensitive (i.e., vulnerable) to small deformations ranging between 0.43 mm and 24.99 mm. Thus, the position of the system (i.e., A, B, or C as presented in Figure 27) can be changed based on the expected soil movement to overcome this issue. However, it should be emphasized that the effectiveness of most warning techniques for predicting landslides is still being researched.

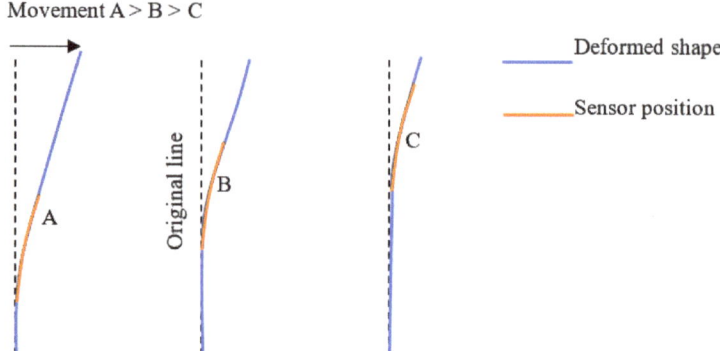

Figure 27. Alternative solutions for movement sensitivity "Reprinted/adapted with permission from Kuang [133], Elsevier".

4.3. Wireless Sensing Network (WSN)

Wired-based systems have apparent disadvantages, such as difficulty in wiring and construction in danger zones, human-caused destruction, and natural catastrophe damage [134]. This significantly increases the effort necessary for installation and operation, both financially and in terms of time. Furthermore, data are often conveyed without any preprocessing, necessitating the storage and delivery of massive packages of redundant data linked to a given node of observation before it can be processed and correlated [7]. Thus, wireless sensor networks have several benefits over traditional techniques, including the following: (1) the ability to gather and analyze multipoint distributed data, (2) the ability to cover a large area with little wiring expenses, (3) they are energy efficient since they can run for months, (4) incorporation with existing equipment [7,28,92,134], (5) installation without preexisting infrastructure, and (6) low vulnerability to environmental impacts [54]. Other appealing characteristics of WSNs include self-organizing and self-healing capabilities, high fault tolerance, and ease of interaction with web-based technologies [6]. Furthermore, unlike human-controlled systems, WSNs use self-governing technologies to limit the risk associated with human workers [135]. It should be emphasized, however, that the base station must be installed in a secure location. The base station consumes energy and must be linked to an electricity network [7]. When several sensors are required for large-area monitoring, it is quite costly [51].

The term "wireless sensor network" (WSN) refers to a wireless network that employs a linked sensor to track the state of physical or environmental factors [111]. The terms "wireless sensor" and "smart transducer" refer to sensors that are outfitted with microcontrollers to give intelligence and network capabilities [107]. It should be noted that WSNs can collect data and move information in real time; however, the precision and accuracy of the measurements are mostly dependent on the monitoring mechanism used [54,134].

The sensor nodes, gateway, and monitoring center comprise the landslide wireless monitoring system. Sensor nodes provide data from the field to the administration of the landslide monitoring center. The gateway is responsible for connecting the node to the internet. The monitoring center is in charge of data storage, processing, and analysis. WSNs are primarily composed of hardware and software systems. The wireless communication modules included in the sensors are commonly long-term evolution (LTE), Bluetooth, ZigBee, Wi-Fi, LoRa, etc. Among these, LoRa modulation technology is an appropriate technological solution for node communication [136]. In a WSN, several sensor nodes structure the linked networks into a certain architecture. The usual network structure is depicted in Figure 28. WSNs primarily use the mesh type, star type, and tree type [92]. The hardware system is made up of four components: (1) a wireless transceiver unit that is in a position to establish wireless connections, (2) a control unit that is responsible for data processing,

(3) a data acquisition module that is in control of collecting data from various sensors, and (4) a background monitoring unit that contains real-time multitasking operating management systems. The software system has the role of arranging programming applications (refer to Figure 28) [107,134].

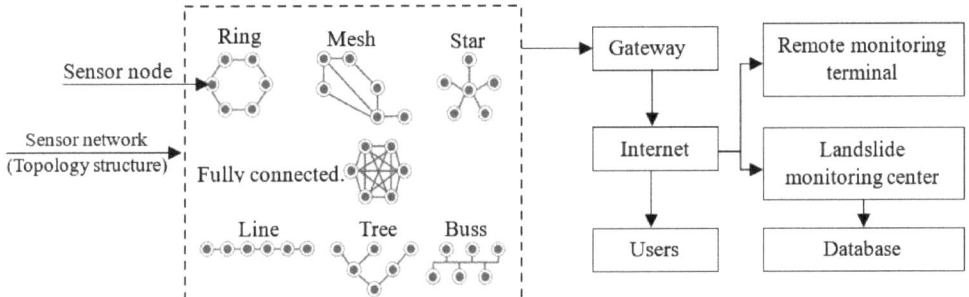

Figure 28. The structure of the wireless sensor network and topology structure (Modified from Yueshun & Wei and Jeong et al. [92,134].

However, wireless sensing networks have some challenging issues, such as energy consumption, memory size, and communication issues [54,137]. To illustrate, the monitoring activity is more accurate if sensor nodes are regularly awakened to sample data, but it has a significant impact on the sensor node lifetime. As a result, it is necessary to develop a flexible system that considers the detection performance, cost, and energy savings [127]. Kumar et al. [137] succeeded in constructing an effective wireless network capable of overcoming the aforementioned drawbacks. This network was built over a 7-acre (approximately 28,328 m^2) rough landscape with 350 sensors, and data was transferred over 320 km to a data center. This system has been functioning for a decade. It has shown itself to be capable of handling heterogeneous sensor readings at rates of up to 1700/s while providing data to the data center with a latency of 10 s.

4.3.1. Energy Consumption Issues

The energy consumption issues are directly related to the amount of transmitted data, sampling rate, and number of sensors, and they are indirectly related to the adopted threshold and prediction accuracy (please see Table 9) [102]. There are three approaches to preserving the system's energy: (1) lowering the frequency at which data are collected, (2) limiting the number of active sensors [103], and (3) improving the self-rechargeability of the power system. As a result, it is critical to comprehend, assess, and construct a threshold that minimizes the sampling rate while maintaining high accuracy. For example, WSNs are capable of making decisions themselves, and data transmission can be minimized during dry seasons [6]. During the rainy season, solar power tends to decline rapidly owing to the increase in the data frequency rate. As a result, limiting energy consumption becomes an overriding concern for the network's long-term operation, particularly when landslides are imminent (for example, heavy-rainfall-induced landslides [4,5,102]).

Table 9. A comparison between different monitoring systems for energy minimization.

	Threshold 1		Threshold 2		Threshold 3		Threshold 4	
Sampling rate	1/s	1/h	1/s	1/h	1/s	1/h	1/s	1/h
Battery lifetime	17 min	26.29 days	9 min	14 days	5 min	6 days	3.8 min	4.5 days
Cost (USD), Kerala, India	150		380		1050 to 3720		2550 to 5220	
Prediction accuracy (%)	50		70		80		>80	
Prediction thresholds	Rainfall (R) (120 mm/day)		R and subsurface moisture state (W_c)		R, W_c, and pore water pressure (PWP)		R, W_c, PWP, and soil movement	

According to Prabha et al. [103], the power consumptions by the sensor nodes, communication system, and processing system were 77.5%, 22%, and 0.45%, respectively. Thus, all attention should be given to the minimization of the sensing power consumption. Ramesh and Rangan [102] studied energy reduction using a prototype field system that consisted of four different types of sensors to measure rainfall, moisture, pore pressure, and movement. Table 9 compares the four alternatives that were used.

Regarding innovative thresholds that are responsible for data frequency lowering, Rosi et al. [7] adopted a threshold that consisted of four stages (quiet stage, quiet-to-motion stage, motion stage, and motion-to-quiet stage), where the sensor starts to collect, store, and send data in the second and third stages only. In the first and last stages, the system shuts down the connection to save energy and maintain accuracy. Ramesh and Rangan [102] established a threshold system with four levels: rainfall (mm), moisture (%), and pore pressure (kPa). The lowest level was (20, 0, 0), while (0, 100, 60) was the highest. At the lowest and highest thresholds, the threshold increased the battery lifespan to 43 days and 63 days, respectively. Another approach was used to reduce further energy use, in which data collection for moisture and pore pressure began after the threshold for precipitation was reached. For the lower and higher thresholds, this threshold could prolong life to 150 days and 400 days, respectively. Additional thresholds can be used, wherein only the sensor with the highest value continues to function while the others go offline. Prabha et al. [103] adopted two thresholds named context-aware data management (CAD) and context-aware energy management (CAE) that can improve the lifetime by six times and twenty times, respectively. To illustrate, the sampling rate of the rain gauge can be modified based on the present rainfall pattern because a significant rise in the rain rate is highly improbable. Sensors for detecting movements, such as strain gauges and tiltmeters, should be detected regularly since their behavior might alter quickly based on specific triggers.

For electromechanical low-power usage sensors, Yang et al. [49] developed a MEMS system that adopted a temporal resolution of 10 min on rainy days and 1 h in dry seasons using four Standard Power 7 Alkaline batteries that can power a single sensor device for more than a year. Abraham et al. [93,94] used a MEMS system with four C-size alkaline batteries and a sensor that sleeps for 10 min after transmitting a signal, extending the battery life in the field. Marino et al. [126] used a technique in which the sensor is turned off when the evaporation rate is very low; otherwise, the data frequency is set to every 2 h. The weight loss between readings was used to estimate the evaporated water. Wang et al. [127] invented the dynamic node cycle. In the absence of unusual movement, this system can be put to sleep; nevertheless, if the sensor node (accelerometer) detects possibly damaging movement, a camera sensor is activated to conduct object recognition and compression transmission. Giri et al. [28] incorporated WSNs with inertial measurement unit (IMU) sensors based on MEMS, which have the advantage of automatically transitioning from a passive state to an active state to save power when no activity is seen for a certain amount of time.

Solar cell systems have been widely used in power-monitoring systems [63,126]. The sensor unit may run semi-permanently without changing the batteries by installing an optional solar battery, which costs approximately USD 5 [49,52,101]. Lin et al. [132] used a unique self-powered wireless sensing method called a zigzag-structured triboelectric nanogenerator (Z-TENG), which has an open-circuit voltage of 2058 V and a short-circuit current of 154 μA. This system has the benefit of using the energy from moving vehicles to power the TTEFBS system. Wireless power transfer (WPT), as a breakthrough method for charging electronic devices, has drawn a significant deal of attention since Tesla's initial WPT experiment at the beginning of the twentieth century to eliminate constant battery changes and charging using plugs. Magnetic resonance wireless power transfer (MR-WPT) offers several benefits, including long coupling distances, high output power, high transfer efficiency, minimal influence from nonferromagnetic barriers, and minimal impact on the human body [138].

Sharma et al. [139] designed WOATCA, which is a revolutionary trust-based energy-efficient protocol based on a whale optimization algorithm that outperforms previous algorithms, such as Adoptive LEACH Mobile (ALM), Topology Control Algorithm for node mobility (TCM), Q12, and secure CH selection protocols. The primary idea is to reduce energy usage by grouping comparable nodes into small disjoint groups (clustering). Ragnoli et al. [136] suggested that LPWAN (LoRa) be utilized in instances where a limited quantity of data is sent at regular intervals. This frequently results in less sophisticated transceiver devices, resulting in lower prices and power. A comparative study of different LPWAN technologies is mentioned in [37]. However, LoRa has several restrictions in terms of data transfer rates. Bagwari et al. [140] integrated LoRa with Wi-Fi architecture and customized the sensor node and gateway node to regularly monitor changes with low energy power consumption.

Hemalatha et al. [57] developed an innovative virtual sensor system based on artificial intelligence models. To illustrate, a machine learning model was created to learn from various sensors over a few years, after which specific sensors were maintained and others were removed. The gained information can be utilized as a virtual sensor for those that were removed. This strategy can both save energy and lower system costs. The sensors that were removed can be employed to gather data in other locations, allowing the system to monitor large regions at a minimal cost and power usage. Jeong et al. [92] developed an innovative technique for optimizing the number of sensors used to decrease both cost and power usage. To demonstrate, a geotechnical investigation was conducted to develop a susceptibility model, and then sensor nodes were placed in areas where the factor of safety was less than unity.

4.3.2. Communication Issues

The system precision is affected by the distance between nodes; the shorter the distance is, the higher the precision. To clarify, the precisions for 110 m, 60 m, and 10 m internode distances were 0.2 m, 0.03 m, and 0.009 m, respectively [54]. Latupapua et al. [111] concluded that the response time rises with the increase in the distance between nodes and station, where the response times were 1.91, 2.98, 3.09, and 4.47 s for distances of 20, 40, 60, and 80 m, respectively, while the monitoring center did not gather any data for distances of 100 m. Rosi et al. [7] adopted a new antenna capable of connecting nodes up to 80 m apart. Mucchi et al. [54] developed a WI-GIM wireless MEMS system with an internode distance between 60 and 90 m. Yang et al. [49] developed a wireless device that can transmit data up to 300 m. Jeong et al. [92] implemented a self-organizing mesh network topology and a time-synchronized mesh protocol (TSMP) to overcome the communication environment of a hilly region; it was found to be more dependable and adaptable than the star network design. Wireless underground sensor networks (WUSNs) cannot be implemented using the current electromagnetic (EM)-based wireless communication technology because it does not match the application requirements of the underground environment (refer to Figure 29). Thus, Wang et al. [141] developed the MIS125-III, which is a magnetic induction communication transceiver that can be buried up to 5.28 m into the ground, and this system is stable without multipath loss, as shown in the comparison between the two systems in Figure 29. The transmitter array technique has the following advantages: (1) a cost-effective method compared with the wireless sensor network, (2) low noise displacement, (3) can be used for ranges up to 250 m, (4) can monitor near-real-time data, and (5) can achieve a centimeter data range. Wang [51] undertook a theoretical study to determine the displacement based on the relative phase difference from two demodulated signals by installing a transmitter (Tx) at the area of interest (AOI). The transmitter is either hardwired or designed to send signals in a coordinated order. The two receivers are spaced close together and demodulate the received signals separately yet coherently.

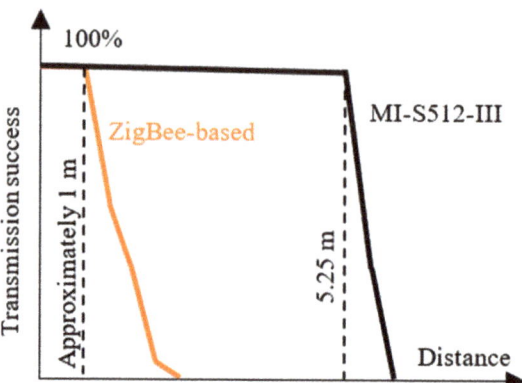

Figure 29. The transmission success rate of MI versus EM (ZigBee). "Reprinted/adapted with per-mission from Wang et al. [141], Wiley".

4.3.3. Data Loss and Size Issues

A large amount of environmental and geophysical data collected by a variety of sensors and systems suffers from high levels of ambiguity, noise, and missing data. To illustrate, the nature of the observed monitoring data fluctuates according to external triggering (rainfall, earthquakes, etc.), and missing records are highly expected to occur throughout the monitoring. Data loss may indicate that the program's goals were not achieved, which is more than just a negative situation. Blahůt et al. [142] revealed that while measuring displacement, missing measurements accounted for approximately 24.6%. To address the aforementioned shortcomings, time-series analysis was used, which included various statistical approaches, such as regression models, to comprehend the underlying context of data points or to make predictions based on prior behavior. A second-order polynomial can be used to approximate trend data representing creep behavior: it should be noted that the displacement can be divided into creep displacement (simple trend) and periodic displacement (complex trend) [4]. Paired adaptive regressors for cumulative sum (PARCS) were utilized for periodic data. Sumathi and Anitha [143] designed a lossless landslide-monitoring (LLM) system. During the data collection and processing phases, two algorithms were used. A modified gray wolf optimization method was employed in the first phase, and an iterative dichotomize-3 (ID-3)-based decision-making strategy was applied in the second phase. This method boosted the delivery ratio by 30%. de Souza and Ebecken [144] adopted artificial intelligence models to predict missing data using principal component analysis (PCA) combined with artificial neural networks (ANNs). Shentu et al. [145] analyzed the monitoring data using a small Feedback Optimizing Background Gray Model (FOBGM (1, N)). Wang and Zhao [146] employed time-series analysis using mean-based low-rank autoregressive tensor completion (MLATC). Li et al. [147] adopted cubic spline interpolation to estimate the missing data. Long et al. [148] employed multi-feature fusion transfer learning (MFTL), assuming that landslides with similar geographical and geological characteristics are comparable but different in magnitude. Practically, De Graff [1] recommended using a parallel landslide monitoring system to overcome the issue of data loss. In other words, when a sensor suffers a data loss issue, the parallel one can help to predict the missing data.

To reduce the amount of data, Wang et al. [127] used an efficient symbolic approach to transform a time series of sensor data into an ordered symbol string, which solved the data volume problem. This method was discovered to keep the critical aspects of the data while reducing its dimension from 128 to 16. Gian et al. [96] developed a novel compressed sensing (CS) technique to provide a novel technique for reducing the data size and power usage. The Fourier transform was employed to turn time-domain data into frequency-domain data, with the transmission based on Fourier coefficients, and a nonlinear method was used to recreate the original data. The optimal compression ratio was found to be 0.55.

4.4. Physical and Prototype Systems

It is challenging to see any purpose for implementing a monitoring program if the devices being used cannot record data with the required frequency, accuracy, or precision stated by its objectives. To illustrate, the difference between precision and accuracy is visualized (the bull's eye targets) in Figure 30 [149]. In the following sections, both experimental and prototype modeling are discussed for the better simulation and investigation of landslide monitoring systems.

   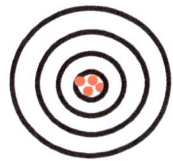

Not accurate and not precise Accurate and not precise Not accurate and precise Accurate and precise

Figure 30. Bull's eye visualization for accuracy and precision conditions.

4.4.1. Experimental Models

Laboratory model testing is a powerful technique that plays a vital part in landslide engineering studies. Although time-consuming, scale-model testing has helped to advance our understanding of landslide causes and processes. The most accurate way to analyze landslides is via laboratory model studies. This is due to the possibility of continuous monitoring of the water content of the soil slope, as well as subsequent deformation, which allows for the management of the soil characteristics and boundary conditions [97,99]. Abraham et al. [93,94] recommended laboratory-scale research that would resemble several types of landslides and identify different criteria for each instance because field testing is expensive, and failure may not occur or occur at a slow rate. Ivanov et al. [82] proposed using experimental-scale modeling to avoid concerns with field testing, such as temperature effects and the harsh environment.

Iai [33] proposed a law for simulation in order to recreate the prototype circumstances in terms of geometry, material properties, beginning state, and boundary conditions. The Buckingham π theorem [150] provides the scaling parameters between the prototype and model, as listed in Table 10, where the length, cohesion, and elastic modulus can be scaled by a constant factor λ; the permeability scaling factor can be $\lambda^{0.5}$; and the density, friction angle, and gravity has a scale factor of 1. According to Iverson [36], the larger the experimental apparatus is, the fewer the scale effects concerning the velocity of a sliding landslide body. Ivanov et al. [82] emphasized the influence of the temporal scaling factor between small-scale experiments and full-scale phenomena, which may be reasonably expected to be greater than 10 based on his model. It is crucial to note that all laboratory tests were conducted in a controlled setting with little outside noise or vibration at a steady room temperature. As a result, these systems need to be revised and established for field monitoring [8].

Table 10. Simulation law for prototype and physical modeling.

Parameter	Length, Cohesion, and Elastic Modulus	Density, Friction Angle, and Gravity	Permeability
Scale factor	λ	1	$\lambda^{0.5}$

4.4.2. Prototype Working Process

The installation of the subsurface monitoring system was illustrated by Chuan et al. [9]. The process consists of (1) hole drilling, (2) monitoring system installation, (3) drilling pipe installation, (4) sensor checkup, (5) powering the system, (6) data analysis, and (7) data processing. First, the depth of the borehole (i.e., sensor tip) is determined using drilling machinery based on the depth of the sliding mass and geological soil profile. The drill pipe is then removed, and the sensors are mounted in accordance with the design objectives. An initial examination is required to ensure that the sensor is linked to the subsurface soil. The next step is to turn on the system and begin data collection and storage. The data are then processed and displayed before being examined. Adopting a probable prediction model based on the processed data is the last stage, as illustrated in Figure 31.

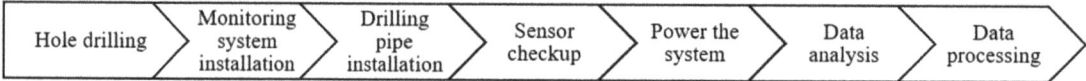

Figure 31. Subsurface monitoring system flowchart.

Zheng et al. [59,60] utilized the aforementioned procedure while installing FBG-based inclinometers. Zheng et al. [89] installed a prototype COFT system in which a borehole was drilled, the OFS was installed, and then cement mortar was injected into the gap between the borehole and the sensor. Similarly, Kumar M and Ramesh [10] installed the SG-DEP sensor from the soil surface to the underlying bedrock using the previous approach. Digging holes in unreachable high mountains is also impossible. As a result, remotely controlling and installing subsurface monitoring systems is critical, especially in harsh outdoor environments. Thus, Molfino et al. [151] developed an innovative robot named Roboclimber that is entirely operated using wireless links. Roboclimber is an autonomous mobile drilling device that can drill boreholes up to 20 m deep and climb (i.e., provide a mobile robotic platform) slopes up to 85 degrees for difficult terrain and rocky landslides (please see Figure 32).

However, one of the most critical aspects of landslide monitoring is the deployment of such monitoring devices in the field. While installing monitoring systems, laborers' health and lives are put at risk by dust, vibrations, accidents caused by falling rocks, etc. These systems are sometimes expensive to construct and operate, restricting their application to well-funded projects. Therefore, a MEMS was recently designed in which sensor modules can be embedded in the ground using a small hammer, which is suitable for shallow landslide monitoring [49]. Qiao et al. [52] installed different wireless MEMS tilt and temperature sensors with different rod lengths for shallow and rotational landslides. These sensors are small in size, have a small weight, and can provide multivariate parameters at the same point. Regarding the installation time of such sensors, Mucchi et al. [54] installed a cluster of 11 wireless MEMS sensors in less than two hours. Figure 33 describes the installation process of the small-sized sensor [101]. Following the subsurface monitoring system flowchart, this process are as follows: (1) soil removal; (2) installing a borehole; (3) removal of the borehole case; (4) inserting a steel rod; (5) mounting the tilt sensor; (6) powering and cabling the system.

Figure 32. Roboclimber field deployment "Reprinted/adapted with permission from Molfino et al. [151], Elsevier".

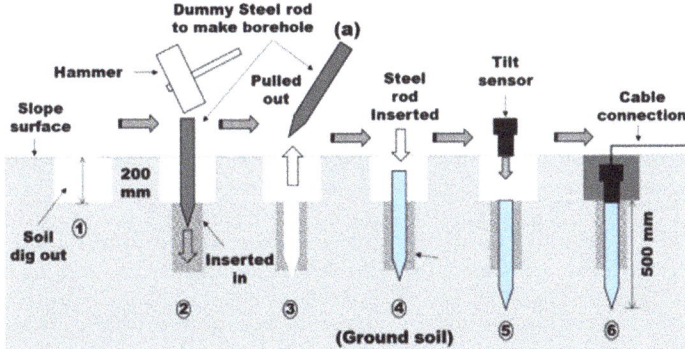

Figure 33. The installation process of small sensors (From Sheikh et al. [101]).

However, because of the nature of target terrains, the required target placements are not always easily accessible to people. Therefore, robotic solutions and unmanned ground vehicles (UGVs) are the sole options for deploying a wireless sensor network, repairing malfunctioning nodes, and charging the batteries of previously placed nodes. Patané [152] created a bioinspired robotic system that combines wheeled and legged robots to deploy a succession of smart sensors at specific sites.

4.4.3. Field Systems

Before implementing any monitoring system, it should be noted that field investigation and laboratory testing are required [92,153]. A site study can offer basic information regarding landslide classification, soil profile and features, sliding surface location, etc. The field investigation program includes a (1) surface geological survey, (2) borehole survey, (3) test pit, (4) standard penetration test (SPT), (5) field density test, (6) field permeability test, (7) surface permeability test, (8) cone penetration test (CPT), (9) refraction seismic survey, and (10) multichannel analysis of surface waves (MASW). Furthermore, laboratory tests for assessing soil attributes include (1) soil classification, (2) water content, (3) Atterberg limits, (4) grain-size distribution, and (5) soil water characteristic curves (SWCC) [92]. Because the previous research strategy is time consuming, subsurface studies employing the geoelectrical resistivity method may be a feasible option. Geoelectrical resistivity is calculated by passing an electric current through a current electrode into the ground and measuring the differential potential of a region. Hasan et al. [153] investigated subsurface soil properties based on the distribution of resistivity values of the soil using the Schlumberger geoelectrical resistivity technique of eight locations with 1 m electrode spacing.

The monitoring locations can be selected using four spatial distribution methods: random, matrix, vulnerable, and hybrid. The monitoring locations in a random method are installed at nonspecific random places on a landslide-prone slope. The whole area of deployment is split into a matrix of NxN cells in the matrix method, and one monitoring probe is placed in each cell of the matrix. Monitoring stations are placed in vulnerable (i.e., critical zones) zones identified during the site investigation, topography mapping, and soil testing in the vulnerability technique. In the hybrid technique, both the matrix and vulnerable approaches are used, i.e., start with the matrix and then adjust the locations of the monitoring devices based on the most critical (i.e., vulnerable) locations [10].

Most of the preceding methods can offer a single measurement (displacement, soil moisture, etc.); however, the possibility of high false alarms limits its usage. Thus, such data should be correlated with other monitoring data, such as rainfall or soil moisture, to reduce such effects. As a result, multimonitoring systems are strongly advised. Multi-monitoring systems can be produced by combining the individual systems shown above or by designing a single sensor node with many functionalities. Multifunctional sensor devices that make use of MEMS sensors and WSNs are now commercially available. Chuan et al. [9] created a system that includes pore water pressure sensors, a stress sensor, and a displacement sensor. However, this system does not support wireless data transmission, and data is stored on an SD card. Ramesh and Vasudevan [6] adopted one of the prototype WSNs by incorporating a variety of subsurface sensors (piezometers, dielectric moisture sensors, strain gauges, tiltmeters, and weather stations). Yunus et al. [107] used a system called wireless sensor network for landslide monitoring (WSNLM) that includes soil moisture, vibration on land, slope angle, soil temperature, air temperature, humidity, and atmospheric pressure. Gian et al. [96] utilized a cost-effective wireless monitoring unit consisting of soil moisture, temperature, tilt meter, geophone sensors, and weather station to monitor rainfall and wind speed and direction. Jeong et al. [92] built a wireless sensor network in which a sensor node consists of a rain gauge, tensiometer, soil moisture sensor, and inclinometer. Yang et al. [49] used a multivariate wireless monitoring sensor (MEMS) that includes soil moisture, soil matric suction, ground vibration, tilt, and rainfall sensors. This device, which can offer real-time data, costs approximately 1500 USD per point. Similarly, Abraham et al. [93] used a MEMS tilt sensor and volumetric water content adopted by Abraham et al. [94], which had the following features: this system is appropriate for shallow landslides where the tilt sensor depth was 1 m and the volumetric water content sensor was 3 cm below ground level. Sheikh et al. [101] built prototype field experiments to investigate the relationship between the tilt angle, displacement, strain, ground level, and rainfall using wireless sensors (tilt sensor, pipe strain gauge, water level gauge, and rain gauge) (refer to Figure 34). Tables 11 and 12 list the physical and prototype system characteristics.

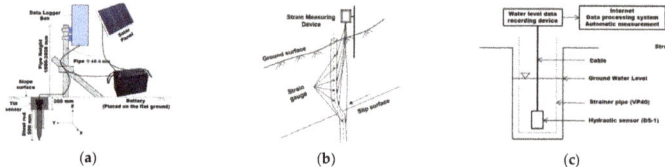

Figure 34. Schematic view of the multifunction monitoring node: (**a**) tilt sensor; (**b**) pipe strain gauge; (**c**) groundwater sensor (from Sheikh et al. [101]).

Table 11. Physical monitoring systems.

Study	Adopted Monitoring System	Model Dimensions (B × L × H) cm	Soil Type and Thickness	(λ)
Schenato et al. [65]	Strain sensor: BRUsens© strain v9 cable (Brugg Kabel AG, Brugg, Switzerland); measurement interval of 1% strain; spatial resolution of 10 mm. Optical frequency domain reflectometer: OBR4600 from Luna Innovation Inc. (University of Padova, Padova, Italy). to measure the strain. Tensiometers: to measure pore water pressure. Water content reflectometer: to measure volumetric soil moisture. Temperature probes: to measure soil temperature. Tipping bucket flow gauges: to measure rainfall intensity.	(200 × 600 × 350) Slope angle 31.14°	A shallow sand layer (60 cm) overlies the clay layer.	–
Ma et al. [53]	3D laser scanner: RIEGL VZ-400 for continuous surface movement measurement. Video camera: for continuous surface movement measurement. Earth PC: Model XTR-2030 to measure earth pressure; capacity of (0–500) kPa. TIR camera: FLIR SC660 to measure surface temperature; temperature sensitivity of 0.03 °C.	(90 × 200 × 74) Varied slope angle	A 4 cm thick sliding layer. Stiff material of clay, sand, bentonite, and water. The soft material of clay, glass beads, and water.	40 (1 g)
Zheng et al. [84]	Combined optic fiber transducer (COFT): the minimum error was achieved using a cement mortar ratio of 1:5 and EPS material (average value of 4.12%). Initial measurement of 1 mm. Maximum sliding distance of 26.5 mm. Dynamic range of 0–23.2 mm. Unit price of 0.2 USD/m.	(200 × 450 × 160) Slope angle 60°	A predefined circular failure surface. A mixture of clay and river sand.	–
Giri et al. [8,29]	3-axis accelerometers BNO055 sensor devices (IMU sensors) with 3-axis accelerometers and 3-axis gyroscopes to measure gravitational acceleration, linear acceleration, and angular velocity. Suitable for translational landslides.	(149 × 183 × 30) Slope angle 35°	Test 1: 12.7 cm sand underlying a 6.35 cm layer of sandy gravelly clay. Test 2: 25.4 cm of homogeneous sand slope.	–
Askarinejad & Springman [105]	SDS sensor: developed at the Institute for Geotechnical Engineering at ETH Zurich to measure horizontal displacement. Initial measurement is <1 mm; data frequency 10 Hz; bending stiffness less than PVC inclinometer by 300 times. Suitable for sand and silt rapid landslides.	(100 × 100 × 750)	A predefined forced failure surface using a hydraulic jack. Poorly graded sand.	–

Table 11. *Cont.*

Study	Adopted Monitoring System	Model Dimensions (B × L × H) cm	Soil Type and Thickness	(λ)
Kuang [133]	Glowstick (chemiluminescence) Powerless and low-cost system to produce early signs based on soil deformations. Extremely sensitive to small motions in the mm range.	(80 × 80 × 80)	A predefined failure surfaces at the mid-elevation. Test 1: the soil was sand. Test 2: the topsoil was clay, and the bottom was sand.	—
Chen et al. [97,99]	Soil moisture sensors: EC-5 by Decagon Devices, Inc. (Pullman, WA, USA), to measure volumetric water content; data frequency of 1 s. Tiltmeter sensor: microelectromechanical system (MEMS) sensor to measure tilt angle; data frequency of 1 s. The elastic wave monitoring system: consists of an exciter, receiver, microcontroller, and data acquisition unit; the exciter is a solenoid (ZHO–1040 L/S by Zonhen Electric Appliances HK Co, Ltd., Shenzhen, China); the receiver is a piezoelectric vibration sensor (VS–BV201 by NEC TOKIN Corporation, Sendai, Japan). Artificial rainfall: spray nozzle (SSXP series by H. IKEUCHI & Co., Ltd, Wuhan, China). This system is suitable for shallow infinite landslides.	(1) Small, fixed test (30 × 70 × 40) with slope angle 45°; (2) small test with varied slope angle; (3) large-scale model (—×790 × 500) with slope angle 45°	Brown natural sand. Three different thicknesses of 5, 10, and 15 cm vertically over the base layer for small tests. Soil thickness of 1 m vertically over concrete for large tests.	—
Chen et al. [98]	Adopted centrifuge tests using the same setup as the small fixed and varied tests listed above by Chen et al. [97,99]			50 g
Zhang et al. [130]	Infrasound–mechanics system Consists of an infrasound sensor (IDS2016), steel tube, pressure sensor, signal transmission line, and signal analysis appliance. IDS2016 infrasonic sensor is small and cost-effective and has the following characteristics: (1) extremely sensitive (50 mV/Pa) to monitor weak signals and (2) high measuring range (0.5 to 200 Hz) to cover a wide frequency range. The seal tube was inserted to a depth of 20 cm from the bottom of the sliding mass to minimize the external noise from the soil mass around the failure surface. ISDAS2016 acquisition device has the following characteristics: (1) low noise, (2) low power consumption, (3) high synchronization accuracy, and (4) sampling rate of 100 Hz.	Small-scale model Slope angle 28°	The failure surface was forced using a hydraulic jack. Seven soil samples (fine sandstone, dolomitic limestone, calcareous mudstone, marl, purple mudstone, calcareous shale, and purple mudstone) with different densities and moistures were adopted.	—

Table 11. *Cont.*

Study	Adopted Monitoring System	Model Dimensions (B × L × H) cm	Soil Type and Thickness	(λ)
Qiao et al. [52]	Tilt sensor: MEMS wireless sensor with a nominal resolution of (0.0025° = 0.04 mm/m); different rod lengths were used (50 mm, 300 mm) through field tests with artificial rainfall.	(45 × 116.5 × 38) Slope angle 43°	Silica sand was used. Test 1: rainfall triggering with fixed slope angle. Test 2: variable slope angle with a defined circular slip surface.	–
Ivanov et al. [82]	TDR: to measure water content. Camera: to visually record the landslide. C-OFTDR—Setup 1: the fibers were perpendicular to sliding directions; setup 2: the fibers were parallel to the sliding direction. Sampling frequency of 20k sample/s; temporal resolution of 15 min; cost approximately EUR 5k. This sensor device is unable to pinpoint the exact location of strain along the line.	(80 × 200 × –) Modifiable angle up to 45°	Homogeneous fine sand with a thickness of 15 cm to simulate shallow landslide. Artificial rainfall was adopted as external triggering.	Temporal scale of 10
Minardo et al. [87]	Tensiometer: to measure soil matric suction. Laser sensor: to measure soil deformation. Camera: to retrieve data using particle image velocimetry (PIV). BOFDA: to measure soil strain; spatial resolution of 5 cm; temporal resolution of 3 min.	(50 ×110 × –) Slope angle 35°	Cohesionless soil with an internal friction angle of 38°. Soil thickness of 13 cm to simulate shallow landslide.	–
Xie et al. [95]	Tilt meter: to measure soil tilting; accuracy of 0.1 degrees. Extensometer: to measure soil displacement; accuracy of 0.1 mm. Digital camera: to monitor soil movement at marked points.	Model 1: (45 × 116.5 × 38); slope angle 40° Model 2: (– × 70 × 30); slope angle 39°	Sandy soil. Model 1: the failure surface was predefined as circular. Model 2: artificial rainfall was applied as external triggering.	–
Xiaochun et al. [118]	High–density electrical instrument: to measure the resistivity with DZD–8 multifunction full waveform DC electrical apparatus (Chongqing Geological Instrument Factory, Chongqing, China). A total of 30 high-density copper electrodes were set on the surface of the model, and the distance between the electrodes was 20 cm. Temporal resolution of 30 min. Soil moisture sensor: to measure soil moisture content (Linde Intelligent Technology Co., Ltd., London, UK).	(80 × 600 × 160) Slope angle 15°	Silty clay was sandwiched with crushed stone and crushed stone soil. Reservoir level change was simulated as external triggering.	–
Liu et al. [11]	Soil moisture sensor: SEN0193. Micropore water pressure transducers: KPG PA. MEMS sensors: 9-axis gyroscope and magnetic meter to measure the deflection angle of soil. The warning signs are based on the values of the factor of safety.	(40 × 80 × 45) Slope angle 45°	Homogeneous slope. Artificial rainfall was adopted.	–

Table 12. Prototype monitoring systems.

Study	Prototype System Components	Notes
Kotta et al. [106]; Rosi et al. [7]	Vibration sensor (accelerometer) to measure the biaxial acceleration change.	Remote real-time system.
Ramesh & Vasudevan [6]	Piezometers: current-based 4-20 mA output piezometers were chosen to eliminate wire length errors; additional filter piezometer tips were installed that could easily be removed for calibration and reinstallation. Dielectric moisture sensors: to measure the volumetric water content; used to quantify a relationship with rainfall infiltration. Strain gauges: the strain gauges were fixed at the outer diameter of an inclinometer casing surface; able to measure four dimensions (X, Y (90°), α (120°), β (240°)). Tiltmeters: attached inside the inclinometer casing to calibrate the strain gauges. Weather station: tipping bucket rain gauges were used to measure (rainfall, humidity, temperature, wind speed, wind direction, pressure).	Remote real-time system.
Chuan et al. [9]	Force sensor: maximum capacity of 500 kPa; voltage of 9 V; output signal 0–5 V; deviation (%) 0.39–2.37; precision of 1%. Pore water pressure meter: maximum capacity of 100 kPa; voltage of 10 VDC; output signal 0–5 V; calibrated using standard test equipment; deviation (%) 0.0–0.26; precision of 0.3%. Displacement sensor: guyed-type displacement sensor; maximum capacity of 200 mm; voltage of 5 VDC; output signal 0–5 V; deviation (%) 0.27–2.22; precision of 0.5%.	Data are recorded on an SD card. Every 5 days the SD card has to be emptied.
He et al. [109]	Stress sensor: measure the sliding force in the anchor cable.	Remote real-time system.
Wang et al. [72]	FBG-based inclinometers: Horizontal displacement with high accuracy in millimeter range; spatial resolution of 1 m; FBG of an internal diameter of 7 cm and thickness of 5 mm; aluminum inclinometers were used where it was expected that the inclinometer casing was consistent with the soil.	Based on wiring and field data collection.
Zheng et al. [59,60]	FBG-based inclinometers: precision of 0.02 mm; maximum deflection up to the damage of the tube; ABS inclinometer from Changzhou Jin Tu Mu Engineering Instrument Co., Ltd. Changzhou, China; the strain collection device TST3826 from Test Electronics Equipment Manufacturing Co. Ltd., Beijing, China.	Based on wiring and field data collection.

Table 12. Cont.

Study	Prototype System Components	Notes
Yunus et al. [107]	Soil moisture sensor and soil temperature sensor. Vibration transducer: to monitor the hill slope. Accelerometer: ADXL 335 accelerometer for slope angle measurement. Seismograph: to measure the seismic vibration using a Visaton FR8 8–ohm loudspeaker. Weather station: to measure temperature, humidity, and atmospheric pressure.	Remote real-time system.
Yang et al. [49]	Inclination sensor: range ±30° resolution 0.0025 degrees. Soil moisture sensor: Decagon EC-5, Decagon Devices, Pullman, WA, USA; up to 30 cm deep; measure volumetric water content using a dielectric constant; range 0–1; accuracy of ±0.01. Soil suction sensor: Tensiomark TS2, Stevens Water Monitoring Systems, Portland, OR, USA; range 0–300 kPa; accuracy ±0.15 kPa. Rain gauge: CAWS 100, Huayun Group, Beijing, China; resolution of 0.2 mm; range 0–4 mm/min.	Remote real-time system. The temporal resolution of 10 min on rainy days and 1 h on dry days. Costs USD 1500.
Prabha et al. [103]	Geophone: to convert the ground movement into voltage. Inclinometer: to measure the slope movement angle. Strain gauges: to measure the micro movement. Dielectric moisture sensor: to measure the volumetric water content. Piezometers: to measure the volumetric water content. Tipping bucket: to measure the rainfall.	Remote real-time system. Two thresholds were adopted to save power consumption.
Askarinejad et al. [73]	Soil deformation sensors (SDS): <1 mm; data frequency 100 Hz; bending stiffness less than PVC inclinometer by 300 times; range 0–25 mm; accuracy = 5%. Earth pressure cells (EPCs): to measure horizontal earth pressure; data frequency 100 Hz; range 0–500 kPa; accuracy = 1 kPa. Piezometer: to measure the groundwater table; range 0–100 kPa; accuracy = 1 kPa. TDR: to measure volumetric water content; range 0–1; accuracy = 0.02. Tensiometer: to measure pore water pressure; range –90 to 100 kPa; accuracy = 0.5 kPa. Strain gauges: to measure bending strain; installed on SDS; range –50 to 20 mε; accuracy = 1 μm. Cameras: multicamera surface monitoring (5 fps).	Remote real-time system. Artificial rainfall was adopted with different intensities and durations.

Table 12. Cont.

Study	Prototype System Components	Notes
Crawford & Bryson [122]; Crawford et al. [62]	Water content reflectometers: Campbell Scientific CS655 to measure volumetric water content, electrical conductivity, dielectric permittivity, and temperature; the sensor was installed at different depths. Porous ceramic disc: MPS-6 sensor to measure the water suction; the sensor was installed at different depths. Cable-extension transducer (CET): to measure the movement; the output signal was voltage and converted to linear displacement. Rain Wise Inc: to measure the rainfall based on a tipping bucket rain gauge; calibrated at 0.25 mm/tip. This system was used to correlate the electrical resistivity measurements (geophysical) with geotechnical measurements.	Based on wiring and field data collection. Data frequency was in 15 min, hourly, and daily average intervals.
Gian et al. [96]	Compressed sensing (CS) was adopted to reduce the amount of data and save power consumption. A multimonitoring system to measure soil moisture, temperature, tilting, and vibration using Geophone, and a weather station to monitor rainfall, wind speed, and wind direction.	Wireless sensor network.
Ho et al. [68]	Inclinometer: accuracy of 2 mm per 25 m; spatial resolution of 0.5 m; range −30° to + 30°. TDR: HL 1101; accuracy of 2 mm; spatial resolution of 0.05 m; range up to 210 −mp reflection coefficients.	Based on wiring and field data collection.
Zheng et al. [78]	COFTI: initial measurement 0.98 mm; maximum range of 36 mm; cost of 0.45 USD/m; consisted of stainless-steel connectors, protective covers, acrylonitrile butadiene styrene (ABS) plastic pipes, capillary steel pipes, and single-module optical fibers.	Based on wiring and field data collection.
Chung & Lin [69]	TDR: RG-8 coaxial cable for soil slopes; P3–500 CA for rock slopes; 75.7 mm in diameter; sand and gravel were suggested to be mixed into the grout cement when grout loss occurs, with water/cement ratio of 1; the spatial resolution was 5 cm; can detect the sliding depth, though displacement quantification is difficult.	Based on wiring and field data collection.
Tao et al. [108]	Constant resistance and large deformation (CRLD) anchor cable: to monitor the sliding force; 900 kN cumulative sliding force was set to be the critical warning level; able to forecast the landslide 4 h before the event.	Based on wiring and field data collection.

Table 12. *Cont.*

Study	Prototype System Components	Notes
Abraham et al. [93,94]	Tilt sensor: accuracy of 0.017°; resolution of 0.003°; sensitivity of 4 V/g; the output was the digital voltage, which was then converted to a tilt angle. Volumetric water content sensor: precision of ±3%; response time of 10 ms; resolution of 0.002 m³/m³. Small dimensions and affordable. The sensor sleeps for 10 min after sending a signal.	Remote real-time system. Suitable for shallow landslides.
Blahůt et al. [142]	3D dilatometer: To measure the movement of slow-moving landslide in (X, Y, Z) directions; TM-71; high precision of ±0.007 mm; temporal resolution of 24 h.	Slow-moving landslide. Automatic data processing.
Zheng et al. [83]	Quasi-distributed fiber-optic displacement sensor (QDFODS): initial measurement of 0.98 mm; maximum value of 36 mm; can determine the sliding surface while the spatial resolution can be determined based on the site investigation studies; used stainless-steel covers to protect the optical fiber and bowknot bending modulator.	Based on wiring and field data collection.
Jeong et al. [92]	Tensiometer: to measure soil water suction; jet fill tensiometer; range 0–100 kPa; accuracy of 1%. Soil moisture sensor: Soil Moisture Equipment Corp. (EC-5); range 0-100%; accuracy of 3%. Rain gauge: KWRG-105 (Wellbian system); accuracy of 3%. Inclinometer: SCA123T-D07 (Murata Electronics); range of −30° to +30°; accuracy of 1.5%. This system is suitable for rainfall-induced landslides, as it can monitor suction stress, soil moisture content, and rainfall.	Remote real-time system. A site investigation was adopted to optimize monitoring locations.
Segui & Veveakis [30]	Thermometer: to monitor the temperature of deep-seated landslides; resolution of 1×10^{-4} °C the sensor was installed at the shear band, which required prior investigation. Piezometer: to monitor water pressure and temperature. Extensometer: To monitor the displacement.	Based on wiring and field data collection.
Wicki & Hauck [116]	ERT: to calculate plot-scale soil moisture fluctuation; spatial resolution of 25 cm; temporal resolution of 2 h during rainy days and daily otherwise; installed approximately 10 cm into the soil. Soil moisture sensor: capacitance-based soil moisture sensors (5TE, METER Group) to verify the ERT method and measure VMC; inserted at different depths (0.15 to 1 m). Tensiometers: T8 Tensiometer, METER Group to measure SWP and verify the ERT method; inserted at different depths (0.15 to 1 m).	Automated ERT system. Can provide spatial resolution instead of point sensors.

Table 12. Cont.

Study	Prototype System Components	Notes
Minardo et al. [87]	BOFDA: The spatial resolution of 5 cm; the temporal resolution of 3 min; monitors rock fall.	Based on wiring and field data collection.
Sheikh et al. [101]	Tilt sensor: can measure tilting in two directions; tilt range ±20°; resolution 1/1000°; precision 10/1000°; service temperature −20 to +60 °C; water pressure resistance 0.5 MPa; temporal resolution of 10 min. Pipe strain gauge: to verify the tilt reading; resolution of 1 microstrain; measuring range of ±20,000 macro strain; temporal resolution of 60 min. Groundwater sensor: measuring range 0–100 m; resolution of 100 mm; temperature compensation range of 0 to −30 °C; temporal resolution of 60 min. Rain gauge: fall mass type; measurement unit of 0.2/1 pulse; temporal resolution of 60 min.	Wireless automatic system. Suitable for shallow landslides. Adopted solar power batteries to overcome power consumption issues.
Chu et al. [125]	Soil moisture sensors: 6 sensors at different depths/nodes, with 3 sensors of STEMMA and 3 sensors of Teros; STEMMA sensors were verified using standard Teros sensors. Accelerometer: to measure ground vibration for early warning; the sensor was turned off after initial development to reduce noise and save battery life. Rainfall: Aerocone tipping bucket rain gauge. Humidity sensor: SHT31D. Piezometric sensor: to measure groundwater level. Pressure and temperature sensor: to measure atmospheric pressure (MS58302).	A wireless automatic system called SitkaNet. Low cost at less than 1000 USD/node. A 5 min temporal resolution.
Xiaochun et al. [118]	High-density electrical instrument: to measure the resistivity with DZD-8 multifunction full waveform DC electrical apparatus (Chongqing Geological Instrument Factory); a total of 40 high-density copper electrodes were set on the surface of the model, and the distance between the electrodes was 2 m. Sample drying method: to measure the soil moisture content.	Based on wiring and field data collection. To test the AI model developed through lab and physical models.
Wielandt et al. [100]	Three-axis accelerometers: ADXL345 from Analog Devices to measure the inclination and deformation of the surrounding soil; the probe is thin and semi-flexible with a length of 1.8 and internal diameter of 6.35 mm; 0.390 mm resolution and a 95% confidence interval of ±0.73 mm per meter of probe length; depth spatial resolution of 100 mm; acceleration range of ±2 g. Temperature sensor: TMP117AIDRVR; high resolution of 0.0078125 °C; accuracy of ±0.1 °C in the −20–50 °C range. Suitable for shallow landslides.	Wireless automatic system. Suitable for shallow landslides.

Table 12. *Cont.*

Study	Prototype System Components	Notes
Setiono et al. [63]	Optical-based wire-extensometers: offers a high resolution of 0.011 ± 0.0083 mm and a speed limit of approximately 36 mm/s.	Wireless automatic system. Suitable for shallow landslides.
Marino et al. [126]	Soil moisture sensor: to measure the volumetric water content; this system is combined with a full meteorological station, tensiometers, and TDR probes; the correlation between the volumetric water content and the sensor output voltage (Vout − 1) reached an R^2 of 0.98.	Wireless automatic system. Suitable for shallow landslides.
Blahůt et al. [142]; Zhang et al. [35]; Zheng et al. [83]	DSS: a novel distributed strain sensing (DSS) cable based on Brillouin frequency; improved soil coupling and developed a new mathematical general model (AIM); depth spatial resolution of 1 m; displacement range based on the field tests up to 12 cm with millimeter range. Quasi-distributed fiber-optic displacement sensor (QDFODS): initial measurement of 0.98 mm; maximum value of 36 mm; can determine the sliding surface while the spatial resolution can be determined based on site investigation studies; used stainless-steel covers to protect the optical fiber and bowknot bending modulator. 3D dilatometer: to measure the movement of slow-moving landslide in (X, Y, Z) directions (TM-71); high precision of ±0.007 mm; temporal resolution of 24 h.	Based on wiring and field data collection. Slow-moving landslide. Automatic data processing.

5. Research Gaps and Future Directions

High-accuracy monitoring can be achieved by considering two main factors: (1) selecting an appropriate monitoring system based on better knowledge of the case study's initial conditions, and (2) selecting a suitable technique to interpret and transfer the data. Choosing the most effective monitoring system necessitates a deep understanding of the triggering conditions, as each case has its distinct features. Thus, in the concluding section, the effective use of each subsurface monitoring system is illustrated (refer to Section 6). However, regardless of the advancement in data transfer and monitoring techniques, some gaps still need to be filled. Methods for interpreting the monitoring data using advanced complex statistical models that better represent such complicated data are missing. Most available techniques are suitable for small regions and are limited by power issues that necessitate developing new systems for wide areas. Some techniques (i.e., warning and subsurface temperature mechanisms) are still undergoing testing and require more investigation to quantify their response to the failure mechanism. It should be noted that a wide range of data issues are considered from the computer science point of view independent of the accuracy of such data from the geotechnical point of view. Installing such systems in a harsh environment is still challenging in terms of both the location and the technique, which necessitates using robotic systems to install such systems considering the vulnerable locations. A common issue about data loss has been considered using statistical models, which neglect the physical slope characteristics where parallel monitoring is missing. Table 13 summarizes the research gaps and provides the recommendations.

Table 13. The research gaps in subsurface landslide monitoring.

Gap	Recommendations
Simple regression analysis was widely utilized to interpret the monitoring results. However, the relationship between subsurface monitoring parameters is complicated and complex.	Using artificial intelligence models is limited in the subsurface monitoring system. Thus, the aforementioned models can provide a possible solution to filling such a gap [4].
Developing a distributed monitoring system that can provide subsurface parameters for wide areas with a large monitoring range, high spatial resolution, suitability for harsh environments, and being self-powered is still a challenging gap to overcome.	Collaboration is needed between different disciplines to design a multi-feature system. To illustrate, triboelectric nanogenerators and wireless power transfer systems can be utilized to power the subsurface monitor system. Moreover, further research is needed to achieve a large monitoring range with high resolution (i.e., the optical fibers).
Warning sign techniques and the subsurface temperature mechanism are still under development and require more research.	More laboratory-scale modeling and prototype field tests are needed to quantify and investigate such techniques.
Data transfer power issues have been widely studied from the perspective of computer science, while considering the accuracy of the data from the geotechnical perspective is still lacking.	A sensitivity analysis considering different frequency rates and different sensor threshold limits is needed to account for the system accuracy considering both power, data size, and accuracy optimization.
Installing the subsurface monitoring system is challenging in terms of (1) accessing the slope and (2) choosing the optimal vulnerable location to be monitored.	(1) A ground vehicle robot can be designed to access places that are very difficult to reach. (2) Statistical or numerical analysis can be used to perform a sensitivity and probability analysis to predict the vulnerable locations [4,5].
Based on the fact that dealing with a harsh environment leads to a high possibility of data loss issues, most studies adopt statistical models to overcome such issues (refer to Section 4.3.3), which neglect the physical and mechanical characteristics of the slope area [4,5].	Designing a parallel system can provide a viable and effective solution. To clarify, using multi-node and multi-feature monitoring systems allows one to obtain different characteristics for the same slope. These data can be correlated with each other, solving data loss issues.

6. Conclusions

This study integrated scientometric and systematic analyses. A scientometric analysis is a potential approach for addressing manual search issues by highlighting the most significant contributions of keywords, authors, organizations, and nations. As a consequence, the key conclusion was that landslide monitoring models have improved over the previous 7 years, indicating growing global concern about preventing the loss of lives and financial resources. This research presented the most recent advancements and state-of-the-art landslide-monitoring technologies. According to the literature, each approach has its own set of pros and limitations.

Surface-monitoring techniques can offer information regarding near-surface movement, moisture content, and other physical information. Such strategies offer the following benefits: (1) they can offer millimeter-level 3D coordinates, and (2) they can provide distributed monitoring data with high spatial resolution across large regions. These studies, however, have the disadvantages of (1) obtaining real-time data is difficult and expensive; (2) they have a coarse resolution; and (3) they are impacted by severe fog, snow/rain, atmospheric delay, dense vegetation, and shadow. As a result, these methodologies are appropriate for creating landslide susceptibility, risk, and vulnerability maps [4,5]. However, such maps cannot provide early warning indications or predict disasters.

These objectives can only be met by a knowledge of the inner mechanism and monitoring of subsurface conditions. Extensometers have a high temporal resolution (36 mm/s) and precision (0.011 ± 0.0083 mm). Nonetheless, this is a single-point surface-movement-monitoring system. These characteristics are appropriate for translational landslides. By detecting subsurface displacement, conventional inclinometers outperform extensometers. The limited spatial vertical resolution (0.5–1 m) restricts its use, particularly for thin shear bandwidth. Unlike traditional inclinometers, TDR can enable exact monitoring of the sliding surface's position (spatial resolution of 0.05 m). When compared with the inclinometer guide enclosure, the coaxial cable costs approximately 55% less. However, measuring the displacement is difficult. The moderate rigidity of inclinometers restricts their use in monitoring minor movements. AE techniques are sensitive to minor deformation and are best suited for slow-moving landslides. Optical-fiber-based inclinometers have recently gained much interest. This technology combines all of the previously mentioned benefits, including high initial measurement (0.98 mm), measuring range (36 mm), low cost (0.45 USD/m), and high spatial resolution of 10 mm. FBG may be coupled with BOTDA to monitor both the strain and temperature across a large region. Because of the restricted monitoring range, this method is best suited for rock landslides. This method is limited in its application since it is based on wire connections.

Tilt sensors have the benefit of being able to determine the direction of a landslide with two-dimensional deformation with an accuracy of $0.0025°$ and a measurement range of $-30°$ to $+30°$. The depth of the sensor rod must be carefully calculated: small and long rods are suited for circular slip surfaces, while long rods should penetrate the rock layer for shallow landslides, as short rods are not effective. Many biaxial tilt sensors may be combined to form a multimodule system (inclinometer) with a spatial resolution of 100 mm, an accuracy of 0.73 mm, and a cost of 70 EUR/m. Tilt sensors, on the other hand, are point sensors and cannot extract deformations in areas where there is no inclination (i.e., translational landslides). Inclinometers based on strain gauges can detect micro-displacement. Soil deformation sensors are excellent for quick landslides since they have a low stiffness when compared with other approaches. SDS can detect micro-displacement (1 mm) throughout a range of 0 to 25 mm. The Strain Gauge Deep Earth Probe (SG-DEP sensor) can give 360-degree directional measurements and is ideal for both shallow and deep landslides, as well as harsh conditions. Acceleration sensors can detect slope movement independently of external triggers. This approach is appropriate for translational quick landslides without tilting, where linear acceleration is the most influential characteristic.

In addition to subsurface monitoring, the best technique to assess the kinematic characteristics of landslides is to monitor the sliding force; however, its installation is complicated. Rainfall monitoring is critical since it is regarded as the primary triggering factor. Based on multiple triboelectric nanogenerator (TENG) units, a self-powered wireless sensor with a high measurement range (0 to 288 mm/d) and resolution (5.5 mm) was recently created. The subsurface moisture state illuminates the antecedent effect of rainfall. The drying technique for determining soil moisture in a laboratory has great accuracy; nonetheless, it is a labor-intensive procedure necessitating massive investigation work for a wide area. It is challenging for AE techniques to link soil moisture with acoustic waves. FBG can detect up to 37% volumetric water content. UHF radio-frequency identification (RFID) sensors can detect soil moisture levels as high as 16%. The smart aggregate (SAs) approach can monitor soil moisture up to 30%. Geophysical methods, such as electrical resistivity tomography (ERT), can offer information about wide areas rather than single spots that provide plot-scale soil moisture variation. The spatial resolution of a region might range from meters to decimeters. This technology can detect soil moisture up to 2 m deep.

MEMS and IoT sensors that can be linked to WSNs can be used to overcome wiring and installation problems. MEMS can be used as an inclinometer, tiltmeter, volumetric water content sensors, etc., with the primary goal of low cost and simple installation and maintenance. These sensors are more suited for shallow landslides. The SitkaNet sensor may represent a realistic solution to construct a deep spatially distributed moisture content sensor for approximately 1000 USD per node. In the shear band, temperature sensitivity is critical for slope stability. Likewise, for shallow strata, the surface temperature can offer an early warning when moving landslides have greater temperatures than stable zones. Multifunction nodes offer a feasible alternative to single-function nodes in terms of cost and false alarm rate.

Regardless of the quantification of subsurface characteristics, warning signs can offer indicators to cope with emergency circumstances. Elastic waves and low-frequency infrasonic signals can provide warning indications when internal mechanisms (such as soil moisture, deformation, matric suction, and effective stresses) change. However, implementing such a strategy is rather difficult. Other warning systems, such as differential capacitors, triboelectric force and bend sensors (TTEFBS), and chemiluminescence-based approaches are currently under development.

Data may be obtained manually; however, critical events may be missed. Natural disasters can cause damage to wire- or cable-based systems. Wireless networks can address the aforementioned limitations by linking several sensors for broad monitoring areas. However, WSNs are limited by power consumption issues, communications issues, and data loss and size issues. For power consumption issues, building a sleep threshold, reducing the number of sensors, and using rechargeable techniques can overcome this dilemma. Regarding communication issues, the communication distance between sensor nodes can affect the precision and the response time for the transmitted data. Available techniques can provide an inter-distance between 90 and 300 m, while the magnetic induction communication transceiver can be buried up to 5.28 m into the ground. Missing data can be obtained using a variety of mathematical methods. Laboratory-scale testing provides an appropriate approach to understanding the mechanism of landslides in a safe and low-cost setting. Prior to the field installation of the monitoring system, a thorough site study is needed. The monitoring system is placed under four conditions: random, matrix, vulnerable, or hybrid. The vulnerable placement allows for reasonable monitoring where the monitoring points are placed in critical locations.

Author Contributions: Conceptualization, K.M.P.E. and T.Z.; methodology, K.M.P.E. and T.Z.; formal analysis, K.M.P.E., S.M.M.H.G. and T.Z.; investigation, K.M.P.E., S.M.M.H.G. and T.Z.; resources, T.Z. and G.A.; data curation, K.M.P.E. and S.M.M.H.G.; writing—original draft preparation, K.M.P.E.; writing—review and editing, S.M.M.H.G., T.Z. and G.A.; supervision, T.Z. All authors have read and agreed to the published version of the manuscript.

Funding: This research was funded by Hong Kong Polytechnic University and partially acknowledge the Innovation and Technology Support Programme (ITSP) of the Hong Kong SAR (grant no. ITS/033/20FP).

Data Availability Statement: Data are contained within the article.

Conflicts of Interest: The authors declare no conflict of interest.

Abbreviations

ABS	Acrylonitrile butadiene styrene	MEMS	Microelectromechanical systems
AE	Acoustic emission	MFTL	Multi-feature fusion transfer learning
AIM	Accumulative integral method	MLATC	Mean-based low-rank autoregressive tensor completion
ANN	Artificial neural networks	MR–WPT	Magnetic resonance wireless power transfer
AOI	Area of interest	OFDR	Optical frequency domain reflectometry
BOCDA	Brillouin optical correlation-domain analysis	OTDR	Optical time domain reflectometry
BOFDA	Brillouin optical frequency-domain analysis	PCA	Principal component analysis
BOTDA	Brillouin optical time-domain analysis	POIS	Position and Inclination Sensor
BOTDR	Brillouin optical time-domain reflectometry	PSCFODS	Parallel-series connected fiber-optic displacement sensor
CAD	Context-aware data management	PS–InSAR	Persistent scatterer interferometry
CAE	Context-aware energy management	PVC	Polyvinyl chloride
CCVDM	Capacitive circuit voltage distribution method	QDFODS	Quasi-distributed fiber-optic displacement sensors
CET	Cable-extension transducer	RDC	Ringdown count
COFT	Combined optical fiber transducer	RTS	Robotized total station
C–OTDR	Coherent optical time domain reflectometry	SAA	ShapeAccelArray
CPT	Cone penetration test	SAAF	ShapeAccelArray/Field
CRLD	Constant resistance and large deformation	SAR	Synthetic aperture radar
CS	Compressed sensing	SAs	Smart aggregates
CS–TENG	Contact–separation mode TENG	SBS	Stimulated Brillouin scattering
DEMs	Digital elevation models	SDSs	Soil deformation sensors
DFOSS	Distributed fiber optical strain sensing	SG–DEP	Strain Gauge Deep Earth Probe
DInSAR	Differential (SAR) interferometry	SOF	Sensing optical fiber
DSS	Distributed strain sensing	SPT	Standard penetration test
EM	Electromagnetic	SSCC	Suction stress characteristic curves
EPCs	Earth pressure cells	SSPDM	Self-structure pressure distribution method
EPS	Expansile polyester ethylene	STFT	Short-time Fourier Transform
ERT	Electrical resistivity tomography	SWCC	Soil water characteristic curve
FBG	Fiber Bragg grating	SWP	Soil water potential
FODSs	Fiber-optic displacement sensors	TBR	Tipping bucket rain gauge
F-TENG	Freestanding TENG	TDR	Time domain reflectometry
GB–InSAR	Ground-based SAR	TENG	Triboelectric nanogenerators
GIS	Global information system	TSMP	Time-synchronized mesh protocol
GNSS	Global navigation satellite system	TTEFBS	Timbo-like triboelectric force and bend sensor
GPR	Ground penetration radar	UGV	Unmanned ground vehicles
GPS	Global positioning system	UHF RFID	Ultrahigh-frequency radio-frequency identification
IMUs	Inertial measuring units	UWB	Ultrawide band
IN	Inclinometers	VMC	Volumetric water content
InSAR	Interferometric synthetic aperture radar	Wi–GIM	Wireless sensor network for ground instability monitoring
IoT	Internet of things	WPT	Wireless power transfer
IPI	In-place inclinometers	WSN	Wireless sensor network
LiDAR	Light detection and ranging	WSNLM	Wireless sensor network for landslide monitoring
LLM	Lossless landslide monitoring	WUSNs	Wireless underground sensor networks
LOS	Line of sight	Z–TENG	Zigzag-structured triboelectric nanogenerator
MASW	Multichannel analysis of surface waves		

References

1. De Graff, J.V. Perspectives for systematic landslide monitoring. *Environ. Eng. Geosci.* **2011**, *17*, 67–76. [CrossRef]
2. Angeli, M.G.; Pasuto, A.; Silvano, S. A critical review of landslide monitoring experiences. *Eng. Geol.* **2000**, *55*, 133–147. [CrossRef]
3. Bicocchi, G.; Tofani, V.; D'Ambrosio, M.; Tacconi-Stefanelli, C.; Vannocci, P.; Casagli, N.; Lavorini, G.; Trevisani, M.; Catani, F. Geotechnical and hydrological characterization of hillslope deposits for regional landslide prediction modeling. *Bull. Eng. Geol. Environ.* **2019**, *78*, 4875–4891. [CrossRef]
4. Ebrahim, K.M.P.; Gomaa, S.M.M.H.; Zayed, T.; Alfalah, G. *Landslide Prediction Models, Part I: Empirical-Statistical and Physically Based Causative Thresholds*; Department of Building and Real Estate, Faculty of Construction and Environment, The Hong Kong Polytechnic University: Hong Kong, China, 2024.
5. Ebrahim, K.M.P.; Gomaa, S.M.M.H.; Zayed, T.; Alfalah, G. *Landslide Prediction Models, Part II: Deterministic Physical and Phenomenologically Models*; Department of Building and Real Estate, Faculty of Construction and Environment, The Hong Kong Polytechnic University: Hong Kong, China, 2024.
6. Ramesh, M.; Vasudevan, N. The deployment of deep-earth sensor probes for landslide detection. *Landslides* **2012**, *9*, 457–474. [CrossRef]
7. Rosi, A.; Berti, M.; Bicocchi, N.; Castelli, G.; Corsini, A.; Mamei, M.; Zambonelli, F. Landslide monitoring with sensor networks: Experiences and lessons learnt from a real-world deployment. *Int. J. Sens. Netw.* **2011**, *10*, 111–122. [CrossRef]
8. Giri, P.; Ng, K.; Phillips, W. Laboratory simulation to understand translational soil slides and establish movement criteria using wireless IMU sensors. *Landslides* **2018**, *15*, 2437–2447. [CrossRef]
9. Chuan, W.; Wen-Qiao, W.; Guo-Jun, W.; Xiao-Ming, W. Multiple parameter monitoring system for landslide. *Int. J. Smart Sens. Intell. Syst.* **2013**, *6*, 1200–1229. [CrossRef]
10. Kumar, N.; Ramesh, M.V. Accurate iot based slope instability sensing system for landslide detection. *IEEE Sens. J.* **2022**, *22*, 17151–17161. [CrossRef]
11. Liu, Y.; Hazarika, H.; Kanaya, H.; Takiguchi, O.; Rohit, D. Landslide prediction based on low-cost and sustainable early warning systems with IoT. *Bull. Eng. Geol. Environ.* **2023**, *82*, 177. [CrossRef]
12. Artese, G.; Perrelli, M.; Artese, S.; Meduri, S.; Brogno, N. POIS, a low cost tilt and position sensor: Design and first tests. *Sensors* **2015**, *15*, 10806–10824. [CrossRef]
13. Shamshi, M.A. Technologies convergence in recent instrumentation for natural disaster monitoring and mitigation. *IETE Tech. Rev.* **2004**, *21*, 277–290. [CrossRef]
14. Eyo, E.E.; Musa, T.A.; Omar, K.M.; MIdris, K.; Bayrak, T.; Onuigbo, I.C.; Opaluwa, Y.D. Application of low-cost GPS tools and techniques for landslide monitoring: A review. *J. Teknol.* **2014**, *71*, 71–78. [CrossRef]
15. Chae, B.G.; Park, H.J.; Catani, F.; Simoni, A.; Berti, M. Landslide prediction, monitoring and early warning: A concise review of state-of-the-art. *Geosci. J.* **2017**, *21*, 1033–1070. [CrossRef]
16. So, A.C.T.; Ho, T.Y.K.; Wong, J.C.F.; Lai, A.C.S.; Leung, W.K.; Kwan, J.S.H. Advancing the Use of Lidar in Geotechnical Applications in Hong Kong-A 10-Year Overview. In Proceedings of the 42nd Asian Conference on Remote Sensing, ACRS 2021, Can Tho City, Vietnam, 22–24 November 2021.
17. Lapenna, V.; Perrone, A. Time-lapse electrical resistivity tomography (TL-ERT) for landslide monitoring: Recent advances and future directions. *Appl. Sci.* **2022**, *12*, 1425. [CrossRef]
18. Breglio, G.; Bernini, R.; Berruti, G.M.; Bruno, F.A.; Buontempo, S.; Campopiano, S.; Catalano, E.; Consales, M.; Coscetta, A.; Cutolo, A.; et al. Innovative Photonic Sensors for Safety and Security, Part III: Environment, Agriculture and Soil Monitoring. *Sensors* **2023**, *23*, 3187. [CrossRef] [PubMed]
19. Huang, G.; Du, S.; Wang, D. GNSS techniques for real-time monitoring of landslides: A review. *Satell. Navig.* **2023**, *4*, 5. [CrossRef]
20. Auflič, M.j.; Herrera, G.; Mateos, R.M.; Poyiadji, E.; Quental, L.; Severine, B.; Peternel, T.; Podolski, L.; Calcaterra, S.; Kociu, A.; et al. Landslide monitoring techniques in the Geological Surveys of Europe. *Landslides* **2023**, *20*, 951–965. [CrossRef]
21. Wuni, I.Y.; Shen, G.Q. Critical success factors for modular integrated construction projects: A review. *Build. Res. Inf.* **2020**, *48*, 763–784. [CrossRef]
22. Yin, X.; Liu, H.; Chen, Y.; Al-Hussein, M. Building information modelling for off-site construction: Review and future directions. *Autom. Constr.* **2019**, *101*, 72–91. [CrossRef]
23. Alshami, A.; Elsayed, M.; Ali, E.; Eltoukhy, A.E.E.; Zayed, T. Harnessing the Power of ChatGPT for Automating Systematic Review Process: Methodology, Case Study, Limitations, and Future Directions. *Systems* **2023**, *11*, 351. [CrossRef]
24. Zou, Y.; Zheng, C. A Scientometric analysis of predicting methods for identifying the environmental risks caused by landslides. *Appl. Sci.* **2022**, *12*, 4333. [CrossRef]
25. Van Eck, N.J.; Waltman, L. VOSviewer: A computer program for bibliometric mapping. In Proceedings of the 12th International Conference on Scientometrics and Informetrics, Nancy, France, 14–17 July 2009; ISSI 2009. pp. 886–897.
26. Moher, D.; Liberati, A.; Tetzlaff, J.; Altman, D.G. Preferred reporting items for systematic reviews and meta-analyses: The PRISMA statement. *J. Clin. Epidemiol.* **2009**, *62*, 1006–1012. [CrossRef] [PubMed]
27. Wohlin, C. Guidelines for snowballing in systematic literature studies and a replication in software engineering. In *ACM International Conference Proceeding Series*; ACM: New York, NY, USA, 2014; pp. 1–10. [CrossRef]
28. Giri, P.; Ng, K.; Phillips, W. Wireless sensor network system for landslide monitoring and warning. *IEEE Trans. Instrum. Meas.* **2018**, *68*, 1210–1220. [CrossRef]

29. Giri, P.; Ng, K.; Phillips, W. Monitoring Soil Slide-Flow Using Wireless Sensor Network-Inertial Measurement Unit System. *Geotech. Geol. Eng.* **2022**, *40*, 367–381. [CrossRef]
30. Seguí, C.; Veveakis, M. Continuous assessment of landslides by measuring their basal temperature. *Landslides* **2021**, *18*, 3953–3961. [CrossRef]
31. Seguí, C.; Veveakis, M. Forecasting and mitigating landslide collapse by fusing physics-based and data-driven approaches. *Geomech. Energy Environ.* **2022**, *32*, 100412. [CrossRef]
32. Huisman, J.A.; Hubbard, S.S.; Redman, J.D.; Annan, A.P. Measuring soil water content with ground penetrating radar: A review. *Vadose Zone J.* **2003**, *2*, 476–491. [CrossRef]
33. Iai, S. Similitude for shaking table tests on soil-structure-fluid model in 1g gravitational field. *Soils Found.* **1989**, *29*, 105–118. [CrossRef]
34. Babaeian, E.; Sadeghi, M.; Jones, S.B.; Montzka, C.; Vereecken, H.; Tuller, M. Ground, proximal, and satellite remote sensing of soil moisture. *Rev. Geophys.* **2019**, *57*, 530–616. [CrossRef]
35. Zhang, L.; Cui, Y.; Zhu, H.; Wu, H.; Han, H.; Yan, Y.; Shi, B. Shear deformation calculation of landslide using distributed strain sensing technology considering the coupling effect. *Landslides* **2023**, *20*, 1583–1597. [CrossRef]
36. Iverson, R.M. Scaling and design of landslide and debris-flow experiments. *Geomorphology* **2015**, *244*, 9–20. [CrossRef]
37. Buurman, B.; Kamruzzaman, J.; Karmakar, G.; Islam, S. Low-power wide-area networks: Design goals, architecture, suitability to use cases and research challenges. *IEEE Access* **2020**, *8*, 17179–17220. [CrossRef]
38. Ma, Y.; Li, F.; Wang, Z.; Zou, X.; An, J.; Li, B. Landslide assessment and monitoring along the Jinsha river, Southwest China, by combining Insar and GPS techniques. *J. Sens.* **2022**, *2022*, 9572937. [CrossRef]
39. Dematteis, N.; Wrzesniak, A.; Allasia, P.; Bertolo, D.; Giordan, D. Integration of robotic total station and digital image correlation to assess the three-dimensional surface kinematics of a landslide. *Eng. Geol.* **2022**, *303*, 106655. [CrossRef]
40. Huang, H.; Ju, S.; Duan, W.; Jiang, D.; Gao, Z.; Liu, H. Landslide Monitoring along the Dadu River in Sichuan Based on Sentinel-1 Multi-Temporal InSAR. *Sensors* **2023**, *23*, 3383. [CrossRef] [PubMed]
41. Refice, A.; Spalluto, L.; Bovenga, F.; Fiore, A.; Miccoli, M.N.; Muzzicato, P.; Nitti, D.O.; Nutricato, R.; Pasquariello, G. Integration of persistent scatterer interferometry and ground data for landslide monitoring: The Pianello landslide (Bovino, Southern Italy). *Landslides* **2019**, *16*, 447–468. [CrossRef]
42. Shirani, K.; Pasandi, M. Landslide monitoring and the inventory map validation by ensemble DInSAR processing of ASAR and PALSAR Images (Case Study: Doab-Samsami Basin in Chaharmahal and Bakhtiari Province, Iran). *Geotech. Geol. Eng.* **2021**, *39*, 1201–1222. [CrossRef]
43. Rebmeister, M.; Auer, S.; Schenk, A.; Hinz, S. Geocoding of ground-based SAR data for infrastructure objects using the Maximum A Posteriori estimation and ray-tracing. *ISPRS J. Photogramm. Remote Sens.* **2022**, *189*, 110–127. [CrossRef]
44. Mayr, A.; Rutzinger, M.; Bremer, M.; Oude Elberink, S.; Stumpf, F.; Geitner, C. Object-based classification of terrestrial laser scanning point clouds for landslide monitoring. *Photogramm. Rec.* **2017**, *32*, 377–397. [CrossRef]
45. Jakopec, I.; Marendić, A.; Grgac, I. Accuracy Analysis of a New Data Processing Method for Landslide Monitoring Based on Unmanned Aerial System Photogrammetry. *Sensors* **2023**, *23*, 3097. [CrossRef] [PubMed]
46. Casagli, N.; Tofani, V.; Ciampalini, A.; Raspini, F.; Lu, P.; Morelli, S. TXT-tool 2.039-3.1: Satellite remote sensing techniques for landslides detection and mapping. In *Landslide Dynamics: ISDR-ICL Landslide Interactive Teaching Tools. Volume 1: Fundamentals, Mapping and Monitoring*; Springer: Cham, Switzerland, 2018; pp. 235–254. [CrossRef]
47. Zhu, Z.W.; Liu, D.Y.; Yuan, Q.Y.; Liu, B.; Liu, J.C. A novel distributed optic fiber transduser for landslides monitoring. *Opt. Lasers Eng.* **2011**, *49*, 1019–1024. [CrossRef]
48. Allil, R.C.; Lima, L.A.; Allil, A.S.; Werneck, M.M. Fbg-based inclinometer for landslide monitoring in tailings dams. *IEEE Sens. J.* **2021**, *21*, 16670–16680. [CrossRef]
49. Yang, Z.; Shao, W.; Qiao, J.; Huang, D.; Tian, H.; Lei, X.; Uchimura, T. A multi-source early warning system of MEMS based wireless monitoring for rainfall-induced landslides. *Appl. Sci.* **2017**, *7*, 1234. [CrossRef]
50. Allasia, P.; Manconi, A.; Giordan, D.; Baldo, M.; Lollino, G. ADVICE: A new approach for near-real-time monitoring of surface displacements in landslide hazard scenarios. *Sensors* **2013**, *13*, 8285–8302. [CrossRef] [PubMed]
51. Wang, B.C. A landslide monitoring technique based on dual-receiver and phase difference measurements. *IEEE Geosci. Remote Sens. Lett.* **2013**, *10*, 1209–1213. [CrossRef]
52. Qiao, S.; Feng, C.; Yu, P.; Tan, J.; Uchimura, T.; Wang, L.; Tang, J.; Shen, Q.; Xie, J. Investigation on surface tilting in the failure process of shallow landslides. *Sensors* **2020**, *20*, 2662. [CrossRef] [PubMed]
53. Ma, J.; Tang, H.; Hu, X.; Bobet, A.; Yong, R.; Ez Eldin, M.A. Model testing of the spatial–temporal evolution of a landslide failure. *Bull. Eng. Geol. Environ.* **2017**, *76*, 323–339. [CrossRef]
54. Mucchi, L.; Jayousi, S.; Martinelli, A.; Caputo, S.; Intrieri, E.; Gigli, G.; Gracchi, T.; Mugnai, F.; Favalli, M.; Fornaciai, A.; et al. A flexible wireless sensor network based on ultra-wide band technology for ground instability monitoring. *Sensors* **2018**, *18*, 2948. [CrossRef]
55. Pawłuszek, K.; Borkowski, A.; Tarolli, P. Towards the optimal Pixel size of dem for automatic mapping of landslide areas. *Int. Arch. Photogramm. Remote Sens. Spat. Inf. Sci.-ISPRS Arch.* **2017**, *42*, 83–90. [CrossRef]
56. Deng, L.; Yuan, H.; Chen, J.; Fu, M.; Chen, Y.; Li, K.; Yu, M.; Chen, T. Experimental Investigation on Integrated Subsurface Monitoring of Soil Slope Using Acoustic Emission and Mechanical Measurement. *Appl. Sci.* **2021**, *11*, 7173. [CrossRef]

57. Hemalatha, T.; Ramesh, M.V.; Rangan, V.P. Effective and accelerated forewarning of landslides using wireless sensor networks and machine learning. *IEEE Sens. J.* **2019**, *19*, 9964–9975. [CrossRef]
58. Zhang, C.C.; Zhu, H.H.; Liu, S.P.; Shi, B.; Zhang, D. A kinematic method for calculating shear displacements of landslides using distributed fiber optic strain measurements. *Eng. Geol.* **2018**, *234*, 83–96. [CrossRef]
59. Zheng, Y.; Huang, D.; Shi, L. A new deflection solution and application of a fiber Bragg grating-based inclinometer for monitoring internal displacements in slopes. *Meas. Sci. Technol.* **2018**, *29*, 055008. [CrossRef]
60. Zheng, Y.; Huang, D.; Zhu, Z.W.; Li, W.J. Experimental study on a parallel-series connected fiber-optic displacement sensor for landslide monitoring. *Opt. Lasers Eng.* **2018**, *111*, 236–245. [CrossRef]
61. Aulakh, N.S.; Chhabra, J.K.; Singh, N.; Jain, S. Microbend resolution enhancing technique for fiber optic based sensing and monitoring of landslides. *Exp. Tech.* **2004**, *28*, 37–42. [CrossRef]
62. Crawford, M.M.; Bryson, L.S.; Woolery, E.W.; Wang, Z. Long-term landslide monitoring using soil-water relationships and electrical data to estimate suction stress. *Eng. Geol.* **2019**, *251*, 146–157. [CrossRef]
63. Setiono, A.; Qomaruddin Afandi, M.I.; Adinanta, H.; Rofianingrum, M.Y.; Suryadi Mulyanto, I.; Bayuwati, D.; Anwar, A.; Widiyatmoko, B. Wire Extensometer Based on Optical Encoder for Translational Landslide Measurement. *Int. J. Adv. Sci. Eng. Inf. Technol.* **2023**, *13*, 17–23. [CrossRef]
64. wiel, G.; Guerriero, L.; Revellino, P.; Guadagno, F.M. A multi-module fixed inclinometer for continuous monitoring of landslides: Design, development, and laboratory testing. *Sensors* **2020**, *20*, 3318. [CrossRef]
65. Schenato, L.; Palmieri, L.; Camporese, M.; Bersan, S.; Cola, S.; Pasuto, A.; Galtarossa, A.; Salandin, P.; Simonini, P. Distributed optical fibre sensing for early detection of shallow landslides triggering. *Sci. Rep.* **2017**, *7*, 14686. [CrossRef]
66. Dunnicliff, J. *Geotechnical Instrumentation for Monitoring Field Performance*; John Wiley & Sons: Hoboken, NJ, USA, 1993.
67. Sargand, S.M.; Sargent, L.; Farrington, S.P. *Inclinometer-Time Domain Reflectometry Comparative Study*; No. FHWA/OH-2004/010; Ohio Research Institute for Transportation and the Environment: Columbus, OH, USA, 2004.
68. Ho, S.C.; Chen, I.H.; Lin, Y.S.; Chen, J.Y.; Su, M.B. Slope deformation monitoring in the Jiufenershan landslide using time domain reflectometry technology. *Landslides* **2019**, *16*, 1141–1151. [CrossRef]
69. Chung, C.C.; Lin, C.P. A comprehensive framework of TDR landslide monitoring and early warning substantiated by field examples. *Eng. Geol.* **2019**, *262*, 105330. [CrossRef]
70. Lin, C.P.; Tang, S.H.; Lin, W.C.; Chung, C.C. Quantification of cable deformation with time domain reflectometry—Implications to landslide monitoring. *J. Geotech. Geoenvironmental. Eng.* **2009**, *135*, 143–152. [CrossRef]
71. Su, M.B.; Chen, I.H.; Liao, C.H. Using TDR cables and GPS for landslide monitoring in high mountain area. *J. Geotech. Geoenvironmental Eng.* **2009**, *135*, 1113–1121. [CrossRef]
72. Wang, Y.L.; Shi, B.; Zhang, T.L.; Zhu, H.H.; Jie, Q.; Sun, Q. Introduction to an FBG-based inclinometer and its application to landslide monitoring. *J. Civ. Struct. Health Monit.* **2015**, *5*, 645–653. [CrossRef]
73. Askarinejad, A.; Akca, D.; Springman, S.M. Precursors of instability in a natural slope due to rainfall: A full-scale experiment. *Landslides* **2018**, *15*, 1745–1759. [CrossRef]
74. Liu, Z.; Cai, G.; Wang, J.; Liu, L.; Zhuang, H. Evaluation of mechanical and electrical properties of a new sensor-enabled piezoelectric geocable for landslide monitoring. *Meas. J. Int. Meas. Confed.* **2023**, *211*, 112667. [CrossRef]
75. Chung, C.C.; Lin, C.P.; Ngui, Y.J.; Lin, W.C.; Yang, C.S. Improved technical guide from physical model tests for TDR landslide monitoring. *Eng. Geol.* **2022**, *296*, 106417. [CrossRef]
76. Kane, W.F.; Beck, T.J.; Hughes, J.J. Applications of time domain reflectometry to landslide and slope monitoring. In Proceedings of the Second International Symposium and Workshop on Time Domain Reflectometry for Innovative Geotechnical Applications, Evanston, IL, USA, 5–7 September 2021; Infrastructure Technology Institute at Northwestern University: Evanston, IL, USA, 2001; pp. 305–314.
77. Chung, C.C.; Tran, V.N.; Azhar, M. Guidelines from direct shear modeling in centrifuge for TDR landslide monitoring. *Eng. Geol.* **2022**, *310*, 106870. [CrossRef]
78. Zheng, Y.; Zhu, Z.W.; Li, W.J.; Gu, D.M.; Xiao, W. Experimental research on a novel optic fiber sensor based on OTDR for landslide monitoring. *Measurement: J. Int. Meas. Confed.* **2019**, *148*, 106926. [CrossRef]
79. Dixon, N.; Smith, A.; Spriggs, M.; Ridley, A.; Meldrum, P.; Haslam, E. Stability monitoring of a rail slope using acoustic emission. *Proc. Inst. Civ. Eng.-Geotech. Eng.* **2015**, *168*, 373–384. [CrossRef]
80. Smith, A.; Moore, I.D.; Dixon, N. Acoustic emission sensing of pipe–soil interaction: Full-scale pipelines subjected to differential ground movements. *J. Geotech. Geoenvironmental Eng.* **2019**, *145*, 04019113. [CrossRef]
81. Dennis, N.D.; Ooi, C.W.; Wong, V.H. Estimating movement of shallow slope failures using time domain reflectometry. In Proceedings of the TDR Conference, Lafayette, IN, USA, 29 October–1 December 2006; Paper ID 41. Purdue University: West Lafayette, IN, USA, 2006; p. 16.
82. Ivanov, V.; Longoni, L.; Ferrario, M.; Brunero, M.; Arosio, D.; Papini, M. Applicability of an interferometric optical fibre sensor for shallow landslide monitoring–Experimental tests. *Eng. Geol.* **2021**, *288*, 106128. [CrossRef]
83. Zheng, Y.; Zhu, Z.W.; Xiao, W.; Gu, D.M.; Deng, Q.X. Investigation of a quasi-distributed displacement sensor using the macro-bending loss of an optical fiber. *Opt. Fiber Technol.* **2020**, *55*, 102140. [CrossRef]
84. Zheng, Y.; Liu, D.Y.; Zhu, Z.W.; Liu, H.L.; Liu, B. Experimental study on slope deformation monitoring based on a combined optical fiber transducer. *J. Sens.* **2017**, *2017*, 7936089. [CrossRef]

85. Zhu, Z.W.; Yuan, Q.Y.; Liu, D.Y.; Liu, B.; Liu, J.C.; Luo, H. New improvement of the combined optical fiber transducer for landslide monitoring. *Nat. Hazards Earth Syst. Sci.* **2014**, *14*, 2079–2088. [CrossRef]
86. Yu, Z.; Dai, H.; Zhang, Q.; Zhang, M.; Liu, L.; Zhang, J.; Jin, X. High-resolution distributed strain sensing system for landslide monitoring. *Optik* **2018**, *158*, 91–96. [CrossRef]
87. Minardo, A.; Zeni, L.; Coscetta, A.; Catalano, E.; Zeni, G.; Damiano, E.; De Cristofaro, M.; Olivares, L. Distributed optical fiber sensor applications in geotechnical monitoring. *Sensors* **2021**, *21*, 7514. [CrossRef]
88. Zeng, B.; Zheng, Y.; Yu, J.; Yang, C. Deformation calculation method based on FBG technology and conjugate beam theory and its application in landslide monitoring. *Opt. Fiber Technol.* **2021**, *63*, 102487. [CrossRef]
89. Zheng, Y.; Zhu, Z.W.; Deng, Q.X.; Xiao, F. Theoretical and experimental study on the fiber Bragg grating-based inclinometer for slope displacement monitoring. *Opt. Fiber Technol.* **2019**, *49*, 28–36. [CrossRef]
90. Li, F.; Zhao, W.; Xu, H.; Wang, S.; Du, Y. A highly integrated BOTDA/XFG sensor on a single fiber for simultaneous multi-parameter monitoring of slopes. *Sensors* **2019**, *19*, 2132. [CrossRef]
91. Guo, T.T.; Yuan, W.Q.; Wu, L.H. Experimental research on distributed fiber sensor for sliding damage monitoring. *Opt. Lasers Eng.* **2009**, *47*, 156–160. [CrossRef]
92. Jeong, S.; Ko, J.; Kim, J. The effectiveness of a wireless sensor network system for landslide monitoring. *IEEE Access* **2019**, *8*, 8073–8086. [CrossRef]
93. Abraham, M.T.; Satyam, N.; Bulzinetti, M.A.; Pradhan, B.; Pham, B.T.; Segoni, S. Using field-based monitoring to enhance the performance of rainfall thresholds for landslide warning. *Water* **2020**, *12*, 3453. [CrossRef]
94. Abraham, M.T.; Satyam, N.; Pradhan, B.; Alamri, A.M. IoT-based geotechnical monitoring of unstable slopes for landslide early warning in the Darjeeling Himalayas. *Sensors* **2020**, *20*, 2611. [CrossRef] [PubMed]
95. Xie, J.; Uchimura, T.; Huang, C.; Maqsood, Z.; Tian, J. Experimental study on the relationship between the velocity of surface movements and tilting rate in pre-failure stage of rainfall-induced landslides. *Sensors* **2021**, *21*, 5988. [CrossRef] [PubMed]
96. Gian Quoc, A.; Nguyen Dinh, C.; Tran Duc, N.; Tran Duc, T.; Kumbesan, S. Wireless technology for monitoring site-specific landslide in Vietnam. *Int. J. Electr. Comput. Eng.* **2018**, *8*, 4448–4455. [CrossRef]
97. Chen, Y.; Irfan, M.; Uchimura, T.; Cheng, G.; Nie, W. Elastic wave velocity monitoring as an emerging technique for rainfall-induced landslide prediction. *Landslides* **2018**, *15*, 1155–1172. [CrossRef]
98. Chen, Y.; Irfan, M.; Uchimura, T.; Wu, Y.; Yu, F. Development of elastic wave velocity threshold for rainfall-induced landslide prediction and early warning. *Landslides* **2019**, *16*, 955–968. [CrossRef]
99. Chen, Y.; Irfan, M.; Uchimura, T.; Zhang, K. Feasibility of using elastic wave velocity monitoring for early warning of rainfall-induced slope failure. *Sensors* **2018**, *18*, 997. [CrossRef]
100. Wielandt, S.; Uhlemann, S.; Fiolleau, S.; Dafflon, B. Low-power, flexible sensor arrays with solderless board-to-board connectors for monitoring soil deformation and temperature. *Sensors* **2022**, *22*, 2814. [CrossRef]
101. Sheikh, M.R.; Nakata, Y.; Shitano, M.; Kaneko, M. Rainfall-induced unstable slope monitoring and early warning through tilt sensors. *Soils Found.* **2021**, *61*, 1033–1053. [CrossRef]
102. Ramesh, M.V.; Rangan, V.P. Data reduction and energy sustenance in multisensor networks for landslide monitoring. *IEEE Sens. J.* **2014**, *14*, 1555–1563. [CrossRef]
103. Prabha, R.; Ramesh, M.V.; Rangan, V.P.; Ushakumari, P.V.; Hemalatha, T. Energy efficient data acquisition techniques using context aware sensing for landslide monitoring systems. *IEEE Sens. J.* **2017**, *17*, 6006–6018. [CrossRef]
104. Askarinejad, A. A method to locate the slip surface and measuring subsurface deformations in slopes. In Proceedings of the Fourth International Young Geotechnical Engineers Conference (4iYGEC), Alexandria, Egypt, 3–6 October 2009; The Egyptian Geotechnical Society: Cairo, Egypt, 2009; pp. 171–174.
105. Askarinejad, A.; Springman, S.M. A novel technique to monitor subsurface movements of landslides. *Can. Geotech. J.* **2018**, *55*, 620–630. [CrossRef]
106. Kotta, H.Z.; Rantelobo, K.; Tena, S.; Klau, G. Wireless sensor network for landslide monitoring in Nusa Tenggara Timur. TELKOMNIKA (Telecommunication Computing Electronics and Control). *J. Mt. Sci.* **2011**, *9*, 9–18. [CrossRef]
107. Yunus, M.A.M.; Ibrahim, S.; Khairi, M.T.M.; Faramarzi, M. The application of WiFi-based wireless sensor network (WSN) in hill slope condition monitoring. *J. Teknol.* **2015**, *73*, 75–84.
108. Tao, Z.; Wang, Y.; Zhu, C.; Xu, H.; Li, G.; He, M. Mechanical evolution of constant resistance and large deformation anchor cables and their application in landslide monitoring. *Bull. Eng. Geol. Environ.* **2019**, *78*, 4787–4803. [CrossRef]
109. He, M.C.; Tao, Z.G.; Zhang, B. Application of remote monitoring technology in landslides in the Luoshan mining area. *Min. Sci. Technol.* **2009**, *19*, 609–614. [CrossRef]
110. Li, M.; Cheng, W.; Chen, J.; Xie, R.; Li, X. A high performance piezoelectric sensor for dynamic force monitoring of landslide. *Sensors* **2017**, *17*, 394. [CrossRef]
111. Latupapua, H.; Latupapua, A.; Wahab, A.; Alaydrus, M. Wireless sensor network design for earthquake's and landslide's early warnings. *Indones. J. Electr. Eng. Comput. Sci.* **2018**, *11*, 437–445. [CrossRef]
112. Hu, Y.; Zhou, J.; Li, J.; Ma, J.; Hu, Y.; Lu, F.; Cheng, T. Tipping-bucket self-powered rain gauge based on triboelectric nanogenerators for rainfall measurement. *Nano Energy* **2022**, *98*, 107234. [CrossRef]
113. Ferraz, E.S.B. *Determining Water Content and Bulk Density of Soil by Gamma Ray Attenuation Methods*; Elsever: Amsterdam, The Netherlands, 1979.

114. Priestley, C.H.B.; Taylor, R.J. On the assessment of surface heat flux and evaporation using large-scale parameters. *Mon. Weather. Rev.* **1972**, *100*, 81–92. [CrossRef]
115. Leone, M.; Principe, S.; Consales, M.; Parente, R.; Laudati, A.; Caliro, S.; Cutolo, A.; Cusano, A. Fiber optic thermo-hygrometers for soil moisture monitoring. *Sensors* **2017**, *17*, 1451. [CrossRef] [PubMed]
116. Wicki, A.; Hauck, C. Monitoring critically saturated conditions for shallow landslide occurrence using electrical resistivity tomography. *Vadose Zone J.* **2022**, *21*, e20204. [CrossRef]
117. Reynolds, S.G. The gravimetric method of soil moisture determination Part IA study of equipment, and methodological problems. *J. Hydrol.* **1970**, *11*, 258–273. [CrossRef]
118. Xiaochun, L.; Xue, C.; Bobo, X.; Bin, T.; Xiaolong, T.; Zhigang, T. Bi-LSTM-GPR algorithms based on a high-density electrical method for inversing the moisture content of landslide. *Bull. Eng. Geol. Environ.* **2022**, *81*, 491. [CrossRef]
119. Lo, W.C.; Yeh, C.L.; Tsai, C.T. Effect of soil texture on the propagation and attenuation of acoustic wave at unsaturated conditions. *J. Hydrol.* **2007**, *338*, 273–284. [CrossRef]
120. Kong, Q.; Chen, H.; Mo, Y.L.; Song, G. Real-time monitoring of water content in sandy soil using shear mode piezoceramic transducers and active sensing—A feasibility study. *Sensors* **2017**, *17*, 2395. [CrossRef]
121. Pichorim, S.F.; Gomes, N.J.; Batchelor, J.C. Two solutions of soil moisture sensing with RFID for landslide monitoring. *Sensors* **2018**, *18*, 452. [CrossRef]
122. Crawford, M.M.; Bryson, L.S. Assessment of active landslides using field electrical measurements. *Eng. Geol.* **2018**, *233*, 146–159. [CrossRef]
123. Binley, A. 11.08-Tools and Techniques: Electrical Methods. In *Treatise on Geophysics*, 2nd ed.; Elsevier: Amsterdam, The Netherlands, 2015; pp. 233–259. [CrossRef]
124. Cheng, Q.; Tao, M.; Chen, X.; Binley, A. Evaluation of electrical resistivity tomography (ERT) for mapping the soil–rock interface in karstic environments. *Environ. Earth Sci.* **2019**, *78*, 439. [CrossRef]
125. Chu, M.; Patton, A.; Roering, J.; Siebert, C.; Selker, J.; Walter, C.; Udell, C. SitkaNet: A low-cost, distributed sensor network for landslide monitoring and study. *HardwareX* **2021**, *9*, e00191. [CrossRef] [PubMed]
126. Marino, P.; Roman Quintero, D.C.; Santonastaso, G.F.; Greco, R. Prototype of an IoT-based low-cost sensor network for the hydrological monitoring of landslide-prone areas. *Sensors* **2023**, *23*, 2299. [CrossRef] [PubMed]
127. Wang, Y.; Liu, Z.; Wang, D.; Li, Y.; Yan, J. Anomaly detection and visual perception for landslide monitoring based on a heterogeneous sensor network. *IEEE Sens. J.* **2017**, *17*, 4248–4257. [CrossRef]
128. Irfan, M.; Uchimura, T.; Chen, Y. Effects of soil deformation and saturation on elastic wave velocities in relation to prediction of rain-induced landslides. *Eng. Geol.* **2017**, *230*, 84–94. [CrossRef]
129. Michlmayr, G.; Or, D.; Cohen, D. Fiber bundle models for stress release and energy bursts during granular shearing. *Phys. Rev. E-Stat. Nonlinear Soft Matter Phys.* **2012**, *86*, 061307. [CrossRef]
130. Zhang, S.; Wang, K.; Hu, K.; Xu, H.; Xu, C.; Liu, D.; Lv, J.; Wei, F. Model test: Infrasonic features of porous soil masses as applied to landslide monitoring. *Eng. Geol.* **2020**, *265*, 105454. [CrossRef]
131. Motakabber, S.M.A.; Ibrahimy, M.I. An approach of differential capacitor sensor for landslide monitoring. *Int. J. Geomate* **2015**, *9*, 1534–1537. [CrossRef]
132. Lin, Z.; He, Q.; Xiao, Y.; Zhu, T.; Yang, J.; Sun, C.; Zhou, Z.; Zhang, H.; Shen, Z.; Yang, J.; et al. Flexible timbo-like triboelectric nanogenerator as self-powered force and bend sensor for wireless and distributed landslide monitoring. *Adv. Mater. Technol.* **2018**, *3*, 1800144. [CrossRef]
133. Kuang, K.S.C. Wireless chemiluminescence-based sensor for soil deformation detection. *Sens. Actuators A Phys.* **2018**, *269*, 70–78. [CrossRef]
134. Yueshun, H.; Zhang, W. The reseach on wireless sensor network for landslide monitoring. *Int. J. Smart Sens. Intell. Syst.* **2013**, *6*, 867. [CrossRef]
135. Van Khoa, V.; Takayama, S. Wireless sensor network in landslide monitoring system with remote data management. *Measurement: J. Int. Meas. Confed.* **2018**, *118*, 214–229. [CrossRef]
136. Ragnoli, M.; Leoni, A.; Barile, G.; Ferri, G.; Stornelli, V. LoRa-Based Wireless Sensors Network for Rockfall and Landslide Monitoring: A Case Study in Pantelleria Island with Portable LoRaWAN Access. *J. Low Power Electron. Appl.* **2022**, *12*, 47. [CrossRef]
137. Kumar, S.; Duttagupta, S.; Rangan, V.P.; Ramesh, M.V. Reliable network connectivity in wireless sensor networks for remote monitoring of landslides. *Wirel. Netw.* **2020**, *26*, 2137–2152. [CrossRef]
138. Xu, Y.; Chen, Q.; Tian, D.; Zhang, Y.; Li, B.; Tang, H. Selection of high transfer stability and optimal power-efficiency tradeoff with respect to distance region for underground wireless power transfer systems. *J. Power Electron.* **2020**, *20*, 1662–1671. [CrossRef]
139. Sharma, R.; Vashisht, V.; Singh, U. WOATCA: A secure and energy aware scheme based on whale optimisation in clustered wireless sensor networks. *IET Commun.* **2020**, *14*, 1199–1208. [CrossRef]
140. Bagwari, S.; Roy, A.; Gehlot, A.; Singh, R.; Priyadarshi, N.; Khan, B. LoRa Based Metrics Evaluation for Real-Time Landslide Monitoring on IoT Platform. *IEEE Access* **2022**, *10*, 46392–46407. [CrossRef]
141. Wang, H.; Zhuo, T.; Zhong, P.; Wei, C.; Zou, D.; Zhong, Y. A novel wireless underground transceiver for landslide internal parameter monitoring based on magnetic induction. *Int. J. Circuit Theory Appl.* **2021**, *49*, 1549–1558. [CrossRef]

142. Blahůt, J.; Balek, J.; Eliaš, M.; Meletlidis, S. 3D Dilatometer time-series analysis for a better understanding of the dynamics of a giant slow-moving landslide. *Appl. Sci.* **2020**, *10*, 5469. [CrossRef]
143. Sumathi, M.S.; Anitha, G.S. Link Aware Routing Protocol for Landslide Monitoring Using Efficient Data Gathering and Handling System. *Wirel. Pers. Commun.* **2020**, *112*, 2663–2684. [CrossRef]
144. de Souza, F.T.; Ebecken, N.F. A data based model to predict landslide induced by rainfall in Rio de Janeiro city. *Geotech. Geol. Eng.* **2012**, *30*, 85–94. [CrossRef]
145. Shentu, N.; Yang, J.; Li, Q.; Qiu, G.; Wang, F. Research on the Landslide Prediction Based on the Dual Mutual-Inductance Deep Displacement 3D Measuring Sensor. *Appl. Sci.* **2022**, *13*, 213. [CrossRef]
146. Wang, C.; Zhao, Y. Time Series Prediction Model of Landslide Displacement Using Mean-Based Low-Rank Autoregressive Tensor Completion. *Appl. Sci.* **2023**, *13*, 5214. [CrossRef]
147. Li, Z.; Cheng, P.; Zheng, J. Prediction of time to slope failure based on a new model. *Bull. Eng. Geol. Environ.* **2021**, *80*, 5279–5291. [CrossRef]
148. Long, J.; Li, C.; Liu, Y.; Feng, P.; Zuo, Q. A multi-feature fusion transfer learning method for displacement prediction of rainfall reservoir-induced landslide with step-like deformation characteristics. *Eng. Geol.* **2022**, *297*, 106494. [CrossRef]
149. Streiner, D.L.; Norman, G.R. "Precision" and "accuracy": Two terms that are neither. *J. Clin. Epidemiol.* **2006**, *59*, 327–330. [CrossRef]
150. Bridgman, P.W. *Dimensional Analysis*; Yale University Press: London, UK, 1922; p. viii+112.
151. Molfino, R.M.; Razzoli, R.P.; Zoppi, M. Autonomous drilling robot for landslide monitoring and consolidation. *Autom. Constr.* **2008**, *17*, 111–121. [CrossRef]
152. Patané, L. Bio-inspired robotic solutions for landslide monitoring. *Energies* **2019**, *12*, 1256. [CrossRef]
153. Hasan, M.F.R.; Salimah, A.; Susilo, A.; Rahmat, A.; Nurtanto, M.; Martina, N. Identification of landslide area using geoelectrical resistivity method as disaster mitigation strategy. *Int. J. Adv. Sci. Eng. Inf. Technol.* **2022**, *12*, 1484–1490. [CrossRef]

Disclaimer/Publisher's Note: The statements, opinions and data contained in all publications are solely those of the individual author(s) and contributor(s) and not of MDPI and/or the editor(s). MDPI and/or the editor(s) disclaim responsibility for any injury to people or property resulting from any ideas, methods, instructions or products referred to in the content.

Article

2D Phase-Based RFID Localization for On-Site Landslide Monitoring

Arthur Charléty [1,*], Mathieu Le Breton [2], Eric Larose [1] and Laurent Baillet [1]

[1] Université Grenoble Alpes, Université Savoie Mont Blanc, CNRS, IRD, Université Gustave Eiffel, ISTerre, 38000 Grenoble, France; eric.larose@univ-grenoble-alpes.fr (E.L.); laurent.baillet@univ-grenoble-alpes.fr (L.B.)
[2] Géolithe, 38920 Crolles, France; mathieu.lebreton@geolithe.com
* Correspondence: arthur.charlety@univ-grenoble-alpes.fr

Abstract: Passive radio-frequency identification (RFID) was recently used to monitor landslide displacement at a high spatio-temporal resolution but only measured 1D displacement. This study demonstrates the tracking of 2D displacements, using an array of antennas connected to an RFID interrogator. Ten tags were deployed on a landslide for 12 months and 2D relative localization was performed using a phase-of-arrival approach. A period of landslide activity was monitored through RFID and displacements were confirmed by reference measurements. The tags showed displacements of up to 1.2 m over the monitored period. The centimeter-scale accuracy of the technique was confirmed experimentally and theoretically for horizontal localization by developing a measurement model that included antenna and tag positions, as well as multipath interference. This study confirms that 2D landslide displacement tracking with RFID is feasible at relatively low instrumental and maintenance cost.

Keywords: phase localization; landslides; RFID; remote sensing; wireless sensor network; early warning

1. Introduction

Ground deformation monitoring with high resolution both in space and time remains a challenge due to the high cost of existing solutions, and to environmental limitations, such as meteorological phenomena, rough terrain or dense vegetation. Amongst several remote sensing methods [1,2], surface monitoring of large landslides can be typically performed through interferometric synthetic aperture radar (InSAR), either by space-borne measurements [3,4] or using ground-based stations [5–9]. Despite the high space resolution of these methods, the station cost remains high and the time resolution can be multiple days in the case of satellite remote sensing. More localized techniques, such as GPS [10–13] and radiofrequency-transponders [14,15], show higher time resolution, but also require on-board power sources which greatly increase initial cost and maintenance.

In this context, radio-rrequency identification (RFID) has shown increasing potential for earth science applications [16,17]. Amongst other applications, it is foreseen as a promising alternative for landslide and civil engineering structure deformation monitoring [18] due to its low cost relative to other solutions, and because it works under rain, snow and vegetation cover conditions [19,20]. It can thus be used as a tool for landslide early-warning [21], forecasting or long-term monitoring [22]. A wide range of solutions exist for tag localization using RFID [23,24], with various possibilities both in measured quantity and in terms of the measuring scheme.

The quantities most used for localization are the received signal strength and the backscattered phase of arrival. Signal-strength-based methods have been widely used for tag localization [25,26,27–29]. However, phase-based methods have shown better precision and reliability in recent years [30–32], primarily because they are less sensitive to environmental

variations and because the phase of the signal varies more rapidly with distance than the received signal strength.

Phase-based localization is divided into multiple schemes, which are extensively presented elsewhere [33–36]. These schemes generally rely on either multistatic stationary antennas and different carrier frequencies [30,37,38], or on a moving antenna with a known trajectory (e.g., the synthetic aperture radar technique) [39–43]. This paper focuses on a monostatic multi-antenna time-domain phase difference (TD-PD)-inspired scheme, as TD-PD has shown the best results for measuring relative displacements outdoors [18], with a precision of about 1 cm over long time periods for 1D displacement tracking. To date, RFID systems deployed to monitor moving ground only provide one-dimensional distance information and are subject to phase unwrapping issues that could be solved by using multiple antennas. In this article, we test the stationary configuration for 2D RFID tag localization using a set of four antennas in a TD-PD relative localization approach, and also discuss 3D localization perspectives. To the best of our knowledge, this is the first attempt at 2D-localization of RFID tags in an outdoor scenario, using a monostatic, monofrequency multi-antenna setup.

In the following section, we present the instrumentation of the experimental site and the methodology for data acquisition and processing. Section 4 provides theoretical background and experimental validation of the RFID measurement error in order to decide on ideal antenna positioning by optimizing the localization accuracy and phase ambiguity. Section 5 reports on an example of 12 months of surface deformation monitoring on the slow-moving Harmalière landslide.

2. Instrumentation and Methods

2.1. Experimental Site: Harmalière Landslide

The Harmalière landslide (Sinard, Isère, France) is located in the Trièves area about 50 km south of Grenoble in the western Prealps (see Figure 1). Trièves appears as a sedimentary plateau eroded by the Drac river; the plateau is formed by Quaternary varved clays and alluvial materials deposited in a glacially dammed lake during the Würm period [44]. Quaternary sediments also include silts, sometimes with a morainic cover, and rest on either interglacial Riss—Würm period glaciofluvial materials (gravels and sands) or on the underlying Jurassic carbonate bedrock. The thickness of the clay deposits can vary from 0 to a maximum of 200 m [45]. The landslide is southeast oriented, 400 m wide at the top, narrowing to 150 m at the toe. It develops from an altitude of 735 m (asl), down to the Monteynard Lake (480 m), over a distance of about 1.5 km. It was abruptly activated in 1981 and has remained active ever since, with new peaks of activity in 2016 and 2017 [46]. The slow moving landslide shows regressive behaviour, the headscarp retreating at an average velocity of 1 m/year, with very strong variations from year-to-year (including almost a decade of rest). The central body of the landslide is moving at velocities ranging from cm/year to m/year, with possible dramatic acceleration phases (m/day). A variety of research subjects are currently investigated in connection with it [46,47].

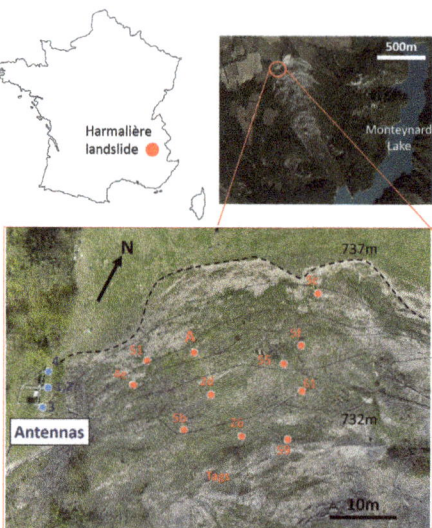

Figure 1. (**Top**) The Harmalière landslide location in France. (**Bottom**) Overview of the Harmalière landslide, with the RFID tag distribution (red points). Blue points : antennas and acquisition system. The dotted black line represents the landslide scar, the gray dotted lines represent 1-meter isolines.

2.2. RFID Instrumentation and Localization

2.2.1. RFID Instrumentation

In February 2020, a section of the landslide was equipped with an RFID system consisting of 32 battery-assisted passive tags and an acquisition station located near the landslide scar (see Figure 1). These tags can last about a decade without maintenance or replacement in the present real-time monitoring scenario. The station includes four antennas, an interrogator (Impinj SR420), a micro-computer (RPI-3B), and a modem to send the data automatically to a remote server, as described by (patent pending FR-17/53739). It is powered by a photovoltaic module and a wind turbine. The station collects RFID data for 3 min every 20 min from every tag and every antenna. The data includes the phase of arrival (termed here "phase") measured at 865.7 Hz, the received signal strength indication and the tag temperature. The tags were placed in pairs on fiber glass stakes 50 cm and 1 m above ground. They were spread out within the antennas reading range in a zone approximately 30 m × 30 m wide (see Figure 1), in such a way as to maximize the line-of-sight readability of each tag by multiple antennas. To validate the RFID localization calculations, the position of the tags was measured with a LEICA TCR805 tacheometer and a handheld target (estimated precision 4 cm), approximately once every month.

2.2.2. RFID Localization Scheme

TD-PD is a relative ranging technique based on phase variation $\delta\phi = \phi_1 - \phi_0$ between two measurements at different points in time. $\delta\phi$ is related to the radial distance variation $\delta r = r_1 - r_0$ between the tag and reader antenna, by the following equation:

$$\delta r = -\frac{c}{4\pi f}\delta\phi \qquad (1)$$

where f is the frequency of the electromagnetic wave (see values above) and c is the speed of light in the propagation medium. It is important to note that Equation (1) is only valid for displacements smaller than $\lambda/4 \approx 8$ cm between two phase measurements because of phase ambiguity. In the present case, this condition is generally fulfilled as the incremental displacements are small compared to the wavelength (usually less than 1 cm between two successive acquisitions). Moreover, a series of phase measurements can generally be unwrapped using well-defined algorithms. In this case, Equation (1) is valid for any unwrapped phase variation.

Section 3 presents a multidimensional localization scheme based on TD-PD.

3. Theoretical Model

In this section, we derive a mathematical model for phase-based RFID localization to compute the localization error of our real experiment. The main goal of this derivation is to study the origins of the localization uncertainty, mainly with respect to the system geometry and the physical measurement process.

From now on, we will consider that all phase measurements are unwrapped, and that Equation (1) is valid for all phase variations. Most presented tags were correctly read and no unwrapping error was detected in the monitored period. The specific case of an unwrapping error is examined separately, and does not fall within the scope of the present study.

In the following, index i describes a series of measurements starting at $i = 0$ and j describes the antenna indexing.

3.1. Localization Model

3.1.1. One Dimensional TD-PD

The localization method presented in this paper is based on the tag phase shift measured by each antenna at different points in time (TD-PD) [36]. In a homogeneous medium, the phase shift $\phi_{i,j} - \phi_{0,j}$ between the initial and the i-th (unwrapped) phase measurement, is directly proportional to the radial displacement $\delta r_{i,j}$ between the tag and antenna j (see Equation (1)).

Assuming an initial radial distance $r_{0,j}$, we obtain a series of radial distances $r_{i,j}$ from a measured phase series $\phi_{i,j}$:

$$r_{i,j} = r_{0,j} + \delta r_{i,j} \quad (2)$$

where $\delta r_{i,j}$ is obtained directly through Equation (1). This localization method is, hence, relative to the initial position, as it does not allow for absolute positioning without further information about the system (e.g., when $r_{0,j}$ is not known).

3.1.2. 2D Relative Displacement Approach

Using the measurements of multiple antennas, we can expand this TD-PD method with spatial considerations. For this purpose, we need both the phase measurements and the geometrical coordinates (x_j, y_j) of each antenna. This derivation focuses on the 2D problem; the 3D case will be briefly discussed at the end.

We define the initial distance $r_{0,j}$ from the antenna j to the tag:

$$r_{0,j} = \sqrt{(x_j - x_0)^2 + (y_j - y_0)^2}$$

where (x_0, y_0) are the initial coordinates of the tag and (x_j, y_j) those of the antenna.

Applying Equation (2), we obtain a series of radial displacements from the phase measurements of each antenna. From these radial distance measurements, a multilateration approach [48] can be applied to estimate the most probable position $(\hat{x}_i, \hat{y}_i)$ for the tag at the ith position. Amongst various possible methods of multilateration, we use an optimization algorithm that minimizes the following cost function Cf for the i-th measurement:

$$Cf_i(x,y) = \sum_{j=1}^{N_a} |r_{i,j} - \sqrt{(x_j - x)^2 + (y_j - y)^2}| \quad (3)$$

$$(\hat{x}_i, \hat{y}_i) = \underset{(x,y) \in \mathbb{R}^2}{argmin}(Cf_i(x,y))$$

where (x,y) are the test point coordinates, N_a is the number of antennas, $r_{i,j}$ is the i-th radial distance from antenna j, and $(\hat{x}_i, \hat{y}_i)$ is the most probable tag position. The minimization of this cost function was performed using the Trust-region optimization algorithm [49] implemented in the Scipy-optimize Python module.

3.2. Geometrical Localization Sensitivity

In this section, we focus on theoretical considerations regarding the localization sensitivity of the geometrical antenna-tag system to compute the value and direction of a displacement error of the tag with respect to a phase measurement error [19]. For a given antenna position (x_j, y_j), the absolute phase accumulated on a linear ray path (line of sight, LOS) between the antenna and a point (x,y) is expressed as follows:

$$\phi_j(x,y) = -\frac{4\pi f}{c} \times \sqrt{(x_j - x)^2 + (y_j - y)^2}$$

Let us define $K_{\phi j}$ as the space gradient of the measured phase ϕ_j, also defined as the phase sensitivity kernel, expressed in the spatial dimension as:

$$K_{\phi j}(x,y) = \begin{bmatrix} \frac{\partial \phi_j}{\partial x} \\ \frac{\partial \phi_j}{\partial y} \end{bmatrix} = \begin{bmatrix} K_{\phi j}^x \\ K_{\phi j}^y \end{bmatrix} \quad (4)$$

For a system consisting of two antennas (A and B) and small phase variations, the relation between the phase variation vector $\delta\phi$ and the true tag displacement δr can then be approximated by the linear matrix system:

$$\begin{bmatrix} \delta\phi_A \\ \delta\phi_B \end{bmatrix} = \begin{bmatrix} K_{\phi A}^x & K_{\phi A}^y \\ K_{\phi B}^x & K_{\phi B}^y \end{bmatrix} \begin{bmatrix} \delta x \\ \delta y \end{bmatrix}$$

That we can simply rewrite:

$$\delta\phi = K\delta r \quad (5)$$

Equation (5) holds for any number of phase measurements (thus any number of antennas N_a), and any number of space dimensions M; in such cases, K will be an $M \times N_a$ matrix. It expresses the direct solution of the phase-based relative localization problem, where K represents the transformation matrix from measured phase space to localization space. For the sake of simplicity, consider now that $N_a = M = 2$, which implies a bijective relationship between phase measurements and tag 2D relative displacement. In this case, the invertibility of the K matrix is almost always possible—the only exceptions are when the point position (x,y) coincides with that of one antenna, or when it is aligned with the two antennas. We exclude these limit cases that have no significance in our experiments. The above equation can then be reversed and gives the theoretical phase sensitivity of the tag position:

$$\delta r = K^{-1} \delta\phi \quad (6)$$

We now consider the linear transformation matrix K^{-1} to which we apply singular value decomposition (SVD). Any real matrix can be decomposed as follows [50]:

$$K^{-1} = U\Sigma V^\top \quad (7)$$

In our model, $V^\top$ represents the eigenvectors in phase space, Σ the diagonal eigenvalue matrix and U the eigenvectors in localization space.

In this derivation, we assume the same variance for all phase measurements; hence, the covariance matrix C_ϕ is defined as follows:

$$C_\phi = \sigma_\phi^2 \cdot I_{N_a} \quad (8)$$

where σ_ϕ is the typical phase standard deviation and I_{N_a} is an identity matrix of size N_a. C_ϕ is, thus, a constant diagonal matrix in our model, with typical values of 0.04 rad. This phase standard deviation is both an experimentally computed value and also corresponds to the modeled approximation of Equation (12) (see next Section).

Considering a given phase measurement uncertainty for each antenna, we can plug any phase distribution into the transformation from Equation (7). The shape and orientation of the resulting spatial distribution around tag position (that we will call localization spot) is described by the localization-space covariance matrix C_r. This matrix can be expressed in the following way, depending on K^{-1} as well as the hypothetical covariance of the phase measurement matrix C_ϕ:

$$C_r = (K^{-1})^\top C_\phi K^{-1} = U\Sigma^2 U^\top \qquad (9)$$

Extracting the eigenvalues and eigenvectors of C_r allows for a completely analytical determination of the localization spot properties (especially the direction of highest error) for a given antenna-tag geometry, as shown in Figure 2. With a phase error of 0.04 rad, and at the given tag position, we expect a localization random error of about 1 cm. Note that in the model, any relative increase in phase error will result in the same relative increase in localization error, as the measurement operator is linear.

The calculation presented above can be extended to a three-antenna system for a 3D localization problem, following Equations (1) to (9) with K a 3×3 matrix. In the case where $N_a > 3$, the system from Equation (5) is overdetermined, and a least-squares solution has to be found [51,52]. Using the pseudo-inverse of K, Equation (6) then gives:

$$\delta r = (K^\top K)^{-1} K^\top \delta\phi \qquad (10)$$

This new system can be solved and the eigenvectors computed by considering the transformation matrix $(K^\top K)^{-1} K^\top$.

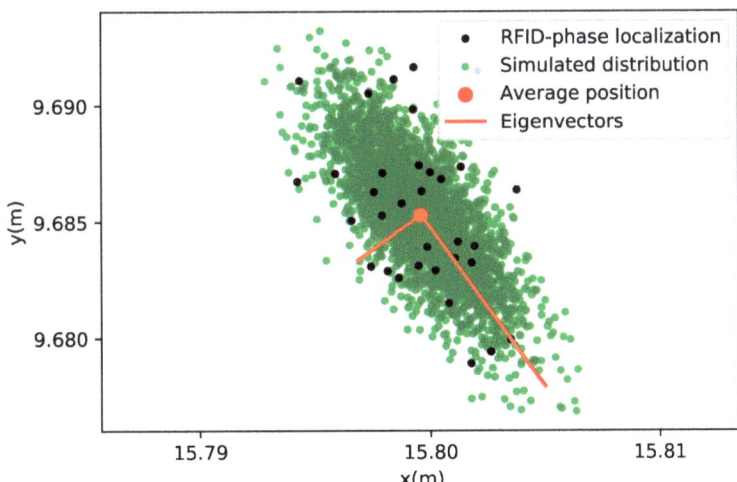

Figure 2. Localization error shape at the position of tag A (see Figure 1) compared with the RFID position estimation during a stable period in the Harmalière (November to December 2020). The green point distribution is computed through the K^{-1} transformation (see Equation (6)), using a Gaussian phase distribution with a standard deviation of 0.04 rad. The eigenvectors of the green distribution (red lines) are scaled up to encompass 97% of the data. The black points correspond to the RFID-phase localization results. The antenna positions are set as in the real experiment (see Figure 1).

3.3. Phase Error Model : Multipath, Phase Standard Deviation and Radiation Pattern

While the previous section focuses on geometrical localization error, we will now incorporate the impact of real-scenario error sources, e.g., antenna radiation pattern and multipath. The following derivation is based on the work of [20].

3.3.1. Multipath Propagation Model

Multipath interference is a major challenge in RFID-localization and several solutions have been proposed to estimate, reduce or mitigate its effect on measurements [53,54]. To start investigating the multipath, we use a simple two-ray model, assuming that the measured signal is a superposition of the line-of-sight ($p = 1$) signal and a signal reflected on the ground ($p = 2$), as shown in Figure 3. The two signals propagate over different path lengths r_p and orientations, which translate in different received power values due to Friis' law:

$$P_p(r) = \left(\frac{\lambda}{4\pi r_p}\right)^2 \times P_t \cdot G_T(i_p) \cdot G_R(i_p) \quad for \ p = \{1,2\}$$

where P_t is the power transmitted by the antenna, P_p is the received power along path p, G_r and G_t are the receiver and transmitter gain which depend on the signal orientation angle i_p and the antenna radiation pattern, λ is the carrier wavelength and r_p is the path propagation distance. We can then define the amplitude gain $A_p(i_p, r_p)$ for the line-of-sight (1) and reflected (2) signals :

$$A_1(i_1, r_1) = \frac{1}{r_1}\sqrt{G_t(i_1) \cdot G_r(i_1)}$$

$$A_2(i_2, r_2) = \frac{1}{r_2} R(i_2) \sqrt{G_t(i_2) \cdot G_r(i_2)}$$

where $R(i_2)$ is the reflection coefficient impacting the reflected ray (which depends on ground relative permittivity). The received signal voltage s_p after normalization by the initial emitted voltage can be expressed by the following phasor:

$$s_p(i_p, r_p) = A_p(i_p, r_p) \cdot \frac{\lambda}{4\pi} \cdot e^{-jkr_p} \quad for \ p = \{1,2\} \tag{11}$$

where k is the wave number. The resulting signal s_{tot} arriving on the tag is the sum of the two phasors:

$$s_{tot} = s_1(i_1, r_1) + s_2(i_2, r_2)$$

After accounting for tag modulation efficiency L_t [55], and due to the reciprocity of all gain values during the back-scattered propagation, the full signal phasor received by the station antenna is finally expressed as follows:

$$s_{full} = s_{tot}^2 \cdot L_t$$

As a reminder, the squared s_{tot} corresponds to the back-and-forth path of the signal.

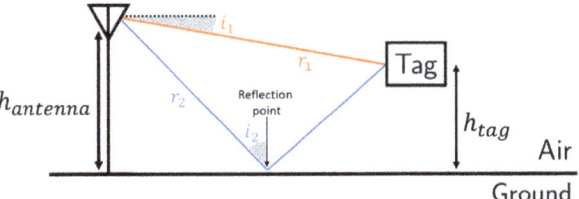

Figure 3. Schematic definition of the two-ray multipath model. The orange line represents the line-of-sight path with angle i_1 and propagation distance r_1. The blue line represents the reflected path with angle i_2 and propagation distance r_2. h_{tag} and $h_{antenna}$ are the tag and antenna heights above ground.

3.3.2. Two Types of Phase Error

We define the phase measurement error as the difference between the ideal LOS phase and the full received phase. This error can be divided into two contributions: the phase random deviation σ_{rdm} and the systematic phase bias ϕ_b, which are both consequences of multipath interference. Let us now consider these two error contributions separately. Previous investigations [18] have shown a direct relationship between antenna received power P(W) and phase random deviation σ_{rdm}(rad), using the same acquisition configuration (tag, interrogator, and communication protocol):

$$\sigma_{rdm} = \frac{4\pi f}{c} \cdot 9.5 \cdot 10^{-9}/\sqrt{P} \qquad (12)$$

where c is the light velocity and f the carrier frequency. This empirical relationship reproduces the phase error value of 0.04 rad used in the previous section. The received power greatly depends on propagation distance, but also on multipath interference, which is why σ_{rdm} is multipath-sensitive. The systematic phase bias ϕ_b is defined as the difference between the ideal LOS phase ϕ_1 and the full received phase ϕ_{full}:

$$\phi_b = arg(s_1^2) - arg(s_{full}) = \phi_1 - \phi_{full} \qquad (13)$$

The phase bias obviously depends on multipath behaviour. In phase space, the two error contributions σ_{rdm} and ϕ_b can be interpreted, respectively, as a scaling and translation operation on an ideal phase measurement distribution. Indeed, σ_{rdm} represents the width of the measurement error distribution, and the bias ϕ_b represents the center of this distribution; compared to the LOS ideal measurement; the true measurement will thus be translated by ϕ_b and scaled to a width of σ_{rdm}. Assuming Gaussian behaviour for the measurement process, each antenna j will, hence, present a measurement distribution ϕ_j following a normal law:

$$\phi_j = \mathcal{N}(\phi_b, \sigma_{rdm}^2)$$

These considerations can be applied in the phase-localization scheme presented in the previous section via a multi-antenna phase distribution.

Let us define the scaling matrix S and the translation vector T as follows:

$$S = \begin{bmatrix} \sigma_1 & 0 \\ 0 & \sigma_2 \end{bmatrix}$$

$$T = \begin{bmatrix} \phi_{b1} & \phi_{b2} \end{bmatrix}^\top$$

The entries of S originate from Equation (12) and the entries of T from Equation (13). They correspond to the values of phase random error and phase bias measured by each antenna ($N_a = 2$ in this simple scenario). Note that the phase random deviation values σ_j are different for each antenna for geometrical reasons; each antenna is in a different location, hence, the multipath and radiation patterns do not yield the same error values. The scaling S in phase space allows for a definition of the phase covariance matrix C_ϕ:

$$C_\phi = S \cdot S^\top$$

C_ϕ can be used in the singular value decomposition to compute the displacement error eigenvectors via the displacement covariance matrix C_r (see Equation (9)). The localization spot dimensions are, hence, fully described by the following covariance matrix in displacement space C_r:

$$C_r = (K^{-1})^\top C_\phi K^{-1} \qquad (14)$$

On the other hand, the translation T induced by the phase bias corresponds to a translation dr_b in displacement space, obtained by:

$$dr_b = K^{-1} \cdot T \qquad (15)$$

Equations (14) and (15) represent our best attempt to model the deviation from an ideal LOS phase measurement, taking into account the various phase measurement errors, and the geometry of the system. Figure 4 presents a 2D schematic view of the measurement distributions from phase space to displacement space. We see that the phase distribution is scaled and translated in phase space, compared to the centered distribution that was set in Equation (8). In displacement space, this gives a specific localization spot with covariance C_r, translated from the true LOS measurement by vector dr_b. The specific values of C_r and dr_b are discussed in Section 4.2.

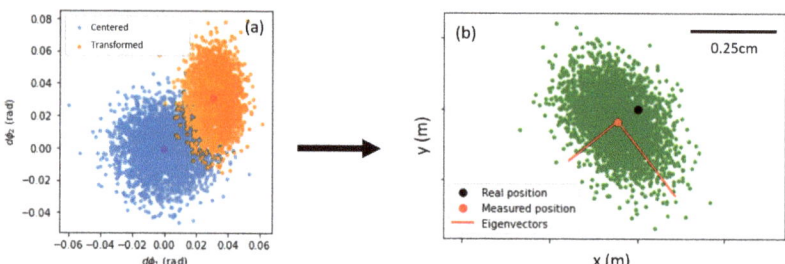

Figure 4. Schematic description of the matrix transformations in phase space towards real 2D space for a 2-antenna system. (**a**) Representation of the simulated multipath-induced phase measurement distribution (orange) compared to the previously assumed centered distribution (blue), highlighting the scaling S and translation T. The translation is illustrated by the shift between the center of the blue distribution and the center of the red distribution. (**b**) True space localization spot obtained by further transformation via the K^{-1} matrix. The antennas are not represented. The systematic bias is again illustrated by the shift between the real position (black point) and the center of the measured distribution (red point).

4. Harmalière Landslide Monitoring

In this section, we will discuss the specific case of the Harmalière landslide RFID system. After presenting the acquired data, the theoretical model will be applied to the real system geometry, then the localization results will be presented.

4.1. Real Phase Data

Among the 32 tags installed in the field, 10 were read almost continuously by more than two antennas for 12 months (January 2021–February 2022). The rest of the tags yielded partial results that could not be used for 2D localization via the present scheme. Two main factors can explain the lack of readability of some tags, namely, the narrow horizontal directivity of the antennas (+/−30° aperture) and signal attenuation—the furthest tags showed the lowest signal quality. Generally speaking, the tags placed 50 cm above ground showed worse results than those placed 1 m above ground, both in terms of data quality and localization accuracy. This observation corresponds to the above theoretical results (see Section 4.2 and Figure 5c), which tend to show that displacements close to the ground are subject to stronger multipath interference. This study will only show the tags read by at least two antennas during the whole period. The unwrapped phase measured during the January 2021–February 2022 time period is presented in Figure 6 for tag A. The data (70 measurements per day) were averaged over 24 h periods before applying the localization algorithm to mitigate the daily phase variations due to humidity and temperature. The missing values correspond to strong weather events that most likely depleted the battery of the acquisition system, or to hardware failures.

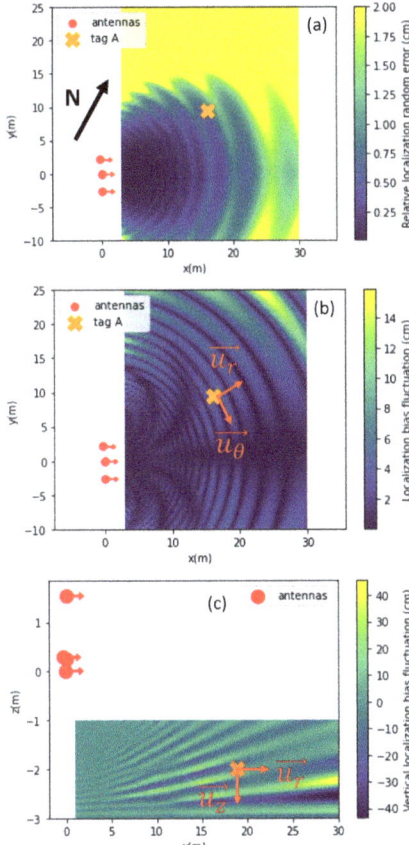

Figure 5. Mapping of the 2D localization error extracted from Equations (14) and (15), simulating the geometry of the Harmalière setup. The red dots represent the reader antennas, and the arrows show the principal antenna directions. The orange cross indicates the position of tag A. The vectors ($\vec{u}_r, \vec{u}_\theta, \vec{u}_z$) define the cylindrical coordinate system used later on. (**a**) The colormap shows the random localization error (maximum dimension of the localization spot) up to 2 cm, related to the phase random deviation σ_{rdm}. The localization bias is not shown. (**b**) Color-mapping of the systematic localization bias (related to ϕ_b) in the xOy plane shows oscillations with meter-order spatial frequency and increasing amplitude with distance from the measurement system. The random localization error is not shown. (**c**) Color-mapping of the systematic localization bias in the xOz plane, with higher oscillation amplitude and frequency. The ground is located at z= −3 m.

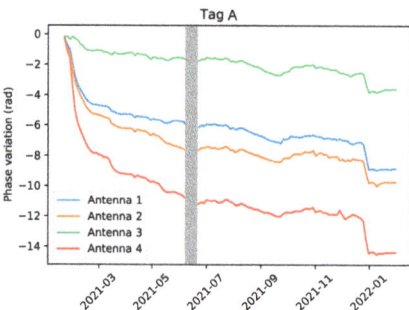

Figure 6. Unwrapped phase variation for tag A, measured by four antennas at a frequency f = 865 MHz, from January 2021 to February 2022. The grey bar shows a period of missing data due to hardware failure. Data was directly available after replacement of the malfunctioning device.

4.2. Application of the Model to a Real Geometry

Before presenting the localization results, we will first apply the previously developed model to the real system geometry. The workflow is presented in Figure 7, showing how the real parameters come together with the geometry and model to compute the localization error mapping.

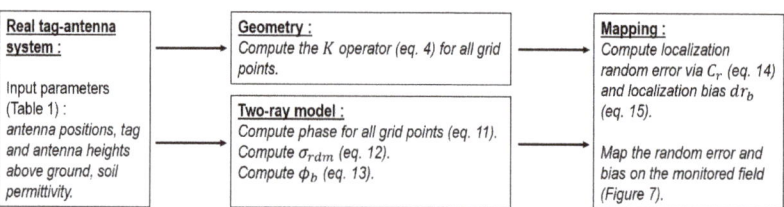

Figure 7. Schematic view of the workflow used to estimate the localization error and bias in the real-scenario Harmalière geometry.

We have set the model geometry according to Table 1, which corresponds to the Harmalière setup geometry. The number of antennas is now set to $N_a = 4$. The ground relative permittivity is set according to the literature for dry soils [56,57], and the following results correspond to this dry soil scenario. In the case of a wet soil, we expect the relative permittivity to reach values around 25. In the model, this turned out generally to increase the phase error (and localization error) values by about 30%, which can represent millimeter to centimeter values depending on the context (see Section 4.2.2).

Table 1. (Up) Geometrical parameters for the positions of the four antennas in the Harmalière setup. (Down) Values of the main variables used in the two-ray model (see Figure 3); the height of the station is relative to the ground at the same position.

Antenna No.	x (m)	y (m)	z (m)
1	0	0	0
2	0.018	−0.034	1.55
3	0.013	−2.608	0.256
4	−0.338	2.148	0.287
$h_{antenna}$ (m)			3
h_{tag} (m)			1
Ground relative permittivity			2.4

4.2.1. Random Localization Error of the Experimental Field

The previous developments (Equations (14) and (15)) have been applied to the geometry installed in the Harmalière landslide, as shown in Figure 5. A mapping of the random localization error (related to σ_{rdm}, Equation (12)) is shown in Figure 5a. We see that the lowest error is obtained when facing the antennas, which are oriented eastward. The plot is separated in two main areas, discriminated by the 2 cm random localization error value. This value was chosen because it reflects the target precision in our application.

4.2.2. Systematic Localization Bias of the Experimental Field

The systematic localization bias (related to ϕ_b, Equation (13)) presented in Figure 5b,c is not to be understood as a raw localization error, but as a varying bias when moving in space; the interference between LOS and the reflected signal changes with tag position.

To better understand the effect of the multipath-induced phase bias on 3D displacement measurements, we propose to consider the typical case of a 1 m displacement along a given spatial direction, starting from the position of tag A. The symmetry of our experiment being mainly cylindrical, we consider a cylindrical coordinate system with its central axis in $(x = 0, y = 0)$. For this displacement, we compute the localization bias fluctuation, and project it on every space direction (along $\vec{u}_\theta$, $\vec{u}_r$, $\vec{u}_z$) to obtain an amplitude value. The displacement length of 1 m was chosen both because it encompasses about one phase bias cycle, and because it corresponds to the actual displacement we measured in the real landslide scenario (see next section).

Table 2 reports the simulated localization bias amplitude in the three space directions, together with real error measurements that were performed on field.

- The direction that produces the least bias variation is a $\vec{u}_\theta$ displacement, which corresponds to the quasi rotational symmetry of the system.
- A horizontal displacement along $\vec{u}_r$ yields a small localization error. This confirms previous studies and demonstrates a centimeter precision for the RFID technique in the horizontal plane [18].
- A vertical displacement along $\vec{u}_z$ undergoes several strong bias oscillations (Figure 5c). The subsequent localization error is a cumulative effect of both the strong multipath interference and the small vertical aperture of the measurement system.

Table 2. Direction-dependent localization bias in the 3 directions (cylindrical coordinates), for a typical 1 m displacement. Each column corresponds to a different direction of displacement. Each line represents the localization bias amplitude along a certain direction, during the 1 m displacement. The *values in italic* correspond to field experiment localization bias measurements.

Bias \ Dir.	$\vec{u}_\theta$	$\vec{u}_r$	$\vec{u}_z$
max. $\vec{u}_\theta$ bias	<1 cm	<1 cm *(1 cm)*	10 cm *(20 cm)*
max. $\vec{u}_r$ bias	1 cm	1 cm *(1 cm)*	2 cm *(15 cm)*
max. $\vec{u}_z$ bias	1 cm	<1 cm *(5 cm)*	70 cm *(110 cm)*

These results tend to show that vertical localization in the current localization scheme cannot be performed with precision. The multipath effect, along with the high system sensitivity in this direction, yield a very high localization bias. This is why we will not present Oz localization results in the following section. This model highlights the importance of the geometrical features of the system, such as antenna position and spacing, tag height and direction of displacement.

4.3. Surface Displacement Monitoring Results

In this section, we present the experimental localization of the tags in the Harmalière landslide. We first focus on the 2D localization of one specific tag (tag A) in Figure 1, then we recapitulate on the whole setup and discuss the results.

4.3.1. 2D Relative Displacement for One Tag

The 2D displacement of tag A, computed from the radial displacements using multilateration and data from the four antennas (see Equation (3)), is shown in Figure 8 against reference tacheometer position measurements. The xOy results are in good agreement with the reference points. Note that, for stable phase periods (for example July 2021), the localization algorithm yields very stable results with a centimeter scale variability, which is in agreement with the theoretical localization error presented in Figure 2. This correspondence between theory and experiment during stable periods is observed for several tags, further validating the measurement error model. Note that Figure 2 does not present any phase bias results, but focuses only on measurement random deviation (dimensions of the localization spot).

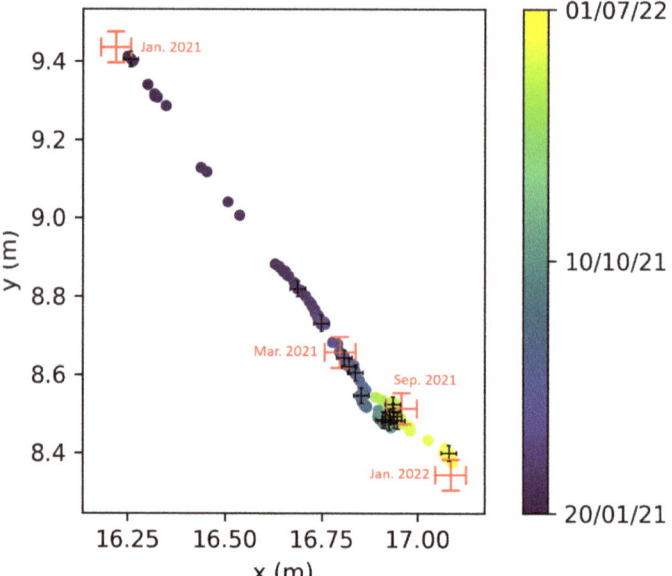

Figure 8. RFID localization in the xOy plane, using phase data for tag A (Figure 6). The total displacement is about 1.6 m. The color plot represents the time evolution of the RFID relative localization. The red crosses represent the reference measurements using a tacheometer, with an estimated error of about 4 cm. The tacheometer measurement of March 2021 is set as an absolute reference for relative localization. The black crosses correspond to the estimated random error bars for TD-phase localization (calculated via the model developed in Section 4.2).

4.3.2. 2D Localization for All Tags

Figure 9 shows an overview of the xOy displacement norm measured by the RFID-phase for all available tags during the measurement period. The total displacement is also shown for every tag in Table 3. All RFID localization results fit with reference measurements, notably for displacements greater than 1 m. The steep displacement increase in January 2022 concerning tags 51, 4e and A, was confirmed by tacheometer measurement. This rapid and localized deformation generated cracks and a landslide retrogression of about two meters in this area. A south-east tendency is clearly validated and corresponds to the landslide main direction, as can be seen in the qualitative vector mapping in Figure 10, with various displacement amplitudes depending on tag location. This opens the way to 2D spatio-temporal monitoring of the landslide surface, offering the possibility to better understand the physical mechanisms at the origin of the landslide activation and propagation, and to build new early warning monitoring systems.

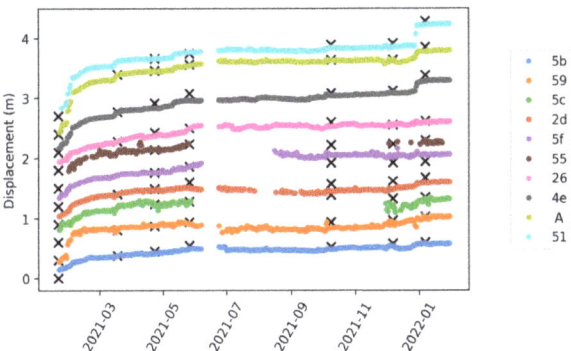

Figure 9. Cumulative 2D displacement norm for each tag, with reference measurements performed via tacheometer (black crosses). An offset was added to every plot to increase readability. The total displacement values are given in Table 3.

Table 3. Total 2D displacement norm for all presented tags computed from the RFID phase, from January 2021 to February 2022. The reference is computed from the tacheometry measurements, with an estimated error of ±4 cm.

Tag	51	A	4e	26	55	5f	2d	5c	59	5b
Total disp (m)	1.54	1.37	1.20	0.81	0.75	0.85	0.69	0.67	0.74	0.56
Reference (m)	1.57	1.45	1.28	0.81	0.79	0.74	0.77	0.74	0.72	0.59

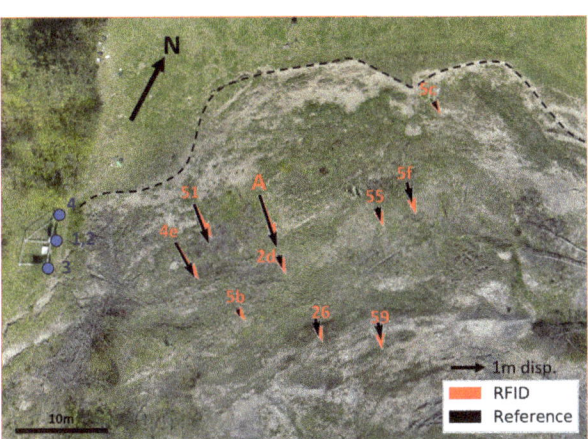

Figure 10. Vector mapping of the total 2D displacement for all available tags from January 2021 to September 2021. The scale is modified for clarity with a 1 m displacement reference (black arrow). The red arrows represent the displacement estimated from the RFID measurements, and the black arrows represent the displacement computed from reference measurements. The blue points numbered 1 to 4 correspond to the reader antennas.

4.4. Discussion

In this section we briefly discuss some of the results presented in this paper, as well as future development of the RFID localization system.

4.5. Localization Error and Reference Measurements

In the context in which RFID localization was performed, absolute reference localization at a centimeter level was a complicated task. For practical reasons, reference positions taken via GPS were not sufficiently accurate to be compared to the RFID localization results. This is why tacheometry was used, which is a relative localization method. A landslide is an ever-changing environment, and using absolute references such as trees or antennas involves several sources of error. For this reason, the tacheometer uncertainty given in Table 3 is ±4 cm. As has been described in previous reports [18], RFID phase outdoor localization can outperform the reference measurements.

4.5.1. Discussion on Antenna Position

The above model (Section 4.2) is a tool for optimizing the antenna positions in a given terrain to minimize localization errors originating from both multipath and geometry. We performed calculations for several geometrical cases in a plane xOy geometry, searching for the lowest random deviation in the monitored zone. As a general rule, we conclude that surrounding the field with antennas yields the best accuracy (lowest localization random deviation). For example, if four antennas are spread around the Harmalière field, the horizontal random localization error is expected to reduce to 1 mm.

Such setups are not always possible in real-environment operational situations—the experimental setup obviously has to be designed taking into account the operational constraints and priorities. In cases where a portion of the field is inaccessible, for example, the distance between antennas (system aperture) should be maximized to obtain the lowest random deviation. This guideline has limitations, such as cable length or station cost, hence the final setup will generally be a compromise between precision and station/maintenance cost. Note that other localization methods, such as angle of arrival techniques [54,58] rely on different system geometries and will not lead to the same optimal antenna disposition. The guidelines provided here only apply to a relative displacement scheme; absolute positioning is a different matter which we do not discuss here.

4.5.2. Perspective for Improving Data Availability

In this investigation, the tags that yielded only partial data (i.e., less than two antenna readings, long time periods without data) were not used, although more complex data assimilation techniques could be of use [59,60]. Exploiting both the knowledge of the landslide mechanics and the redundancy of information that the system yields could allow tag monitoring even in partial data scenarios, which are a common issue in outdoor environments. Such techniques will be implemented in future work.

5. Conclusions

We have derived a phase-based 2D localization error theoretical model that allows for error estimation in a scenario of two to four static interrogator antennas, taking into account the specific setup geometry. The model is based on both the sensitivity kernel of the measurement system and a two-ray propagation model (multipath). Under certain conditions, this model confirms the ability to track centimetric ground displacements. The in-plane horizontal measurements demonstrate much better accuracy than the out-of-plane vertical measurements, due to the preferential horizontal antenna distribution, and to ground-reflection multipath interference.

A set of RFID tags was placed on an active landslide and phase measurements were performed over several months to monitor the tags' displacement. The results show a clear south-east displacement of about 1 m in the horizontal plane over the monitored area. The presented method, inspired by the time-difference phase-difference scheme, has shown very good results for the monitoring of relative displacements in 2D at the centimeter scale. The monitoring of landslides using RFID technology was demonstrated to be a viable solution, with centimeter-scale accuracy over large periods of time. A further step in large scale monitoring could be to deploy a moving antenna (SAR) over greater lengths, and to implement a data assimilation approach to increase data availability.

Author Contributions: Conceptualization, A.C., M.L.B., E.L. and L.B.; Data curation, A.C.; Funding acquisition, M.L.B., E.L. and L.B.; Investigation, A.C.; Methodology, M.L.B., E.L. and L.B.; Supervision, M.L.B., E.L. and L.B.; Validation, A.C., M.L.B., E.L. and L.B.; Writing—original draft, A.C.; Writing—review & editing, M.L.B., E.L. and L.B. All authors have read and agreed to the published version of the manuscript.

Funding: This work was partially funded by the ANR LABCOM GEO3ILAB and by the Region Auvergne Rhone Alpes RISQID project. This work is part of the "Habitability" LABEX coordinated by OSUG. We acknowledge experimental help from B. Vial, M. Langlais, G. Scheiblin, and G. Bièvre from ISTerre.

Conflicts of Interest: The authors declare no conflict of interest.

References

1. Scaioni, M.; Longoni, L.; Melillo, V.; Papini, M. Remote sensing for landslide investigations: An overview of recent achievements and perspectives. *Remote Sens.* **2014**, *6*, 9600–9652. [CrossRef]
2. Zhao, C.; Lu, Z. Remote sensing of landslides—A review. *Remote Sens.* **2018**, *10*, 279. [CrossRef]
3. Colesanti, C.; Wasowski, J. Investigating landslides with space-borne Synthetic Aperture Radar (SAR) interferometry. *Eng. Geol.* **2006**, *88*, 173–199. [CrossRef]
4. Strozzi, T.; Farina, P.; Corsini, A.; Ambrosi, C.; Thüring, M.; Zilger, J.; Wiesmann, A.; Wegmüller, U.; Werner, C. Survey and monitoring of landslide displacements by means of L-band satellite SAR interferometry. *Landslides* **2005**, *2*, 193–201. [CrossRef]
5. Wang, Y.; Hong, W.; Zhang, Y.; Lin, Y.; Li, Y.; Bai, Z.; Zhang, Q.; Lv, S.; Liu, H.; Song, Y. Ground-based differential interferometry SAR: A review. *IEEE Geosci. Remote Sens. Mag.* **2020**, *8*, 43–70. [CrossRef]
6. Tarchi, D.; Casagli, N.; Fanti, R.; Leva, D.D.; Luzi, G.; Pasuto, A.; Pieraccini, M.; Silvano, S. Landslide monitoring by using ground-based SAR interferometry: An example of application to the Tessina landslide in Italy. *Eng. Geol.* **2003**, *68*, 15–30. [CrossRef]
7. Monserrat, O.; Crosetto, M.; Luzi, G. A review of ground-based SAR interferometry for deformation measurement. *ISPRS J. Photogramm. Remote Sens.* **2014**, *93*, 40–48. [CrossRef]
8. Helmstetter, A.; Garambois, S. Seismic monitoring of Séchilienne rockslide (French Alps): Analysis of seismic signals and their correlation with rainfalls. *J. Geophys. Res. Earth Surf.* **2010**, *115*. [CrossRef]
9. Aryal, A.; Brooks, B.A.; Reid, M.E.; Bawden, G.W.; Pawlak, G.R. Displacement fields from point cloud data: Application of particle imaging velocimetry to landslide geodesy. *J. Geophys. Res. Earth Surf.* **2012**, *117*. [CrossRef]
10. Benoit, L.; Briole, P.; Martin, O.; Thom, C.; Malet, J.P.; Ulrich, P. Monitoring landslide displacements with the Geocube wireless network of low-cost GPS. *Eng. Geol.* **2015**, *195*, 111–121. [CrossRef]
11. Li, Y.; Huang, J.; Jiang, S.H.; Huang, F.; Chang, Z. A web-based GPS system for displacement monitoring and failure mechanism analysis of reservoir landslide. *Sci. Rep.* **2017**, *7*, 17171. [CrossRef]
12. Šegina, E.; Peternel, T.; Urbančič, T.; Realini, E.; Zupan, M.; Jež, J.; Caldera, S.; Gatti, A.; Tagliaferro, G.; Consoli, A.; et al. Monitoring Surface Displacement of a Deep-Seated Landslide by a Low-Cost and near Real-Time GNSS System. *Remote Sens.* **2020**, *12*, 3375. [CrossRef]
13. Dong, M.; Wu, H.; Hu, H.; Azzam, R.; Zhang, L.; Zheng, Z.; Gong, X. Deformation prediction of unstable slopes based on real-time monitoring and deepar model. *Sensors* **2020**, *21*, 14. [CrossRef]
14. Intrieri, E.; Gigli, G.; Gracchi, T.; Nocentini, M.; Lombardi, L.; Mugnai, F.; Frodella, W.; Bertolini, G.; Carnevale, E.; Favalli, M.; et al. Application of an ultra-wide band sensor-free wireless network for ground monitoring. *Eng. Geol.* **2018**, *238*, 1–14. [CrossRef]
15. Mucchi, L.; Jayousi, S.; Martinelli, A.; Caputo, S.; Intrieri, E.; Gigli, G.; Gracchi, T.; Mugnai, F.; Favalli, M.; Fornaciai, A.; et al. A flexible wireless sensor network based on ultra-wide band technology for ground instability monitoring. *Sensors* **2018**, *18*, 2948. [CrossRef]
16. Schneider, J.M.; Turowski, J.M.; Rickenmann, D.; Hegglin, R.; Arrigo, S.; Mao, L.; Kirchner, J.W. Scaling relationships between bed load volumes, transport distances, and stream power in steep mountain channels. *J. Geophys. Res. Earth Surf.* **2014**, *119*, 533–549. [CrossRef]
17. Breton, M.L.; Liébault, F.; Baillet, L.; Charléty, A.; Éric Larose.; Tedjini, S. Dense and longdterm monitoring of Earth surface processes with passive RFID—A review. *arXiv* **2021**, arXiv:2112.11965.

18. Le Breton, M.; Baillet, L.; Larose, E.; Rey, E.; Benech, P.; Jongmans, D.; Guyoton, F.; Jaboyedoff, M. Passive radio-frequency identification ranging, a dense and weather-robust technique for landslide displacement monitoring. *Eng. Geol.* **2019**, *250*, 1–10. [CrossRef]
19. Le Breton, M.; Baillet, L.; Larose, E.; Rey, E.; Benech, P.; Jongmans, D.; Guyoton, F. Outdoor uhf rfid: Phase stabilization for real-world applications. *IEEE J. Radio Freq. Identif.* **2017**, *1*, 279–290. [CrossRef]
20. Le Breton, M. Suivi Temporel d'un Glissement de Terrain à l'Aide d'Étiquettes RFID Passives, Couplé à l'Observation de Pluviométrie et de Bruit Sismique Ambiant. Ph.D. Thesis, Université Grenoble Alpes (ComUE), Grenoble, France, 2019.
21. Intrieri, E.; Gigli, G.; Mugnai, F.; Fanti, R.; Casagli, N. Design and implementation of a landslide early warning system. *Eng. Geol.* **2012**, *147*, 124–136. [CrossRef]
22. Intrieri, E.; Carlà, T.; Gigli, G. Forecasting the time of failure of landslides at slope-scale: A literature review. *Earth-Sci. Rev.* **2019**, *193*, 333–349. [CrossRef]
23. Balaji, R.; Malathi, R.; Priya, M.; Kannammal, K. A Comprehensive Nomenclature Of RFID Localization. In Proceedings of the 2020 International Conference on Computer Communication and Informatics (ICCCI), Coimbatore, India, 22-24 January 2020; pp. 1–9.
24. Miesen, R.; Ebelt, R.; Kirsch, F.; Schäfer, T.; Li, G.; Wang, H.; Vossiek, M. Where is the tag? *IEEE Microw. Mag.* **2011**, *12*, S49–S63. [CrossRef]
25. Ni, L.M.; Liu, Y.; Lau, Y.C.; Patil, A.P. LANDMARC: Indoor location sensing using active RFID. In Proceedings of the 1st IEEE International Conference on Pervasive Computing and Communications, 2003.(PerCom 2003), Fort Worth, Texas, USA, 23–26 March 2003; pp. 407–415.
26. Subedi, S.; Pauls, E.; Zhang, Y.D. Accurate localization and tracking of a passive RFID reader based on RSSI measurements. *IEEE J. Radio Freq. Identif.* **2017**, *1*, 144–154. [CrossRef]
27. Rohmat Rose, N.D.; Low, T.J.; Ahmad, M. 3D trilateration localization using RSSI in indoor environment. *Int. J. Adv. Comput. Sci. Appl.* **2020**, *11*, 385–391.
28. Martinelli, F. A robot localization system combining RSSI and phase shift in UHF-RFID signals. *IEEE Trans. Control. Syst. Technol.* **2015**, *23*, 1782–1796. [CrossRef]
29. Shen, L.; Zhang, Q.; Pang, J.; Xu, H.; Li, P. PRDL: relative localization method of RFID tags via phase and RSSI based on deep learning. *IEEE Access* **2019**, *7*, 20249–20261. [CrossRef]
30. Scherhäufl, M.; Pichler, M.; Stelzer, A. UHF RFID localization based on evaluation of backscattered tag signals. *IEEE Trans. Instrum. Meas.* **2015**, *64*, 2889–2899. [CrossRef]
31. Wang, Z.; Ye, N.; Malekian, R.; Xiao, F.; Wang, R. TrackT: Accurate tracking of RFID tags with mm-level accuracy using first-order taylor series approximation. *Ad Hoc Netw.* **2016**, *53*, 132–144. [CrossRef]
32. Zhou, C.; Griffin, J.D. Accurate phase-based ranging measurements for backscatter RFID tags. *IEEE Antennas Wirel. Propag. Lett.* **2012**, *11*, 152–155. [CrossRef]
33. Li, C.; Mo, L.; Zhang, D. Review on UHF RFID localization methods. *IEEE J. Radio Freq. Identif.* **2019**, *3*, 205–215. [CrossRef]
34. Huiting, J.; Flisijn, H.; Kokkeler, A.B.; Smit, G.J. Exploiting phase measurements of EPC Gen2 RFID tags. In Proceedings of the 2013 IEEE International Conference on RFID-Technologies and Applications (RFID-TA), Johor Bahru, Malaysia, 4–5 September 2013; pp. 1–6.
35. Pelka, M.; Bollmeyer, C.; Hellbrück, H. Accurate radio distance estimation by phase measurements with multiple frequencies. In Proceedings of the 2014 International Conference on Indoor Positioning and Indoor Navigation (IPIN), Busan, Korea, 27–30 October 2014; pp. 142–151.
36. Nikitin, P.V.; Martinez, R.; Ramamurthy, S.; Leland, H.; Spiess, G.; Rao, K. Phase based spatial identification of UHF RFID tags. In Proceedings of the 2010 IEEE International Conference on RFID (IEEE RFID 2010), Orlando, FL, USA, 14–16 April 2010; pp. 102–109.
37. Povalac, A.; Sebesta, J. Phase difference of arrival distance estimation for RFID tags in frequency domain. In Proceedings of the 2011 IEEE International Conference on RFID-Technologies and Applications, Sitges, Spain, 15–16 September 2011; pp. 188–193.
38. Scherhäufl, M.; Pichler, M.; Stelzer, A. UHF RFID localization based on phase evaluation of passive tag arrays. *IEEE Trans. Instrum. Meas.* **2014**, *64*, 913–922. [CrossRef]
39. Buffi, A.; Nepa, P.; Cioni, R. SARFID on drone: Drone-based UHF-RFID tag localization. In Proceedings of the 2017 IEEE International Conference on RFID Technology & Application (RFID-TA), Warsaw, Poland, 20-22 September 2017; pp. 40–44.
40. Buffi, A.; Motroni, A.; Nepa, P.; Tellini, B.; Cioni, R. A SAR-based measurement method for passive-tag positioning with a flying UHF-RFID reader. *IEEE Trans. Instrum. Meas.* **2018**, *68*, 845–853. [CrossRef]
41. Motroni, A.; Nepa, P.; Magnago, V.; Buffi, A.; Tellini, B.; Fontanelli, D.; Macii, D. SAR-based indoor localization of UHF-RFID tags via mobile robot. In Proceedings of the 2018 International Conference on Indoor Positioning and Indoor Navigation (IPIN), Nantes, France, 24–27 September 2018; pp. 1–8.
42. Bernardini, F.; Buffi, A.; Motroni, A.; Nepa, P.; Tellini, B.; Tripicchio, P.; Unetti, M. Particle swarm optimization in SAR-based method enabling real-time 3D positioning of UHF-RFID tags. *IEEE J. Radio Freq. Identif.* **2020**, *4*, 300–313. [CrossRef]
43. Gareis, M.; Fenske, P.; Carlowitz, C.; Vossiek, M. Particle filter-based SAR approach and trajectory optimization for real-time 3D UHF-RFID tag localization. In Proceedings of the 2020 IEEE International Conference on RFID (RFID), Orlando, FL, USA, 28 September–16 October 2020; pp. 1–8.

44. Monjuvent, G. La transfluence Durance-Isère Essai de synthèse du Quaternaire du bassin du Drac'(Alpes françaises). *Géol. Alp.* **1973**, *49*, 57–118.
45. Jongmans, D.; Bièvre, G.; Renalier, F.; Schwartz, S.; Beaurez, N.; Orengo, Y. Geophysical investigation of a large landslide in glaciolacustrine clays in the Trièves area (French Alps). *Eng. Geol.* **2009**, *109*, 45–56. [CrossRef]
46. Fiolleau, S.; Borgniet, L.; Jongmans, D.; Bièvre, G.; Chambon, G. Using UAV's imagery and LiDAR to accurately monitor Harmalière (France) landslide evolution. In *Geophysical Research Abstracts*; European Geosciences Union: Munchen, Germany, 2019; Volume 21.
47. Fiolleau, S.; Jongmans, D.; Bièvre, G.; Chambon, G.; Lacroix, P.; Helmstetter, A.; Wathelet, M.; Demierre, M. Multi-method investigation of mass transfer mechanisms in a retrogressive clayey landslide (Harmalière, French Alps). *Landslides* **2021**, *18*, 1981–2000. [CrossRef]
48. Norrdine, A. An algebraic solution to the multilateration problem. In Proceedings of the 15th International Conference on Indoor Positioning and Indoor Navigation, Sydney, Australia, 13–15 November 2012; Volume 1315.
49. Conn, A.R.; Gould, N.I.; Toint, P.L. *Trust Region Methods*; SIAM: Philadelphia, PA, USA, 2000.
50. Van Loan, C.F. Generalizing the singular value decomposition. *SIAM J. Numer. Anal.* **1976**, *13*, 76–83. [CrossRef]
51. Anton, H.; Rorres, C. *Elementary Linear Algebra: Applications Version*; John Wiley & Sons: Hoboken, NJ, USA, 2013.
52. Golub, G.; Kahan, W. Calculating the singular values and pseudo-inverse of a matrix. *J. Soc. Ind. Appl. Math. Ser. Numer. Anal.* **1965**, *2*, 205–224. [CrossRef]
53. Wang, G.; Qian, C.; Cui, K.; Shi, X.; Ding, H.; Xi, W.; Zhao, J.; Han, J. A Universal Method to Combat Multipaths for RFID Sensing. In Proceedings of the IEEE INFOCOM 2020-IEEE Conference on Computer Communications, Toronto, ON, Canada, 6–9 July 2020; pp. 277–286.
54. Faseth, T.; Winkler, M.; Arthaber, H.; Magerl, G. The influence of multipath propagation on phase-based narrowband positioning principles in UHF RFID. In Proceedings of the 2011 IEEE-APS Topical Conference on Antennas and Propagation in Wireless Communications, Torino, Italy, 12–16 September 2011; pp. 1144–1147.
55. Rembold, B. Optimum modulation efficiency and sideband backscatter power response of RFID-tags. *Frequenz* **2009**, *63*, 9–13. [CrossRef]
56. ITU-R P. 523-7; Electrical Characteristics of the Surface of the Earth. ITU-R: Geneva, Switzerland, 1992.
57. Lytle, R.J. Measurement of earth medium electrical characteristics: Techniques, results, and applications. *IEEE Trans. Geosci. Electron.* **1974**, *12*, 81–101. [CrossRef]
58. Azzouzi, S.; Cremer, M.; Dettmar, U.; Kronberger, R.; Knie, T. New measurement results for the localization of uhf rfid transponders using an angle of arrival (aoa) approach. In Proceedings of the 2011 IEEE International Conference on RFID, Sitges, Spain, 15–16 September 2011; pp. 91–97.
59. Sun, S.L.; Deng, Z.L. Multi-sensor optimal information fusion Kalman filter. *Automatica* **2004**, *40*, 1017–1023. [CrossRef]
60. Sarkka, S.; Viikari, V.V.; Huusko, M.; Jaakkola, K. Phase-based UHF RFID tracking with nonlinear Kalman filtering and smoothing. *IEEE Sens. J.* **2011**, *12*, 904–910. [CrossRef]

Article

Superpixel and Supervoxel Segmentation Assessment of Landslides Using UAV-Derived Models

Ioannis Farmakis [1,*], Efstratios Karantanellis [2], D. Jean Hutchinson [1], Nicholas Vlachopoulos [3] and Vassilis Marinos [4]

[1] Department of Geological Sciences and Geological Engineering, Queen's University, Kingston, ON K7L 3N6, Canada
[2] Department of Earth Science and Environmental Sciences, University of Michigan, Ann Arbor, MI 48103, USA
[3] Department of Civil Engineering, Royal Military College of Canada, Kingston, ON K7K 7B4, Canada
[4] Department of Civil Engineering, National Technical University of Athens, 157 80 Athens, Greece
* Correspondence: i.farmakis@queensu.ca; Tel.: +30-698-410-1585

Abstract: Reality capture technologies such as Structure-from-Motion (SfM) photogrammetry have become a state-of-the-art practice within landslide research workflows in recent years. Such technology has been predominantly utilized to provide detailed digital products in landslide assessment where often, for thorough mapping, significant accessibility restrictions must be overcome. UAV photogrammetry produces a set of multi-dimensional digital models to support landslide management, including orthomosaic, digital surface model (DSM), and 3D point cloud. At the same time, the recognition of objects depicted in images has become increasingly possible with the development of various methodologies. Among those, Geographic Object-Based Image Analysis (GEOBIA) has been established as a new paradigm in the geospatial data domain and has also recently found applications in landslide research. However, most of the landslide-related GEOBIA applications focus on large scales based on satellite imagery. In this work, we examine the potential of different UAV photogrammetry product combinations to be used as inputs to image segmentation techniques for the automated extraction of landslide elements at site-specific scales. Image segmentation is the core process within GEOBIA workflows. The objective of this work is to investigate the incorporation of fully 3D data into GEOBIA workflows for the delineation of landslide elements that are often challenging to be identified within typical rasterized models due to the steepness of the terrain. Here, we apply a common unsupervised image segmentation pipeline to 3D grids based on the superpixel/supervoxel and graph cut algorithms. The products of UAV photogrammetry for two landslide cases in Greece are combined and used as 2D (orthomosaic), 2.5D (orthomosaic + DSM), and 3D (point cloud) terrain representations in this research. We provide a detailed quantitative comparative analysis of the different models based on expert-based annotations of the landscapes and conclude that using fully 3D terrain representations as inputs to segmentation algorithms provides consistently better landslide segments.

Keywords: UAV; image segmentation; 3D modelling; landslide; superpixels; supervoxels; graph cut

Citation: Farmakis, I.; Karantanellis, E.; Hutchinson, D.J.; Vlachopoulos, N.; Marinos, V. Superpixel and Supervoxel Segmentation Assessment of Landslides Using UAV-Derived Models. *Remote Sens.* **2022**, *14*, 5668. https://doi.org/10.3390/rs14225668

Academic Editor: Rachid El Hamdouni

Received: 22 September 2022
Accepted: 7 November 2022
Published: 10 November 2022

Publisher's Note: MDPI stays neutral with regard to jurisdictional claims in published maps and institutional affiliations.

Copyright: © 2022 by the authors. Licensee MDPI, Basel, Switzerland. This article is an open access article distributed under the terms and conditions of the Creative Commons Attribution (CC BY) license (https://creativecommons.org/licenses/by/4.0/).

1. Introduction

Landslide disasters constitute a global issue that threatens the sustainability of infrastructure and the environment as well as human lives [1]. Sophisticated digital methods for accurate and efficient assessment based on cutting-edge technologies can enhance traditional protocols. Unmanned Aerial Vehicle (UAV) photogrammetry has become a state-of-the-art technique in landslide research in recent years. The ability to produce high-resolution digital models of inaccessible areas in a short time and with relatively low cost has made UAVs key assets in landslide assessment and they support decision making. Digital landslide models permit landslide assessment and support decision making to mitigate landslide risk. Consequently, there is a need to improve accuracy, efficiency, and automation in landslide modelling and minimize subjectivity and human error in visual interpretations. The products of UAV surveys vary from the 2D orthomosaic and 2.5D digital surface model (DSM) to high-resolution 3D point cloud. These products permit powerful visualizations, precise measurements, and detailed analysis tasks such as change detection to be carried out.

However, there is room to improve the capabilities of our models. To make the models more intelligent and leverage autonomy, strengthen inter-communication, and perform efficient queries to speed up current analysis frameworks, we primarily need to transfer formalized knowledge and inject semantics into the digital model. This mainly comprises a clustering problem. However, the identification of commonalities, in ultra-high-resolution geographic data, constitutes a challenge within several applications. It is also important to structure a large model in a way that it can be manipulated efficiently. This requires the development of semantic segmentation methodologies that are able to recognize different geologic/geomorphologic features within a digital model and efficiently accommodate multi-level conceptualizations [2].

1.1. Motivation and Objectives

The motivation for the work presented in this paper is to assess the effectiveness of UAV-derived terrain representations of varying dimensionality to support the extraction of meaningful landslide objects. In landslide scenes, segmentations have predominantly been applied to 2D or 2.5D images (pixel grids) (see Section 1.2 below). Landslide scenes are often steep terrains where critical elements such as the scarp zones and/or the flanks of a landslide may not be adequately modeled in top-view projections. In such cases, a fully 3D approach might be essential. Although powerful, fully 3D approaches are often constrained by the high computational resources demand, especially at regional scale analyses. However, at site-specific scales, the data load can still be handled by commercial hardware. In this study, the authors thoroughly compare site-specific scale landslide segmentations produced in all 2D, 2.5D, and 3D operational domains using UAV-derived terrain representations as inputs to GEOBIA workflows. The segmentation algorithm used for the assessment of the models was selected based on three criteria:

(a) it is unsupervised (no requirement for large amounts of training data);
(b) it is applicable in the 3D space as well (maintains consistency in the experiments); and
(c) it is tested for the first time in a landslide environment (provides new data for the landslide community).

It is, however, worth mentioning that the definition of the optimum landslide segmentation algorithm for the task is beyond the scope of this research. Only the dimensionality of the landslide representation is assessed as the input, based on the algorithm that fulfills the above criteria, to maintain consistency between the different operational spaces. UAV deployment at site-specific scales enables the acquisition of detailed 3D models. In steep terrains such as landslides, moving from 2D image segmentation to fully 3D analyses may prove to be decisive in the extraction of confident geologic information.

In the research reported here, superpixels and supervoxels are generated under different resolutions and the final segmentation of each model is achieved using a data-driven approach. The supervoxel implementations are modified to operate directly on the voxel grid rather than on stacked pixel grids. The results of two geologically/geomorphologically different landslide cases are evaluated against expert annotated reference data, considering state-of-the-art performance metrics for the task. The methodology followed is illustrated in the flowchart in Figure 1 and aims to investigate the suitability of individual terrain representations of varying spatial dimensionality for automatic unsupervised segmentation towards object-oriented semantic injection in landslide models.

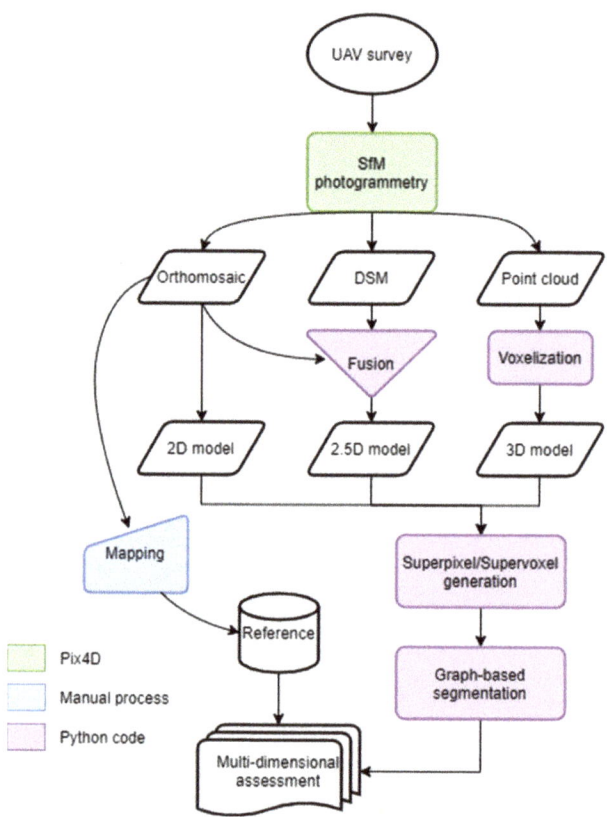

Figure 1. The methodology of the analysis followed for each landslide study site in the current study from raw data capture to segmentation and multi-dimensional assessment of UAV-derived models based on expert-based reference annotations. Each process is coloured by the software/technique used for the implementation while their products are mentioned in white frames.

1.2. Segmentation Background

Image segmentation was initially implemented within Object-Based Image Analysis (OBIA) frameworks for medical imaging, but since then it has been applied to multiple domains such as indoor mapping, traffic detection, and range imaging [3]. It has revolutionized image analysis processes with a move from the traditional pixel-based model to an object-based contextual model that endeavours to emulate the way humans interpret images. Subsequently, the objective of segmentation has been changed from pixel labelling to object partitioning [4]. Since the early 2000s, the same shift has been taking place in remote sensing. The new paradigm named Geographic Object-Based Image Analysis (GEOBIA), promises to change the way geoscientists perceive, analyze, and use remote (or close-range) sensing data [5,6]. In remote sensing, the main objectives are to detect, identify, analyze, and monitor the dynamics of natural phenomena and several studies have highlighted the advantages of GEOBIA compared to the pixel-based paradigm [7,8]. However, there are specific challenges in using the object-oriented data model, such as the under-segmentation and over-segmentation error associated with the segmentation scale. This occurs when the object entities include more than one object, or the objects are unnecessarily broken apart in the segmentation process [9], respectively.

Image segmentation is a critical step in GEOBIA. There are various methods for semi-automatic or automatic object detection that are based on the application of various segmentation algorithms to Earth Observation (EO) data [10]. Many methods for object recognition and segmentation rely on the tessellation of an image into "superpixels". A superpixel is an image patch which is better aligned with object boundaries than a rectangular patch. The objective is to group pixels into perceptually meaningful regions to replace the rigid structure of the raster [11]. Superpixels aim at balancing the conflicting goals of reducing image complexity and yielding new properties, while avoiding under-segmentation. The extensive state-of-the-art review on segmentation methods for GEOBIA conducted by [4] reveals that MRS [12] is the most cited segmentation algorithm and is often used as a reference for benchmarking. However, Ref. [13] demonstrates that similar or even better accuracy can be achieved using superpixels instead of MRS-derived segments as objects. The accuracy and precision of a segmentation technique refer to the degree by which the segmentation results agree with the reference data extent. Image segmentation literature is extensive, including different algorithms and approaches. Furthermore, Ref. [14] have demonstrated a detailed review of recent trends in object detection with the use of deep learning (e.g., [15]). Because segmentation is the initial phase in the data analysis, an accurate, precise, and efficient approach must be implemented to minimize inappropriate results.

The segmentation process defines the entities for object-oriented analysis which should aim to create meaningfully delineated real-world objects. Entities can be generated based on different attributes, such as size, texture, shape, spatial and spectral distribution. The precision and accuracy of object boundary delineation is significant due to its impact on the subsequent classification phase. Subsequently, the segmentation can be completed treating these newly formed entities, with their respective attributes, as the image unit within the desirable clustering approach. This is essentially the principle of object-oriented models which have proven to yield better performance compared to traditional pixel-based analyses of 2D and 2.5D images [16].

The same concepts apply to 3D space where the analogous to pixels, voxels (volumetric pixels) are grouped into supervoxels as discussed in [2]. The generation of the voxel grid requires the point cloud to be recursively subdivided into eight child voxels until the desired resolution is reached. The voxel colour and/or other layers are inherited from the point cloud and the points included within each voxel. In contrast to 2D image over-segmentation, little work has been carried out in supervoxel algorithms operating directly on 3D space (i.e., [17,18]). Despite the usefulness in digitizing the real world, 3D point clouds are tedious to work with and conventional point cloud tools are limited. The vast majority of the advances in computer vision and machine learning today deal with 2D images [19]. That is the reason why point clouds are often written off as the raw product with the focus being put on processed data formats such as rasters. Three-dimensional scenes are usually treated as a stack of slices along one of the dimensions, each processed as an individual image. We use point clouds as visualizations for virtual inspections and limit our work with them to measuring simple distances. However, given the ever-increasing availability of high-resolution point clouds, fully 3D object-oriented operations might add substantial value to object-based landslide semantic segmentation.

In the landslide domain, GEOBIA has become a popular method for semantic image analysis due to the integration of different data types in very high spatial resolutions. The automatic identification of specific landslide and geomorphologic features [20,21] have been studied in the literature and GEOBIA has already proven its strength (Martha et al., 2010). Many studies have used spectral information combined with Digital Elevation Model (DEM) derivatives to delineate landslide bodies within satellite images utilizing a non-seeded region growing segmentation algorithm in commercial software [22–25]. Furthermore, the study by [26] proposed a workflow for landslide mapping from satellite images using open-source tools with a mean shift segmentation. Earlier, Ref. [27] had successfully segmented glacier surfaces form Airborne Laser Scanning (ALS) data using a seeded region growing segmentation based on DEM derivatives and laser intensity. Other landform applications of GEOBIA include the mapping of drumlins [28] and gully-affected areas [29].

However, landslide segmentation for object-oriented analysis can be even more detailed, when utilizing UAV-derived data, to segment individual landslide elements. Recently, Ref. [30] utilized UAV photogrammetry to produce 2.5D representations of two landslide cases at a site-specific scale. The authors used GEOBIA to segment and classify distinct landslide features such as the scarp, depletion and accumulation zones as well as anthropogenic structures that constitute elements at risk within the landslide scene. The segmentation performed in a multi-dimensional feature space consisted of spectral, topographic, and geometric features using a non-seeded region growing algorithm. Their approach aimed to propose a framework for targeted landslide mapping and quantified characterization of specific semantic zones.

Given the advantages of GEOBIA and the ever-increasing availability of high-resolution 3D point clouds, a new approach to object-oriented analyses of landslide scenes, based on direct point cloud segmentation is proposed by [31]. The development of distinct 3D object entities represents an advantageous approach for analyzing natural scenes because points can be meaningfully partitioned into networked homogeneous entities that carry their full semantic information. Nonetheless, direct 3D data processing is more complicated than image processing and requires proficiency. To date, very few studies have used Object-Oriented Point cloud Analysis (OBPA) of landslide scenes. Preliminary efforts using LiDAR point clouds include: (a) segmentation of rotational landslides into individual geomorphologic zones and vegetation based on a seeded region growing algorithm using point-based geometric descriptors [32], and (b) segmentation of a multi-hazard railway rock slope including debris channels, steep rock outcrops, rock benches, and transportation infrastructure using voxel-based features and a non-seeded region growing segmentation in 3D [2].

2. Material and Methods

In this study, the derivatives of close-range UAV photogrammetry are employed to produce 2-, 2.5-, and 3-dimensional coloured terrain representations of landslide cases at different settings. Superpixel/supervoxel segmentation is applied to each model (depending on the dimensionality) examining different configurations of the algorithm used. The resulting over-segmentations are subsequently partitioned into homogenous segments utilizing an unsupervised colour-based clustering technique while different threshold values are considered. The whole segmentation pipeline is illustrated in Figure 2. Finally, all three dimensionality input models, accompanied by the varying algorithm configurations, are evaluated against expert annotated ground truth images of the sites.

This section is organized as follows: Section 2.1 includes a description of the examined sites aiming to provide the reader with a detailed overview of the geologic and geometric setting as well as the specificities associated with each landslide mechanism. Section 2.2 provides the technical details of the UAV survey and the techniques used for the generation of the multi-dimensional models, while in Sections 2.3 and 2.4 the superpixel/supervoxel generation and unsupervised segmentation algorithms are explained, respectively, together with the examined parameter settings. Section 2.5 introduces the evaluation methodology.

2.1. Study Sites

Case 1 is located adjacent to the main road network from Karpenisi to Proussos Monastery in the Evritania region, Greece (Figure 3A). The slope has an average width of 10 m in the scarp zone and almost 70 m in the depositional zone crossed by the road. The relative elevation from the top to the toe is approximately 90 m while the hillslope is approximately 70° steep, facing north-east (NE) and located at a 780 m elevation. Geologically the test site is located in the flysch environment of the Pindos geotectonic unit, in the central part of Greece. Case 2 is located at the Red Beach of Santorini, Greece, a volcanic island in the Cyclades Archipelago, and it represents a coastal failure site (Figure 3B). The site of Red Beach constitutes part of Cape Mavrorachidi and is geologically situated in the Akrotiri Volcanic Complex [33]. The coastal erosion and the nature of the geomaterial created very steep volcanic slopes which dip up to 80° and are up to 45 m in height. The geomaterials examined here consist of medium to well-cemented scoria and compact lavas. Scoria formation is composed of alternating coarse-grained, medium cemented volcanic breccias and fine-grained, well-cemented volcanic breccias. The geotechnical properties of scoria are difficult to assess due to the friable character of rocks that contain air bubbles [34]. The brittle nature of the material in combination with the erosion processes due to enhanced wave energy and wind guts results in the area being susceptible to rockslides and rockfalls.

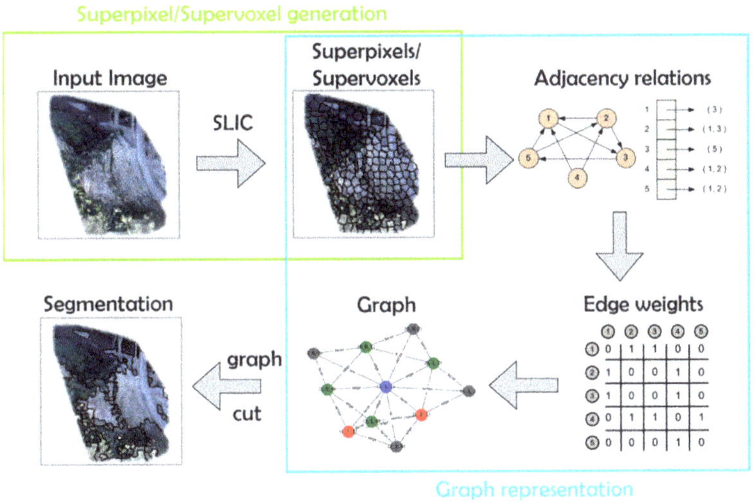

Figure 2. Step-by-step illustration of the segmentation pipeline developed for the segmentation of the multi-dimensional UAV-derived models.

Figure 3. Overview of the landslide sites examined in the current comparative study for automatic segmentation via multi-dimensional UAV-derived digital models. (**A**): Proussos landslide and (**B**): the active Red Beach cliffs.

2.2. Digital Model Creation

For the creation of the multi-dimensional digital models, a set of UAV-derived photographs was collected for each study site. Case study research with a low-cost UAV platform was carried out at the two landslide sites. The photogrammetric procedure was performed in four distinct stages: (1) flight planning, (2) flight execution and imagery collection, (3) Structure-from-Motion processing, and (4) Orthomosaic and digital surface model creation (Figure 4) [35]. The initial stage includes the flight route planning, which must ensure the best coverage of the target area with an adequate image frontlap and sidelap, considering the camera footprint and flight altitude. The UAV platform was programmed to hover at a constant altitude along the landslide sites to assure optimal conditions for tie-point identification and bundle adjustment [36].

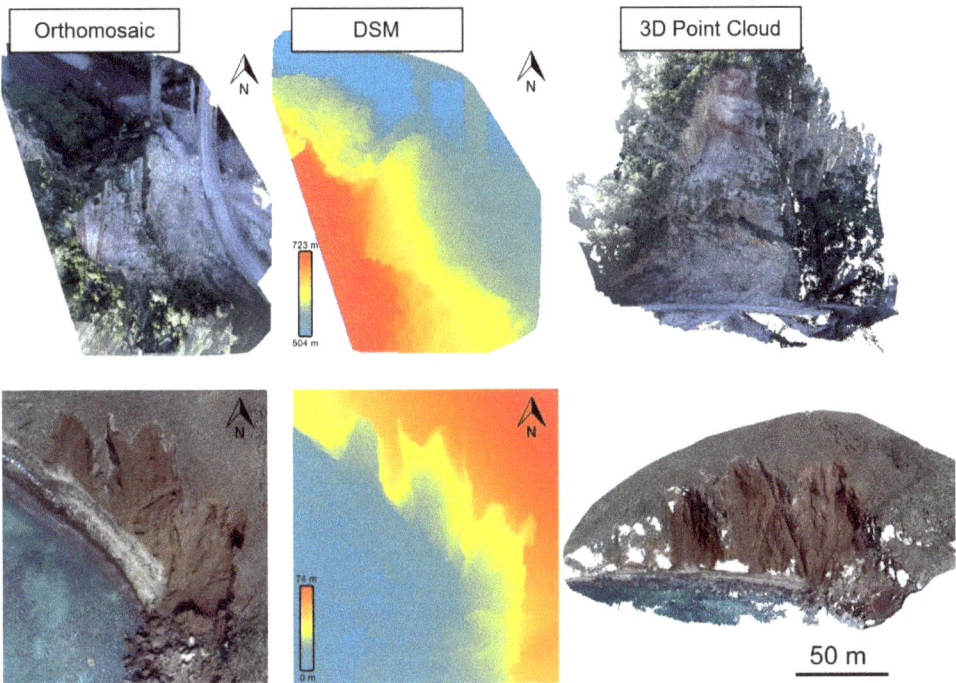

Figure 4. The digital models (upper: Proussos case, lower: Red Beach case) generated by SfM photogrammetry and used as inputs to the developed segmentation pipeline. From left to right, the columns depict: the orthomosaic, the DSM showing lower elevation in cooler colours and higher elevation in warmer colours, and 3D point cloud.

At the Proussos case site, field investigations and UAV photogrammetric surveys were performed on 24 September 2020. In total, 114 photos were collected during the flight surveys with a commercial UAV platform (DJI Phantom 4 Pro) with a median of 5243 keypoints per image. Regarding the Red Beach case, 104 images were collected during field investigations during May 2020, with a median of 1485 keypoints per image. All the images were processed to the orthophoto and the DSM with 0.5 m resolution by using Pix4D photogrammetric software with WGS84 as the coordinate reference system for the data. Information on the specific photogrammetric surveys can be found in Table 1.

Table 1. Technical details of the UAV missions for the acquisition of each dataset.

	Case 1	Case 2
Number of images	114	104
Average flying altitude (m)	50	70
Image overlap (sidelap–frontlap) (%)	70	70
Ground resolution (m/pixel)	0.034	0.024
Area extent (m^2)	16,000	3000
Point density (points/m^2)	348.85	213.99
Point cloud points	3,164,848	16,888,932
Orthomosaic resolution (m/pixel)	0.5	0.5
DSM resolution (m/pixel)	0.5	0.5
Overall error X, Y (m)	0.02	0.06
Overall error Z (m)	0.07	0.09

The orthomosaic of each site was chosen to represent the 2D geographic model which includes a Red-Green-Blue (RGB) description for each pixel in the rasterized format. Fusion of the orthomosaic with the digital surface model (DSM) adds the elevation as a fourth dimension to the pixel description, thereby creating a 2.5D model. The point cloud which retains direct 3D information of a given area was used to create the 3D model. However, due to the nature and the inherent lack of structure in the point cloud's specific data type, a voxelization process was first implemented to assign structure in a grid-like format (for further information regarding voxelization please refer to [2]). The product of the voxelization process is a voxel grid, which is technically the 3D analogue to the pixel grid. Each voxel (volumetric pixel) represents a cube assigned the mean RBG value of the points it encloses. Similar to the pixel grid, the resolution of a voxel grid is defined as the length of the edge of the voxel. The software for the raster operations was developed in Python 3.8 programming language making use of the Python Image Library (PIL) and Rasterio, while the point cloud and voxel grid manipulation is based on Open3D [37], networkx [38], and Numerical Python (NumPy) [39] modules.

2.3. Superpixel/Supervoxel Generation

Having each study site represented by both 2-, 2.5-, and 3-dimensional coloured geographic models (Figure 4) in grid-like formats enables their tessellation into superpixels or supervoxels, depending on the grid dimensionality. The Simple Linear Iterative Clustering (SLIC) algorithm is used for this task. SLIC is a state-of-the-art superpixel generation algorithm discussed in detail by [11]. SLIC includes a local implementation of the *k*-means clustering algorithm to the image, thereby offering the following two advantages:

(a) The search area is limited to a specified extent around each cluster centre rather than searching the whole image. This simplifies the complexity and makes the execution time linear in the number of pixels and independent of the value of *k*.
(b) It introduces a distance measure which accounts for colour and spatial similarity simultaneously and controls the size and compactness of the superpixels.

The process starts with the initialization of the cluster centres, with spatial location controlled by the step parameter(s). Step controls the arrangement of the superpixels along a regular grid with a resolution s times coarser than the original model resolution. The pixel gradient in each cluster is then calculated within a 3 × 3 window and the cluster centre is moved to the lowest gradient position. The gradient is defined as the normalized sum of distances in feature space from all the neighbours and computed as:

$$G(i) = \sum_{j=1}^{k_{adj}} \frac{|k(i) - k(j)|}{N_{adj}}, \qquad (1)$$

where, k is the notation for a pixel (or voxel in the 3D), i is the index of each processed element, j is the index of each element adjacent to the processed element within the defined window, and N is the total number of adjacent elements.

In SLIC, it is usually recommended to transform the RGB image into the non-linear L*a*b* colour space (L* for lightness, and a* and b* for the position between red-green and blue-yellow, respectively) where a given numerical change corresponds to a similar perceived change in colour. To keep the methodology consistent between the different models and to not introduce bias into the comparison, the authors applied this transformation to all the applications. In the case of the 2D model where the pixel is characterized by only the colour values, each cluster centre is initialized as $C_i = [L_i\ a_i\ b_i\ x_i\ y_i]^T$ while in the 2.5D model case, the elevation dimension (E) is added to the clustering space as $C_i = [L_i\ a_i\ b_i\ E_i\ x_i\ y_i]^T$. The elevation range across the whole model is standardized in the range [0:100] which is the exact same range the parameter L (luminosity) fluctuates within.

For the adaptation of the above process in the 3D space and the generation of supervoxels, the authors developed an extension of the algorithm for application on voxel grids based on the Voxel Cloud Connectivity Segmentation (VCCS) proposed by Papon et al. (2013). Since the bounding box of a point cloud is almost never entirely occupied by points, there are several cluster seeds within the initial supervoxel grid that remain empty of points. The step parameter is still linked to the seed grid resolution and affects the supervoxel segmentation result. In this case, the centre of each non-empty supervoxel is moved to the nearest voxel while the rest are rejected. Each cluster centre is then moved to the lower gradient position calculated within a 3 × 3 × 3 window, like in the previous two cases, using Equation (1).

The colour is again defined in the CIELAB space and the cluster centres initialized as $C_i = [L_i\ a_i\ b_i\ x_i\ y_i\ z_i]^T$ with the third dimension being directly integrated in the spatial component of the distance measure which is calculated as:

$$D = \sqrt{d_f^2 + \left(\frac{d_s}{s}\right)^2 c^2} \qquad (2)$$

where, d_f and d_s denote the distance in feature space and spatial distance, respectively, s defines the step parameter, and c weights the relative importance between feature space similarity (colour or colour + elevation) and spatial proximity. The larger the value of c, the more important the spatial proximity is and so the resulting superpixels/supervoxels are more compact in shape. In contrast, lower c values lead to superpixels/supervoxels more flexible in shape. As such, c stands for compactness which is mathematically expressed as the area to perimeter ratio. In general, in CIELAB space operation, the compactness can be in the range [1:40] [11]. The c values were defined after an exploratory trial-and-error analysis for each site in a way that allows enough flexibility during the formation of superpixels/supervoxels (between 20 and 30).

2.4. Graph-Based Segmentation

Once the homogenization of the scene and the complexity reduction have been achieved, the superpixels and supervoxels represent the image unit for further processing within an object-based conceptualization. For the organization of them into meaningful objects, the authors employ a graph cut approach to assess the implementation of an unsupervised data-driven segmentation that does not require empirical knowledge.

Graph cut describes a set of edges whose removal makes the different graphs disconnected. The set of models, in their corresponding feature space, are represented by a weighted undirected graph structure $G = (V, E)$, where the nodes of the graph are associated with the superpixel or supervoxels and the edges connect adjacent nodes. Each edge $(v_i, v_j) \in E$ is weighted by the dissimilarity between nodes v_i and v_j. In this implementation, colour dissimilarity is used as the edge weight $w(v_i, v_j)$ [40,41]. This stage of the methodology aims at providing an initial estimate of the representativity of the final segments and their potential within unsupervised object-based semantic segmentation workflows from UAV models.

The normalized cut (Ncut) function is used to avoid min cut bias. This approach was developed to solve the perceptual grouping problem in vision by normalizing for the size of each segment. The normalized cut criterion accounts for both the total dissimilarity between different clusters and the total similarity within the clusters (Jianbo Shi and Malik, 2000). In this way, the methodology extracts the global impression of a scene rather than only assessing local information and its consistency in the model. The formulation of the normalized cut criterion is given in Equation (3).

$$Ncut(A, B) = \frac{cut(A,B)}{\sum_{u \in A, t \in V} w(u,t)} + \frac{cut(A,B)}{\sum_{u \in B, t \in V} w(u,t)} \qquad (3)$$

where, $\sum_{u \in A, t \in V} w(u,t)$ is the total weight from edges connecting the nodes in A to all the nodes in the graph, and $\sum_{u \in B, t \in V} w(u,t)$ is for all the nodes in B.

2.5. Multi-Dimensional Assessments

The landslide sites described in Section 3.1 are both used in the evaluation phase as they introduce diversity in slope geometry, failure mechanism, and rock types. The evaluation of the multi-dimensional UAV-derived terrain representations for each case is carried out in both stages of the segmentation pipeline presented in Figure 2. Expert-based annotations were prepared for the sites to be used as references for the comparisons. These annotations include geologic/geomorphologic landslide features such as the scarp, depletion zone, accumulation zone, coastline, and rockfalls as well as anthropogenic features such as roads threatened by the landslide activity. The authors project the 3D segment boundaries to the top-view plane for the comparisons with the raster-based segmentations using the methodology by [17].

Initially, the superpixels or supervoxels generated by the SLIC algorithm are evaluated for different resolutions ranging from 5 to 20 m. This aims to assess the suitability of the models as inputs to over-segmentation methods for properly generating meaningful image objects, while simultaneously reducing the complexity of a scene. Subsequently, the result of the unsupervised graph cut segmentation, considering all the different SLIC outputs, is evaluated separately. The results of different cut cost values are examined representing three orders of magnitude (0.01, 0.1, and 1) to assess the sensitivity of the outputs in complex natural terrains such as these active landslide slopes.

3. Results

This section details the segmentation results of the experimental analyses conducted in the two different landslide scenes introduced in 3.1. Each of the two steps of the segmentation pipeline (Figure 2) is thoroughly analyzed using the multi-dimensional UAV-derived models (Figure 4) as inputs, respectively, and expert-labeled ground-truth segmentations as reference. The authors first provide insights on the generated superpixels/supervoxels under different resolutions of the SLIC algorithm. Afterwards, each of the superpixel/supervoxel assemblies is used to derive the final segmentations through normalized cut analysis.

3.1. Evaluation Metrics

The most important property of a segmentation algorithm is the ability to generate segments able to adhere to, and not cross, real object boundaries. To assess the results quantitatively, two standard metrics called boundary recall [42] and under-segmentation error [43] are widely used in the literature [11,17,18].

However, these metrics will always provide the best scores for scenes segmented into many small segments, which is undesirable [18]. The objective of any segmentation technique is to provide the best scores with the least possible number of segments, and it is commonly preferred for the metric to be plotted against the number of segments. However, due to the fact that the number of segments increases exponentially by adding the third dimension of elevation and to keep the comparison unbiased, the authors plot the scores against the over-segmentation resolution which is directly related to the number of formed segments. The lower the resolution, the fewer the SLIC segments and the higher the cut cost, the fewer the final segments.

3.1.1. Boundary Recall

Boundary recall (BR) measures what fraction of the ground truth edges fall within at least two pixels of a segment boundary (Equation (4)). High boundary recall indicates that the segments properly follow the edges of objects defined in the ground truth labeling.

$$BR = \frac{True\ Positives}{True\ Positives + False\ Negatives} \quad (4)$$

3.1.2. Under-Segmentation Error

Under-segmentation error (UE) estimates the area that corresponds to erroneous overlaps. If A is a ground truth segment and B is a generated segment that intersects A, Bin and Bout refer to the portions of B that do and do not overlap with A, respectively. Thus, the UE is calculated as follows:

$$UE = \frac{1}{N}\left[\sum_A \left(\sum_{B \cap A} min(|B_{in}|, |B_{out}|)\right)\right] \quad (5)$$

where N defines the total number of pixels in the reference model.

3.2. Superpixel/Supervoxel Evaluation

The SLIC algorithm was applied to the three different terrain representations (2D, 2.5D, and 3D) for each case site. Due to the fact that appropriate segmentations depend on the suitable scale of analysis, and the goal is to achieve the best boundary adherence with the least possible number of segments, a range of different superpixel/supervoxel resolutions was examined (5 to 20 m, with an interval of 2.5 m). The authors investigated the superpixel/supervoxel generation for landslide models using the multi-dimensional processing methods discussed in 3.3. Figure 5 illustrates over-segmentation results of the 2D, 2.5D, and 3D processing methods, respectively, with both a fine and coarse resolution for each. It is note-worthy that the 3D results, especially for the Red Beach case, depict a few intersected segment boundaries. This is due to three factors: (a) the noise introduced to the point cloud through the SfM process due to the sea waves which is reflected when projected to the top-down view, (b) shadowed or occluded areas that interrupt the continuity of the voxel grid and lead the supervoxel seeds to small, disconnected clusters, and (c) areas of negative slope where segments that are at different elevations intersect with each other when projected.

It was found that the 3D supervoxel algorithm generates over-segmentations with over 80% boundary recall (Figure 6) and below 20% under-segmentation error (Figure 7) for almost all the examined SLIC resolutions for both study sites. In particular, for the finer resolutions, the boundary recall exceeds 90% while the under-segmentation errors do not exceed 12%. In contrast, the boundary recall for the respective 2D and 2.5D superpixel algorithms does not exceed 80%, while it fluctuates between 30–60% for the coarser settings. The same trend is observed for the under-segmentation error too. Two-dimensional and 2.5D models produce consistently higher under-segmentation error scores than the 3D applications.

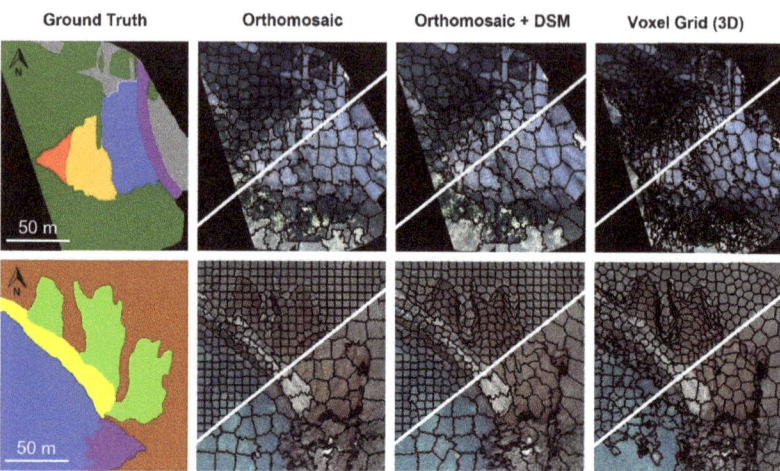

Figure 5. Examples of superpixel/supervoxel output. From left to right: the ground truth annotation, 2D, 2.5D, and 3D input model. Each is shown with two different superpixel/supervoxel resolutions. In the Proussos case, above, elements such as the scarp (red), depletion zone (yellow), accumulation zone (blue), road (purple), and non-affected area (green) are delineated, while in the Red Beach case, below, the annotated elements include landslide areas (green), rockfall deposits (purple), the beach (yellow), waterbody (blue), and the non-affected area (brown).

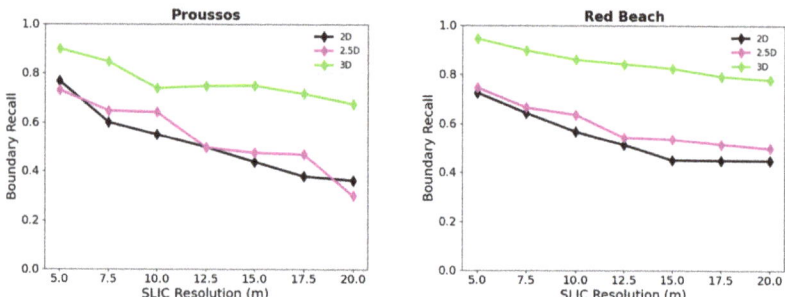

Figure 6. Quantitative over-segmentation performance evaluation based on the boundary recall using SLIC for each dimensionality of the UAV-derived input digital models.

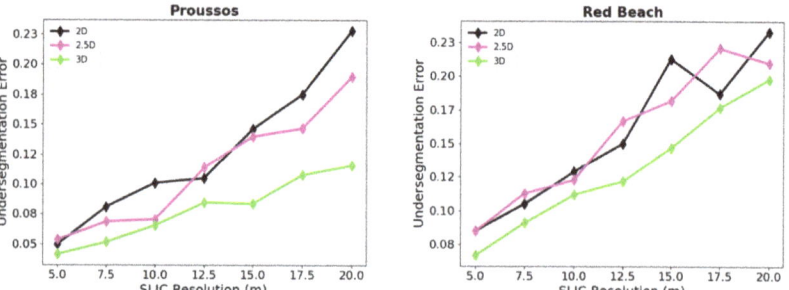

Figure 7. Quantitative over-segmentation performance evaluation based on the under-segmentation error using SLIC for each dimensionality of the UAV-derived input digital models.

3.3. Final Segmentation Evaluation

The seven superpixel/supervoxel assemblies of each model generated in the previous step were subsequently clustered into larger segments based on the normalized graph cut segmentation (Figure 8).

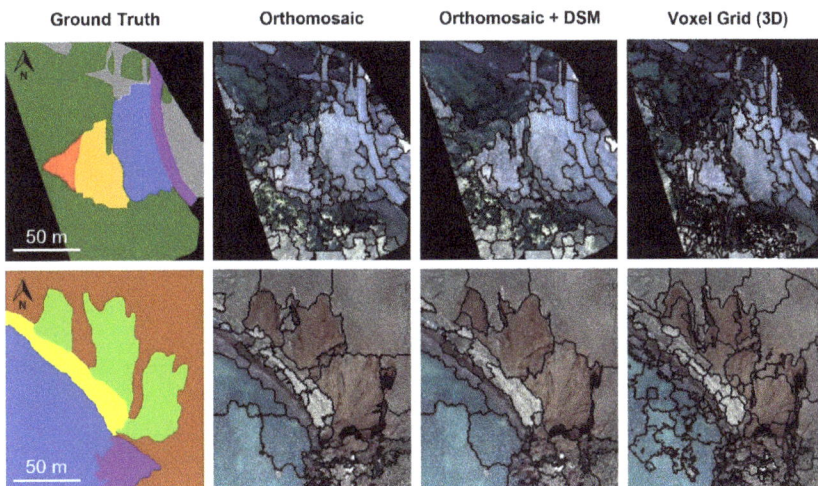

Figure 8. Examples of superpixel/supervoxel-based normalized cut segmentation outputs. From left to right: the ground truth annotation, 2D, 2.5D, and 3D input model. In the Proussos case, above, elements such as the scarp (red), depletion zone (yellow), accumulation zone (blue), road (purple), and non-affected area (green) are delineated, while in the Red Beach case, below, the annotated elements include landslide areas (green), rockfall deposits (purple), the beach (yellow), waterbody (blue), and the non-affected area (brown).

To obtain a better estimate of the different models, the authors performed the segmentation with three different orders of homogeneity magnitude (i.e., 0.01, 0.1, and 1). The same trend of the superpixel/supervoxel evaluation is propagated to the output of the final segmentation. In both case studies, it was found that the 3D segmentation algorithm produces better segments for almost all the resolutions. It turns out that the incorporation of the third dimension into the 2.5D model by means of a spectral feature provides subtle improvement in the segmentation output. An example is illustrated in Figure 9 where even sub-elements such as the flanks and the crest of the landslide are separated with the 3D operations, compared to the other two models with the same configuration, maintaining a comparable number of total segments. To achieve a similar decomposition of the landslide into segments that adhere well to the sub-elements' boundaries using either the 2D or 2.5D model, a lower resolution would be required resulting in an exponentially higher number of segments, which is undesirable.

The best segmentation scores are observed for all the models at the higher cut cost (1), which is expected since it does not allow merges to happen easily. However, great boundary recall and low under-segmentation error values are recorded for Ncut = 0.1 (Figures 10 and 11). This is found in the Red Beach case especially, where the boundary recall scores fluctuate between 70–80% for all the configurations. In Figure 11, it can be observed that direct 3D operations for the segmentation of both the landslide sites decreases under-segmentation error for the Ncut = 0.1 model, which seems to be the most effective order of magnitude for these specific cases.

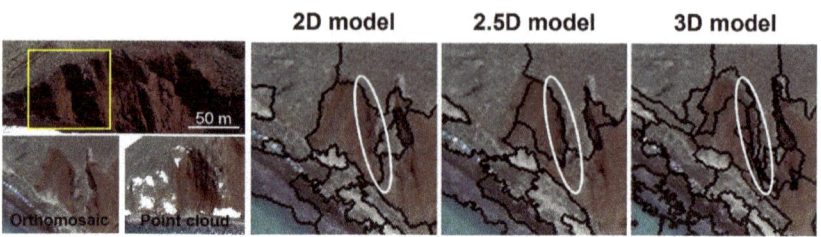

Figure 9. A characteristic example of the different outputs produced by 2D, 2.5D, and 3D segmentations for the delineation of steep sub-elements such as the flanks of a landslide. The white ellipse delineates the flank and part of the scarp, and the increasing level of important detail defined as the dimensions of the analysis increase.

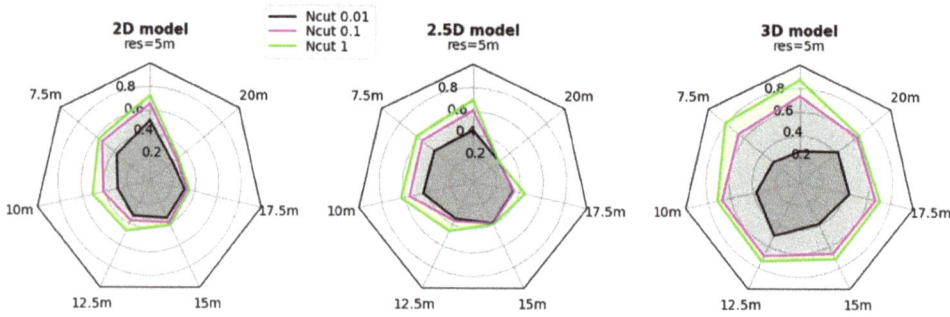

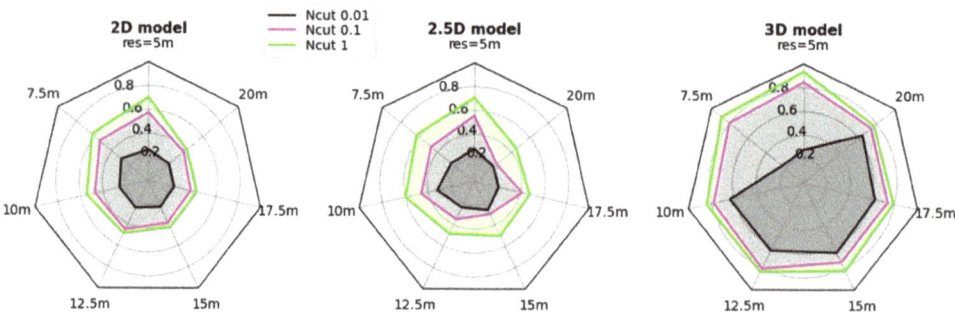

Figure 10. The boundary recall of the superpixel/supervoxel-based graph-based segmentation for each dimensionality of the UAV-derived input digital models.

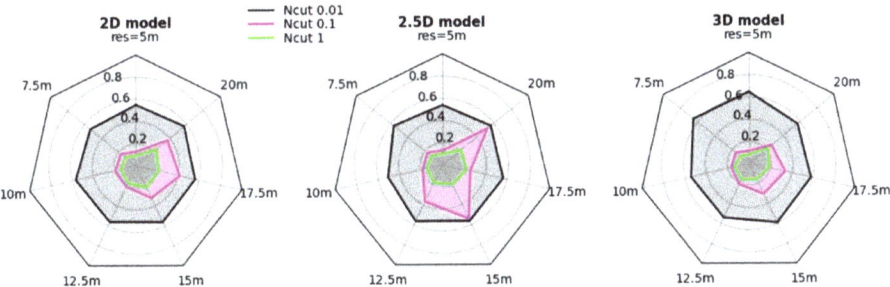

Figure 11. The under-segmentation error of the superpixel/supervoxel-based graph-based segmentation for each dimensionality of the UAV-derived input digital models.

4. Discussion

UAVs constitute a valuable, low-cost tool for data collection in landslide research. Their ability to produce high-resolution imagery of even inaccessible areas along rock slopes in a short time places them among the most essential tools in landslide risk assessment. SfM techniques are able to generate detailed 3D terrain representations from the acquired set of images, the so-called 3D point clouds, which provide supplementary geometric and topographic information. However, point cloud processing is often tedious and conventional tools are limited. Due to this reason, processed data formats such as rasters are commonly used as 2.5D DSMs to augment the image information. With this rich information in hand, efforts towards the utilization of semi-automated or automated image analysis methods of landslide scenes have been initiated. Image analysis tasks such as segmentation have been successfully implemented within GEOBIA workflows for landslide mapping from satellite images and DEM derivatives. However, although the great advances in image segmentation have been adopted by the geoscience community for landslide mapping at the regional scale, little work has been carried out in image segmentation of landslide scenes at the site-specific scale.

In this transition from a regional to site-specific scale, the use of UAVs supports the acquisition of 3D terrain representations. This increases the amount of available information and the complexity of the scene while raising the question regarding the adequacy of simplified 2.5D models compared to fully 3D terrain representations. This study provides insights about the potential of integrating advanced 3D point cloud segmentation methods for geographic OBPA of landslide scenes. A 3D modification of a state-of-the-art image segmentation pipeline was implemented by the authors and compared to the image segmentations by both 2D and 2.5D models. It incorporates the concept of voxelization, which generates a 3D grid that can facilitate neighbourhood and metric operations, thereby maintaining the 3D character of the data throughout the segmentation process. It highlights the advantages of direct 3D point cloud processing for landslide scene segmentation in two complex cases where different site-specific elements exist.

The supervoxels formed by the point cloud-based 3D modification of the SLIC algorithm are shown to segment the two landslide scenes more appropriately than the superpixels generated with either the 2D or 2.5D model as the input raster. In contrast, no significant difference was observed between the 2D and 2.5D superpixels for any of the examined resolutions. Subsequently, the graph-based segmentation performed at the superpixel/supervoxel level was proven quite effective for fully unsupervised landslide scene segmentation with Ncut values of 0.1 (Figures 10 and 11). Supervoxel-based segmentations of almost 80% boundary recall (Figure 6) and below 15% under-segmentation error were achieved (Figure 7). However, the future investigation of image segmentation algorithms applied directly on 3D space will provide a more confident estimate regarding the suitability of multi-dimensional UAV-derived models for landslide scene segmentation.

Although the complexity and lack of specialized tools may demotivate landslide experts to work with 3D point clouds directly, the results of this study encourage familiarization with advanced point cloud processing. The integration of the great advances in point cloud processing into landslide research can provide strong support to object-oriented landslide assessment at the site-specific scale. The segmentation results are proven to better agree with the expert annotations and the resulting point cloud segments can provide further information for the subsequent classification phase. For instance, areas of negative slope and overhangs become "visible". The authors believe that object-oriented analysis using supervoxels has the potential to support intelligent landslide modelling.

5. Conclusions

UAV-derived models can be used effectively for object-oriented landslide scene analysis at site-specific scales. The detail of the obtained information coupled with advanced processing methods supports semi-automated or automated identification of landslide elements within digital models to efficiently support enhanced landslide risk management.

The experimental analysis conducted in this study shows that performing landslide scene segmentation directly on UAV-derived 3D point clouds is more effective than leveraging 2D or 2.5D images by means of the orthomosaic and DSM. Modelling the landslide scene by adding the elevation information as a fourth image band does not add much value to the segmentation compared to using only spectral information. In contrast, the ability to directly segment the 3D point cloud seems to provide a promising opening to 3D GEOBIA or GEOBPA (Geographic Object-Based Point cloud Analysis) in landslide research.

In steep terrains, such as rock slopes prone to landslides, rasterized projections of the terrain may not always capture the landscape adequately. The analysis has revealed that this type of 2.5D representation could limit the effectiveness of the segmentations of the steep boundary features of a landslide. Typical examples of such elements include the scarp and the flanks of a landslide which have proven to be more precisely segmented by directly utilizing fully 3D terrain representations and operations. The ability to operate in the 3D space may provide an opportunity to unlock the full potential of UAV surveys in site-specific object-oriented analysis of landslide scenes.

The improvement imposed to landslide scene segmentations by utilizing fully 3D modelling and processing can enhance landslide mapping and assessment procedures that employ UAVs in emergency response situations. Furthermore, future research is encouraged to build upon the findings of this study and orient their efforts towards the integration of advanced 3D segmentation methods.

Author Contributions: Conceptualization, I.F.; methodology, I.F. and E.K.; software, I.F.; formal analysis, I.F. and E.K.; investigation, I.F. and E.K.; resources, N.V., D.J.H. and V.M.; data curation, I.F. and E.K.; writing—original draft preparation, I.F.; writing—review and editing, E.K., D.J.H. and N.V.; visualization, I.F.; supervision, D.J.H., N.V. and V.M. All authors have read and agreed to the published version of the manuscript.

Funding: This research was funded by Natural Sciences and Engineering Research Council (NSERC) of Canada.

Data Availability Statement: The data presented in this study are available on request from the corresponding author.

Acknowledgments: The authors acknowledge the RMC Green Team and the Canadian Department of National Defense as well as the State Scholarships Foundation of Greece for supporting this research.

Conflicts of Interest: The authors declare no conflict of interest. The funders had no role in the design of the study; in the collection, analyses, or interpretation of data; in the writing of the manuscript; or in the decision to publish the results.

References

1. Froude, M.J.; Petley, D.N. Global Fatal Landslide Occurrence from 2004 to 2016. *Nat. Hazards Earth Syst. Sci.* **2018**, *18*, 2161–2181. [CrossRef]
2. Farmakis, I.; Bonneau, D.; Hutchinson, D.J.; Vlachopoulos, N. Targeted Rock Slope Assessment Using Voxels and Object-Oriented Classification. *Remote Sens.* **2021**, *13*, 1354. [CrossRef]
3. Pham, T.D. Computing with Words in Formal Methods. *Int. J. Intell. Syst.* **2000**, *15*, 801–810. [CrossRef]
4. Hossain, M.D.; Chen, D. Segmentation for Object-Based Image Analysis (OBIA): A Review of Algorithms and Challenges from Remote Sensing Perspective. *ISPRS J. Photogramm. Remote Sens.* **2019**, *150*, 115–134. [CrossRef]
5. Blaschke, T.; Hay, G.J.; Kelly, M.; Lang, S.; Hofmann, P.; Addink, E.; Queiroz Feitosa, R.; van der Meer, F.; van der Werff, H.; van Coillie, F.; et al. Geographic Object-Based Image Analysis—Towards a New Paradigm. *ISPRS J. Photogramm. Remote Sens.* **2014**, *87*, 180–191. [CrossRef]
6. Hay, G.J.; Castilla, G. Geographic Object-Based Image Analysis (GEOBIA): A New Name for a New Discipline. In *Lecture Notes in Geoinformation and Cartography*; Springer: Berlin/Heidelberg, Germany, 2008; pp. 75–89. [CrossRef]
7. Radoux, J.; Bogaert, P. Good Practices for Object-Based Accuracy Assessment. *Remote Sens.* **2017**, *9*, 646. [CrossRef]
8. Ye, S.; Pontius, R.G.; Rakshit, R. A Review of Accuracy Assessment for Object-Based Image Analysis: From per-Pixel to per-Polygon Approaches. *ISPRS J. Photogramm. Remote Sens.* **2018**, *141*, 137–147. [CrossRef]
9. Kohntopp, D.; Lehmann, B.; Kraus, D.; Birk, A. Segmentation and Classification Using Active Contours Based Superellipse Fitting on Side Scan Sonar Images for Marine Demining. In Proceedings of the 2015 IEEE International Conference on Robotics and Automation (ICRA), Seattle, WA, USA, 26–30 May 2015; pp. 3380–3387. [CrossRef]
10. Mueller, N.; Lewis, A.; Roberts, D.; Ring, S.; Melrose, R.; Sixsmith, J.; Lymburner, L.; McIntyre, A.; Tan, P.; Curnow, S.; et al. Water Observations from Space: Mapping Surface Water from 25 years of Landsat Imagery across Australia. *Remote Sens. Environ.* **2016**, *174*, 341–352. [CrossRef]
11. Achanta, R.; Shaji, A.; Smith, K.; Lucchi, A.; Fua, P.; Süsstrunk, S. SLIC Superpixels Compared to State-of-the-Art Superpixel Methods. *IEEE Trans. Pattern Anal. Mach. Intell.* **2012**, *34*, 2274–2282. [CrossRef]
12. Baatz, M.; Schape, A. Multiresolution Segmentation: An Optimization Approach for High Quality Multi-Scale Image Segmentation. In *Angewandte Geographische Informations—Verarbeitung XII*; Strobl, T., Blaschke, T., Griesebner, G., Eds.; Wichmann Verlag: Karlsruhe, Germany, 2000; pp. 12–23.
13. Csillik, O. Fast Segmentation and Classification of Very High Resolution Remote Sensing Data Using SLIC Superpixels. *Remote Sens.* **2017**, *9*, 243. [CrossRef]
14. Hoeser, T.; Bachofer, F.; Kuenzer, C. Object Detection and Image Segmentation with Deep Learning on Earth Observation Data: A Review-Part II: Applications. *Remote Sens.* **2020**, *12*, 53. [CrossRef]
15. Liu, S.; Chen, P.; Woźniak, M. Image Enhancement-Based Detection with Small Infrared Targets. *Remote Sens.* **2022**, *14*, 3232. [CrossRef]
16. Keyport, R.N.; Oommen, T.; Martha, T.R.; Sajinkumar, K.S.; Gierke, J.S. A Comparative Analysis of Pixel- and Object-Based Detection of Landslides from Very High-Resolution Images. *Int. J. Appl. Earth Obs. Geoinf.* **2018**, *64*, 1–11. [CrossRef]
17. Papon, J.; Abramov, A.; Schoeler, M.; Worgotter, F. Voxel Cloud Connectivity Segmentation—Supervoxels for Point Clouds. In Proceedings of the IEEE Computer Society Conference on Computer Vision and Pattern Recognition, Portland, OR, USA, 23–28 June 2013; pp. 2027–2034. [CrossRef]
18. Ben-Shabat, Y.; Avraham, T.; Lindenbaum, M.; Fischer, A. Graph Based Over-Segmentation Methods for 3D Point Clouds. *Comput. Vis. Image Underst.* **2018**, *174*, 12–23. [CrossRef]
19. Guo, Y.; Wang, H.; Hu, Q.; Liu, H.; Liu, L.; Bennamoun, M. Deep Learning for 3D Point Clouds: A Survey. *IEEE Trans. Pattern Anal. Mach. Intell.* **2020**, *43*, 4338–4364. [CrossRef]

20. Hölbling, D.; Friedl, B.; Eisank, C. An Object-Based Approach for Semi-Automated Landslide Change Detection and Attribution of Changes to Landslide Classes in Northern Taiwan. *Earth Sci. Inform.* **2015**, *8*, 327–335. [CrossRef]
21. Yang, Y.; Song, S.; Yue, F.; He, W.; Shao, W.; Zhao, K.; Nie, W. Superpixel-Based Automatic Image Recognition for Landslide Deformation Areas. *Eng. Geol.* **2019**, *259*, 105166. [CrossRef]
22. Feizizadeh, B.; Blaschke, T. A Semi-Automated Object Based Image Analysis Approach for Landslide Delineation. In Proceedings of the 2013 European Space Agency Living Planet Symposium, Edinburgh, UK, 9–13 September 2013; pp. 9–13. [CrossRef]
23. Hölbling, D.; Abad, L.; Dabiri, Z.; Prasicek, G.; Tsai, T.T.; Argentin, A.L. Mapping and Analyzing the Evolution of the Butangbunasi Landslide Using Landsat Time Series with Respect to Heavy Rainfall Events during Typhoons. *Appl. Sci.* **2020**, *10*, 630. [CrossRef]
24. Hölbling, D.; Füreder, P.; Antolini, F.; Cigna, F.; Casagli, N.; Lang, S. A Semi-Automated Object-Based Approach for Landslide Detection Validated by Persistent Scatterer Interferometry Measures and Landslide Inventories. *Remote Sens.* **2012**, *4*, 1310–1336. [CrossRef]
25. Lahousse, T.; Chang, K.T.; Lin, Y.H. Landslide Mapping with Multi-Scale Object-Based Image Analysis—A Case Study in the Baichi Watershed, Taiwan. *Nat. Hazards Earth Syst. Sci.* **2011**, *11*, 2715–2726. [CrossRef]
26. Amatya, P.; Kirschbaum, D.; Stanley, T.; Tanyas, H. Landslide Mapping Using Object-Based Image Analysis and Open Source Tools. *Eng. Geol.* **2021**, *282*, 106000. [CrossRef]
27. Höfle, B.; Geist, T.; Rutzinger, M.; Pfeifer, N. Glacier Surface Segmentation Using Airborne Laser Scanning Point Cloud and Intensity Data. *ISPRS Workshop Laser Scanning* **2007**, *XXXVI*, 195–200.
28. Eisank, C.; Smith, M.; Hillier, J. Assessment of Multiresolution Segmentation for Delimiting Drumlins in Digital Elevation Models. *Geomorphology* **2014**, *214*, 452–464. [CrossRef] [PubMed]
29. d'Oleire-Oltmanns, S.; Marzolff, I.; Tiede, D.; Blaschke, T. Detection of Gully-Affected Areas by Applying Object-Based Image Analysis (OBIA) in the Region of Taroudannt, Morocco. *Remote Sens.* **2014**, *6*, 8287–8309. [CrossRef]
30. Karantanellis, E.; Marinos, V.; Vassilakis, E.; Christaras, B. Object-Based Analysis Using Unmanned Aerial Vehicles (UAVs) for Site-Specific Landslide Assessment. *Remote Sens.* **2020**, *12*, 1711. [CrossRef]
31. Farmakis, I.; Bonneau, D.; Hutchinson, D.J.; Vlachopoulos, N. Supervoxel-Based Multi-Scale Point Cloud Segmentation Using Fnea for Object-Oriented Rock Slope Classification Using Tls. *ISPRS—Int. Arch.Photogramm. Remote Sens. Spat. Inf. Sci.* **2020**, *XLIII-B2-2*, 1049–1056. [CrossRef]
32. Mayr, A.; Rutzinger, M.; Bremer, M.; Oude Elberink, S.; Stumpf, F.; Geitner, C. Object-Based Classification of Terrestrial Laser Scanning Point Clouds for Landslide Monitoring. *Photogramm. Rec.* **2017**, *32*, 377–397. [CrossRef]
33. Druitt, T.H.; Edwards, L.; Mellors, R.M.; Pyle, D.M.; Sparks, R.S.J.; Lanphere, M.; Davies, M.; Barreirio, B. *Santorini Volcano*; Geological Society of London: London, UK, 1999; Volume 19.
34. Marinos, V.; Prountzopoulos, G.; Asteriou, P.; Papathanassiou, G.; Kaklis, T.; Pantazis, G.; Lambrou, E.; Grendas, N.; Karantanellis, E.; Pavlides, S. Beyond the Boundaries of Feasible Engineering Geological Solutions: Stability Considerations of the Spectacular Red Beach Cliffs on Santorini Island, Greece. *Environ. Earth Sci.* **2017**, *76*, 513. [CrossRef]
35. Westoby, M.J.; Brasington, J.; Glasser, N.F.; Hambrey, M.J.; Reynolds, J.M. "Structure-from-Motion" Photogrammetry: A Low-Cost, Effective Tool for Geoscience Applications. *Geomorphology* **2012**, *179*, 300–314. [CrossRef]
36. Triggs, B.; McLauchlan, P.F.; Hartley, R.I.; Fitzgibbon, A.W. Bundle Adjustment—A Modern Synthesis. In *Vision Algorithms: Theory and Practice. IWVA 1999. Lecture Notes in Computer Science*; Triggs, B., Zisserman, A., Szeliski, R., Eds.; Springer: Berlin/Heidelberg, Germany, 2000; pp. 298–372.
37. Zhou, Q.-Y.; Park, J.; Koltun, V. Open3D: A Modern Library for 3D Data Processing. *arXiv* **2018**, arXiv:1801.09847.
38. Hagberg, A.A.; Schult, D.A.; Swart, P.J. Exploring Network Structure, Dynamics, and Function Using NetworkX. In Proceedings of the 7th Python in Science Conference (SciPy 2008), Pasadena, CA, USA, 19–24 August 2008; pp. 11–15.
39. Harris, C.R.; Millman, K.J.; van der Walt, S.J.; Gommers, R.; Virtanen, P.; Cournapeau, D.; Wieser, E.; Taylor, J.; Berg, S.; Smith, N.J.; et al. Array Programming with NumPy. *Nature* **2020**, *585*, 357–362. [CrossRef] [PubMed]
40. Felzenszwalb, P.F.; Huttenlocher, D.P. Efficient Graph-Based Image Segmentation. *Int. J. Comput. Vis.* **2004**, *59*, 167–181. [CrossRef]
41. Shi, J.; Malik, J. Normalized Cuts and Image Segmentation. *IEEE Trans. Pattern Anal. Mach. Intell.* **2000**, *22*, 888–905. [CrossRef]
42. Martin, D.; Fowlkes, C.; Tal, D.; Malik, J. A Database of Human Segmented Natural Images and Its Application to Evaluating Segmentation Algorithms and Measuring Ecological Statistics. In Proceedings of the Eighth IEEE International Conference on Computer Vision. ICCV 2001, Vancouver, BC, Canada, 7–14 July 2001; Voume 2, pp. 416–423.
43. Silberman, N.; Hoiem, D.; Kohli, P.; Fergus, R. Indoor Segmentation and Support Inference from RGBD Images. In Proceedings of the European Conference on Computer Vision, Florence, Italy, 7–13 October 2012; Springer: Berlin/Heidelberg, Germany, 2012; pp. 746–760.

Article

Comparison of Three Mixed-Effects Models for Mass Movement Susceptibility Mapping Based on Incomplete Inventory in China

Yifei He [1,2] and Yaonan Zhang [1,2,3,*]

1. Northwest Institute of Eco-Environment and Resources, CAS, Lanzhou 730000, China
2. National Cryosphere Desert Data Center, Lanzhou 730000, China
3. Gansu Data Engineering and Technology Research Center for Resources and Environment, Lanzhou 730000, China
* Correspondence: yaonan@lzb.ac.cn

Abstract: Generating an unbiased inventory of mass movements is challenging, particularly in a large region such as China. However, due to the enormous threat to human life and property caused by the increasing number of mass movements, it is imperative to develop a reliable nationwide mass movement susceptibility model to identify mass movement-prone regions and formulate appropriate disaster prevention strategies. In recent years, the mixed-effects models have shown their unique advantages in dealing with the biased mass movement inventory, yet there are no relevant studies to compare different mixed-effects models. This research compared three mixed-effects models to explore the most plausible and robust susceptibility mapping model, considering the inherently heterogeneously complete mass movement information. Based on a preliminary data analysis, eight critical factors influencing mass movements were selected as basis predictors: the slope, aspect, profile curvature, plan curvature, road density, river density, soil moisture, and lithology. Two additional factors, namely, the land use and geological environment division, representing the inventory bias were selected as random intercepts. Subsequently, three mixed-effects models—Statistical-based generalized linear mixed-effects model (GLMM), generalized additive mixed-effects model (GAMM), and machine learning-based tree-boosted mixed-effects model (TBMM)—were adopted. These models were used to evaluate the susceptibility of three distinct types of mass movements (i.e., 28,814 debris flows, 54,586 rockfalls and 108,432 landslides), respectively. The results were compared both from quantitative and qualitative perspectives. The results showed that TBMM performed best in all three cases with AUROCs (Area Under the Receiver Operating Characteristic curve) of cross-validation, spatial cross-validation, and predictions on simulated highly biased inventory, all exceeding 0.8. In addition, the spatial prediction patterns of TBMM were more in line with the natural geomorphological underlying process, indicating that TBMM can better reduce the impact of inventory bias than GLMM and GAMM. Finally, factor contribution analysis showed the key role of topographic factors in predicting the occurrence of mass movements, followed by road density and soil moisture. This study contributes to assessing China's overall mass movement susceptibility situation and assisting policymakers in master planning for risk mitigation. Further, it demonstrates the tremendous potential of TBMM for mass movement susceptibility assessment, despite inherent biases in the inventory.

Keywords: nationwide; susceptibility mapping; mass movement; inventory bias; tree-boosted; mixed-effects models

---

1. Introduction

China is the country that experiences the most frequent natural disasters globally, as it witnesses numerous hazards such as floods, droughts, earthquakes, sandstorms and wildfires every year [1,2]. The landscape of China is characterized by widespread mountainous areas, which makes it highly prone to mass movements [3]. Mass movements such as debris flows, rockfalls and landslides are major geological disasters that have devastating effects on property, human life and the country's ecological environment [4,5]. It has been recognized that mass movements may pose an even more serious threat in the future due to the potential effects of rapid urbanization, compounded by aggravated climate change [6]. Therefore, a credible nationwide mass movement susceptibility map is paramount for providing a generalized overview of potential mass movement propagation areas in China.

Over the past few decades, an increasing number of studies on mass movement susceptibility assessments have been published, most of which have been performed on local areas [7–9]. However, in order to improve the overall perception of mass movement risk, some studies have begun to assess susceptibility in very large areas, including national-scale evaluations [10–15], continental-scale analyses [16–19], and global-scale assessments [20–22]. It can be found that such evaluations for large areas are mostly based on statistical or machine learning models, as the lack of detailed geotechnical data limits the application of physical-based models [23]. Both statistical models and machine learning models assume that the conditions that caused mass movements in the past may lead to future mass movements; thus, the correlations between controlling factors and mass movement inventories of past events were fitted by models to determine the probability of future mass movement occurrence [24]. So far, many statistical models have been successfully applied to mass movement susceptibility mapping, such as weight of evidence [25], frequency ratio [26], logistic regression [27], information value [28] and generalized additive model [29]. Recently, many machine learning models have demonstrated excellent performance, such as support vector machine [30], artificial neural network [31], maximum entropy model [32,33], naïve Bayes [34], decision tree [35], random forest [36] and gradient boosted trees [37]. Each of these approaches has its own pros and cons. For example, as black-box models, random forest and gradient-boosted trees tend to show better performance but low interpretability.

During mass movement susceptibility modeling, many factors will affect the final assessment outcome, including the quality of the mass movement inventory [12,38,39], the selection of spatial mapping units and their resolutions [40], the sampling strategy for mass movement-free units [41], the selection of conditional factors and their quality [42], the choice of the susceptibility algorithm [8,43], the optimization of model parameters [44] and the model validation metrics [45]. Among them, a representative mass movement inventory is the key prerequisite to getting a reliable susceptibility map [46–49]. However, the available mass movement databases are often biased and incomplete. In general, the reported mass movements tend to be more representative in economically advanced areas with a large population [4,50]. Such areas have more abundant detection methods with programs for detailed investigations of mass movements. Thus, mass movements in densely populated or trafficked areas are more likely to be observed and reported. Suppose the data on mass movement is derived from the interpretation of optical remote sensing or LiDAR. In that case, it is typically overrepresented within forested areas and underestimated in regions with intense human activity, such as the presence of cultivated land [51]. Because of the distinctive morphological characteristics of mass movements in forest areas, they are easily identified. On the other hand, for the mass movements occurring within arable land, their topographical features are easily blurred or eliminated or altered by human activities. Thus, for a study region as large as China, with its complex topographic conditions and unbalanced population and economic development [52], the available mass movement inventory is usually heterogeneously complete.

The spatially heterogeneous completeness of mass movement information has an enormous impact on susceptibility assessment. Several studies have shown that if inventory bias is directly ignored, even models that perform well on quantitative metrics such as accuracy, F1-score, or AUROC will propagate this bias into the model results, leading to geomorphological implausibility in final spatial prediction patterns [48,53]. For example, Steger et al. [54] simulated a highly incomplete landslide inventory in forested areas. They found that when simply ignored this bias, the models would predict landslide susceptibility to be low in forested areas, as opposed to predictions when the inventory was complete.

Some researchers have proposed using mixed-effects models, considered beneficial where measurements are repeated or are statistically relevant. Popular in medicine, ecology, and economy because of their unique advantages in analyzing hierarchical or longitudinal data [55,56], they are also helpful in mass movement modelling studies. In a study by Steger et al. [54], mixed-effects models successfully reduced the effects of inventory bias in mass movement susceptibility mapping. Unlike the traditional mass movement susceptibility models that use only fixed effects to predict susceptibility, mixed-effects models also incorporate random effects components. Introducing these additional components can account for spatially heterogeneous completeness of mass movement inventory, thereby counterbalancing the associated propagation of biases in the data. In addition, the fixed-effects term of the mixed effects model can also be implemented in many ways to achieve the desired results. The generalized linear mixed-effects model (GLMM) [53,54], based on the GLM, and the generalized additive mixed-effects model (GAMM) [57,58], based on the GAM, are examples of statistical-based mixed models that have performed well in this regard. Recently, Sigrist [59] has innovatively developed Tree-boosted mixed-effects model (TBMM) and demonstrated its excellent performance. However, no relevant research articles illustrate the suitability of this machine learning-based mixed model in the field of mass movement susceptibility assessment and its comparison with statistical-based mixed models.

Against the backdrop of the above discussions, this paper has used three mixed-effects models (i.e., statistical-based GLMM, GAMM and machine learning-based TBMM) to assess the mass movement susceptibility in selected areas of China. This study addresses how to account for the inherent spatial incompleteness of the inventory and compares the performance of the models, both from quantitative and qualitative perspectives and explores which of them show superior performance. The comparison is based on quantitative analysis of AUROCs of cross-validation, spatial cross-validation, and predictions on simulated highly biased inventory, as well as qualitative perspectives using spatial patterns of susceptibility maps.

2. Study Area and Materials

2.1. Characterization of the Study Area

Located in the eastern part of Eurasia along the Pacific Ocean coastline, China is the third largest country in the world, occupying a territory of about 9.6 million km^2. Its geographic coverage is approximately 73°–135°E longitude and 18°–54°N latitude (Figure 1). There are widely varying landscapes and climate zones in China. The mountainous areas (including mountains, hills and rugged plateaus) are vast and cover about 70% of the country's land area [3]. The terrain is generally characterized by a high in the west and a low in the east, with a "staircase"-type distribution consisting of three steps. The first step of the staircase is the Qinghai–Tibet Plateau (Tibet region), with an average height of over 4000 m. The Kunlun, Qilian and Hengduan Mountains are located on its northern and eastern edges, which mark the boundary between the first and second steps of the terrain. The second step (including the NW, Loess and SW regions) has huge basins and plateaus with an average elevation of 1000–2000 m. The Greater Khingan, Taihang, and Xuefeng Mountains are located to the east, forming a dividing line between the second and third steps of the terrain. The third step (including NE, Yangtze and SC regions) is dominated by vast plains and hills, most of which occur at elevations less than 500 m above sea level.

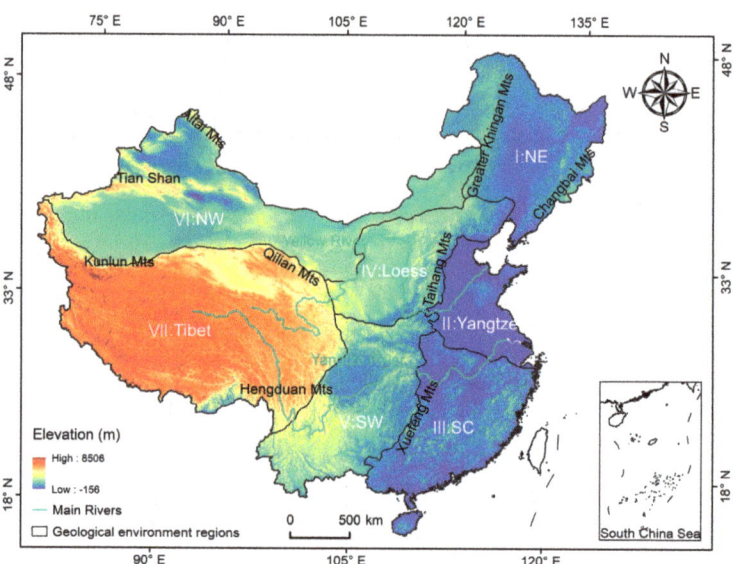

Figure 1. Overview map of China. The seven geological environment regions are shown as the Northeastern plain and mountain region (I: NE); the Huang–Huai–Hai–Yangtze River Delta plain region (II:Yangtze); the South China low mountain and hill region (III:SC); the North China Loess Plateau region (IV:Loess); the Southwest karst mountain region (V:SW); the Northwest mountain and basin region (VI:NW); and the Qinhai–Tibet Plateau region (VII:Tibet).

Due to the vastness of its territory and complex terrain characteristics, China exhibits variable climatic conditions. Its eastern part is significantly affected by the monsoon. This region exhibits tropical and subtropical climates (SW and SC region) and temperate monsoon climates (NE region) distributed from south to north. The western part is located inland, mainly with mountainous highland climate (Tibet region) and temperate continental climate (NW region) [60]. Precipitation generally declines from southeast to northwest in China. As for earthquakes, China lies between two of the world's most extensive seismic belts, i.e., the Circum-Pacific belt and the Alpide belt, with intense seismic activity [61]. Due to the widespread mountainous areas, rugged terrain, active seismicity and intense monsoons, China is prone to different geological disasters. Numerous landslides, rockfalls, and debris flows that occur here cause immeasurable casualties and economic damage [4].

2.2. Spatial Database
2.2.1. Inventory of Mass Movement

A spatial dataset representing former mass movements is essential for carrying out mass movement susceptibility mapping and hazard assessment [62]. Since 2005, the China Geological Survey has been conducting detailed investigations on six types of geological disasters, namely, landslide, rockfall, debris flow, ground subsidence, ground collapse and ground fissure. The survey consists of three main procedures: (i) interpretation of high-resolution remote sensing imagery; (ii) field verification; (iii) collating and correcting of obtaining data. The mass movement catalog used in this paper is credited to the National Geological Hazard Detailed Survey (https://geocloud.cgs.gov.cn/#/home accessed on 2 April 2022). In this work, the three most widely distributed and profound mass movements, including landslide (all slide-type movements), rockfall (all fall-type movements) and debris flow (all flow-type movements) were selected for analysis. There is a total of 108,432, 54,586 and 28,814 reported landslides, rockfalls and debris flows points, respectively. The inventory maps and kernel density maps for the three mass movements are shown in Figure 2. It was apparent that the completeness of the mass movement in different regions is heterogeneous. For example, the very low mass movement density in the northwest China may be related to sparse populations and tough surveys.

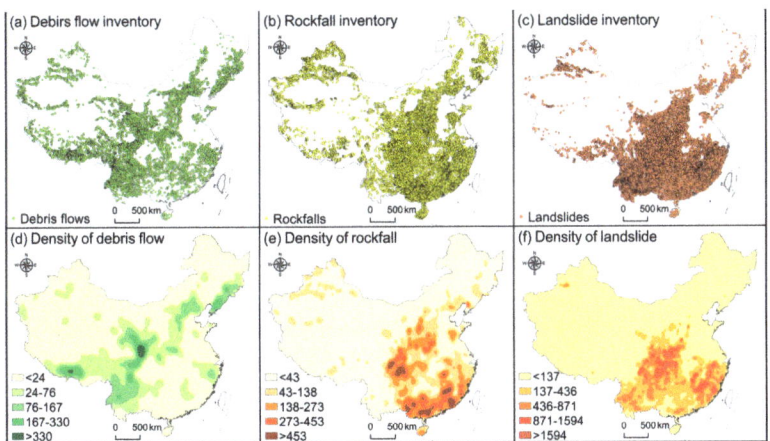

Figure 2. Spatial distribution maps and kernel density maps of three types of mass movements.

2.2.2. Mass Movement Influencing Factors

Mass movements' occurrence mechanism is extremely intricate and is affected by various conditional factors [7,63]. In the light of the relevant literature, six common categories of influencing factors, including topography (slope, aspect, profile curvature, plan curvature), human activities (road density), hydrology (river density, soil moisture), geology (lithology), land use and geological environment division are chosen as primary factors [4,10,57]. It should be emphasized that the mass movement data used in this paper are primarily triggered by heavy rainfall events, floods, earthquakes, or a combination thereof. Thus, considering the diversity of mass movement triggering conditions, we finally chose the common influencing factors of these mass movement as evaluation indicators [64,65]. The information of types and sources for mass movement inventory and influencing factors are shown in Table 1. Among these factors, slope, aspect, profile curvature, plan curvature, road density, river density, soil moisture and lithology are considered basic predictors (fixed-effect factors). In contrast, land use and geological environment division are considered as factors that are linked to the incompleteness of mass movement data (random intercept factors). The relationships between mass movement occurrence and conditional factors are described below. The fixed-effect and random intercept factors are also discussed in some detail.

Table 1. Data type and sources of mass movements inventory and influencing factors.

Data	Original Data Type	Data Sources
Mass movements inventory	Point	China geological survey
Slope	Grid (90 m)	Derived from DEM https://srtm.csi.cgiar.org (accessed on 5 April 2022)
Aspect	Grid (90 m)	Derived from DEM
Profile curvature	Grid (90 m)	Derived from DEM
Plan curvature	Grid (90 m)	Derived from DEM
Road density	Line	https://www.tianditu.gov.cn (accessed on 9 April 2022)
River density	Line	https://www.tianditu.gov.cn (accessed on 9 April 2022)
Soil moisture	Grid (1 km)	https://csidotinfo.wordpress.com/data/global-high-resolution-soil-water-balance (accessed on 10 April 2022)
Lithology	Polygon	https://www.uni-hamburg.de (accessed on 6 April 2022)
Land use	Grid (1 km)	https://www.resdc.cn (accessed on 6 April 2022)
Geological environment division	Polygon	https://geocloud.cgs.gov.cn/#/home (accessed on 15 April 2022)

(1) Fixed-Effect Factors

Topography, road density, hydrology, and geological properties are the fixed-effect factors discussed here. The slope is the most widely adopted parameter for mass movement susceptibility mapping [7]. It is not only a prerequisite for the occurrence of a mass movement but also affects the infiltration process and the resulting field distribution [66]. Aspect represents the orientation of slope, which will affect radiation absorption, rainfall runoff and weathering conditions, thus indirectly influencing the occurrence of mass movement [67]. Profile curvature and plan curvature indicate the change rate of slope along and perpendicular to slope gradient, which primarily influences soil erosion and surface runoff [68]. These terrain factors were generated from Shuttle Radar Topography Mission (SRTM) DEM at 90m resolution [69].

Road density is important, as many mass movements tend to occur along the roads in mountainous areas. This is primarily due to the instability of the slope caused by the destruction of mountains for road construction. Thus, road density is a commonly utilized anthropogenic variable for assessing mass movement susceptibility [70].

Hydrological conditions, including river density and soil moisture, are important in mass movements. Banks of rivers may collapse due to infiltration of pore water and erosion of slopes [71]. The soil moisture plays a crucial role in soil cohesion and permeation, leading to changes in soil shear strength [72]. Data on roads and rivers are generated from the National Platform for Common Geospatial Information Services. Data on soil moisture is derived from the Global High-Resolution Soil-Water dataset at 1km resolution and calculated as an annual average value [73].

Concerning the geological properties, lithology was selected to represent the physical and chemical properties of rocks. There are 16 types of lithology in China, namely, Basic Volcanic Rocks (VB), Intermediate Volcanic Rocks (VI), Acid Volcanic Rocks (VA), Basic Plutonic Rocks (PB), Intermediate Plutonic Rocks (PI), Acid Plutonic Rocks (PA), Metamorphic Rocks (MT), Evaporites (EV), Pyroclastic (PY), Carbonate Sedimentary Rocks (SC), Siliciclastic Sedimentary Rocks (SS), Mixed Sedimentary Rocks (SM), Unconsolidated Sediments (SU), Ice and Glaciers (IG), Water Bodies (WB), and No Data (ND). The lithology data was derived from the Global lithological map (GLiM) developed by Hartmann and Moosdorf [74].

(2) Random intercept factors

Next, we discuss random intercept factors associated with spatial heterogeneity of mass movement completeness, introduced into mixed-effects models as random intercept terms. Based on previous research, land use and division of the geological environment were chosen in China to account for mass movement incompleteness [57]. As for land use, mass movements in agricultural land or along transportation infrastructure are more likely to be blurred or removed by human activities. Those in forest areas are easily detected due to their distinct characteristics that differentiate them from the surrounding environment [54]. Thus, the mass movement inventory is expected to be underrepresented in arable land and overrepresented in forests. In this study, the land use data is derived from the Remote Sensing Monitoring Database of China's Land use/Cover in 2005 with 1km resolution, which was reclassified into five categories to better account for the bias of mass movement inventory. These are classified as Arable land (Ar), Forest land (Fo), Meadowland (Me), Settlements and Artificial land (SA), and Unutilized land (Un).

The China Geological Survey describes the geological environment as seven divisions based on geologic structure, geographical conditions, and geomorphology. These are the Northeastern plain and mountain region (I.NE), the Huang–Huai–Hai–Yangtze River Delta plain region (Yangtze), the South China low mountain and hill region (III.SC), the North China Loess Plateau region (IV.Loess), the Southwest karst mountain region (V.SW), the Northwest mountain and basin region (VI.NW) and the Qinhai–Tibet Plateau region (VII. Tibet) (Figure 1). Due to differences in topographical conditions and economic development, the completeness of mass movement inventory vary from region to region [57]. For example, mass movement investigations are generally more detailed in the economically developed SC region. However, the Tibet region is too high to reach, resulting in poorer availability of mass movement [75].

There is no universal guideline on the choice of mapping unit for mass movement susceptibility assessment [76]. This paper selected the most popular grid with 1 km resolution as the primary mapping unit. All the thematic layers were resampled or converted to 1 km resolution for consistency. Since the mass movement inventory was stored as points, a grid was considered as one with mass movement if it contained at least one event of mass movement. If the grid unit did not include any mass movements, it was considered as one without any mass movements. Ultimately, a total of 90,558, 47,057 and 25,425 grid units were confirmed, which included landslide, rockfall and debris flow, respectively. The distribution maps of all influencing factors are shown in Figure 3. Further, it has been noted that the selection of ratio between absence data and presence of data and the sampling method will affect the accuracy of the mass movement susceptibility model [41]. Based on previous research, this study adopted a usual 1:1 ratio and performed random sampling on absence grid units [42,77].

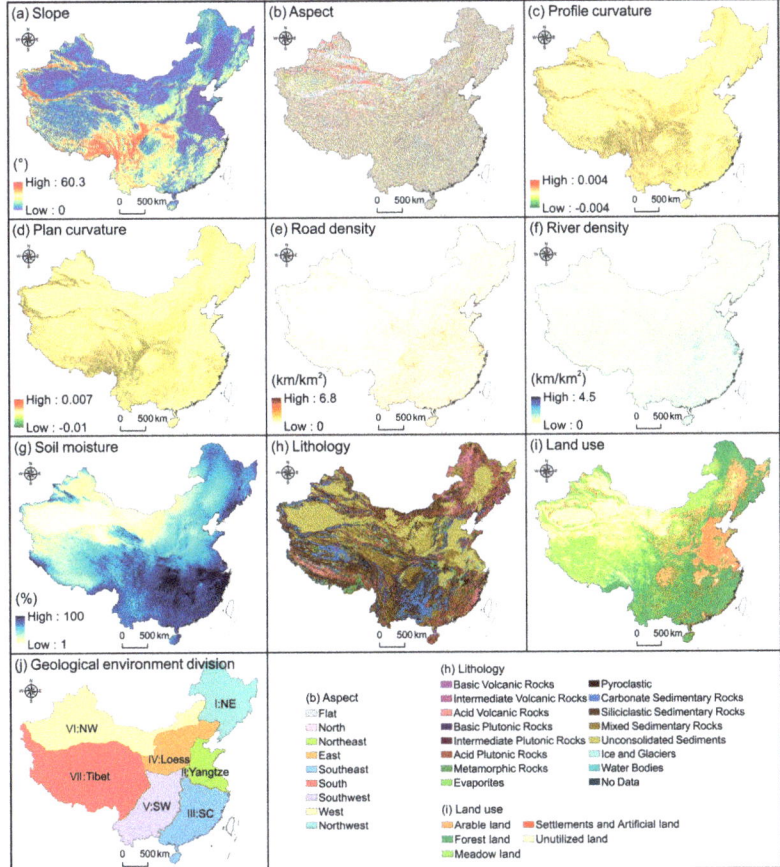

Figure 3. Spatial distribution map of influencing factors. (**a**) slope; (**b**) aspect (**c**) profile curvature; (**d**) plan curvature; (**e**) road density; (**f**) river density; (**g**) soil moisture; (**h**) lithology; (**i**) land use (**j**) geological environment division.

3. Methodology

The methodological flowchart of this study, including four main phases, is shown in Figure 4. In step 1, a spatial database was constructed with three mass movement inventories as response variables, eight basic predictors as fixed effects and two factors closely related to incompleteness of inventories as random intercepts. In step 2, preliminary data analysis was carried out to study the correlations between influencing factors and their relationships with mass movements, and to further confirm the factors that describe the incompleteness of the inventories. In step 3, three mixed-effects models were implemented for all the mass movements based on the established spatial database. It should be noted that the model fitting is based on fixed and random effects. In contrast, model prediction uses only fixed-effect factors, and the random effects that account for mass movement incompleteness are zeroed (i.e., averaged-out). In step 4, the prediction results of the three mixed models are compared and analyzed from both quantitative and qualitative perspectives.

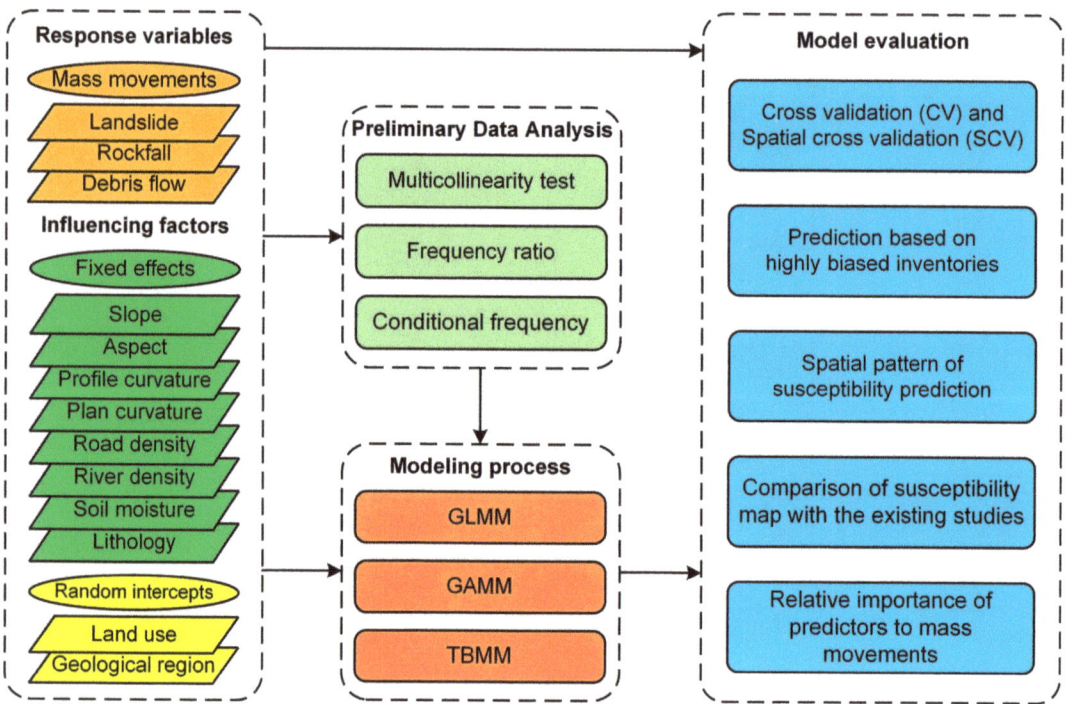

Figure 4. Workflow performed in this study.

3.1. Generalized Linear Mixed-Effects Model

The generalized linear mixed-effects model (GLMM) [78] combines and extends the characteristics of the linear mixed-effects model (LMM) and generalized linear model (GLM). Its dependent variable is no longer required to follow a Gaussian distribution (from GLM), and independent variables can contain both fixed and random effects (from LMM). GLMM has distinct advantages for analyzing the clustered, longitudinal and hierarchical data that are grouped at different levels [79]. In recent years, GLMM has been used to address data incompleteness in mass movement susceptibility mapping with considerable success [54]. Specifically, while modeling mass movement susceptibility, the Logit Link Function is adopted to represent the probability of mass movement occurrence P (Y=1) while specifying variables related to mass movement incompleteness as random intercept terms and the other basic predictors as fixed-effect terms. The GLMM for mass movement susceptibility can be expressed as follows:

$$logit(P(Y=1)) = \beta_0 + \beta_1 x_1 + \beta_2 x_2 + \cdots + \beta_m x_m + \gamma \quad (1)$$

where β_0 is the coefficient for the intercept, $\beta_1 \cdots \beta_m$ are the fixed-effect coefficients of associated influencing factors, $x_1 \cdots x_m$, and γ is the random intercept component that is presumed to be normally distributed with mean zero and variance σ^2 [80]. For more description on the application of GLMM in mass movement susceptibility, see Steger et al. [54].

3.2. Generalized Additive Mixed-Effects Model

The generalized additive mixed-effects model (GAMM) [81,82] extends the properties of the GLMM approach by replacing linear functions with smoothing functions to allow nonlinear associations between the dependent and independent variables while maintaining additivity [83]. GAMM is highly flexible and can easily control overfitting because it employs non-parametric additive functions to model linear or nonlinear covariate effects [84]. To evaluate a binary outcome (i.e., mass movement presence/absence) in mass movement susceptibility modeling, the GAMM has the form:

$$logit(P(Y=1)) = \beta_0 + f_1(x_1) + f_2(x_2) + \cdots + f_m(x_m) + \gamma \quad (2)$$

where the feature function, $f_1 \cdots f_m$, is constructed by a non-parametric smoothing spline, which can automatically model nonlinear associations without manually trying out many different transformations on each factor. Other parameters are similar to those in GLMM (Section 3.1). See Lin et al. [57] for more details on the methodology.

3.3. Tree-Boosted Mixed-Effects Model

Tree-boosted mixed-effects model (TBMM) [59] is a novel machine learning-based mixed-effects model that combines gradient-boosted trees with random effects. The model allows the relaxation of the linearity assumption of the response variable in a flexible nonparametric manner, and it can handle both continuous and discrete independent variables. The boosted trees have recently attracted significant attention in mass movement susceptibility mapping because of their state-of-the-art prediction performance and higher flexibility than other machine learning methods [8,85]. TBMM also has these advantages and can combine random effects to analyze grouped data; thus, it has great potential to solve the problem of incompleteness in mass movement susceptibility mapping. The equation of TBMM in this research is as follows:

$$logit(P(Y=1)) = TB(X) + \gamma \quad (3)$$

where X is the m-dimensional fixed-effects design matrix, e.g., there are m predictor variables. $TB()$ is a boosted tree. More specifically, the GPBoost algorithm is adopted to train the model, which iteratively learns the (co)variance parameters of the random effects and uses a gradient boosting step to add a tree to the ensemble of trees. In particular, the LightGBM [86] library is used to learn tree-boosting. LightGBM is very suitable for handling large-scale data due to its higher efficiency than other gradient boosting trees. More detailed principles of TBMM can be found in Sigrist [59]. Regarding the hyperparameter settings of this study, we used the built-in grid search function ('gpb.grid_search_tune_parameters') of the "gpboost" package to select the optimal algorithm parameter. The final parameters settings for the three types of mass movements are shown in Table 2.

Table 2. Parameter settings of TBMM for three types of mass movement.

Parameter	Debris Flow	Rockfall	Landslide
learning_rate	0.8	0.5	0.8
max_depth	10	6	5
min_data_in_leaf	80	30	30
num_boost_round	200	200	300

3.4. Model Evaluation

To quantitatively assess the predictive ability of all the mixed-effects models, the Area Under the Receiver Operating Characteristic curve (AUROC) was adopted [87]. The value ranges from 0.5 to 1, where 0.5 is a random prediction and 1 is a perfect prediction. The AUROCs for all models were calculated by repeated Spatial Cross Validation (SCV) and non-spatial Cross Validation (CV) approaches [88]. In this context, 10-times-repeated 10-fold partitioning of training and testing sets were used for both SCV and CV. The final spatial mass movement susceptibility maps produced from each model have been checked for plausibility, considering the incompleteness of the data on spatial variation of the mass movement and the comparative analysis of the spatial pattern observed from different models.

All the data preprocessing was performed in ArcMap software. Before introducing the data into the model, a multicollinearity test and frequency ratio analysis was performed using python and ArcMap software. The conditional frequency plots were based on the cdplot function in the R. GLMM and GAMM were achieved with "lme4" [89] and "mgcv" [90] packages in software R, respectively. TBMM is built based on the package "gpboost" [59] in python. SCV and CV for GLMM and GAMM were estimated using R package "sperrorest" [88], and for TBMM, they were implemented with the "sklearn" [91] package in python.

4. Results

The results of this study mainly include analysis of influencing factors, quantitative performance comparison of different mixed models, spatial pattern comparison of susceptibility map, and evaluation of the relative importance of influencing factors to three types of movements.

4.1. Preliminary Data Analysis

4.1.1. Multicollinearity Test

Multicollinearity is a common issue in model evaluation. It occurs when there are high correlations among predictors, which can reduce the stability of the model or even cause the model to fail. Thus, it is essential to perform a multicollinearity calculation before the predictors are entered into the model. The tolerance ($TOL = 1 - R^2 j$) and variance inflation factor ($VIF = 1/TOL$) are typical indicators for testing multicollinearity. If the value of $TOL < 0.1$ or $VIF > 10$, it means serious multicollinearity [92]. The results demonstrate that the VIF values of all predictors are < 5, and the TOL values are > 0.2, among which the soil moisture achieved the highest VIF of 4.3645 and the lowest TOL of 0.2291 (Table 3). These results indicate that there is no multicollinearity problem for all selected predictors.

Table 3. Results of multicollinearity test for predictors.

Influencing Factors	VIF	TOL
Slope	2.3926	0.4179
Aspect	3.1363	0.3188
Profile curvature	1.1988	0.8342
Plan curvature	1.1532	0.8672
Road density	1.2351	0.8087
River density	1.3272	0.7535
Soil moisture	4.3645	0.2291
Lithology	2.3306	0.4291

4.1.2. Correlation Analysis between Mass Movements and Influencing Factors

The Frequency Ratio (FR) approach was adopted to analyze the association between inventoried mass movements and each conditional factor [93]. A FR value below 1 reveals a weak relationship, and a value above 1 indicates a high probability of mass movement occurrence [94]. Since FR can handle only categorical variables, the Jenks natural breaks method is adopted to classify continuous factors. Figure 5 depicts the relationship between the three types of movements and each influencing factor. It can be concluded that the slope of the land illustrated a positive association with debris flows. However, for rockfalls and landslides, there was an initial increasing trend followed by a decreasing trend. Except for the first class of all mass movements and the last class of landslides, all other FR values exceed 1, indicating that the slope is crucial to the occurrence of mass movements. For aspect, the frequencies of three types of mass movements are slightly higher in east and southeast. For profile curvature, all movements have a high frequency of occurrence in the first three categories. Rockfalls also have a higher incidence in the class of 0.002–0.02, and landslides additionally have a higher incidence in the categories of 0.002–0.02 and 0.02–0.03, indicating that most mass movements are more abundant in concave and some convex areas. Regarding the plan curvature, mass movements are more frequently distributed within concave and convex classes. The results of FR values for road density and river density show that the occurrence frequency of mass movements is positively correlated with the density of roads and rivers, indicating their critical influence on mass movement occurrence. As for soil moisture, the FR values of debris flows initially increased and then decreased, while the frequency of rockfall and landslide occurrence is proportional to soil moisture. At the same time, debris flows have a higher frequency in areas with moderate humidity (categories 3 to 6). In comparison, rockfalls and landslides have a higher frequency in areas with high humidity (the last two categories). The lithology results show that Intermediate Volcanic Rocks, Basic Plutonic Rocks, Acid Plutonic Rocks, Metamorphic Rocks, Pyroclastic, Carbonate Sedimentary Rocks, Siliciclastic Sedimentary Rocks and Mixed Sedimentary Rocks have higher FR values, demonstrating a high likelihood of mass movement in these lithological units.

The three types of movements are more concentrated in arable and forest land for the land use factor. In addition, debris flows are abundant on meadowland. Landslides are also more distributed in settlements and artificial land. This phenomenon may be related to the discrepancies in the completeness of the inventory of mass movements for different land use types reported by Petschko et al. [95]. In addition, as can be found in Figure 6a, the distribution of land use varies between different slopes. Both arable land and settlements and artificial land are more frequently spread over gentler areas, while forest land and meadowland tend to be concentrated in steeper terrains. Therefore, this confounding relationship needs to be avoided when modeling. In the case of geological environment division, the distribution of mass movements in different regions varies greatly. Debris flows are reported in more detail in SW, Loess and Tibet regions, rockfalls are predominantly reported in SC and Loess regions, and landslides are more concentrated in SW and SC regions. The distribution difference in mass movements across the geological environment division is believed to be tightly associated with the difference in resource investment for mass movement investigations and the availability of mass movement information [57]. In addition, geological environment division and soil moisture are also spatially correlated (Figure 6b). For example, SC and SW are mainly distributed in humid areas, while NW is primarily located in arid areas. Therefore, this discrepancy in the representation of movements data within different geological environment subregions could have biased effects on soil moisture.

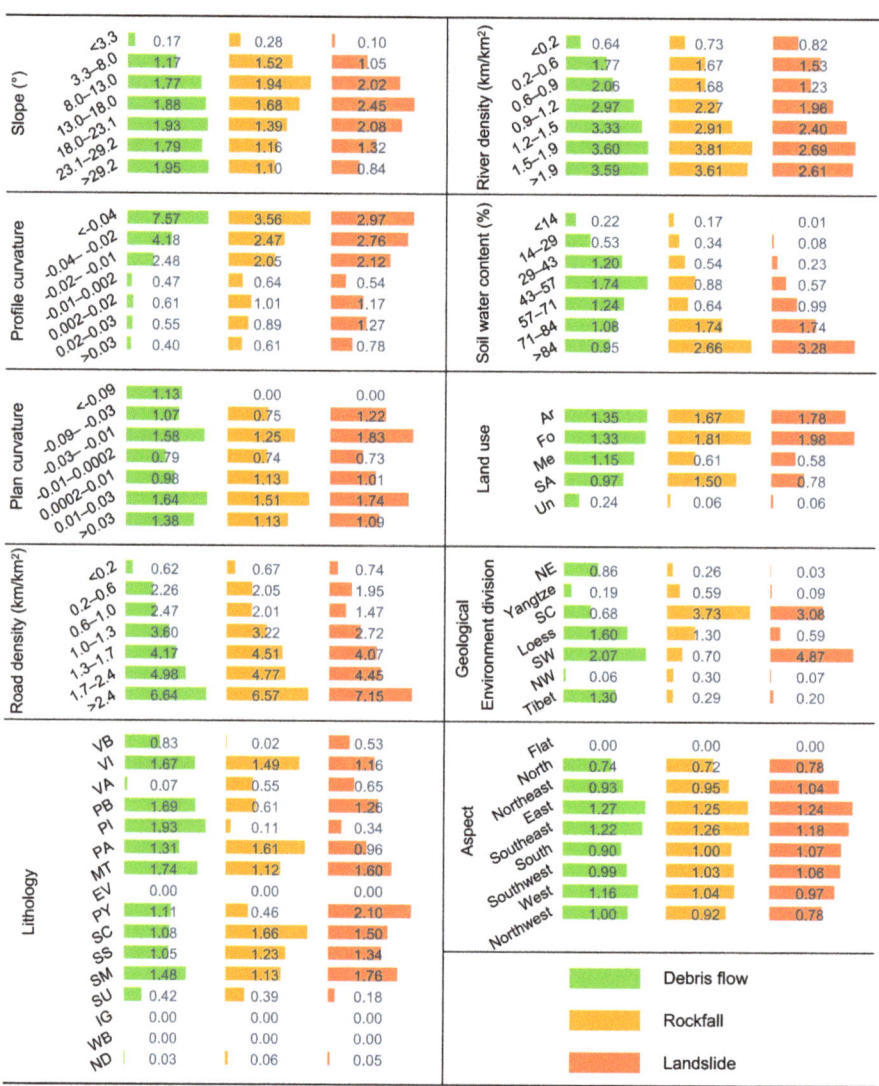

Figure 5. Correlations between three types of mass movements and influencing factors.

In summary, we find that land use and geological environment division are directly related to the incompleteness of data on mass movements. They are also associated with other predictors (e.g., slope and soil moisture). Therefore, mixed-effects models are needed to account for these biases [54,57].

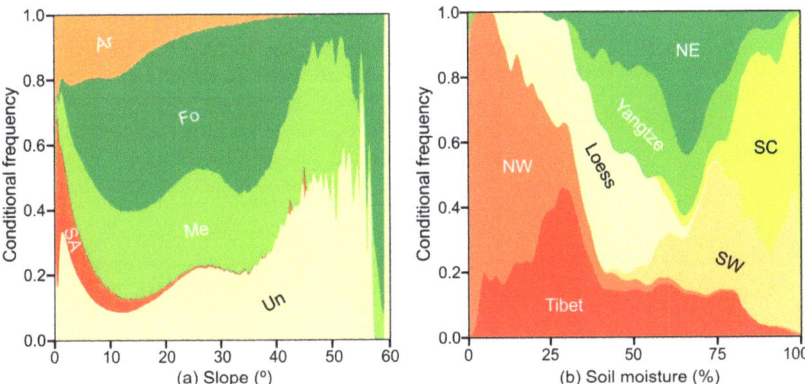

Figure 6. Conditional frequency plots between (**a**) land use and slope; (**b**) geological environment division and soil moisture.

4.2. Quantitative Performance Comparison

4.2.1. Cross-Validation Results

Table 4 presents the comparison results of three mixed-effects models for different types of movements. For debris flow, the median AUROCs of non-spatial cross-validation (CV) for all models are higher than 0.8. For GAMM and TBMM, the median AUROCs of spatial cross-validation (SCV) are above 0.8. This value is lower than 0.8 for GLMM, indicating the better spatial and non-spatial performance of GAMM and TBMM. In addition, the median AUROCs of all models decrease (SCV vs. CV) are below 0.02, with all interquartile ranges below 0.1, demonstrating the robustness of spatial predictions. Regarding the rockfall, the median AUROCs of SCV results for all models are above 0.8, and the median AUROCs of CV results for GAMM and TBMM are higher than 0.8 but only 0.773 for GLMM, suggesting that GAMM and TBMM perform better. In addition, the median AUROCs of the SCV for GLMM and GAMM are abnormally greater than that of CV, and the SCV of TBMM is only 0.013 lower than CV with the interquartile ranges less than 0.15, indicating that the prediction performance of TBMM is more stable. In the case of landslide, the median AUROCs of CV results for all models are above 0.8, while the median AUROCs of SCV results for GLMM and GAMM are below 0.8, indicating their unstable spatial predictive performance. For TBMM, the median AUROC value dropped by (SCV vs. CV) only 0.018, indicating that its predictive power is more spatially robust compared to GLMM (0.039) and GAMM (0.056). Regarding the differences in interquartile, GAMM has the best performance with an interquartile of 0.055, and both GLMM and TBMM performed somewhat poorly, with interquartile ranges of about 0.19.

Overall, TBMM consistently produced SCV and CV results above 0.8 and higher than GLMM and GAMM for all types of movements, and its reduced values (SCV vs. CV) and interquartile ranges were acceptable, which are generally considered to reflect superior validation performance [54].

Table 4. Cross-validation results of three mixed models for mass movements.

Model	AUROC Median (1st–3rd Quantile)	Debris Flow	Rockfall	Landslide
GLMM	Non-spatial Cross Validation	0.816 (0.813–0.819)	0.773 (0.769–0.775)	0.827 (0.825–0.830)
	Spatial Cross Validation	0.799 (0.760–0.832)	0.805 (0.679–0.817)	0.788 (0.654–0.845)
GAMM	Non-spatial Cross Validation	0.848 (0.844–0.852)	0.801 (0.799–0.805)	0.839 (0.836–0.842)
	Spatial Cross Validation	0.844 (0.781–0.855)	0.805 (0.734–0.846)	0.783 (0.751–0.806)
TBMM	Non-spatial Cross Validation	0.866 (0.863–0.868)	0.830 (0.826–0.833)	0.841 (0.837–0.844)
	Spatial Cross Validation	0.848 (0.800–0.858)	0.817 (0.733–0.865)	0.823 (0.678–0.867)

4.2.2. Predictions Based on Highly Biased Inventories

According to Steger et al. [54], an excellent model could maintain higher predictive performance even when the inventory is severely incomplete. This study simulated several inventory data that were highly biased by randomly deleting 80% of debris flows, rockfalls and landslides in the SC and SW regions, NW and Tibet regions, forest land, and arable land, respectively. Finally, SC-and SW-related, NW- and Tibet-related, forest-related, and arable-related biased inventories were obtained (Table 5).

Table 5. The number of mass movements in different highly biased inventories.

Movements	Original Data	SC and SW Regions	NW and Tibet Regions	Forest Land	Arable Land
Debris flow	25,425	19,307	17,252	19,393	20,218
Rockfall	47,057	22,479	41,697	31,632	35,192
Landslide	90,558	28,187	85,622	58,112	66,092

Then, all mixed models were trained based on these highly biased data and were used to predict the unmodified mass movements; the results are shown in Figure 7. A conclusion can be drawn that TBMM exhibits the best predictive performance for all the cases. At the same time, the range of variation of the prediction results based on different biased data was compared. For debris flow, all models showed a small range of variations from 0.002 for GAMM, 0.003 TBMM and 0.006 for GLMM. In the case of rockfall, the value for TBMM remained stable with a range of 0.007, while the GAMM and GLMM performed poorly with values, respectively, in the ranges of 0.020 and 0.023. Regarding landslide, GLMM and TBMM performed best with a range of 0.006, while GAMM was the worse, with a range of 0.025. Overall, TBMM showed the best score with a stable variation range (less than 0.01) across all types of land movements, indicating its advanced and robust prediction performance based on highly biased inventories.

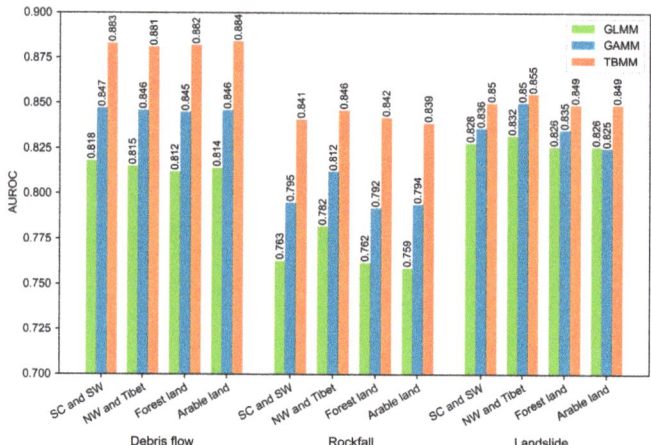

Figure 7. Comparison of model predictions (AUROC) for different types of mass movements based on highly incomplete data.

4.3. Spatial Pattern Comparison of the Susceptibility Map

Susceptibility maps for three movements were generated by predicting the probability of occurrence for each grid using three mixed-effects models separately. Then, the susceptibility index was classified into five categories using the Jenks natural break strategy in ArcMap 10.8 [96]. Figures 8–10 demonstrate the spatial patterns of debris flow, rockfall and landslide susceptibility based on the three mixed models and detailed local area comparisons. Regarding the differences between susceptibility maps, we first quantified susceptibility levels from very low to very high as one to five, and then the differences are obtained by subtracting one susceptibility map from the other through Raster Calculator tool in ArcMap. The results of debris flow show that three susceptibility maps (Figure 8a–c) are generally consistent. Areas that are highly prone to mass movements are abundant in the hilly areas of the Changbai Mountains (NE region), the southeast hills (SC region), and the Taihang Mountains (Loess region). The mountains around the Sichuan Basin (SW region), the Altai Mountains and the Tian Shan Mountains (NW region) and the mountains of the Qinghai–Tibetan Plateau (Tibet region) are also highly prone to mass movements. These results show that the three mixed-effects models can effectively reduce the impact of inventory bias, so they also have better predictive capability in the NW and Tibet regions, where there is a significant lack of data.

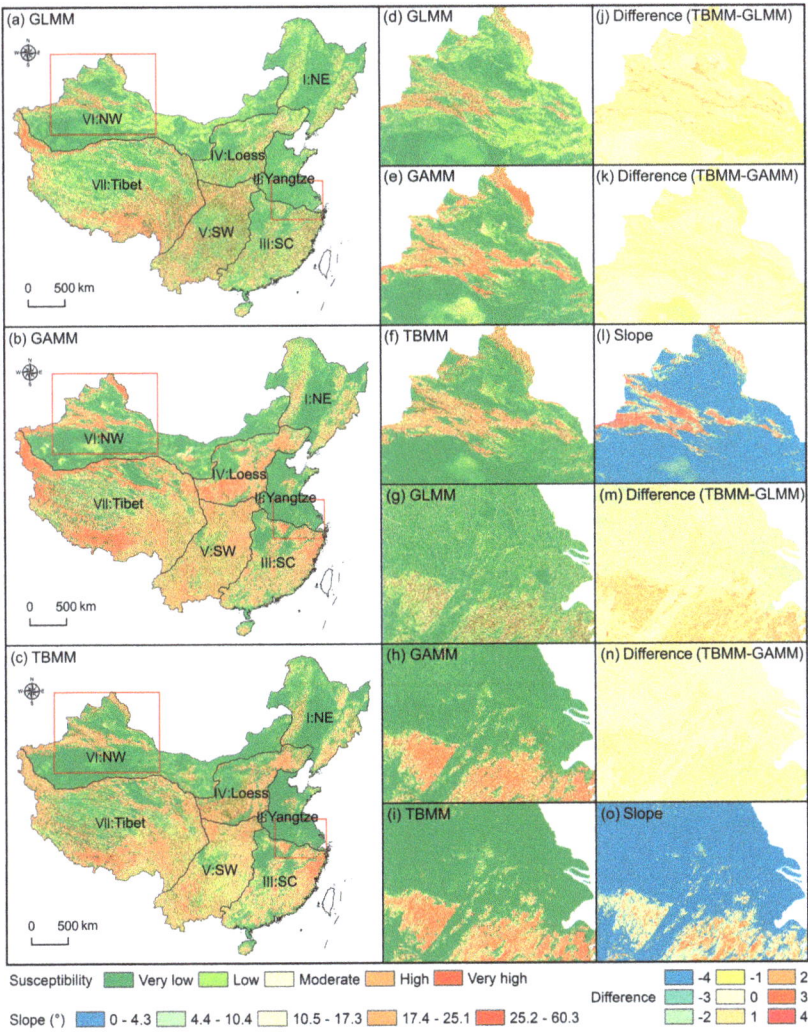

Figure 8. Debris flow susceptibility maps. (**a–i**) Susceptibility maps for China and local areas yield by three mixed models; (**j,k,m,n**) the difference between TBMM and other models in local areas; (**l,o**) the topography slope in local areas.

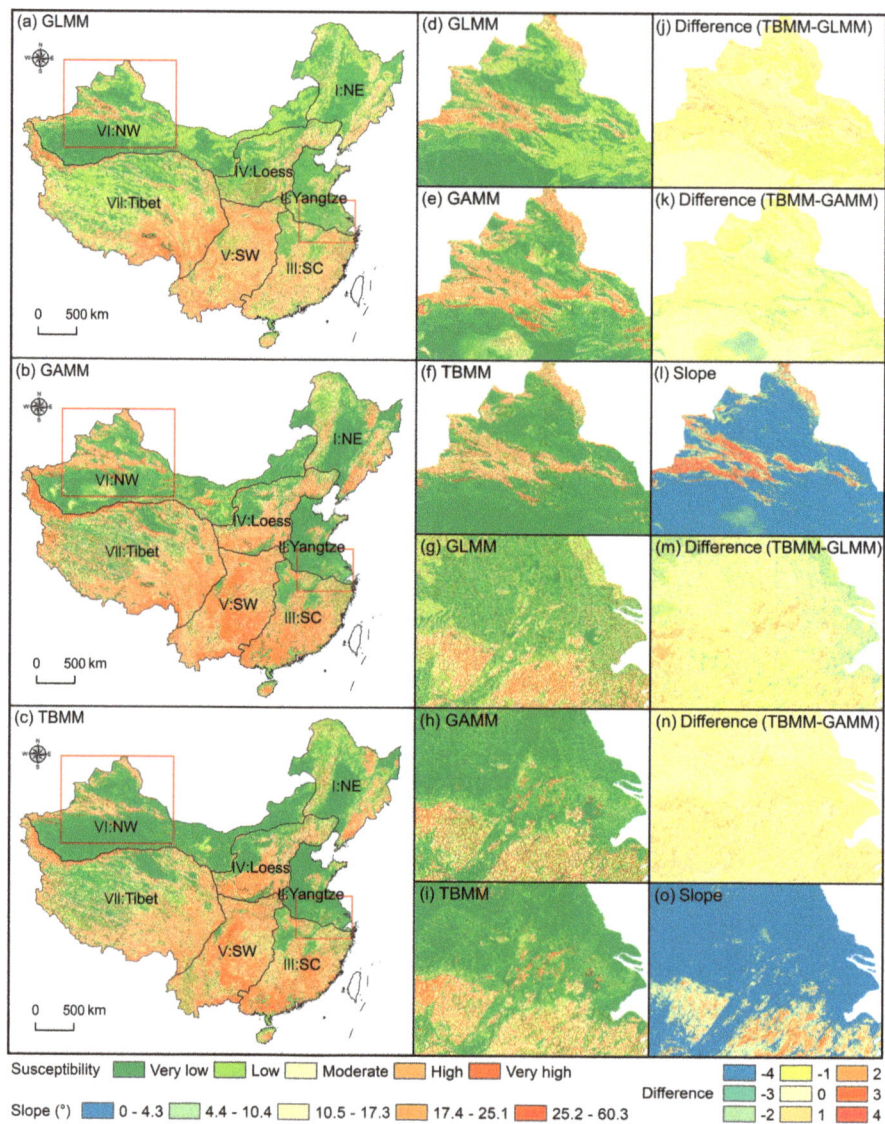

Figure 9. Rockfall susceptibility maps. (**a**–**i**) Susceptibility maps for China and local areas yielded by three mixed models; (**j**,**k**,**m**,**n**) the difference between TBMM and other models in local areas; (**l**,**o**) the topography slope in local areas.

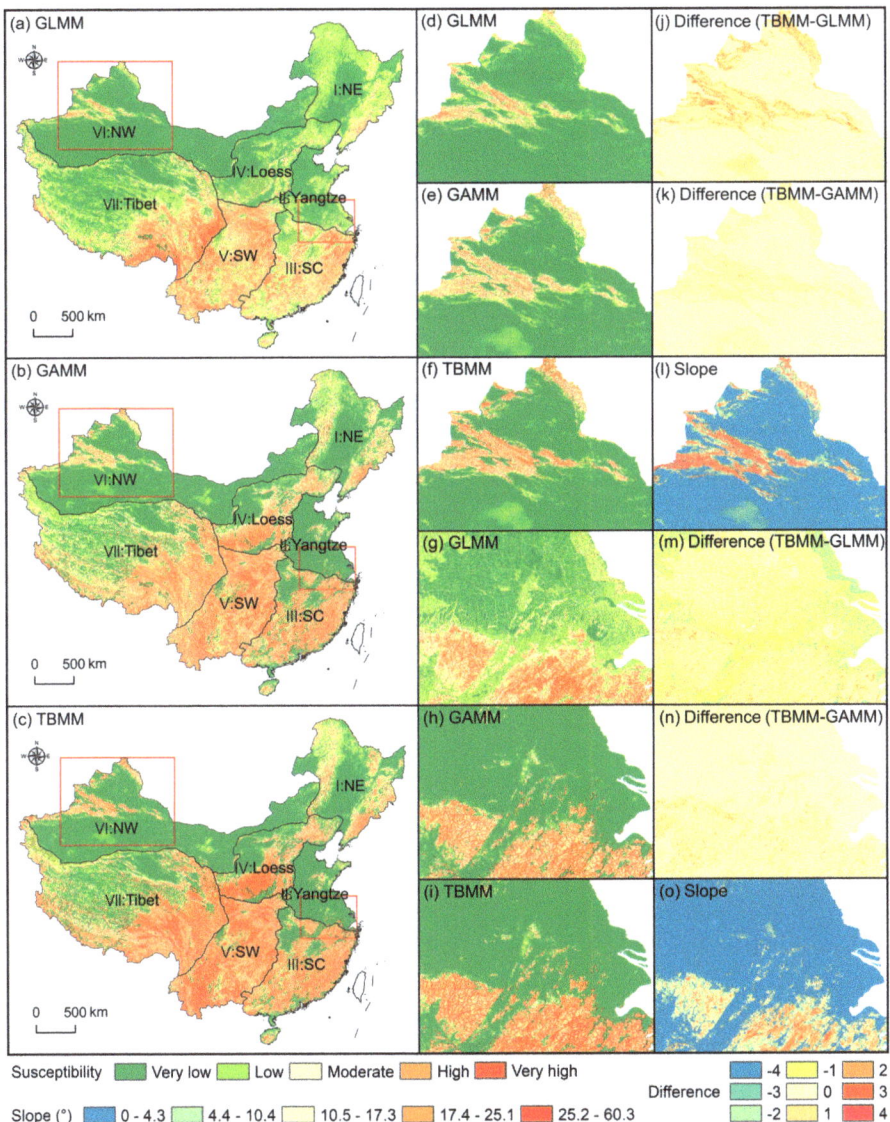

Figure 10. Landslide susceptibility maps. (**a–i**) Susceptibility maps for China and local areas generated by three mixed models; (**j,k,m,n**) the difference between TBMM and other models in local areas; (**l,o**) the topography slope in local areas.

Further, it has been observed that the susceptibility maps produced by GLMM are poor representations of the ground truth. For example, the slope angles in the Yangtze River Delta Plain are almost below 5 degrees. Still, the susceptibility map obtained by GLMM is greatly affected by roads in the plain. Many grids near roads are classified as medium or high susceptibility (see Figure 8o,g). The difference between TBMM and GAMM is lower than that of GLMM, with most values ranging between −1 and 1 (Figure 8j,k,m,n). Their predicted susceptibility in the plains was generally very low and significantly lower than that of GLMM (Figure 8h,i,m). The susceptibility was higher in mountainous areas than in

GLMM (Figure 8j,m), suggesting that the prediction results of TBMM and GAMM are more reasonable. Besides, some areas in the middle of the Taklimakan desert region (southern NW region) are predicted by GAMM to be moderate to high susceptibility (Figure 8e). In fact, the incidence of debris flow in arid regions is very low, and there are no debris flow points in the region (Figure 2a). In contrast, the susceptibility based on TBMM is lower than that of GAMM in this region (Figure 8k), indicating that the prediction of TBMM is better.

For rockfall, the susceptibility results of the GLMM (Figure 9a) performed relatively poorly. For example, in the Yangtze River Delta Plain, GLMM is severely affected by roads, and many areas are classified into medium to very high levels (Figure 9g). Overall, TBMM (Figure 9c) and GAMM (Figure 9b) are similar and perform better than GLMM in both mountains (Figure 9j,m) and plains (Figure 9m). TBMM also performed better than GAMM in desert areas (Figure 9k). The results indicate that GAMM may have overestimated the susceptibility in many areas, while GLMM underestimated the results.

Regarding the landslide, both TBMM (Figure 10c) and GAMM (Figure 10b) showed better performance. GLMM classified many areas in the Yangtze delta plain as moderate-to-high landslide susceptibility types (Figure 10g), which obviously do not match the actual topographic features. In addition, it can be observed that TBMM generates higher susceptibility indices than GAMM in the Tianshan and Altai mountains and the southeastern hills (Figure 10k,n). These results are similar to a previous study that used a representative landslide inventory from the local area [97]. The comparability of results suggests that TBMM has a more robust predictive capability.

To understand the overall pattern of mass movement distribution and mass movement susceptible areas, we performed summary statistics on the distribution of mass movement in each susceptibility class, distribution of susceptibility classes and the frequency ratio of mass movements in each susceptibility class (Figure 11). In general, models with more historical mass movement points concentrated in predicted high-prone areas are considered to have better performance [98]. From Figure 11, it can be found that for the three types of mass movements, only TBMM's results have more than 80% of the mass movement points concentrated in the areas with high and very high susceptibility (81.6%, 80.4%, and 85.1% for debris flow, rockfall and landslide, respectively), and the least mass movements were classified to low and very low-prone areas (6.9%, 5.8%, and 2.5%, respectively). Additionally, according to the results of Figure 11c, with the improvement of the susceptibility grade, the frequency ratios of debris flow, rockfall and landslide points increased by: 199 (0.033–6.557), 123 (0.039–4.813) and 247 (0.017–4.192) times, respectively, in TBMM; 104 (0.048–4.993), 52 (0.077–4.015) and 190 (0.027–5.139) times, respectively, in GAMM; and 51 (0.135–6.934), 37 (0.123–4.493) and 134 (0.037–4.962) times, respectively, in GLMM. Therefore, the susceptibility maps produced by TBMM showed the most reliable results. In order for readers to accurately determine the susceptibility of any location, we provide the original data of three susceptibility maps generated by TBMM in the Supplementary Materials.

Almost all previous literature has only evaluated landslide susceptibility in China. Therefore, we compare the TBMM-generated landslide susceptibility map from this paper with three previous studies carried out in recent years (Figure 12). In general, Figure 12a is very similar to Figure 12b, which also employed a mixed-effects model. The results in both these figures predicted higher susceptibility in the NW and Tibet region's mountainous areas compared to those shown in Figure 12c,d. Both of these are the results from the application of traditional non-mixed models. Due to the high altitude and sparse population of Tibet and NW regions, the survey of landslides is relatively rough and possibly not representative of the actual situation. This example demonstrates how the lack of data significantly influences the model's outcome and confirms the mixed-effects models' superiority.

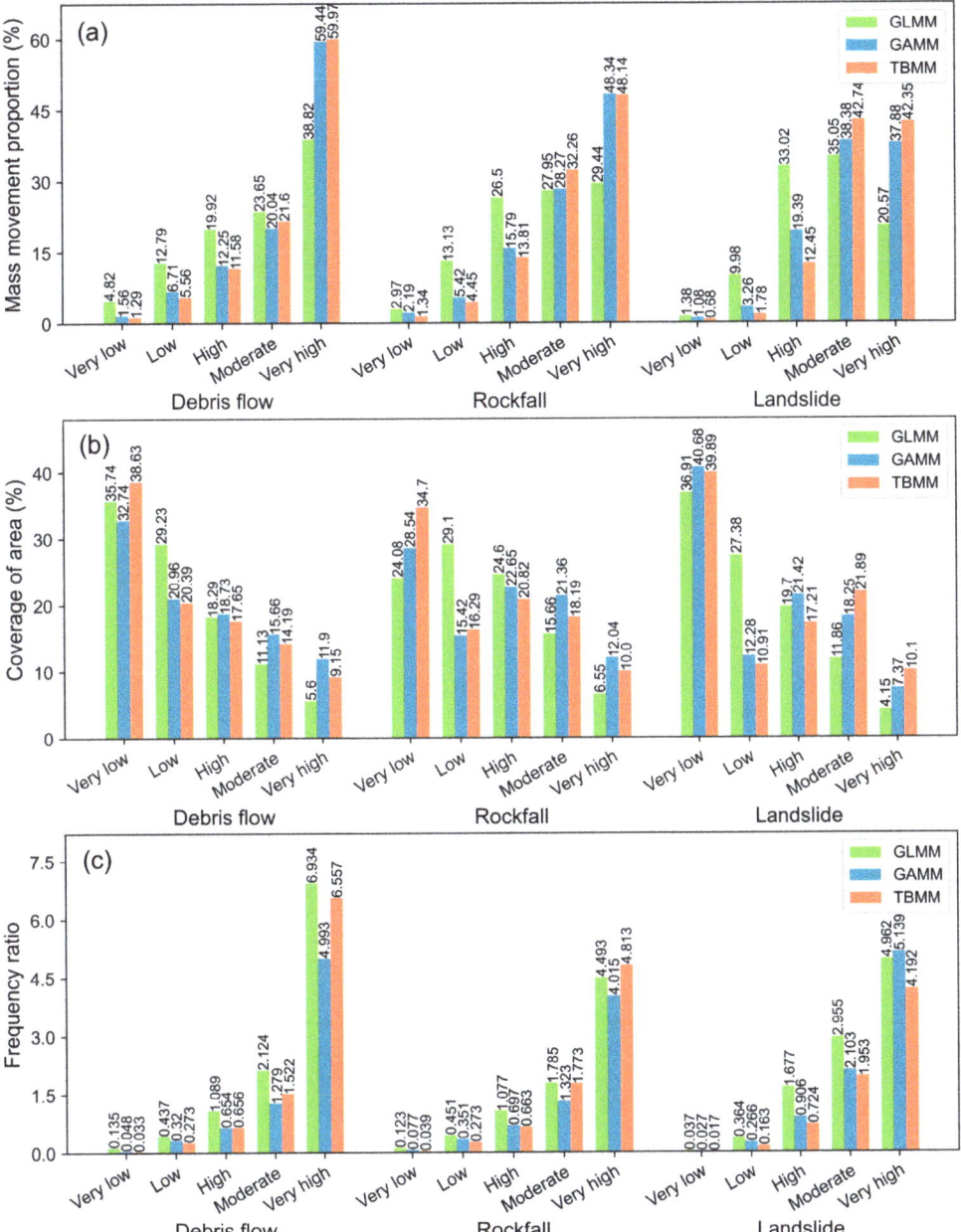

Figure 11. Statistics on the susceptibility maps of three type of mass movements. (**a**) distribution of mass movements in each susceptibility class; (**b**) distribution of susceptibility classes; (**c**) frequency ratio of mass movements in each susceptibility class.

Further, the results presented in Figure 12a show lower susceptibility in the Yangtze River Delta plain compared to the situation presented in Figure 12b. This result indicates that TBMM performs better in areas where landslides are unlikely to occur. It is to be noted that between these two cases, there is a notable difference in the basic unit used for the study. While the spatial mapping unit used in Figure 12a is a grid with an area of 1 km^2, that for Figure 12b is a much larger sub-watershed, with an average area of 129.1 km^2. Thus, in the latter case, almost all the mountainous areas in the SW and SC regions are predicted to be more vulnerable, which can be misguiding and, therefore, not appropriate to be recommended as guidelines for the government for decision-making and evolving disaster prevention measures. On the contrary, the TBMM-based results that presented in Figure 12a are finer and more realistic, which allows for more targeted development of disaster prevention and resource investment strategies.

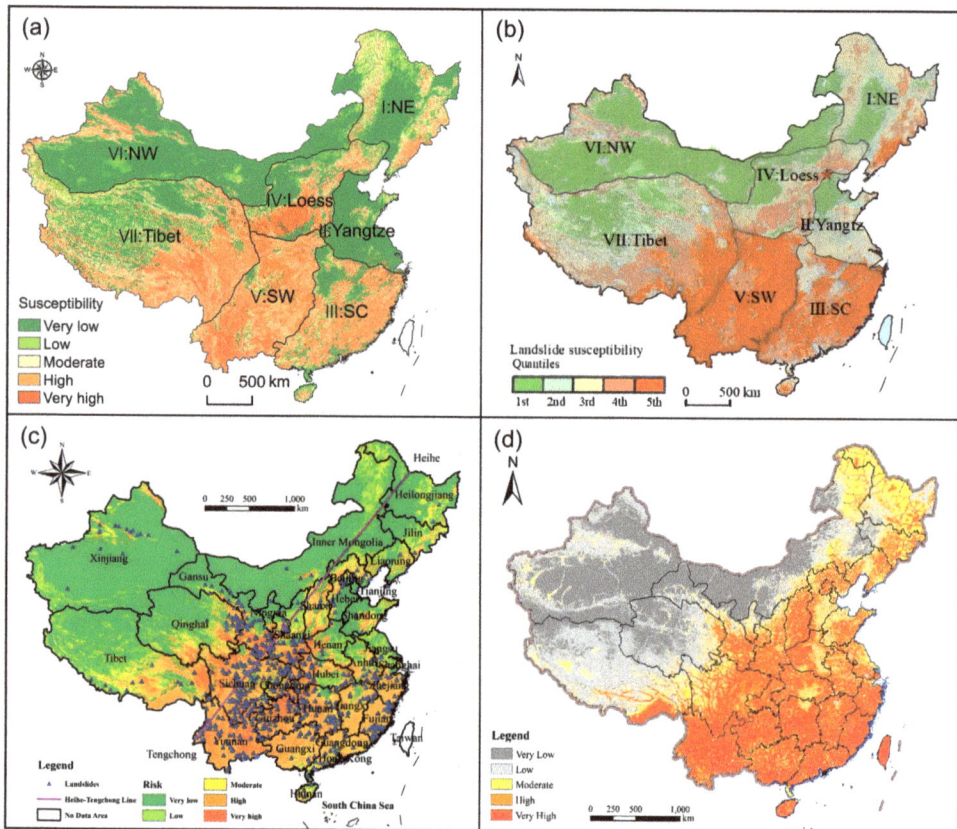

Figure 12. Comparison of results from this study with that of previous research. (**a**) Landslide susceptibility map generated by this paper based on TBMM; similar maps from (**b**) Lin et al. [57]; (**c**) Liu and Miao [99]; (**d**) Wang et al. [10].

4.4. Factor Contribution Analysis

Since the fixed part of TBMM is fitted with LightGBM, the relative importance of each predictor for susceptibility modeling can be obtained. Results illustrating the mean and standard deviation (error bar) of the relative importance of each factor based on both spatial and non-spatial cross-validations are shown in Figure 13. It can be found that profile curvature, slope, road density and soil moisture are the four significant factors with relative importance greater than 0.1 for susceptibility modeling of debris flow. With relative importance of less than 0.05, aspect and plan curvature are factors that seem to have the least importance. In the case of rockfall, the slope, road density and soil moisture contributed more to predicting rockfall risks with relative importance above 0.1. With a relative importance factor of less than 0.05, aspect, plan curvature and lithology have little influence on rock fall. In the case of landslide susceptibility modeling, the slope, road density and soil moisture have more influence, with their relative importance of more than 0.1. In contrast, the aspect and plan curvature seem relatively insignificant predictors.

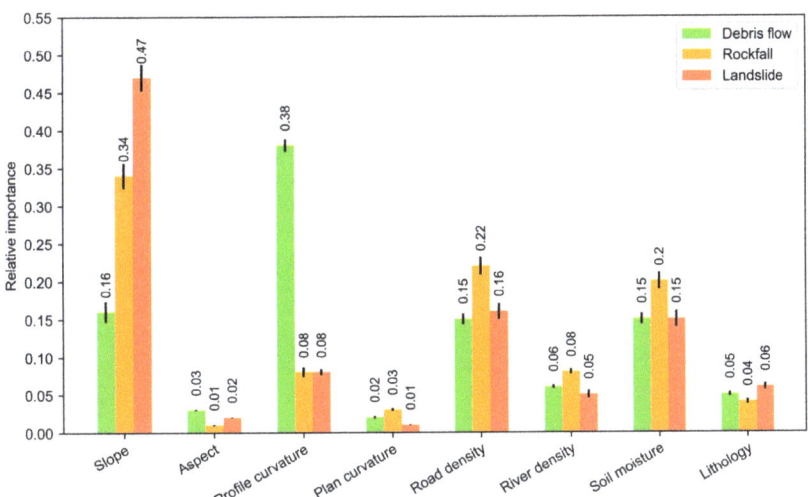

Figure 13. The relative importance of predictors in different mass movements.

The influencing parameters in the three types of movements have varying levels of importance. While slope, road density and soil moisture contribute significantly to the generation of debris flow, rockfall and landslide susceptibility, plan curvature and aspect appear to have the least importance in triggering all these movements. Additionally, it is noted that profile curvature has the highest importance in debris flow but is less critical for rockfall and landslide.

5. Discussion

With the development of artificial intelligence and the improvement in computing power, significant progress has been made in mass movement susceptibility modeling. An increasing number of susceptibility assessment models and optimization algorithms are being developed, and they are becoming increasingly complex [7–9]. However, reducing the bias propagation effect originating from the unavoidable incompleteness of mass movement inventory remains a fundamental and important challenge for susceptibility assessment, especially on a large scale such as that of China [57].

To address the issue of incomplete inventory in modeling mass movement susceptibility and to make realistic predictions, this research proposes three mixed-effects models, namely, GLMM, GAMM and TBMM. These models combine basic predictors (topography, road density, hydrology, geological properties, etc.) and random effect factors that account for biases in the mass movement inventory to improve the reliability of susceptibility evaluation in China. It is worth noting that all mixed models use both fixed and random effects to train the algorithm while only using fixed effects to predict the final susceptibility index. This step ensures that all models can counterbalance the adverse effects of biases in the inventory. However, there are some fundamental differences between the structure of the three models. The fixed part of GLMM is fitted by a parametric linear Generalized linear model (GLM). On the other hand, GAMM uses a semi-parametric nonlinear Generalized additive model (GAM) to fit the fixed effects term. Deviating from GLM and GAM, the fixed effects of TBMM are implemented by a non-parametric nonlinear LightGBM model.

Our study indicates that the forms of GAMM and TBMM are more flexible and may better fit the relationship between basic predictors and response variables. The empirical results also confirm this speculation. Based on the results presented in Table 4, GLMM had the worst performance in Cross-validation (CV) and Spatial cross-validation (SCV) for all types of mass movements. It is to be noted that with only half (3/6) of median AUROCs above 0.8, the level of performance of GLMM was low. While the median AUROCs produced by GAMM were mostly (5/6) above 0.8. TBMM performed best, with all values above 0.8. As for the difference between median AUROCs of SCV and CV, TBMM performed best, with the largest variation of 0.018, followed by GLMM with the largest difference of 0.039, and GAMM performed worst with the largest variation of 0.056. These differences are significantly more minor than those of the traditional non-mixed-effects model in the results of Lin et al. [57], and are close to the results of the mixed-effects model it used, which reflects the robustness of the mixed-effects model. Furthermore, according to Steger et al. [54], a good model can still have a high prediction score for the original mass movement data when the training data is highly biased. This paper simulated several types of highly biased inventory data to train all models and then predict the position of the original mass movements (Figure 7). It was found that TBMM still outperformed the other mixed models with all AUROCs above 0.8 and with the smallest fluctuations, further demonstrating the excellent performance of TBMM.

For the final mass movement susceptibility map generated for China, it can be found that both the mixed-effects models (Figures 8–10 and 12a,b) and the traditional non-mixed models (Figure 12c,d) yield similar high to very high susceptibilities in the SC, SW and Loess regions of China. It shows that the mass movement inventory in these areas is well represented; thus, all the models can yield good results. However, too low susceptibility levels for the northwestern mountainous areas (Tian Shan Mountainous and Altai Mountainous) obtained from traditional models are not credible, as they propagate the biases of mass movement inventory directly into the final results. On the other hand, results of all mixed-effects models classified the susceptibility of mountainous areas in the NW region as moderate to very high, which are also consistent with the results based on a representative inventory [97]. In addition, from the comparison of the susceptibility maps from different mixed effects models, it is found that TBMM not only maintained its excellent performance in mountainous areas but also did well in plain and desert areas, where mass movements are unlikely to occur. All these results indicate the superior quality of TBMM. Finally, the factor contribution analysis shows that the specific dominant factors of three type of mass movements are different (Figure 13). However, in general, topographic factors are the most important, followed by human activities and hydrology, indicating that topographical conditions are very crucial to the occurrence of mass movements. This observation is consistent with the views of most published results [42,64,65,100].

The results of this research emphasize the necessity to account for the effects of inherent inventory bias for susceptibility mapping, especially where the study area is as large as China. Considering this aspect of modelling, we implemented three mixed-effects models to account for the incompleteness of mass movement data, and our studies have suggested the superiority of TBMM. We find that this model can fit the relationship between the basic predictor and the response variables quite well while reducing the effect of the inventory bias. Therefore, this superior mass movement susceptibility evaluation model will provide a solid foundation for subsequent risk analysis and disaster prevention. In recent years, several other novel mixed-effects models have emerged. Among them are the MERF and BiMM combined with random forests [101,102]; the GMET and GMERT that combined decision trees [103,104]; and the MeNets and LMMNN combined with neural networks [105,106]. These novel mixed models have great potential to be explored in the field of mass movement susceptibility mapping. Moreover, there are some other strategies to deal with the incompleteness of the inventory data, such as using a fuzzy logic model to simulate the distribution characteristics of the entire mass movement based on a small portion of the mass movement inventory [107]. Training the susceptibility model for a small area, where the inventory was considered relatively complete, and then making predictions for the whole research region based on the calibrated model [18]. Combining the results of remote sensing interpretation, heuristic and multinomial statistical models to compensate for the uncertainty caused by limited inventory data [75]. Applying the maximum entropy model to deal with limited data due to its advantage of not requiring negative samples [108,109]. A comparison of mixed-effects models with these methods should be considered in future studies.

Although this study found a better mixed-effects model to be more efficient in dealing with biased inventory, limitations still exist, starting with the size of the mapping unit. Due to the computational inefficiency of mixed-effects models, we need to find a compromise between resolution and computational efficiency in such a large study area. The 1 km × 1 km grid used in this study is relatively coarse, which may result in some grids containing more than one mass movement. More reasonably sized mapping units or intensity mapping instead of susceptibility mapping can be explored in future studies, as has been suggested by some previous researchers [40,110]. Secondly, implementing mass movement susceptibility mapping over such a large area makes it impossible to obtain an unbiased input data. Although the mixed-effects models we used can minimize this biasing effect, the results remained difficult to interpret and validate. Thirdly, this paper employed grid search to optimize the parameters of TBMM, but this method is still rough and extremely time-consuming [111]. Exploring better parameter optimization strategies is a focus of future research. Fourthly, Some studies have found that recent mass movements are more likely to affect future mass movements than earlier ones [112]. Since the mass movement inventory used in this paper have been collected since 2005, there may be some older mass movements that have little impact on the future and therefore affect the predictive performance of the model. Finally, we used road and river data in 2017 and land use data in 2005 due to limitations of data acquisition, which may influence the plausibility of the model predictions and need to be improved in future work. Given these limitations, we emphasize that the three types of mass movement susceptibility maps in this paper provide a general situational awareness of mass movement-prone areas in China, but they are not recommended for local decision making.

6. Conclusions

The incompleteness of inventory data is inevitable while performing the susceptibility mapping for mass movements in large areas such as China. The mixed-effects model proposed in recent years can solve this problem well, but there are also many ways to achieve it. In this paper, three mixed-effects models are implemented to evaluate the susceptibility of mass movements in China, and several important conclusions can be drawn.

(i) From a quantitative point of view, the tree-boosted mixed-effects model (TBMM) performs best in both spatial and non-spatial cross-validation for all mass movements. In addition, when further reducing the completeness of inventory data in different categories of land use or geological environment division, TBMM maintained the best AUROC scores with little variation among the different highly biased types.

(ii) From a qualitative point of view, the derived TBMM yielded more plausible spatial susceptibility patterns than the other two mixed models and conventional methods discussed in the existing literature.

(iii) Through the factor contribution analysis, it was found that the profile curvature and slope contribute significantly to the evaluation of debris flow. For rockfall, slope, soil moisture and road density had more significant contribution. Regarding the landslide, slope and road density were the most critical factors.

In general, this paper aims to explore the best mixed-effects model to counterbalance the undesired effects of incomplete inventory data in susceptibility mapping and finally confirm the superiority of TBMM from both quantitative and qualitative perspectives. The susceptibility maps obtained in this study serve as a foundation for disaster prevention and spatial planning. Further, this model provides a reference for future susceptibility mapping of other regions.

Supplementary Materials: The following supporting information can be downloaded at: https://www.mdpi.com/article/10.3390/rs14236068/s1, The original format of the three types of mass movements susceptibility maps generated by TBMM.

Author Contributions: Conceptualization, Y.Z.; methodology, Y.H.; validation, Y.Z. and Y.H.; formal analysis, Y.H.; resources, Y.Z.; data curation, Y.Z.; writing—original draft preparation, Y.H.; writing—review and editing, Y.H. and Y.Z.; visualization, Y.H.; supervision, Y.Z.; funding acquisition, Y.Z. All authors have read and agreed to the published version of the manuscript.

Funding: This research was funded by the National key research and development program of Ministry of Science and Technology (No. 2022YFF0711704), the National Cryosphere Desert Data Center (No. E01Z790201) and the Capacity Building for Cryosphere Desert Data Center, Chinese Academy of Sciences (No. Y929830201).

Data Availability Statement: All the data are available in the public domain at the links provided in the texts.

Acknowledgments: Thanks to the scholars who provided the data used in this work. Thanks to the editors and reviewers for their comments.

Conflicts of Interest: The authors declare no conflict of interest.

References

1. Wang, Q.; Zhang, Q.-P.; Liu, Y.-Y.; Tong, L.-J.; Zhang, Y.-Z.; Li, X.-Y.; Li, J.-L. Characterizing the spatial distribution of typical natural disaster vulnerability in China from 2010 to 2017. *Nat. Hazards* **2020**, *100*, 3–15. [CrossRef]
2. Xu, L.; Meng, X.; Xu, X. Natural hazard chain research in China: A review. *Nat. Hazards* **2014**, *70*, 1631–1659. [CrossRef]
3. Wang, J.A.; Xiao, H.; Hartmann, R.; Yue, Y. Physical Geography of China and the US. In *A Comparative Geography of China and the US*; Springer: Berlin/Heidelberg, Germany, 2014; pp. 23–81.
4. Lin, Q.; Wang, Y. Spatial and temporal analysis of a fatal landslide inventory in China from 1950 to 2016. *Landslides* **2018**, *15*, 2357–2372. [CrossRef]
5. Froude, M.J.; Petley, D.N. Global fatal landslide occurrence from 2004 to 2016. *Nat. Hazards Earth Syst. Sci.* **2018**, *18*, 2161–2181. [CrossRef]
6. Lin, Q.; Wang, Y.; Glade, T.; Zhang, J.; Zhang, Y. Assessing the spatiotemporal impact of climate change on event rainfall characteristics influencing landslide occurrences based on multiple GCM projections in China. *Clim. Chang.* **2020**, *162*, 761–779. [CrossRef]
7. Reichenbach, P.; Rossi, M.; Malamud, B.; Mihir, M.; Guzzetti, F. A review of statistically-based landslide susceptibility models. *Earth-Sci. Rev.* **2018**, *180*, 60–91. [CrossRef]
8. Merghadi, A.; Yunus, A.P.; Dou, J.; Whiteley, J.; ThaiPham, B.; Bui, D.T.; Avtar, R.; Abderrahmane, B. Machine learning methods for landslide susceptibility studies: A comparative overview of algorithm performance. *Earth-Sci. Rev.* **2020**, *207*, 103225. [CrossRef]
9. Pourghasemi, H.R.; Teimoori Yansari, Z.; Panagos, P.; Pradhan, B. Analysis and evaluation of landslide susceptibility: A review on articles published during 2005–2016 (periods of 2005–2012 and 2013–2016). *Arab. J. Geosci.* **2018**, *11*, 193. [CrossRef]
10. Wang, D.; Hao, M.; Chen, S.; Meng, Z.; Jiang, D.; Ding, F. Assessment of landslide susceptibility and risk factors in China. *Nat. Hazards* **2021**, *108*, 3045–3059. [CrossRef]
11. Huang, H.; Wang, Y.; Li, Y.; Zhou, Y.; Zeng, Z. Debris-Flow Susceptibility Assessment in China: A Comparison between Traditional Statistical and Machine Learning Methods. *Remote Sens.* **2022**, *14*, 4475. [CrossRef]
12. Gaprindashvili, G.; Van Westen, C.J. Generation of a national landslide hazard and risk map for the country of Georgia. *Nat. Hazards* **2016**, *80*, 69–101. [CrossRef]
13. Ngo, P.T.T.; Panahi, M.; Khosravi, K.; Ghorbanzadeh, O.; Kariminejad, N.; Cerda, A.; Lee, S. Evaluation of deep learning algorithms for national scale landslide susceptibility mapping of Iran. *Geosci. Front.* **2021**, *12*, 505–519. [CrossRef]
14. Komac, M.; Ribičič, M. Landslide susceptibility map of Slovenia at scale 1: 250,000. *Geologija* **2006**, *49*, 295–309. [CrossRef]
15. Saroglou, C. GIS-based rockfall susceptibility zoning in Greece. *Geosciences* **2019**, *9*, 163. [CrossRef]
16. Titti, G.; Borgatti, L.; Zou, Q.; Cui, P.; Pasuto, A. Landslide susceptibility in the Belt and Road Countries: Continental step of a multi-scale approach. *Environ. Earth Sci.* **2021**, *80*, 630. [CrossRef]
17. Broeckx, J.; Vanmaercke, M.; Duchateau, R.; Poesen, J. A data-based landslide susceptibility map of Africa. *Earth-Sci. Rev.* **2018**, *185*, 102–121. [CrossRef]
18. Van Den Eeckhaut, M.; Hervás, J.; Jaedicke, C.; Malet, J.-P.; Montanarella, L.; Nadim, F. Statistical modelling of Europe-wide landslide susceptibility using limited landslide inventory data. *Landslides* **2012**, *9*, 357–369. [CrossRef]

19. Günther, A.; Reichenbach, P.; Malet, J.-P.; Van Den Eeckhaut, M.; Hervás, J.; Dashwood, C.; Guzzetti, F. Tier-based approaches for landslide susceptibility assessment in Europe. *Landslides* **2013**, *10*, 529–546. [CrossRef]
20. Lin, L.; Lin, Q.; Wang, Y. Landslide susceptibility mapping on a global scale using the method of logistic regression. *Nat. Hazards Earth Syst. Sci.* **2017**, *17*, 1411–1424. [CrossRef]
21. Hong, Y.; Adler, R.; Huffman, G. Use of satellite remote sensing data in the mapping of global landslide susceptibility. *Nat. Hazards* **2007**, *43*, 245–256. [CrossRef]
22. Jia, G.; Alvioli, M.; Gariano, S.L.; Marchesini, I.; Guzzetti, F.; Tang, Q. A global landslide non-susceptibility map. *Geomorphology* **2021**, *389*, 107804. [CrossRef]
23. Corominas, J.; van Westen, C.; Frattini, P.; Cascini, L.; Malet, J.-P.; Fotopoulou, S.; Catani, F.; Van Den Eeckhaut, M.; Mavrouli, O.; Agliardi, F. Recommendations for the quantitative analysis of landslide risk. *Bull. Eng. Geol. Environ.* **2014**, *73*, 209–263. [CrossRef]
24. Tien Bui, D.; Tuan, T.A.; Klempe, H.; Pradhan, B.; Revhaug, I. Spatial prediction models for shallow landslide hazards: A comparative assessment of the efficacy of support vector machines, artificial neural networks, kernel logistic regression, and logistic model tree. *Landslides* **2016**, *13*, 361–378. [CrossRef]
25. Ilia, I.; Tsangaratos, P. Applying weight of evidence method and sensitivity analysis to produce a landslide susceptibility map. *Landslides* **2016**, *13*, 379–397. [CrossRef]
26. Li, L.; Lan, H.; Guo, C.; Zhang, Y.; Li, Q.; Wu, Y. A modified frequency ratio method for landslide susceptibility assessment. *Landslides* **2017**, *14*, 727–741. [CrossRef]
27. Xiong, K.; Adhikari, B.R.; Stamatopoulos, C.A.; Zhan, Y.; Wu, S.; Dong, Z.; Di, B. Comparison of different machine learning methods for debris flow susceptibility mapping: A case study in the Sichuan Province, China. *Remote Sens.* **2020**, *12*, 295. [CrossRef]
28. Melo, R.; Zêzere, J.L. Modeling debris flow initiation and run-out in recently burned areas using data-driven methods. *Nat. Hazards* **2017**, *88*, 1373–1407. [CrossRef]
29. Goetz, J.; Brenning, A.; Petschko, H.; Leopold, P. Evaluating machine learning and statistical prediction techniques for landslide susceptibility modeling. *Comput. Geosci.* **2015**, *81*, 1–11. [CrossRef]
30. Huang, Y.; Zhao, L. Review on landslide susceptibility mapping using support vector machines. *Catena* **2018**, *165*, 520–529. [CrossRef]
31. Lee, D.-H.; Kim, Y.-T.; Lee, S.-R. Shallow landslide susceptibility models based on artificial neural networks considering the factor selection method and various non-linear activation functions. *Remote Sens.* **2020**, *12*, 1194. [CrossRef]
32. Di Napoli, M.; Marsiglia, P.; Di Martire, D.; Ramondini, M.; Ullo, S.L.; Calcaterra, D. Landslide susceptibility assessment of wildfire burnt areas through earth-observation techniques and a machine learning-based approach. *Remote Sens.* **2020**, *12*, 2505. [CrossRef]
33. Lin, J.; He, P.; Yang, L.; He, X.; Lu, S.; Liu, D. Predicting future urban waterlogging-prone areas by coupling the maximum entropy and FLUS model. *Sustain. Cities Soc.* **2022**, *80*, 103812. [CrossRef]
34. Tsangaratos, P.; Ilia, I. Comparison of a logistic regression and Naïve Bayes classifier in landslide susceptibility assessments: The influence of models complexity and training dataset size. *Catena* **2016**, *145*, 164–179. [CrossRef]
35. Park, S.-J.; Lee, C.-W.; Lee, S.; Lee, M.-J. Landslide susceptibility mapping and comparison using decision tree models: A Case Study of Jumunjin Area, Korea. *Remote Sens.* **2018**, *10*, 1545. [CrossRef]
36. Shirvani, Z. A holistic analysis for landslide susceptibility mapping applying geographic object-based random forest: A comparison between protected and non-protected forests. *Remote Sens.* **2020**, *12*, 434. [CrossRef]
37. Sahin, E.K. Comparative analysis of gradient boosting algorithms for landslide susceptibility mapping. *Geocarto Int.* **2020**, *37*, 2441–2465. [CrossRef]
38. Cascini, L. Applicability of landslide susceptibility and hazard zoning at different scales. *Eng. Geol.* **2008**, *102*, 164–177. [CrossRef]
39. Steger, S.; Brenning, A.; Bell, R.; Glade, T. The propagation of inventory-based positional errors into statistical landslide susceptibility models. *Nat. Hazards Earth Syst. Sci.* **2016**, *16*, 2729–2745. [CrossRef]
40. Erener, A.; Düzgün, H. Landslide susceptibility assessment: What are the effects of mapping unit and mapping method? *Environ. Earth Sci.* **2012**, *66*, 859–877. [CrossRef]
41. Hong, H.; Miao, Y.; Liu, J.; Zhu, A.-X. Exploring the effects of the design and quantity of absence data on the performance of random forest-based landslide susceptibility mapping. *Catena* **2019**, *176*, 45–64. [CrossRef]
42. Sun, D.; Shi, S.; Wen, H.; Xu, J.; Zhou, X.; Wu, J. A hybrid optimization method of factor screening predicated on GeoDetector and Random Forest for Landslide Susceptibility Mapping. *Geomorphology* **2021**, *379*, 107623. [CrossRef]
43. Cheng, J.; Dai, X.; Wang, Z.; Li, J.; Qu, G.; Li, W.; She, J.; Wang, Y. Landslide Susceptibility Assessment Model Construction Using Typical Machine Learning for the Three Gorges Reservoir Area in China. *Remote Sens.* **2022**, *14*, 2257. [CrossRef]
44. Kang, L.; Chen, R.-S.; Xiong, N.; Chen, Y.-C.; Hu, Y.-X.; Chen, C.-M. Selecting hyper-parameters of Gaussian process regression based on non-inertial particle swarm optimization in Internet of Things. *IEEE Access* **2019**, *7*, 59504–59513. [CrossRef]
45. Frattini, P.; Crosta, G.; Carrara, A. Techniques for evaluating the performance of landslide susceptibility models. *Eng. Geol.* **2010**, *111*, 62–72. [CrossRef]

46. Steger, S.; Mair, V.; Kofler, C.; Pittore, M.; Zebisch, M.; Schneiderbauer, S. Correlation does not imply geomorphic causation in data-driven landslide susceptibility modelling–Benefits of exploring landslide data collection effects. *Sci. Total Environ.* **2021**, *776*, 145935. [CrossRef]
47. Zêzere, J.; Pereira, S.; Melo, R.; Oliveira, S.; Garcia, R.A. Mapping landslide susceptibility using data-driven methods. *Sci. Total Environ.* **2017**, *589*, 250–267. [CrossRef]
48. Steger, S.; Brenning, A.; Bell, R.; Petschko, H.; Glade, T. Exploring discrepancies between quantitative validation results and the geomorphic plausibility of statistical landslide susceptibility maps. *Geomorphology* **2016**, *262*, 8–23. [CrossRef]
49. Harp, E.L.; Keefer, D.K.; Sato, H.P.; Yagi, H. Landslide inventories: The essential part of seismic landslide hazard analyses. *Eng. Geol.* **2011**, *122*, 9–21. [CrossRef]
50. Kirschbaum, D.; Stanley, T.; Zhou, Y. Spatial and temporal analysis of a global landslide catalog. *Geomorphology* **2015**, *249*, 4–15. [CrossRef]
51. Steger, S.; Schmaltz, E.; Glade, T. The (f) utility to account for pre-failure topography in data-driven landslide susceptibility modelling. *Geomorphology* **2020**, *354*, 107041. [CrossRef]
52. Yanar, T.; Kocaman, S.; Gokceoglu, C. Use of Mamdani fuzzy algorithm for multi-hazard susceptibility assessment in a developing urban settlement (Mamak, Ankara, Turkey). *ISPRS Int. J. Geo-Inf.* **2020**, *9*, 114. [CrossRef]
53. Lima, P.; Steger, S.; Glade, T. Counteracting flawed landslide data in statistically based landslide susceptibility modelling for very large areas: A national-scale assessment for Austria. *Landslides* **2021**, *18*, 3531–3546. [CrossRef]
54. Steger, S.; Brenning, A.; Bell, R.; Glade, T. The influence of systematically incomplete shallow landslide inventories on statistical susceptibility models and suggestions for improvements. *Landslides* **2017**, *14*, 1767–1781. [CrossRef]
55. Dingemanse, N.J.; Dochtermann, N.A. Quantifying individual variation in behaviour: Mixed-effect modelling approaches. *J. Anim. Ecol.* **2013**, *82*, 39–54. [CrossRef] [PubMed]
56. Ngufor, C.; Van Houten, H.; Caffo, B.S.; Shah, N.D.; McCoy, R.G. Mixed Effect Machine Learning: A framework for predicting longitudinal change in hemoglobin A1c. *J. Biomed. Inform.* **2019**, *89*, 56–67. [CrossRef]
57. Lin, Q.; Lima, P.; Steger, S.; Glade, T.; Jiang, T.; Zhang, J.; Liu, T.; Wang, Y. National-scale data-driven rainfall induced landslide susceptibility mapping for China by accounting for incomplete landslide data. *Geosci. Front.* **2021**, *12*, 101248. [CrossRef]
58. Steger, S.; Mair, V.; Kofler, C.; Schneiderbauer, S.; Zebisch, M. The necessity to consider the landslide data origin in statistically-based spatial predictive modelling-A landslide intervention index for South Tyrol (Italy). In *EGU General Assembly Conference Abstracts*; EGU: Munich, Germany, 2020; p. 3440.
59. Sigrist, F. Gaussian process boosting. *arXiv* **2020**, arXiv:2004.02653. [CrossRef]
60. Domrös, M.; Peng, G. *The Climate of China*; Springer Science & Business Media: Berlin/Heidelberg, Germany, 2012.
61. Chen, G.; Magistrale, H.; Rong, Y.; Cheng, J.; Binselam, S.A.; Xu, X. Seismic site condition of Mainland China from geology. *Seismol. Res. Lett.* **2021**, *92*, 998–1010. [CrossRef]
62. Galli, M.; Ardizzone, F.; Cardinali, M.; Guzzetti, F.; Reichenbach, P. Comparing landslide inventory maps. *Geomorphology* **2008**, *94*, 268–289. [CrossRef]
63. Pellicani, R.; Van Westen, C.J.; Spilotro, G. Assessing landslide exposure in areas with limited landslide information. *Landslides* **2014**, *11*, 463–480. [CrossRef]
64. Luo, W.; Liu, C.-C. Innovative landslide susceptibility mapping supported by geomorphon and geographical detector methods. *Landslides* **2018**, *15*, 465–474. [CrossRef]
65. Yang, Y.; Yang, J.; Xu, C.; Xu, C.; Song, C. Local-scale landslide susceptibility mapping using the B-GeoSVC model. *Landslides* **2019**, *16*, 1301–1312. [CrossRef]
66. Zhou, X.; Wen, H.; Zhang, Y.; Xu, J.; Zhang, W. Landslide susceptibility mapping using hybrid random forest with GeoDetector and RFE for factor optimization. *Geosci. Front.* **2021**, *12*, 101211. [CrossRef]
67. Pourghasemi, H.R.; Rossi, M. Landslide susceptibility modeling in a landslide prone area in Mazandarn Province, north of Iran: A comparison between GLM, GAM, MARS, and M-AHP methods. *Theor. Appl. Climatol.* **2017**, *130*, 609–633. [CrossRef]
68. Burrough, P.A.; McDonnell, R.A.; Lloyd, C.D. *Principles of Geographical Information Systems*; Oxford University Press: Oxford, UK, 2015.
69. Farr, T.G.; Kobrick, M. Shuttle Radar Topography Mission produces a wealth of data. *Eos Trans. Am. Geophys. Union* **2000**, *81*, 583–585. [CrossRef]
70. Huang, F.; Pan, L.; Fan, X.; Jiang, S.-H.; Huang, J.; Zhou, C. The uncertainty of landslide susceptibility prediction modeling: Suitability of linear conditioning factors. *Bull. Eng. Geol. Environ.* **2022**, *81*, 182. [CrossRef]
71. Bui, D.T.; Lofman, O.; Revhaug, I.; Dick, O. Landslide susceptibility analysis in the Hoa Binh province of Vietnam using statistical index and logistic regression. *Nat. Hazards* **2011**, *59*, 1413–1444. [CrossRef]
72. Hua, Y.; Wang, X.; Li, Y.; Xu, P.; Xia, W. Dynamic development of landslide susceptibility based on slope unit and deep neural networks. *Landslides* **2021**, *18*, 281–302. [CrossRef]
73. Trabucco, A.; Zomer, R. Global High-Resolution Soil-Water Balance. *Figshare. Fileset* **2019**, *10*, m9. [CrossRef]
74. Hartmann, J.; Moosdorf, N. The new global lithological map database GLiM: A representation of rock properties at the Earth surface. *Geochem. Geophys. Geosyst.* **2012**, *13*, 12. [CrossRef]
75. Du, J.; Glade, T.; Woldai, T.; Chai, B.; Zeng, B. Landslide susceptibility assessment based on an incomplete landslide inventory in the Jilong Valley, Tibet, Chinese Himalayas. *Eng. Geol.* **2020**, *270*, 105572. [CrossRef]

76. Zhao, S.; Zhao, Z. A comparative study of landslide susceptibility mapping using SVM and PSO-SVM models based on Grid and Slope Units. *Math. Probl. Eng.* **2021**, *2021*, 8854606. [CrossRef]
77. Hussin, H.Y.; Zumpano, V.; Reichenbach, P.; Sterlacchini, S.; Micu, M.; van Westen, C.; Bălteanu, D. Different landslide sampling strategies in a grid-based bi-variate statistical susceptibility model. *Geomorphology* **2016**, *253*, 508–523. [CrossRef]
78. Breslow, N.E.; Clayton, D.G. Approximate inference in generalized linear mixed models. *J. Am. Stat. Assoc.* **1993**, *88*, 9–25. [CrossRef]
79. Stroup, W.W. *Generalized Linear Mixed Models: Modern Concepts, Methods and Applications*; CRC Press: Boca Raton, FL, USA, 2012.
80. Rao, C.R.; Miller, J.P.; Rao, D.C. Essential statistical methods for medical statistics. In *Handbook of Statistics: Epidemiology and Medical Statistics*; Elsevier Inc.: Amsterdam, The Netherlands, 2011; pp. 1–351.
81. Lin, X.; Zhang, D. Inference in generalized additive mixed modelsby using smoothing splines. *J. R. Stat. Soc. Ser. B (Stat. Methodol.)* **1999**, *61*, 381–400. [CrossRef]
82. Wang, Y. Mixed effects smoothing spline analysis of variance. *J. R. Stat. Soc. Ser. B (Stat. Methodol.)* **1998**, *60*, 159–174. [CrossRef]
83. Mullah, M.A.S.; Hanley, J.A.; Benedetti, A. Modeling perinatal mortality in twins via generalized additive mixed models: A comparison of estimation approaches. *BMC Med. Res. Methodol.* **2019**, *19*, 209. [CrossRef]
84. Iddrisu, W.A.; Nokoe, K.S.; Luguterah, A.; Antwi, E.O. Generalized Additive Mixed Modelling of River Discharge in the Black Volta River. *Open J. Stat.* **2017**, *7*, 621. [CrossRef]
85. Sahin, E.K. Assessing the predictive capability of ensemble tree methods for landslide susceptibility mapping using XGBoost, gradient boosting machine, and random forest. *SN Appl. Sci.* **2020**, *2*, 1308. [CrossRef]
86. Ke, G.; Meng, Q.; Finley, T.; Wang, T.; Chen, W.; Ma, W.; Ye, Q.; Liu, T.-Y. Lightgbm: A highly efficient gradient boosting decision tree. *Adv. Neural Inf. Process. Syst.* **2017**, *30*, 3146–3154.
87. Swets, J.A. Measuring the accuracy of diagnostic systems. *Science* **1988**, *240*, 1285–1293. [CrossRef] [PubMed]
88. Brenning, A. Spatial cross-validation and bootstrap for the assessment of prediction rules in remote sensing: The R package sperrorest. In Proceedings of the 2012 IEEE International Geoscience and Remote Sensing Symposium, Munich, Germany, 22–27 July 2012; pp. 5372–5375.
89. Bates, D.; Mächler, M.; Bolker, B.; Walker, S. Fitting linear mixed-effects models using lme4. *arXiv* **2014**, arXiv:1406.5823. [CrossRef]
90. Wood, S.N. *Generalized Additive Models: An Introduction with R*, 2nd ed.; Chapman and Hall/CRC: Boca Raton, FL, USA, 2017; pp. 1–496.
91. Buitinck, L.; Louppe, G.; Blondel, M.; Pedregosa, F.; Mueller, A.; Grisel, O.; Niculae, V.; Prettenhofer, P.; Gramfort, A.; Grobler, J. API design for machine learning software: Experiences from the scikit-learn project. *arXiv* **2013**, arXiv:1309.0238. [CrossRef]
92. Mandal, K.; Saha, S.; Mandal, S. Applying deep learning and benchmark machine learning algorithms for landslide susceptibility modelling in Rorachu river basin of Sikkim Himalaya, India. *Geosci. Front.* **2021**, *12*, 101203. [CrossRef]
93. Korup, O.; Stolle, A. Landslide prediction from machine learning. *Geol. Today* **2014**, *30*, 26–33. [CrossRef]
94. Chen, W.; Chai, H.; Sun, X.; Wang, Q.; Ding, X.; Hong, H. A GIS-based comparative study of frequency ratio, statistical index and weights-of-evidence models in landslide susceptibility mapping. *Arab. J. Geosci.* **2016**, *9*, 204. [CrossRef]
95. Petschko, H.; Bell, R.; Glade, T. Effectiveness of visually analyzing LiDAR DTM derivatives for earth and debris slide inventory mapping for statistical susceptibility modeling. *Landslides* **2016**, *13*, 857–872. [CrossRef]
96. Youssef, A.M.; Pourghasemi, H.R. Landslide susceptibility mapping using machine learning algorithms and comparison of their performance at Abha Basin, Asir Region, Saudi Arabia. *Geosci. Front.* **2021**, *12*, 639–655. [CrossRef]
97. Liu, L.; Li, S.; Li, X.; Jiang, Y.; Wei, W.; Wang, Z.; Bai, Y. An integrated approach for landslide susceptibility mapping by considering spatial correlation and fractal distribution of clustered landslide data. *Landslides* **2019**, *16*, 715–728. [CrossRef]
98. Chang, L.; Zhang, R.; Wang, C. Evaluation and Prediction of Landslide Susceptibility in Yichang Section of Yangtze River Basin Based on Integrated Deep Learning Algorithm. *Remote Sens.* **2022**, *14*, 2717. [CrossRef]
99. Liu, X.; Miao, C. Large-scale assessment of landslide hazard, vulnerability and risk in China. *Geomat. Nat. Hazards Risk* **2018**, *9*, 1037–1052. [CrossRef]
100. Shirzadi, A.; Bui, D.T.; Pham, B.T.; Solaimani, K.; Chapi, K.; Kavian, A.; Shahabi, H.; Revhaug, I. Shallow landslide susceptibility assessment using a novel hybrid intelligence approach. *Environ. Earth Sci.* **2017**, *76*, 60. [CrossRef]
101. Hajjem, A.; Bellavance, F.; Larocque, D. Mixed-effects random forest for clustered data. *J. Stat. Comput. Simul.* **2014**, *84*, 1313–1328. [CrossRef]
102. Speiser, J.L.; Wolf, B.J.; Chung, D.; Karvellas, C.J.; Koch, D.G.; Durkalski, V.L. BiMM forest: A random forest method for modeling clustered and longitudinal binary outcomes. *Chemom. Intell. Lab. Syst.* **2019**, *185*, 122–134. [CrossRef]
103. Hajjem, A.; Larocque, D.; Bellavance, F. Generalized mixed effects regression trees. *Stat. Probab. Lett.* **2017**, *126*, 114–118. [CrossRef]
104. Fontana, L.; Masci, C.; Ieva, F.; Paganoni, A.M. Performing Learning Analytics via Generalised Mixed-Effects Trees. *Data* **2021**, *6*, 74. [CrossRef]
105. Xiong, Y.; Kim, H.J.; Singh, V. Mixed effects neural networks (menets) with applications to gaze estimation. In Proceedings of the IEEE/CVF Conference on Computer Vision and Pattern Recognition, Long Beach, CA, USA, 15–20 June 2019; pp. 7743–7752.
106. Simchoni, G.; Rosset, S. Integrating Random Effects in Deep Neural Networks. *arXiv* **2022**, arXiv:2206.03314. [CrossRef]

107. Robinson, T.R.; Rosser, N.J.; Densmore, A.L.; Williams, J.G.; Kincey, M.E.; Benjamin, J.; Bell, H.J. Rapid post-earthquake modelling of coseismic landslide intensity and distribution for emergency response decision support. *Nat. Hazards Earth Syst. Sci.* **2017**, *17*, 1521–1540. [CrossRef]
108. Javidan, N.; Kavian, A.; Pourghasemi, H.R.; Conoscenti, C.; Jafarian, Z.; Rodrigo-Comino, J. Evaluation of multi-hazard map produced using MaxEnt machine learning technique. *Sci. Rep.* **2021**, *11*, 6496. [CrossRef] [PubMed]
109. Rahmati, O.; Golkarian, A.; Biggs, T.; Keesstra, S.; Mohammadi, F.; Daliakopoulos, I.N. Land subsidence hazard modeling: Machine learning to identify predictors and the role of human activities. *J. Environ. Manag.* **2019**, *236*, 466–480. [CrossRef]
110. Lombardo, L.; Opitz, T.; Huser, R. Point process-based modeling of multiple debris flow landslides using INLA: An application to the 2009 Messina disaster. *Stoch. Environ. Res. Risk Assess.* **2018**, *32*, 2179–2198. [CrossRef]
111. Sameen, M.I.; Pradhan, B.; Lee, S. Application of convolutional neural networks featuring Bayesian optimization for landslide susceptibility assessment. *Catena* **2020**, *186*, 104249. [CrossRef]
112. Samia, J.; Temme, A.; Bregt, A.; Wallinga, J.; Guzzetti, F.; Ardizzone, F.; Rossi, M. Do landslides follow landslides? Insights in path dependency from a multi-temporal landslide inventory. *Landslides* **2017**, *14*, 547–558. [CrossRef]

Article

The Suitability of UAV-Derived DSMs and the Impact of DEM Resolutions on Rockfall Numerical Simulations: A Case Study of the Bouanane Active Scarp, Tétouan, Northern Morocco

Ali Bounab [1], Younes El Kharim [1] and Rachid El Hamdouni [2,*]

[1] GERN, Department of Geology, Faculty of Sciences, Abdelmalek Essaadi University, 93030 Tetouan, Morocco
[2] Civil Engineering Department, E.T.S. Ingenieros de Caminos, Canales y Puertos, Campus Universitario de Fuentenueva, s/n Granada University, 18071 Granada, Spain
* Correspondence: rachidej@ugr.es

Abstract: Rockfall simulations constitute the first step toward hazard assessments and can guide future rockfall prevention efforts. In this work, we assess the impact of digital elevation model (DEM) resolution on the accuracy of numerical rockfall simulation outputs. For this purpose, we compared the simulation output obtained using 1 m, 2 m and 3 m resolution UAV-derived DEMs, to two other models based on coarser topographic data (a 5 m resolution DEM obtained through interpolating elevation contours and the Shuttle Radar Topographic Mission 30m DEM). To generate the validation data, we conducted field surveys in order to map the real trajectories of three boulders that were detached during a rockfall event that occurred on 1 December 2018. Our findings suggest that the use of low to medium-resolution DEMs translated into large errors in the shape of the simulated trajectories as well as the computed runout distances, which appeared to be exaggerated by such models. The geometry of the runout area and the targets of the potential rockfall events also appeared to be different from those mapped on the field. This hindered the efficiency of any prevention or correction measures. On the other hand, the 1m UAV-derived model produced more accurate results relative to the field data. Therefore, it is accurate enough for rockfall simulations and hazard research applications. Although such remote sensing techniques may require additional expenses, our results suggest that the enhanced accuracy of the models is worth the investment.

Keywords: rockfall simulation; Rif; DEM resolution; UAV; back analysis

1. Introduction

Rockfalls and rock avalanches are characterized by high velocities and important runout distances [1], which could lead to injuries or even casualties in populated areas. In the western Mediterranean region, the population growth and its consequential urban expansion have led to the exploration of hazardous rocky cliffs, which resulted in many rockfall occurrences in the last few decades [2–8]. In some cases, this natural hazard can threaten the lives and goods of people residing in high-risk areas, which subsequently affects their socio-economic development.

To assess rockfall hazards, different laboratory and field tests (e.g., [9]) as well as numerical techniques are being used by hazard scientists (e.g., [10,11]). For the latter approach, topographic data are needed to predict the trajectory of potential future events. Such data can be freely downloaded with different spatial resolutions depending on the location of the study area or generated by landslide researchers. For instance, airborne LiDAR data, which provide more accurate digital elevation models (DEMs) compared to other DEM generation techniques, cover only a small portion of the globe [12]. In other parts of the world, airborne imagery techniques can be deployed to generate LiDAR scenes for local and regional applications. However, the cost is often high making it inaccessible to many researchers worldwide [13–15]. Other techniques such as structure-from-motion

(SFM), also known as digital aerial photogrammetry (DAP), are less expensive but are difficult to use in densely vegetated areas [16]. Therefore, the resolution and accuracy of DEMs that are used for rockfall trajectory simulations depend on the availability of the data for a given study area as well as the financial resources dedicated to the research project. Such challenges may affect the usability and reliability of the output spatial hazard maps and hinder the ability to compare the results coming from different areas of the world. In fact, previous research efforts attempted to assess the effects of DEM resolution on rockfall numerical simulations. Generally, the researchers agreed that, in addition to the physical parameters of the soil layers, DEM resolution is equally significant in determining the output of the said models (e.g., [17–20]). Despite this obvious effect, some authors (e.g., [21–23]) used coarse DEMs to build rockfall hazard assessment models and based their conclusions and assumptions on the said models. Since the latter constituted the basis for all hazard prevention and mitigation attempts, it is important to assess the significance and potential influence of simulation errors on prevention scenarios and strategies. If such investigations show that coarser DEMs are practically unusable, the investment into acquiring more detailed topographic data becomes more justified.

Based on this and given the popularity of numerical modeling techniques, the accuracy and usability of rockfall simulation models may be compromised in various applications. In fact, incorrect models can decrease the value of property in low-hazard areas while limiting the ability of decision makers and authorities to prevent future occurrences in high-hazard locations. To avoid this, validation data needs to be generated through either field investigations and the documentation of previous occurrences if possible, or by conducting laboratory tests that simulate such processes [24–29] in order to assess the accuracy of the simulations. However, most efforts have mostly used laboratory tests and rarely field data. Though the former can be rightly used for such purposes, the latter is more objective and, therefore, excludes any experimental biases. However, it is difficult to map rockfall trajectories on a field, given their spatio-temporal unpredictability.

Therefore, a well-documented rockfall occurrence in the Bouanane cliff in Northern Morocco, is used in this paper as the validation data to assess the performance of five DEMs used for the rockfall simulations. The event that took place on 1 December 2018, offers the opportunity to evaluate the accuracy of rockfall simulation models that were produced using a 1 m, 2 m, 3 m, 5 m and 30 m DEM through conducting a back-analysis of this event. The results will be analyzed and an attempt to explain the variability of the results will be presented.

2. Study Area
2.1. Geological and Geomorphological Setting

The Bouanane cliff is located immediately to the South of Tetouan (Figure 1a). From a morphological point of view, it is characterized by a subvertical topography with an elevation difference of about 120 m. The geological material outcropping at the site is attributed to the Dorsale Calcaire structural unit, which is mainly formed by thick layers of Triasic and Jurassic carbonate rocks with a fairly developed Tertiary sedimentary cover [30–33]. This structural unit is considered a unique morphostructural domain that is essentially constructed by carbonate rock ridges consequential to the thrust sheet structure of the Rif Cordillera [34].

However, the structure of the Bouanane cliff is not attributed to these N-S oriented thrust faults. It is, in fact, the result of late Miocene to Pliocene extensional deformation, which reactivated the E-W striking, right-lateral, strike-slip Tetouan accident into a normal faulting system (Figure 1b). This system of strike-slip faults, mainly N50° to N90°, is believed to be responsible for the morphogenesis of the Tetouan water-gap [35,36]. The field observations and visual interpretation of the aerial photographs allow for the mapping such structures to the south of the Bouanane site. These aerial photographs show that the Bouanane ridge is formed by thick limestone and dolostone strata that are affected by the E-W strike-slip faults (Figure 1). Using the thick strata as a reference mark, the right-lateral motion of the faults can be clearly distinguished. The Bouanane cliff is also considered to be the result of complex right-lateral normal faulting that cuts the Bouanane ridge short to the north. On the field, one can also observe a dense tectonic joint network striking mainly N130° to N160°, which split the massive carbonate rocks into multiple metric blocs.

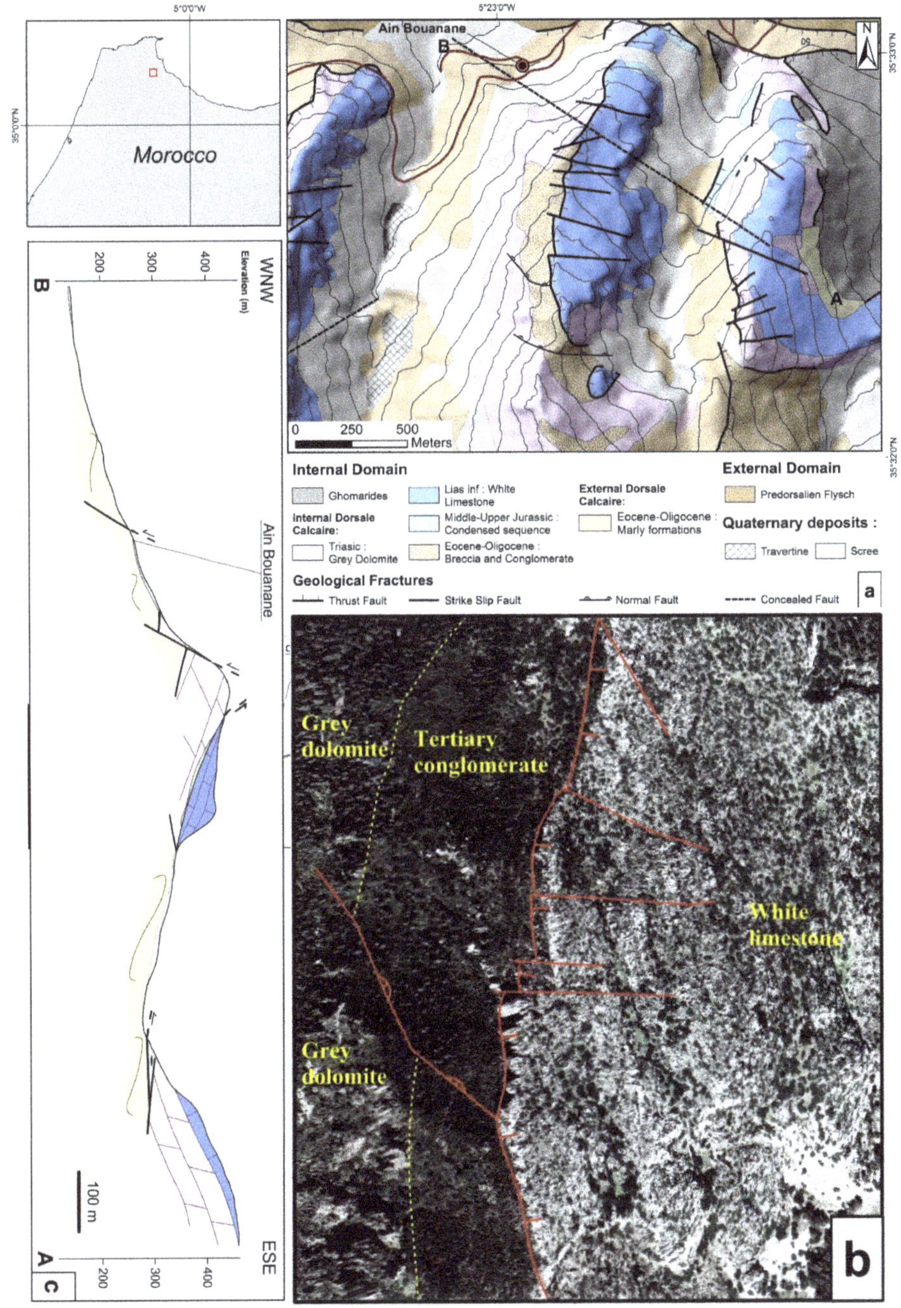

Figure 1. (**a**) Geological map of the study area. (**b**) Aerial photographs showing strike-slip faults affecting the Bouanane massif. (**c**) Geological cross section of the Dorsale Calcaire thrust sheets south of Tetouan.

2.2. Rockfall Occurences in the Bouanane Site

In Northern Morocco, which is part of the Mediterranean peripheral chains, several rockfall and rock avalanche occurrences have had pronounced consequences. The sites subject to such phenomena include the village of Ametrass [37], El Onsar village north of Tetouan [38] and the cliff of Bouanane which is the subject of the present paper. The common characteristics between the above-mentioned cases is the damage caused due to the presence of human dwellings and infrastructures in the piedmont of these cliffs and the lack of protection structures against rockfalls. In Bouanane, three previous events have caused material damage without any casualties. The first, which dates back to 1994, did not cause any damage despite the large size of the detached boulder. The second, unlike the first, damaged two parked vehicles in 2011, despite the relatively smaller size of the boulders involved. The most recent occurrence, which took place on 1 December 2018, caused damage to the decorative fence surrounding a car parking spot and stopped at the doorstep of a coffee shop. The latter event will be investigated in detail. The trajectory and some of the impact points will also be used as validation data to assess the accuracy of the simulation models produced using the three DEMs with different resolutions.

3. Materials and Methods

3.1. Stability of the Bouanane Cliff

Before presenting the simulation models, we will attempt to study the detachment mechanisms and assess the stability of the Bouanane cliff. To do so, the strike and dip of 56 tectonic joints were measured during several field surveys conducted in 2017 and 2018. This helps to identify the major joint directions/families in the area and their geometrical association with the cliff's morphology.

Regarding the stability of the Bouanane cliff, the SMR (slope mass rating) index [39] was calculated using the automated SMRTool beta 1.10 [40]. Although this classification constitutes a good tool for assessing the stability of the rocky slopes, several modifications were developed to further enhance its performance in heterogeneous or anisotropic slopes [41]. Given the important elevation difference of the Bouanane cliff, we opted for the Chinese adaptation CSMR [42], which integrates the elevation difference of the investigated slope into the calculation Formula (1)

$$CSMR = E \cdot RMR + L (F1 \cdot F2 \cdot F3) + F4 \qquad (1)$$

where $E = 0.43 + 0.57$ (80/elevation difference), RMR (rock mass rating) is the Beniawski index [43], F1, F2, F3 and F4 are the correction parameters and L is an index that reflects the state of the geological fractures affecting the cliff [39].

L (between 0.8 and 1) is the index that reflects the state of the geological fractures affecting the cliff. A value of 0.8 is generally assigned where the involved geological fractures are large-scale metric or decametric joints, and a value of 1 in the opposite case. In this study, the former value is adopted. The height of the slope used in this calculation was 65 m, while its dip and dip direction were 50° and N315, respectively. As the general slope was natural, the F4 parameter value used was 15.

3.2. Detachment Mechanisms

To decide which detachment mechanism controls the rockfall dynamics at the study area, dip and dip direction measurements were used to graphically estimate the probability of toppling (T), planar sliding (P) and wedge sliding (W) in Bouanane, using the Goodman method [44]. To estimate the risk of toppling, we counted the number of joint planes poles that were projected inside the zone and delimited by the friction cone (trend = slope dip direction + 90, plunge = 0; and angle = 60) and the sliding limit surface (dip = slope dip-friction angle; dip direction = slope dip direction). Such joints have a high potential for producing rockfalls through toppling. The more poles there are, the greater the risk. To determine which joint planes may cause planar sliding, the density of the poles located outside the friction cone (trend = 0, plunge = 90 and angle = friction angles) and delimited by the daylight envelope of the slope, were visually estimated, which reflected the probability of rockfall occurrence by planar sliding. For the last mechanism, the intersection points between the average planes of each joint family, located within the zone delimited by the slope and the friction cone (trend = 0; plunge = 90; and angle = 90 friction angle), indicated the potentially unstable planes that may produce wedge sliding. The quantification of this risk was also achieved through a graphical estimation of the density of unstable joint families.

3.3. Preparation of Topographic Data

In this study, we will use five DEMs with varying spatial resolution and vertical accuracies. The first was downloaded from the USGS's Earth Explorer platform (https://earthexplorer.usgs.gov/, accessed on 3 December 2022). It is a 1 arc second (approximately 30 meter) resolution DEM produced by the Shuttle Radar Topography Mission (SRTM). This dataset has the lowest resolution. The second DEM (5 m resolution) was generated using elevation contours derived from the 1:25,000 topographic map of the study area. The elevation contours were digitized manually and an interpolation of the data was performed.

Another high-resolution digital surface model (DSM) (11 cm) was generated using a set of UAV-derived aerial photographs. The UAV used was a DJI Phantom 4 drone equipped with a Global Positioning System (GPS) module, with a vertical accuracy of 0.1 m and a horizontal accuracy of 0.3 m. The camera used for capturing the aerial photographs is a 1/2.3″ CMOS camera with a total number of effective pixels of 12.4 M. Further information regarding the hardware specifications and acquisition parameters is presented in Supplementary Table S1.

The processing chain that was followed to obtain this model first consisted of eliminating the bad quality tie points and introducing 310 ground control points (GCPs) (Supplementary Figure S1) to optimise the initial point cloud before generating the 3D dense points cloud and the digital surface model (DSM). The GCPs that were exploited to enhance the model were acquired by the Urban Agency of Tetouan using a dual frequency SpectraSP60 D-GPS in static mode, with a spatial accuracy of +/− 5 cm. The tool used to perform the structure-from-motion (SFM) analyses in this study was the open source MicMac photogrammetry software that is freely available at https://micmac.ensg.eu, accessed on 3 December 2022.

One of the main difficulties that we encountered in this study, was the presence of a dense forest canopy in the surveyed area, which needed to be filtered out in order to avoid creating artificial barriers in the simulation model. To do so, several post-processing steps were performed to correct the altitude values of the forest canopy tie points. The correction methods frequently deployed were based on the use of GCPs that were measured using precise ground positioning techniques. These were subsequently used to correct the elevation values for the areas with dense vegetation [45], a near-infrared based filtration of the points corresponding to the forest canopy [46] and the use of a ground-based scanner (TLS) [47]. LiDAR imagery was an alternative, but its cost was higher and it was consequently not considered in this study for budgetary reasons. In addition, obtaining permits for the use of this technique in our study area was not possible, so we opted for the SFM alternative.

To correct the forest canopy points, we used the first approach where 186 of the 310 GCPs, provided by the Urban Agency of Tetouan (Supplementary Figure S1) and measured using differential GPS, were used to calculate the mean elevation of the pine tree canopy in the study area. This was achieved through subtracting the interpolated elevation surface, calculated using the GCPs, from the canopy elevation raster provided by the UAV-derived model. The average tree height was found to be around 9.23 m with a standard deviation of 2.74 m. This low variance was due to the fact that all the trees at the site were planted in the same year (1969) as part of a reforestation effort by the Moroccan government and, therefore, have a more or less similar size. After the subtraction was done, the surface was smoothed using a majority filtering algorithm that replaced each cell value based on the twelve contiguous neighboring cells (Figure 2).

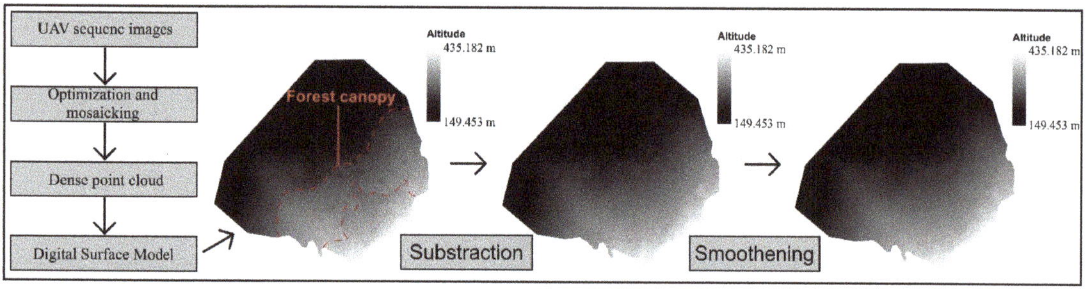

Figure 2. Processing chain of the UAV images.

The resolution of the UAV model was downgraded to 1 m to avoid memory allocation errors in the Rocpro 3D simulation software (Figure 3). Two other models were derived from the UAV DEM with spatial resolutions of 2 m and 3 m. Along with the other two DEMs (Figure 3), the topographic data was prepared for simulation purposes. An additional 102 GCPs (Supplementary Figure S1B) acquired from the same agency were used to assess the vertical accuracy of the used DEMs and compare them to one another. The vertical error distribution presented in Figure 3D shows that the 1 m and 3 m DEMs are very similar in terms of the mean error values, the standard deviation and the overall distribution. Surprisingly, the 2 m DEM was less accurate than the 3 m one. Finally, the 5 m and 30 m DEMs were shown to present the highest mean error values and also presented a more uniform error distribution compared to the UAV-derived DEM.

3.4. D trajectory Simulation

The software deployed in this study (RocPro3D) used physical parameters such as rebound (restitution coefficient and lateral deviation), rolling (friction coefficient, limit velocity) and transition parameters of the soil and boulders in order to estimate the possible propagation trajectories, energy, velocity and impact points. The approximate values of these parameters for bare and densely vegetated dolomite scree deposits and loose soils were taken from the typical values table available in the Help section of the website (www.rocscience.com, accessed on 3 December 2022), which is based on previous experimental research [21,23]. This research estimated the parameter range for soils that are similar to the ones present at Bouanane. Although the individual trees were not integrated into the 3D mesh, their effects on the soil parameters were also included in these estimation efforts. The soils covered by the forest canopy in our study area were considered densely vegetated soils and, therefore, their corresponding parameters were adopted. With respect to the trajectory estimation, we chose the rigid body approach of the RocPro3D simulator for a maximum output of 50 simulated trajectories. The boulder diameter used was that of the biggest dolomite boulder that fell in 2018. This was done in order to validate our simulation results using the field data, which cannot be achieved if the boulder sizes are different.

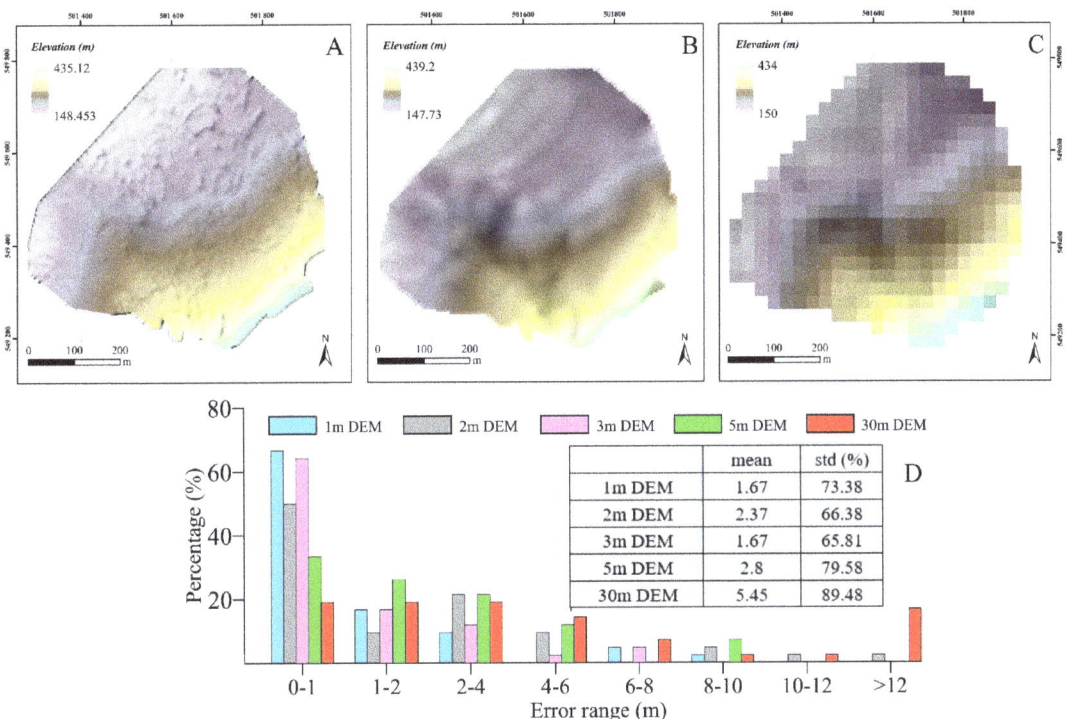

Figure 3. DEMs used for simulating the 2018 event. (**A**) UAV-derived DEM. (**B**) 5 m DEM. (**C**) SRTM 30 m DEM. (**D**) Vertical error distribution for the three DEMs used in this study.

3.5. Statistical Analyses

While visual interpretation of the results allowed for a comparison the three models used in this study, numerically deploying statistical techniques quantified the difference, which provided more solid evidence regarding the degree of significance of our findings. Therefore, we first attempted to assess the horizontal error distribution through a histogram plot of the distance to real trajectory of each of the five simulated models. To do so, the polyline shapefiles representing the simulated trajectories were first rasterized using a GIS tool. Then, the latter were used to compute the Euclidean distance separating the centre of each pixel from the field trajectories. Generally speaking, an error range of 0 to 2 m is considered good since all of the potential rockfall targets at Bouanane are objects wider than 4 m (e.g., buildings, cars, café terraces, etc.). Therefore, the probability that the simulated boulder missed its real target is low for such an error range.

Additionally, we prepared a box plot of the runout simulation results in order to graphically represent the simulated samples and compare them to the field reference values. Subsequently, the Kruskal–Wallis non-parametric test [48] was performed to compare the three produced simulations, since the data were clearly non-normally distributed [48]. The input variables introduced to the Kruskal–Wallis algorithm were the simulated velocity, energy and bounce height, obtained from our 3D simulation effort. The null hypothesis for such a test was that all five models would be equal and that the observed difference would not be statistically significant. Dunn's (PostHoc) test [49] was also performed to point out which of the models were similar and which were different based on a pairwise comparison approach.

3.6. Morphometry of Scree Pebbles and Boulders

To study the morphometry of the piedmont deposits at Bouanane, the length (a), width (b) and thickness (c) of 1749 pebbles were measured at 11 sampling stations during several field investigations (Supplementary Tables S2 and S3 and Figure S2). The same measurements were taken for 55 boulders, the locations of which were determined using a GPS tool (Supplementary Tables S4 and S5). For the boulders, the measured values were projected onto the Sneed and Folk ternary diagram [50] using the Hockey coordinate system [51]. The latter was adopted for its simplicity. The computer tool used in this study was "Tri-plot" (https://www.lboro.ac.uk/microsites/research/phys-geog/tri-plot/index.html, accessed on 3 December 2022). The results should allow for the studying of the morphology of the pebbles and boulders and consequently the characterization of their size and shape distributions. For the scree pebbles, the value of the 'b' axis was used to generate a cumulative frequency plot in order to determine the statistical mode of the samples. This allowed for the determination of the physical soil parameters of the scree deposits layers using the above-mentioned reference values table.

4. Results

4.1. Stability of the Bouanane Cliff and Detachment Mechanisms

The geometric attributes and geomechanical measurements of the seven tectonic joint families identified in the study area were used in CSMR calculations. Our results yielded a value of around 70 in the study area (Table 1), which means that the Bouanane cliff can be considered as a class II slope according to the Romana classification [39]. In terms of the stability, this class was deemed stable with few rockfall occurrences. Their temporal probability should not exceed 0.2 according to the same classification, which was in accordance with the testimonies and accounts of the local people.

Table 1. CSMR analysis results for the Bouanane cliff.

Joints Family *	RMR	α (j)	β (j)	α (s)	β (s)	H (m)	L	E	F1	F2	F3	F1*F2*F3	CSMR
F1	40.12	225	55	315	50	65	0.85	2.23	0.15	1	0.65	0	89.5
F2	27.32	270	66	315	50	65	1	2.23	0.22	0.98	1.19	−0.25	61
F2'	27.32	90	66	315	50	65	1	2.23	0.22	1	−2.15	−0.47	60
F3	34.6	315	54	315	50	65	1	2.23	1	0.96	−4.68	−4	73
F4	39.4	180	65	315	50	65	0.9	2.23	1	0.96	−1.76	−0.39	88
F5'	27.1	135	80	315	50	65	1	2.23	1	1	−25	−25	35
F5	27.1	315	80	315	50	65	1	2.23	1	0.99	−0.64	−0.63	60
F6	39.33	225	85	315	50	65	0.8	2.23	0.15	1	−25.31	−4	85
F7	39.3	0	63	315	50	65	1	2.23	0.22	0.97	−1.47	−0.31	87
												Average =	70

* RMR (rock mass rating). α (j) joint dip direction. β (j) joint dip. α (s) slope dip direction. β (sj) slope dip. H (m) height of the slope. L (index that reflects the state of the geological fractures affecting the cliff). E = 0.43 + 0.57 (80/elevation difference). F1, F2 and F3 are the correction parameters of Romana.

According to our field surveys and the UAV high-resolution images, the main joints family responsible for the segmentation of the cliff into smaller metric blocks is the conjugated WNW-ESE to NW-SE tectonic joints (Figure 4). Together, they form X-shaped lines that intersect the cliff at a 60° to 70° angles.

Figure 4. Major joint directions obtained from the field investigations and UAV images.

Regarding the boulder detachment mechanisms, the Goodman analysis [44] performed in the Bouanane cliff (Figure 5) showed that wedge sliding processes had a high occurrence probability. This was explained by the presence of the above-described dense joint network, the orientation of which was quasi-orthogonal to the main scarp direction. As for the other possible detachment styles, the probability of planar sliding and toppling were found to be low (<5%) due to the low frequency of NE-SW oriented tectonic joints (Figures 4 and 5).

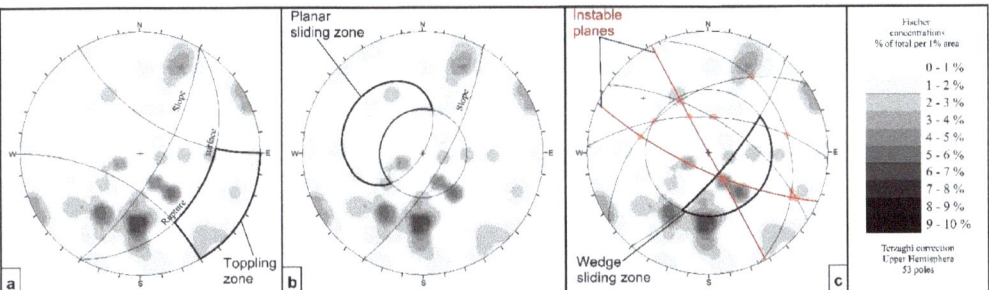

Figure 5. Result of the Goodman analysis conducted on the Bouanane cliff. (**a**) Toppling analysis, (**b**) Planar slide analysis. (**c**) Wedge slide analysis.

4.2. Investigation of the December 2018 Event

Our field surveys that were conducted immediately after the rockfall event of 1 December 2018, allowed us to draw the trajectory of three detached boulders (Figure 6a). At the source area, fresh impact holes were identified (Figure 6b). These holes were not very deep but were large enough to suggest a significant impact force that exposed the dark quaternary soil layers. Immediately below the source area, fresh mechanical injuries were seen on some trees that did not show any signs of impact prior to the event (Figure 6c,d). The fresh yellowish color of the impacts on the Pinus halepenesis trunks also indicated the freshness of the injuries, since old impacts on these pine trees tend to darken very quickly due to the hardening of the resin secretions [52]. At the bottom of the steep slope, the three detached boulders landed separately. The two that reached the inhabited zone were easy to locate since they left behind clear impacts on manmade structures (Figure 6e,f). However, the third boulder was more difficult to find since it was stopped by an older and bigger buried boulder (Figure 6g). The latter boulder was distinguished from the older boulders by the presence of reddish wood and resin stains on its surface. The size of all three boulders was measured on the field. Their diameters were found to be 1.2 m, 1.8 m and 1.5 m, respectively, indicating a more or less similar size.

Figure 6. (**a**) Trajectories of the 2018 event projected over the UAV-derived orthoimage. (**b**) Boulder impact near the source area. (**c**) and (**d**) Mechanical injuries on the tree stems. (**e**) and (**f**) impact observed on manmade structures. (**g**) photograph showing a recently detached boulder with fresh resin and live wood stains on its surface and its supposed impact on a nearby tree.

4.3. Significance of the 2018 Event

The diameter-frequency distribution (Figure 7a) for the 55 boulders measured below the source area, showed that 0.5-to-2-meter large boulders constituted the dominant category in the study area (Figure 7a). Given that 26 of the 55 boulders belonged to the 1 to 1.5 m category (Figure 7a), the rockfall event of December 2018 can be considered typical of the Bouanane site. In fact, with a boulder diameter ranging from 1.2 to 1.8 m, the detached boulders were "coarse" according to the [53] classification, with this category being the most dominant in the study area.

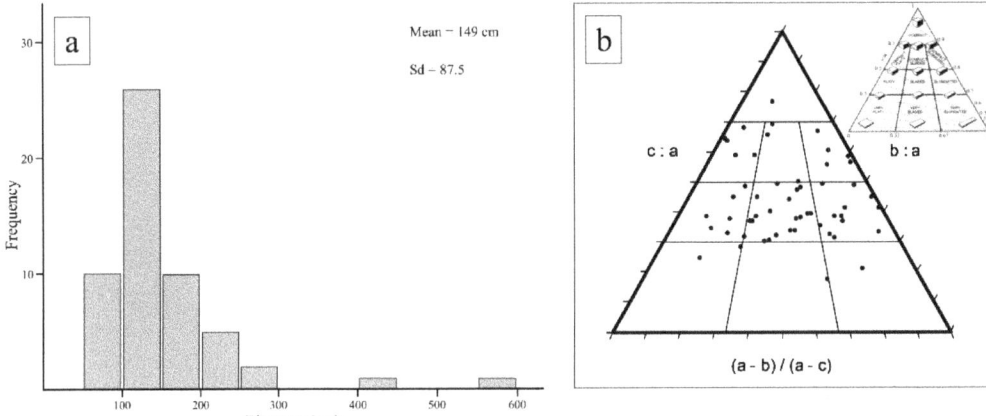

Figure 7. (a) Size-frequency distribution of boulders in the Bouanane site. (b) Boulders measured in the Bouanane site projected onto the Sneed and Folks (1958) diagram. The used classification for the boulders investigated in the study area is shown in Table S6 of the Supplementary Materials.

4.4. Rockfall Trajectory Simulation and Back-Analysis

The trajectory simulations produced using the five above-mentioned DEMs are shown in Figure 8. The first model, which used the 1 m resolution DEM as the topography input, produced complex trajectories that corresponded more or less to the field observations (Figure 8a). Of the different calculated trajectories, a significant portion did not reach manmade structures, while the rest either stopped at the coffee shop impacted by the 2018 event or continued further to impact other buildings downhill. As can be seen in Figure 8, the runout distances rarely exceed that of the longest documented path (Figure 8f), suggesting a reasonably accurate depiction of the real event. The 2 m (Figure 8b) and 3 m (Figure 8c) DEM simulations were more or less similar trajectories with slightly wider invasion zones and longer runout distances. These differences, although small, produced a significant runout calculation error according to Figure 8f, since the runout value range did not intersect that of the real trajectories.

The fourth and fifth models, which used the 5 m and 30 m DEMs as the topographic data, respectively, produce smoother and longer trajectories that did not follow the paths of the 2018 rockfall occurrence (Figure 8d,e). For the mid-resolution model, most if not all calculated runouts significantly exceeded the observed values with no boulders stopping before they reached the inhabited area (Figure 8f). As for the low-resolution simulation, the runout values were even bigger, with all the boulders reaching areas that presented no signs of any recent occurrences (Figure 8e,f). The simulated trajectories of the fifth model also missed the coffee shop that was affected by the recorded 2018 rockfall, yielding a deformed representation of the real event. The consequences of such errors are discussed in Section 5.

In terms of the horizontal error distribution, the distance to real trajectory histogram presented in Figure 9 confirmed the visual interpretation results. According to this histogram, the percentage of the 0 to 2 m error range (good error range) negatively correlated with the DEM resolution, which means that the lower the resolution, the greater the overlap between the simulated and real invasion zones. Conversely, the >25 m error range was dominated by the 30 m model (Figure 9), which also correlated well with our visual interpretation of the results. Consequently, the high spatial errors were reverse proportional to the resolution of the data where coarser models produced significantly worse results relative to the real trajectories.

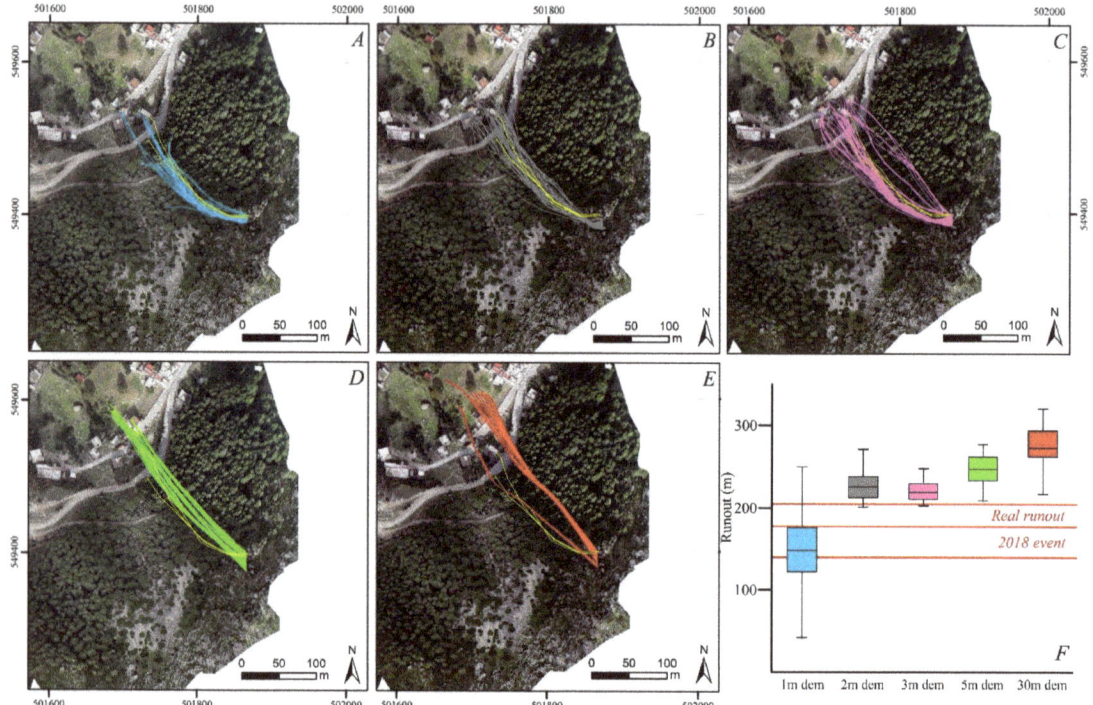

Figure 8. Simulated rockfall trajectories using the 1 m UAV-derived DEM (**A**), the 2 m DEM (**B**), the 3 m DEM (**C**), the 5 m DEM (**D**) and the 30 m DEM (**E**). (**F**) Box plot of the runout distances obtained from the simulation models compared to the observed runouts of the 2018 event.

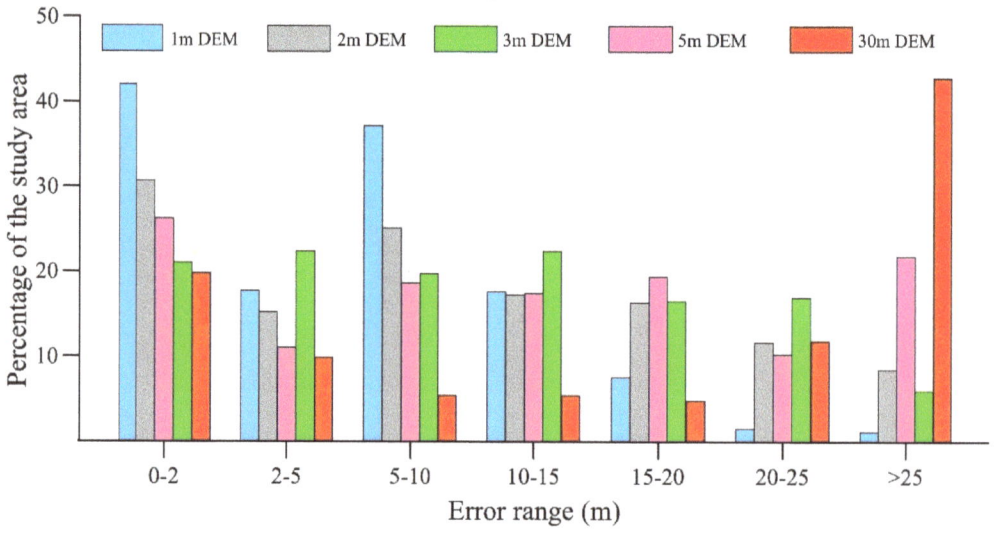

Figure 9. Statistical distribution of the horizontal error for all five simulated models.

4.5. Energy and Velocity Simulation

The simulated velocity values showed that the 3 m resolution DEM produced the widest range of values. It also yielded the highest velocity with a maximum of 33.96 m/s compared to a maximum of only 29.28 m/s, 28.17 m/s, 26.69 m/s and 21.56 m/s for the other models (Figure 10). However, when the simulated boulders reached the manmade structures, the velocity for the first and third models (Figure 10a,c) was low in comparison to the rest, which presented higher velocities reaching up to 20 m/s. By comparing these values to the impacts observed on the field, the latter appeared to correspond more to the high-resolution simulation since the damage was mainly aesthetic and indicated low-velocity impacts. As for the 2 m and the medium and low-resolution models, they were found to exaggerate the true velocity downhill. Similar remarks can be given regarding the spatial distribution of the energy values (Figure 11).

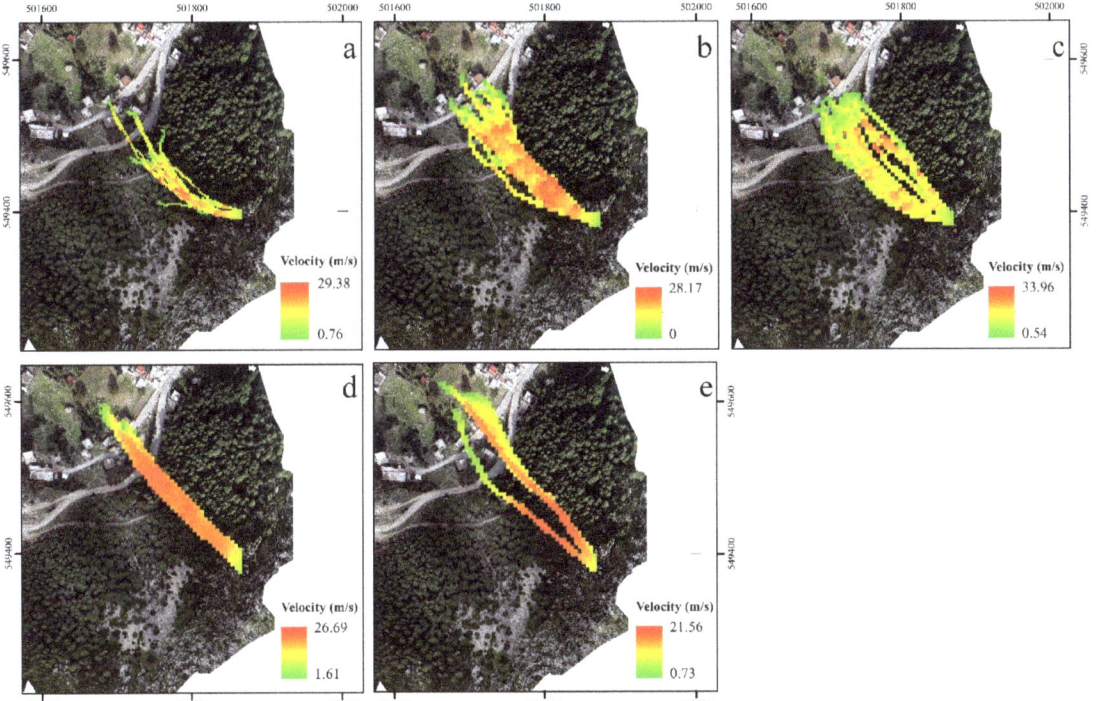

Figure 10. Simulated rockfall velocity simulation in the Bouanane cliff obtained using the 1 m DEM (**a**), the 2 m DEM (**b**), the 3 m DEM (**c**), the 5 m DEM (**d**) and the 30 m DEM (**e**).

In terms of the percentage, the spatial distribution of both the velocity and energy values differs, where low to medium values tend to cover the most area in the 1 m and 3 m models, while high values dominate in the other models, especially the 5 m model. The probable reasons for this are explained in Section 5.

This spatial distribution difference was well attested by the Kruskal–Wallis test results (Table 2) that showed a low P-value output (<0.0001). Given a threshold of 0.05, these findings proved that the observed spatial distribution difference was statistically significant. The pairwise Dunn test also yielded a similar output for the velocity and energy, with P-values largely inferior to 0.05 (<0.001) (Table 3), with the exception of the 2 m and 3 m DEMs, which appeared to have similar velocity distributions.

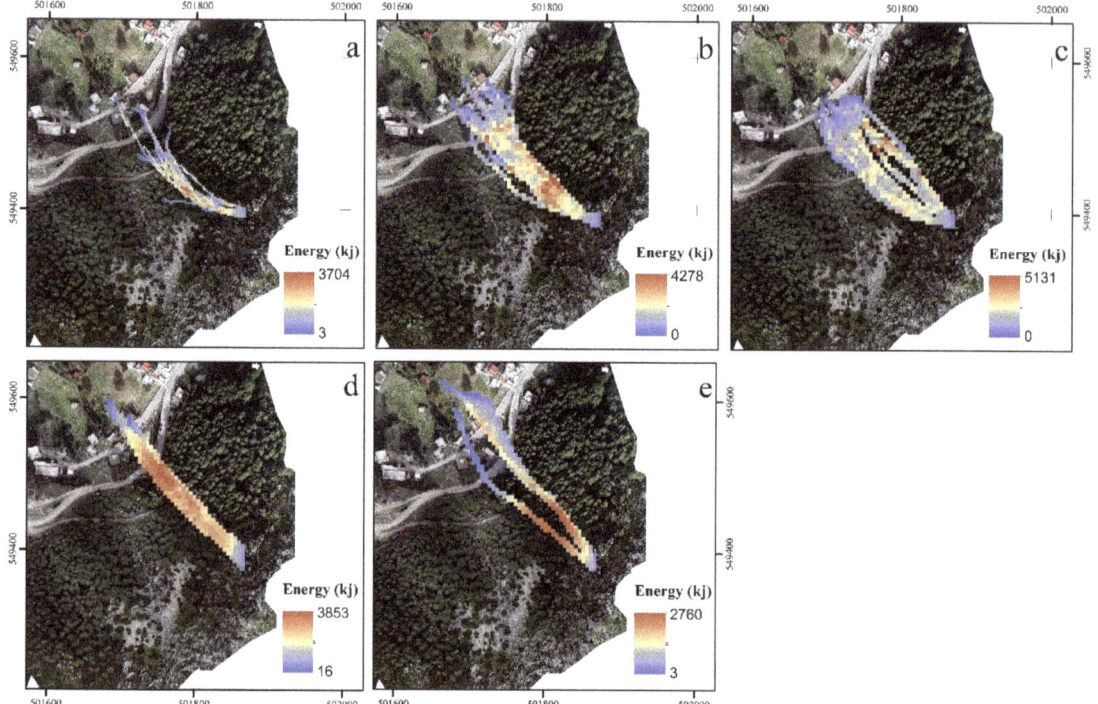

Figure 11. Simulated rockfall energy simulation in the Bouanane cliff obtained using the 1 m DEM (**a**), the 2 m DEM (**b**), the 3 m DEM (**c**), the 5 m DEM (**d**) and the 30 m DEM (**e**).

Table 2. Kruskal–Wallis test results.

Dependent Variable	Kruskal–Wallis Test	
	H	p-Value
Velocity	470.035	<0.001
Energy	1387.988	<0.001
Bouncing height	498.55	<0.001

Table 3. Dunn's (Post Hoc) test results.

Dependent Variable	DEMs	p-Value	Dependent Variable	DEMs	p-Value
Velocity	1–2 m	<0.001	Bouncing height	1–2 m	<0.001
	1–3 m	<0.001		1–3 m	<0.001
	1–5 m	<0.001		1–5 m	<0.001
	1–30 m	0.002		1–30 m	<0.001
	2–30 m	<0.001		2–3 m	<0.001

Table 3. Cont.

Dependent Variable	DEMs	p-Value	Dependent Variable	DEMs	p-Value
	2–3 m	0.237		2–5 m	<0.001
	2–5 m	<0.001		2–30 m	<0.001
	3–5 m	<0.001		3–5 m	<0.001
	3–30 m	<0.001		3–30 m	<0.001
	5–30 m	<0.001		5–30 m	<0.001
	1–2 m	<0.001			
	1–3 m	<0.001			
	1–5 m	<0.001			
	1–30 m	<0.001			
Energy	2–3 m	<0.001			
	2–5 m	<0.001			
	2–30 m	<0.001			
	3–5 m	<0.001			
	3–30 m	<0.001			
	5–30 m	<0.001			

5. Discussion

5.1. How the DSM Resolution Impacts the Rockfall Numerical Simulations

In the rockfall simulations, the researchers sought to model the geometry of the possible boulder trajectories, which allowed for distinguishing between hazardous and safe areas, the runout values that determine how far a detached boulder can travel and the velocity and energy of boulders once they descend the slope. The DEM resolution influenced all the above-mentioned aspects of the rockfall research. In fact, [18] declared that the use of coarser topographic grids translated into a decreased variability in the computed trajectories, higher mean velocity values and lower bounce heights due to the smoother geometry of the DEM. Similar remarks were reported in a more recent study by [19] who showed that higher resolution input data exhibited more complex trajectory shapes that agreed more with historical inventories. In our case study, the UAV-derived 1 m DEM yielded more complex geometries that agreed well with our field observations (Figure 11a). The invasion zone produced by this high-resolution model was also smaller but had more variability in terms of the shape and length of the simulated trajectories, which was similar to the previous findings.

In the previous research, the runout extent was also shown to change significantly in some cases [19,20] and stayed the same in others [18]. This effect can be explained by the roughness of the terrain which was demonstrated to be proportional to the runout variability using anecdotal evidence [20]. In this work, the mean runout for the high-resolution model was lower than those obtained using coarser DEMs. In fact, it appeared that the lower the resolution of the topographic data, the longer a boulder will travel downhill. Such results were due (in our study as well) to the rugged and bumpy topography of the site, which translated into large differences between the DEMs used in this study. However, the effects of the slope roughness on the travel mode of rockfalls depends on the slope ratio as well. [54] showed that while rough surfaces tend to promote higher velocity rockfalls in steep slopes, their influence is reversed in gentler slopes due to a loss of energy on impact (Figure 10). This explains why the highest velocity calculated in this case study was given by the higher resolution DEMs near the source area, despite them presenting the shortest runouts. Consequently, the appropriateness of a DEM for use in a simulation can be summarized in its ability to depict the micro-topographic geometries, such as micro terraces, small holes and micro-ridges. Such small but significant details regarding the topography of a given slope can change the outcome of a rockfall simulation depending on the steepness of the hillslopes and its real shape. In Bouanane, the topography was dominated by small convex uphill surfaces, which explained why the simulated velocity in this segment of the slope was higher in the high-resolution model compared to the medium and low-resolution models (Figure 12a). Larger concave downhill surfaces were accurately represented by the 1 m DEM but appeared smoother in the coarser DEM, resulting in shorter runouts in the former and longer trajectories in the latter (Figure 12b,c).

Figure 12. Theoretical 2D profiles of a rockfall event, drawn on a topographic profile extracted from the 1 m resolution DEM (**a**), the 5 m resolution DEM (**b**) and the 30 m resolution DEM (**c**).

In addition to the roughness, the elevation of the unstable cliff/slope determined how big the effect of the DEM resolution was on the simulation results, especially the runout distance. The experimental results published by [18] showed that the runout did not change significantly when using different DEM resolutions. However, in cases where the source area belongs to the domain of middle slopes and ridges, such as the case of the Bouanane cliff, the resolution and detailed depiction of the topography determined mostly where the boulders stop. This was because the valley floor was a flat surface, the shape of which does not vary from one DEM to the other. Therefore, the falling boulder stopped at approximately the same area due to an absence of a horizontal component of acceleration. In addition, the moving object lost energy quickly regardless of the initial impact force. Nevertheless, when the topography was inclined, the stopping position was more sensitive to the acceleration component and friction forces that mainly depended on the shape of the impact surface. As such, the existence of micro topography features became more relevant to the results. For instance, the 1 m, 2 m and 3 m DEMs adopted in this study produced significantly different invasion zones and runout distributions despite being derived from the same techniques. The reason for this was the loss of the microtopographic details mentioned above.

5.2. Impact of the Simulation Results on the Hazard Assessment and Prevention Efforts

When an area is revealed to be hazardous, various solutions are implemented to reduce the vulnerability of constructions to rockfalls. Such protection countermeasures are either natural (e.g., tree barrier effect, etc.), quasi-natural (e.g., building embankments and ditches), structural (e.g., monolithic rockfall protection galleries) or flexible (e.g., fences and other flexible barriers). The proper design and dimensions of such protection structures varies depending on the geometry of the slope, the simulated energy of the boulders, the impact load and the bouncing height in the runout area [55]. Since such values are generally obtained from numerical models, the above-presented effects of the topography on the simulation results plays a major role in rockfall prevention efforts. For instance, the construction of rockfall protection fences in Bouanane (which are deemed the most appropriate technique due to the geometry of the slope) required information regarding the runout, bouncing height and the most probable targets for potential future events. In the coarse resolution simulations, the output was shown to differ significantly from the real trajectory drawn on the field. For the 5 m DEM model, the inaccuracies were mainly due to the shape of the modeled trajectories, which deviated significantly from field data. In addition, the velocity and energy values (Figures 10 and 11) were exaggerated near the real target area compared to the slight impacts observed (Figure 6). As for the 30 m model, it was shown to miss the most probable trajectory, stopping far away from the observed rockfall path (Figure 6). Therefore, if such models were to be used, they would induce significant positioning errors. Conversely, the 1 m model presented a reasonably accurate invasion zone that corresponded to the reality of the field, which could be used for hazard zonation mapping and rockfall prevention efforts.

Regarding the bouncing height distribution for the modeled trajectories, the 1 m and 2 m models appeared to present the highest uphill values (Figure 13a,b) compared to the coarse models where the boulders kept rolling downhill (Figure 13d,e). Although the detached boulders of the 2018 event did not leave behind evidence of high bouncing, old broken tree stems were hit at a ground elevation of 2 to 3 m (Figure 13g). The area where these trees were found corresponded to a clear land in the middle of the forest (Figure 13f), where the modeled trajectories revealed the highest bouncing heights. The implications for such variation are significant, given that a future protection fence should absorb the energy of the falling rock and, therefore, needs to be as high as or higher than the bouncing boulder itself. As such, the underestimation of the bouncing by the medium and coarse resolution models could hinder the effectiveness of the rockfall protection structures if the engineers were to rely solely on such data.

5.3. How the DEM Resolution Improve the Results

Although it is obvious that higher resolution data improves the simulation results, the question of a threshold resolution for the rockfall simulations remains valid. In this study, we tested five DEMs with a resolution range of 1 to 30 metres, with some being practically unusable for achieving accurate rockfall modeling. While the improvement is very significant going from the lowest to the highest resolutions, the difference between the 1 m and 2 m DEMs is less obvious, especially in terms of the trajectory shape and invasion zone geometry. Therefore, we wonder if exceeding the 1 m resolution limit would have produced any significant improvement in our case study. Some results presented by [20], who worked on two different sites, clearly showed that the geomorphological setting of the study area determined how the resolution will influence the output. Therefore, such investigations need to be conducted in a variety of locations around the world to determine a "universal" DEM threshold for rockfall hazard modeling under different geomorphological circumstances. Failling to do so may lead to large errors in many models that still use coarse DEMs. This matter will be more significant in developing countries where high-resolution topographic data is scarce and the investment in data acquisition technology is low.

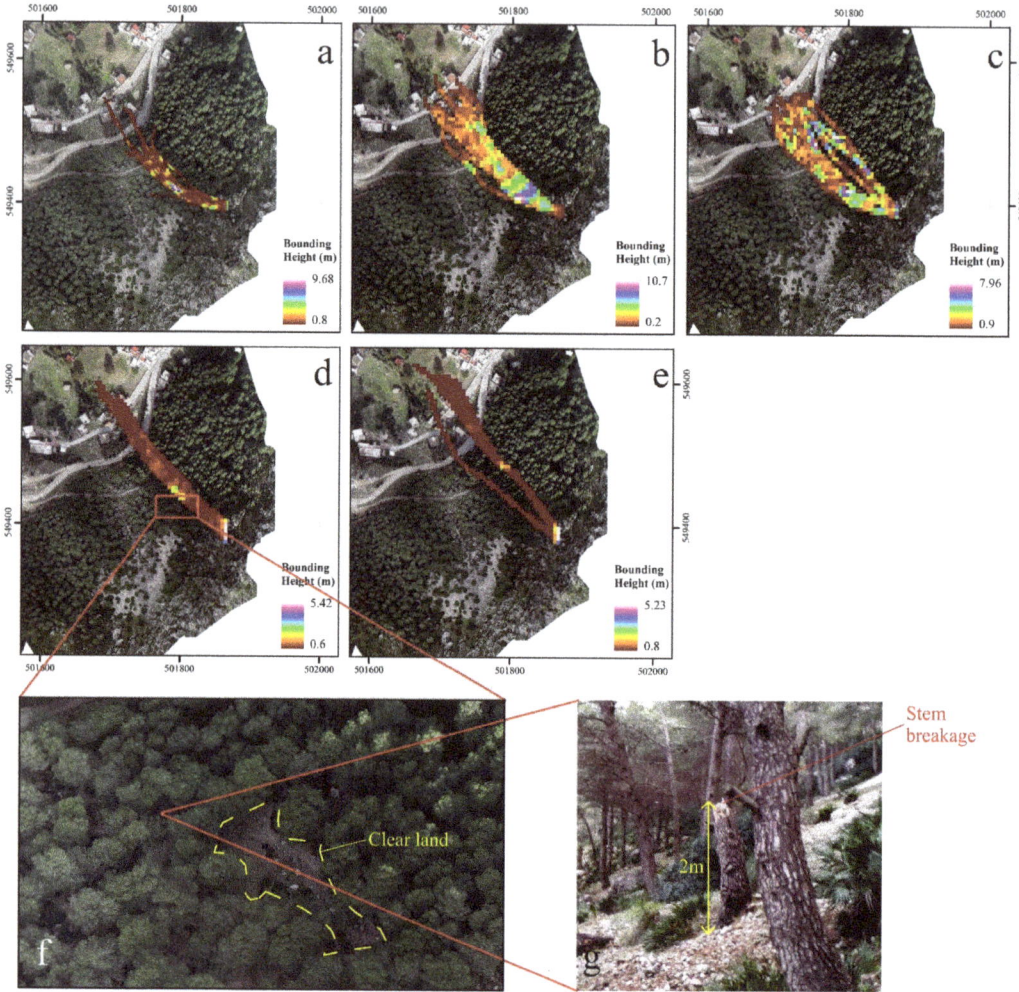

Figure 13. Bouncing height simulations produced using the 1 m DEM (**a**), the 5 m DEM (**b**) and the 30 m DEM (**c**). (**d**) UAV-derived orthoimage showing clear land (**f**) caused by repetitive rockfall events. (**e**) stem breakage (**g**) caused by an old rockfall occurrence.

6. Conclusions

Based on our findings, we believe that low-resolution DEMs are unsuitable for site specific rockfall simulations. Moreover, medium and coarse-resolution topographic data also induce large positioning and geometric errors that degrade the quality of a simulation and hinder its usability for prevention and protection efforts. The results of this paper suggest that high-resolution DEMs are required to produce reasonably accurate simulation models, with trajectory shapes, invasion zones and runout areas that agree well with field observations. This is especially true for areas belonging to the domain of middle slopes and ridges, such as the case of the Bouanane cliff, where the microtopographic features determine the simulation output. Although such data are not available worldwide, the use of new remote-sensing technology such UAVs can solve this problem and provide the accurate data needed for such studies. This technology is low cost, which should encourage its use especially when we consider the significant difference it has with respect to the accuracy of the output. However, fieldwork must always constitute the reference data, which allows for the calibration and validation of the simulation no matter the accuracy of the input.

Supplementary Materials: The following supporting information can be downloaded at: https://www.mdpi.com/article/10.3390/rs14246205/s1, Figure S1: Spatial distribution of GCPs used for constructing the UAV-derived DEM (A) and for testing the accuracy of all DEMs used in this study (B). Figure S2: Photographs taken during field missions; A- the measured rock clast axes (after [55]). B- and C- are photographed taken in measurement stations 1 and 8; Table S1: Hardware specifications and UAV acquisition parameters.; Table S2: b-axis measurements (cm) for pebbles forming the scree deposits downhill of the Bouanane cliff.; Table S3: WGS84 coordinates for pebble measurement stations; Table S4: EPSG:26191 coordinates of the boulders subject to morphometric analyses; Table S5: Morphometric measurements of boulders in the Bouanane site; Table S6: Sneed and Folks classification for the boulders investigated in the study area. Ref. [56] is cited in the Supplementary Materials.

Author Contributions: Conceptualization, A.B., Y.E.K. and R.E.H.; data curation, A.B.; formal analysis, A.B.; funding acquisition, Y.E.K. and R.E.H.; investigation, A.B., Y.E.K. and R.E.H.; methodology, Y.E.K. and R.E.H.; project administration, Y.E.K.; resources, Y.E.K. and R.E.H.; software, A.B.; supervision, Y.E.K. and R.E.H.; validation, A.B., Y.E.K. and R.E.H.; visualization, A.B.; writing—original draft, A.B.; writing—review and editing, Y.E.K. and R.E.H. All authors have read and agreed to the published version of the manuscript.

Funding: This research was funded by CNRST, the "Centre National de Recherche Scientifique et Technique" of Morocco, as part of the PPR2/205/65 project.

Data Availability Statement: Not applicable. All data generated or analyzed during this study are included in this article and its Supplementary Materials.

Acknowledgments: The authors of this paper wish to express their sincere appreciation for the financial support received from CNRST. We also thank Tetouan Urban Agency for providing the ground control points that we used to correct the forest canopy points in the UAV derived DSM obtained in the study area. We would also like to thank Jean-Dominique Barnichon for sharing the RocPro 3D license with us free of charge.

Conflicts of Interest: The authors declare no conflict of interest. The funders had no role in the design of the study; in the collection, analyses, or interpretation of the data; in the writing of the manuscript; or in the decision to publish the results.

References

1. Cruden, D.M.; Varnes, D.J. Landslide Types and Processes, Special Report, Transportation Research Board, National Academy of Sciences. *U. S. Geol. Surv.* **1996**, *247*, 36–75.
2. Guzzetti, F.; Reichenbach, P.; Ghigi, S. Rockfall Hazard and Risk Assessment Along a Transportation Corridor in the Nera Valley, Central Italy. *Environ. Manag.* **2004**, *34*, 191–208. [CrossRef]
3. Schweigl, J.; Ferretti, C.; Nössing, L. Geotechnical Characterization and Rockfall Simulation of a Slope: A Practical Case Study from South Tyrol (Italy). *Eng. Geol.* **2003**, *67*, 281–296. [CrossRef]
4. Gunzburger, Y.; Merrien-Soukatchoff, V.; Guglielmi, Y. Influence of Daily Surface Temperature Fluctuations on Rock Slope Stability: Case Study of the Rochers de Valabres Slope (France). *Int. J. Rock Mech. Min. Sci.* **2005**, *42*, 331–349. [CrossRef]
5. Abellán, A.; Vilaplana, J.M.; Calvet, J.; García-Sellés, D.; Asensio, E. Rockfall Monitoring by Terrestrial Laser Scanning-Case Study of the Basaltic Rock Face at Castellfollit de La Roca (Catalonia, Spain). *Nat. Hazards Earth Syst. Sci.* **2011**, *11*, 829–841. [CrossRef]
6. Sarro, R.; Mateos, R.M.; García-Moreno, I.; Herrera, G.; Reichenbach, P.; Laín, L.; Paredes, C. The Son Poc Rockfall (Mallorca, Spain) on the 6th of March 2013: 3D Simulation. *Landslides* **2014**, *11*, 493–503. [CrossRef]
7. Mateos, R.M.; Azañón, J.M.; Roldán, F.J.; Notti, D.; Pérez-Peña, V.; Galve, J.P.; Pérez-García, J.L.; Colomo, C.M.; Gómez-López, J.M.; Montserrat, O.; et al. The Combined Use of PSInSAR and UAV Photogrammetry Techniques for the Analysis of the Kinematics of a Coastal Landslide Affecting an Urban Area (SE Spain). *Landslides* **2017**, *14*, 743–754. [CrossRef]
8. Dellero, H.; El Kharim, Y. Rockfall Hazard in an Old Abandoned Aggregate Quarry in the City of Tetouan, Morocco. *Int. J. Geosci.* **2013**, *4*, 1228–1232. [CrossRef]
9. Asteriou, P.; Tsiambaos, G. Empirical Model for Predicting Rockfall Trajectory Direction. *Rock Mech. Rock Eng.* **2016**, *49*, 927–941. [CrossRef]
10. Chen, G.; Zheng, L.; Zhang, Y.; Wu, J. Numerical Simulation in Rockfall Analysis: A Close Comparison of 2-D and 3-D DDA. *Rock Mech. Rock Eng.* **2013**, *46*, 527–541. [CrossRef]

11. Glover, J.; Schweizer, A.; Christen, M.; Gerber, W.; Leine, R.; Bartelt, P. Numerical Investigation of the Influence of Rock Shape on Rockfall Trajectory. In Proceedings of the EGU General Assembly Conference Abstracts, Vienna, Austria, 22–27 April 2012; p. 11022.
12. Tarquini, S.; Vinci, S.; Favalli, M.; Doumaz, F.; Fornaciai, A.; Nannipieri, L. Release of a 10-m-Resolution DEM for the Italian Territory: Comparison with Global-Coverage DEMs and Anaglyph-Mode Exploration via the Web. *Comput. Geosci.* **2012**, *38*, 168–170. [CrossRef]
13. Cao, L.; Liu, H.; Fu, X.; Zhang, Z.; Shen, X.; Ruan, H. Comparison of UAV LiDAR and Digital Aerial Photogrammetry Point Clouds for Estimating Forest Structural Attributes in Subtropical Planted Forests. *Forests* **2019**, *10*, 145. [CrossRef]
14. Goodbody, T.R.H.; Coops, N.C.; White, J.C. Digital Aerial Photogrammetry for Updating Area-Based Forest Inventories: A Review of Opportunities, Challenges, and Future Directions. *Curr. For. Rep.* **2019**, *5*, 55–75. [CrossRef]
15. Jiang, S.; Jiang, C.; Jiang, W. Efficient Structure from Motion for Large-Scale UAV Images: A Review and a Comparison of SfM Tools. *ISPRS J. Photogramm. Remote Sens.* **2020**, *167*, 230–251. [CrossRef]
16. Westoby, M.J.; Brasington, J.; Glasser, N.F.; Hambrey, M.J.; Reynolds, J.M. 'Structure-from-Motion'Photogrammetry: A Low-Cost, Effective Tool for Geoscience Applications. *Geomorphology* **2012**, *179*, 300–314. [CrossRef]
17. Žabota, B.; Repe, B.; Kobal, M. Influence of Digital Elevation Model Resolution on Rockfall Modelling. *Geomorphology* **2019**, *328*, 183–195. [CrossRef]
18. Agliardi, F.; Crosta, G.B. High Resolution Three-Dimensional Numerical Modelling of Rockfalls. *Int. J. Rock Mech. Min. Sci.* **2003**, *40*, 455–471. [CrossRef]
19. Lan, H.; Martin, C.D.; Zhou, C.; Lim, C.H. Rockfall Hazard Analysis Using LiDAR and Spatial Modeling. *Geomorphology* **2010**, *118*, 213–223. [CrossRef]
20. Bühler, Y.; Christen, M.; Glover, J.; Christen, M.; Bartelt, P. Significance of Digital Elevation Model Resolution for Numerical Rockfall Simulations. In Proceedings of the 3rd International Symposium Rock Slope Stability C2ROP RSS 2016, Lyon, France, 15–17 November 2016; pp. 15–17.
21. PFEIFFER, T.J.; BOWEN, T.D. Computer Simulation of Rockfalls. *Bull. Assoc. Eng. Geol.* **1989**, *26*, 135–146. [CrossRef]
22. Fisher, R.A. The Correlation between Relatives on the Supposition of Mendelian Inheritance. *Earth Environ. Sci. Trans. R. Soc. Edinburgh* **1919**, *52*, 399–433. [CrossRef]
23. Rammer, W.; Brauner, M.; Dorren, L.K.A.; Berger, F.; Lexer, M.J. Evaluation of a 3-D Rockfall Module within a Forest Patch Model. *Nat. Hazards Earth Syst. Sci.* **2010**, *10*, 699–711. [CrossRef]
24. Žabota, B.; Kobal, M. A New Methodology for Mapping Past Rockfall Events: From Mobile Crowdsourcing to Rockfall Simulation Validation. *ISPRS Int. J. Geo-Inf.* **2020**, *9*, 514. [CrossRef]
25. Lambert, S.; Bourrier, F. Design of Rockfall Protection Embankments: A Review. *Eng. Geol.* **2013**, *154*, 77–88. [CrossRef]
26. Schober, A.; Bannwart, C.; Keuschnig, M. Rockfall Modelling in High Alpine Terrain—Validation and Limitations/Steinschlagsimulation in Hochalpinem Raum—Validierung Und Limitationen. *Geomech. Tunn.* **2012**, *5*, 368–378. [CrossRef]
27. Pellicani, R.; Spilotro, G.; Van Westen, C.J. Rockfall Trajectory Modeling Combined with Heuristic Analysis for Assessing the Rockfall Hazard along the Maratea SS18 Coastal Road (Basilicata, Southern Italy). *Landslides* **2016**, *13*, 985–1003. [CrossRef]
28. Bonneau, D.A.; Hutchinson, D.J.; DiFrancesco, P.-M.; Coombs, M.; Sala, Z. Three-Dimensional Rockfall Shape Back Analysis: Methods and Implications. *Nat. Hazards Earth Syst. Sci.* **2019**, *19*, 2745–2765. [CrossRef]
29. Saroglou, C.; Asteriou, P.; Zekkos, D.; Tsiambaos, G.; Clark, M.; Manousakis, J. UAV-Based Mapping, Back Analysis and Trajectory Modeling of a Coseismic Rockfall in Lefkada Island, Greece. *Nat. Hazards Earth Syst. Sci.* **2018**, *18*, 321–333. [CrossRef]
30. Fallot, P. *Essai Sur La Géologie Du Rif Septentrional*; Imprimerie officielle: Rabat, Morocco, 1937; 553p.
31. Durand-Delga, M.; Hottinger, L.; Marcais, J.; Mattauer, M.; Milliard, Y.; Suter, C. *Données Actuelles sur la Structure du Rif. Livre a la Mémoire du Professeur Paul Fallot*; Société Géologique de France: Paris, France, 1961; pp. 339–422.
32. Didon, J.; Durand-Delga, M.; Kornprobst, J. Homologies Géologiques Entre Les Deux Rives Du Détroit de Gibraltar. *Bull. Soc. Géologique Fr.* **1973**, *7*, 77–105. [CrossRef]
33. Nold, M.; Uttinger, J.; Wildi, W. Géologie de La Dorsale Calcaire Entre Tétouan et Assifane (Rif Interne, Maroc). *Notes Mémoires Serv. Géologique Maroc* **1981**, *233*, 1–233.
34. El Gharbaoui, A. Note Preliminaire Sur l'evolution Geomorphologique de La Peninsule de Tanger. *Bull. Société Géologique Fr.* **1977**, *7*, 615–622. [CrossRef]
35. Romagny, A. Evolution des Mouvements Verticaux Néogènes de La Chaîne du Rif (Nord-Maroc): Apports d'une Analyse Structurale et Thermochronologique. Doctoral Dissertation, Université Nice Sophia Antipolis, Nice, France, 2014.
36. Benmakhlouf, M. Genèse et Évolution de l'accident de Tétouan et Son Rôle Transformant Au Niveau Du Rif Septentrional (Maroc) (Depuis l'oligocène Jusqu'à l'actuel). Ph.D. Thesis, Université Mohammed V, Faculté des Sciences, Rabat, Morocco, 1990.
37. Mastere, M. La Susceptibilité Aux Mouvements de Terrain Dans La Province de Chefchaouen: Analyse Spatiale, Modélisation Probabiliste Multi-Échelle et Impacts Sur l'aménagement & l'urbanisme. Ph.D. Thesis, Université de Bretagne Occidentale, Brest, France, 2011.
38. El Kharim, Y.; Darraz, C.; Hlila, R.; El Hajjaji, K. Écroulements et Mouvements de Versants Associés Au Niveau Du Col de Onsar (Rif, Maroc) Dans Un Contexte Géologique de Décrochement. *Rev. Française Géotechnique* **2003**, *103*, 3–11. [CrossRef]
39. Romana, M.R. A Geomechanical Classification for Slopes: Slope Mass Rating. In *Rock Testing and Site Characterization*; Elsevier: Amsterdam, The Netherlands, 1993; pp. 575–600. [CrossRef]

40. Riquelme, A.; Tomás, R.; Abellán, A. SMRTool Beta. A Calculator for Determining Slope Mass Rating (SMR). Universidad de Alicante. License: Creative Commons BY-NC-SA. 2014. Available online: http://personal.ua.es/es/ariquelme/smrtool.html (accessed on 2 May 2022).
41. Romana, M.; Tomás, R.; Serón, J.B. Slope Mass Rating (SMR) Geomechanics Classification: Thirty Years Review. In Proceedings of the 13th ISRM International Congress of Rock Mechanics, Montreal, QC, Canada, 10–13 May 2015; Volume 2015-MAY.
42. Chen, Z. Recent Developments in Slope Stability Analysis. In Proceedings of the 8th ISRM Congress, Tokyo, Japan, 25–29 September 1995.
43. Beniawski, Z.T. Rock Mass Classification in Rock Engineering Applications. In Proceedings of the a Symposium on Exploration for Rock Engineering 12, Johannesburg, South Africa, 1–5 November 1976; pp. 97–106.
44. Goodman, R.E. *Introduction to Rock Mechanics*; Wiley: New York, NY, USA, 1980; pp. 254–287.
45. Meng, X.; Shang, N.; Zhang, X.; Li, C.; Zhao, K.; Qiu, X.; Weeks, E. Photogrammetric UAV Mapping of Terrain under Dense Coastal Vegetation: An Object-Oriented Classification Ensemble Algorithm for Classification and Terrain Correction. *Remote Sens.* **2017**, *9*, 1187. [CrossRef]
46. Skarlatos, D.; Vlachos, M. Vegetation Removal from UAV Derived DSMS, Using Combination of RGB and NIR IMAGERY. In Proceedings of the ISPRS Annals of the Photogrammetry, Remote Sensing and Spatial Information Sciences, Riva del Garda, Italy, 4–7 June 2018; Volume 2.
47. Prokop, A.; Panholzer, H. Assessing the Capability of Terrestrial Laser Scanning for Monitoring Slow Moving Landslides. *Nat. Hazards Earth Syst. Sci.* **2009**, *9*, 1921–1928. [CrossRef]
48. Blanca Mena, M.J.; Alarcón Postigo, R.; Arnau Gras, J.; Bono Cabré, R.; Bendayan, R. Non-Normal Data: Is ANOVA Still a Valid Option? *Psicothema* **2017**, *29*, 552–557.
49. Dunn, O.J. Multiple Comparisons Using Rank Sums. *Technometrics* **1964**, *6*, 241–252. [CrossRef]
50. Sneed, E.D.; Folk, R.L. Pebbles in the Lower Colorado River, Texas a Study in Particle Morphogenesis. *J. Geol.* **1958**, *66*, 114–150. [CrossRef]
51. Hockey, B. An Improved Co_Ordinate System for Particle Shape Representation: NOTES. *J. Sediment. Res.* **1970**, *40*, 1054–1056. [CrossRef]
52. Perret, S.; Baumgartner, M.; Kienholz, H. Inventory and Analysis of Tree Injuries in a Rockfall-Damaged Forest Stand. *Eur. J. For. Res.* **2006**, *125*, 101–110. [CrossRef]
53. Blair, T.C.; McPherson, J.G. Grain-Size and Textural Classification of Coarse Sedimentary Particles. *J. Sediment. Res.* **1999**, *69*, 6–19. [CrossRef]
54. Wang, I.-T.; Lee, C.-Y. Influence of Slope Shape and Surface Roughness on the Moving Paths of a Single Rockfall. *Int. J. Civ. Environ. Eng.* **2010**, *4*, 122–128.
55. Abramson, L.W.; Lee, T.S.; Sharma, S.; Boyce, G.M. *Slope Stability and Stabilization Methods*; John Wiley & Sons, INC.: Hoboken, NJ, USA, 2001; Volume 706.
56. Rosenberg, D.; Shtober-Zisu, N. The Stone Components of the Pits and Pavements. In *An Early Pottery Neolithic Occurrence at Beisamoun, the Hula Valley, Northern Israel*; BAR International Series: Oxford, UK, 2007; pp. 19–34.

Article

Monitoring Seasonal Movement Characteristics of the Landslide Based on Time-Series InSAR Technology: The Cheyiping Landslide Case Study, China

Yiting Gou [1,2,3], Lu Zhang [1,2,3,*], Yu Chen [1,2,3], Heng Zhou [1,2,3], Qi Zhu [1,2,3], Xuting Liu [1,2,3] and Jiahui Lin [1,2,3]

[1] Key Laboratory of Digital Earth Science, Aerospace Information Research Institute, Chinese Academy of Sciences, Beijing 100094, China; gouyiting21@mails.ucas.ac.cn (Y.G.); chenyu@radi.ac.cn (Y.C.); zhouheng20@mails.ucas.ac.cn (H.Z.); zhuqi20@mails.ucas.ac.cn (Q.Z.); liuxuting20@mails.ucas.ac.cn (X.L.); linjiahui20@mails.ucas.ac.cn (J.L.)
[2] International Research Center of Big Data for Sustainable Development Goals, Beijing 100094, China
[3] University of Chinese Academy of Sciences, Beijing 100049, China
* Correspondence: zhanglu@radi.ac.cn

Citation: Gou, Y.; Zhang, L.; Chen, Y.; Zhou, H.; Zhu, Q.; Liu, X.; Lin, J. Monitoring Seasonal Movement Characteristics of the Landslide Based on Time-Series InSAR Technology: The Cheyiping Landslide Case Study, China. *Remote Sens.* 2023, 15, 51. https://doi.org/10.3390/rs15010051

Academic Editor: Francesca Ardizzone

Received: 28 September 2022
Revised: 17 December 2022
Accepted: 19 December 2022
Published: 22 December 2022

Copyright: © 2022 by the authors. Licensee MDPI, Basel, Switzerland. This article is an open access article distributed under the terms and conditions of the Creative Commons Attribution (CC BY) license (https://creativecommons.org/licenses/by/4.0/).

Abstract: Landslides are one of the extremely high-incidence and serious-loss geological disasters in the world, and the early monitoring and warning of landslides are of great importance. The Cheyiping landslide, located in western Yunnan Province, China, added many cracks and dislocations to the surface of the slope due to the severe seasonal rainfall and rise of the water level, which seriously threaten the safety of residents and roads located on the body and foot of the slope. To investigate the movement of the landslide, this paper used Sentinel-1A SAR data processed by time-series interferometric synthetic aperture radar (InSAR) technology to monitor the long-time surface deformation. The landslide boundary was defined, then the spatial distribution of landslide surface deformation from 5 January 2018 to 27 December 2021 was obtained. According to the monthly rainfall data and the temporal deformation results, the movement of the landslide was highly correlated with seasonal rainfall, and the Cheyiping landslide underwent seasonal sectional accelerated deformation. Moreover, the water level change of the Lancang River caused by the water storage of the hydropower station and seasonal rainfall accelerates the deformation of the landslide. This case study contributes to the interpretation of the slow deformation mechanism of the Cheyiping landslide and early hazard warning.

Keywords: Cheyiping landslide; seasonal movement; time-series InSAR technology; deformation monitoring

1. Introduction

The landslide is the movement of a large amount of rock, mud, or debris along a slope [1], usually triggered by external factors such as earthquakes, heavy rainfall, water level change, typhoons, floods, etc. [2]. A total of 4862 fatal landslide disasters were recorded from 2004 to 2016 around the world, most of which were located in Central America, the Caribbean islands, South America, East Africa, Asia, Turkey, Iran, and the European Alps [3]. The majority of fatal landslides are caused by intense rainfall around the world, and most disasters occur from June to September in Asia because of the summer monsoon [4,5]. China suffered numerous disasters compared to other countries. For example, in 2010, 87% of the landslides triggered by rainfall in Asia occurred in China, especially during the peak of rainfall in July and August [6]. Due to huge potential energy, landslides carry on a high-speed dangerous geological body after breaking away from the parent rock, causing serious loss of life and property [7,8]. There were 55,997 fatalities caused by landslides between 2004 and 2016. From 1950 to 2016, 1911 non-earthquake landslides caused 28,139 deaths in China [9]. Some landslide events were extremely hazardous. On 22 March 2014, a landslide near Oso, Washington, USA caused a great catastrophe. The mud and debris crossed a floodplain for more than 1 km and then demolished the Steelhead Haven community, killing 43 people and destroying 35 houses [10,11]. On 23 July 2019, a landslide occurred at Jichang Town in Shuicheng, Liupanshui City, Guizhou Province, resulting in 43 deaths, 9 missing, 11 injuries, and a direct economic loss of 190 million yuan [12,13]. It can be seen that landslide disasters pose serious harm to people; therefore, landslide detection and early warning are extremely necessary.

Among the quantifiable parameters in landslide monitoring (volume, position, activity status, etc.), the surface deformation caused by slope movement is the most direct physical quality reflecting the current stability and movement condition of the landslide [14]. Traditional methods such as manual field investigation, Globe Positioning System, real-time monitoring, photogrammetry, distributed fiber optic sensing, and geodetic methods have high monitoring accuracy in field measurements [15–18]. While in some places where the terrain is steep or landslides have already occurred, it is difficult for people to reach the sites, making field monitoring and rescue operations difficult [19,20]. Because of the all-weather, all-time, and strong penetrability qualities of the data, Synthetic Aperture Radar Interferometry (InSAR) developed in the past 30 years can detect micro-deformation in the early stage of landslide disasters with large space coverage, high monitoring accuracy [21,22]. In 1990, Gabriel et al. first proposed the differential interferometric synthetic aperture radar technique (D-InSAR) and validated its application in surface deformation monitoring [23,24]. Subsequently, D-InSAR technology has been successively used to monitor land subsidence [25], earthquakes [26,27], landslide movement [28], etc.

Nevertheless, the D-InSAR method seriously interfered with atmospheric factors, and the change in the scattering characteristics of ground objects when the observation time becomes longer leads to a decrease in image coherence, which means that the accuracy of D-InSAR results often fails to meet expectations [29,30]. To solve this problem, scientists have proposed time-series InSAR technology [31]. In 2000, Ferretti et al. proposed the permanent scatterer interferometry technique (PS-InSAR) [32], and Berardino et al. proposed the short baseline set differential interferometry technique (SBAS-InSAR) in 2002 [33]. Time-series InSAR technology extracts coherent points with stable scattering characteristics in multi-scene SAR data for deformation analysis, reducing the decoherence effect caused by long-term baselines, removing atmospheric effects through statistical methods, and achieving considerable monitoring for slow and long-term landslides [34,35].

The Cheyiping landslide is an ancient landslide, located in the high-altitude geological disaster area in northwest Yunnan Province. As early as the 1920s, the residents moved out because of the severe surface deformation of the landslide. In the 1980s, the overall situation became stable, so the residents moved back to the original site one after another. After the rainy season in 2017, there were gaps and cracks at the front, middle, and trailing edges of the mountain. As the Huangdeng Hydropower Station downstream began to impound in May 2018, the changes in the water level of the Lancang River impaired the stability of the landslide. Moreover, Preliminary ground investigations show that the speed of the landslide is about 1m/a. The slow-moving landslide tends to persist for several years to decades, and once it occurs, it can cause damage to infrastructure or even serious casualties in a short period of time [36], which poses a serious threat to the safety of people's lives and property. However, the local monitoring of the Cheyiping landslide is still mainly based on field geological surveys, and there is no detailed and long-term observation record to clarify the motion of the landslide. Therefore, it is significant to monitor the movement patterns of the Cheyiping landslide following its resurrection in recent years, especially based on the time-series InSAR technology, which can obtain monitoring results in a wide coverage, high resolution, and long time series.

In this paper, 60 scenes of Sentinel-1A data were collected to monitor the Cheyiping landslide from January 2018 to December 2021 by using PS-InSAR and SBAS-InSAR techniques to get long-time series, high-precision, and high-density surface deformation of the study area. This study gives a case study to analyze the changes in the time series of the surface deformation speed and the accumulated settlement of the landslide, detect the characteristics of landslide movements and deformation, and explore the inducement of the landslide based on the geological and geomorphological conditions, the seasonal rainfall and the fluctuation of the Lancang River water level caused by the hydropower station. This study reveals the evolution process of the Cheyiping landslide, which could provide data support for the early warning of landslide disasters, thus, reducing loss of life and property, and setting a case example for the geological hazard in the nearby region suffering a similar external environmental condition.

2. Study Area and Data

2.1. Study Area

The Cheyiping landslide is a medium-sized, slow-moving planar sliding landslide composed of clay and sand [37], which is located in Shideng Township, Lanping County, Lisu Autonomous Prefecture of Nujiang, Yunnan Province. The study area has little cultivated land, a large elevation difference, and plenty of deeply cut valleys, in which several geological disasters are distributed. Figure 1a shows the topography and geographical location of the study area. The center of landslide is located at 26.7928°N, 99.1863°E, with an altitude range of 1796 to 1855 m, a slope aspect of approximately 250°, and a terrain slope of around 25°. Located along the Lancang River, 155 households with a total of 535 people live in Cheyiping Village and the primary school on the landslide. In addition, the Bao–Tibet Highway was built in the middle of it. There have been lots of cracks appearing on the road and walls in the village because of movement in recent years. The location of the highway and the village on the landslide are shown in Figure 1b.

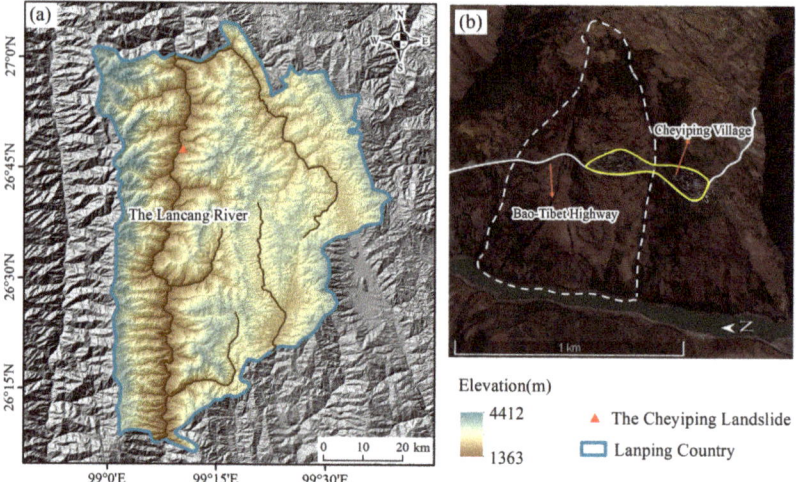

Figure 1. Overview of the study area. (**a**) The geographical location of the Cheyiping landslide (the red triangle). (**b**) The Google map of the Cheyiping landslide, labeled with Bao–Tibet highway, Cheyiping village, and the boundary of the Cheyiping landslide.

2.2. Data

This paper uses 60 scenes of Sentinel-1A SAR ascending data of orbit 172 acquired from 5 January 2018 to 27 December 2021. The imaging mode is IW (Interferometric Wideswath) SLC (Single Look Complex), and the central incident angle is 39.28°. The resolution is 13.94 m in azimuth and 2.32 m in the slant range. The Sentinel-1 satellite operates in the C-band with an orbital height of about 7000 km, with a 12-day revisit period and a large-scale spatial coverage of 250 km × 250 km. It can perform all-weather and all-day high-resolution monitoring of the global land and sea surface in multi-polarization.

Therefore, these characteristics of Sentinel-1A data could meet the requirements of the landslide observation. The time information of the data used is shown in Table 1.

Table 1. Acquisition time of SAR data.

Number	Date	Number	Date	Number	Date	Number	Date
1	5 January 2018	16	31 December 2018	31	7 January 2020	46	13 January 2021
2	29 January 2018	17	24 January 2019	32	31 January 2020	47	6 February 2021
3	22 February 2018	18	17 February 2019	33	24 February 2020	48	14 March 2021
4	18 March 2018	19	13 March 2019	34	19 March 2020	49	7 April 2021
5	11 April 2018	20	6 April 2019	35	12 April 2020	50	1 May 2021
6	5 May 2018	21	12 May 2019	36	6 May 2020	51	25 May 2021
7	29 May 2018	22	5 June 2019	37	30 May 2020	52	18 June 2021
8	22 June 2018	23	29 June 2019	38	23 June 2020	53	12 July 2021
9	16 July 2018	24	23 July 2019	39	17 July 2020	54	5 August 2021
10	9 August 2018	25	16 August 2019	40	10 August 2020	55	29 August 2021
11	2 September 2018	26	9 September 2019	41	3 September 2020	56	22 September 2021
12	26 September 2018	27	3 October 2019	42	27 September 2020	57	16 October 2021
13	20 October 2018	28	27 October 2019	43	21 October 2020	58	9 November 2021
14	13 November 2018	29	20 November 2019	44	14 November 2020	59	3 December 2021
15	7 December 2018	30	14 December 2019	45	20 December 2020	60	27 December 2021

It is necessary to remove terrain phase errors from the satellite orbit information during the process of image registration and differential interference. Therefore, the POD precise orbit data was used for orbit refinement when importing data [38,39]. The image of the study area was cropped out to improve processing efficiency (Figure 2). The SRTM1 30 m elevation data jointly measured by NASA and the Department of Defense's National Mapping Agency (NIMA) were used in the interferometric processing to remove the topographic phase [40].

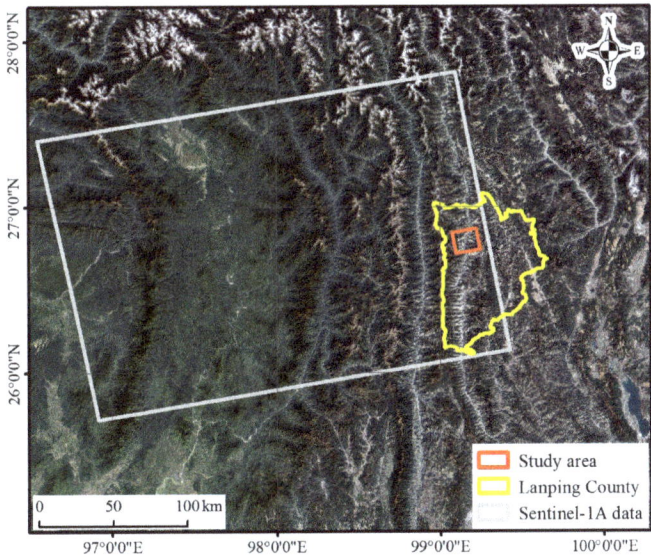

Figure 2. The Sentinel-1A SAR data coverage.

3. Methodology

The workflow of this paper is shown in Figure 3, which is mainly divided into datasets, data process, results, deformation analysis, and inducement analysis of landslide. The data processing method is described in detail here.

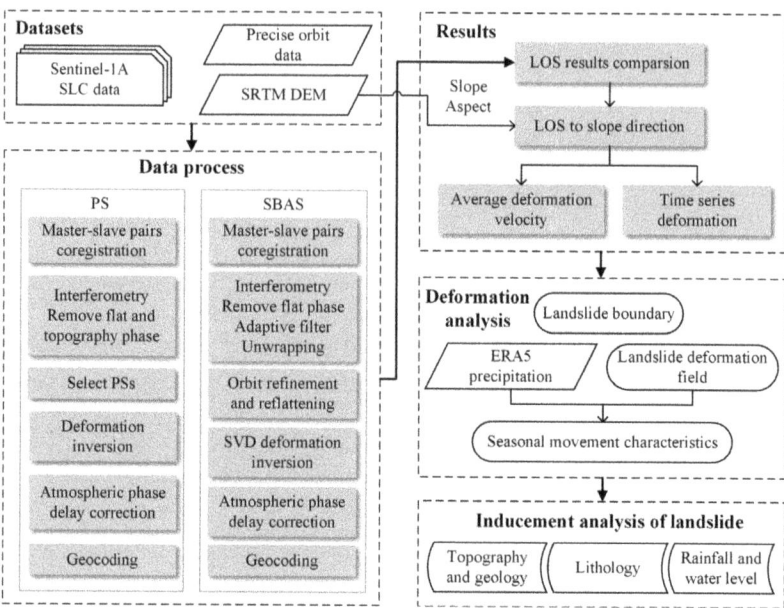

Figure 3. The workflow of the study.

3.1. The Principle of PS-InSAR

The process of PS-InSAR uses multi-scene SAR images to detect highly coherent persistent scatterers (PSs) that are not affected by time and space baseline decorrelation based on a statistical analysis of the stability of amplitude and phase information in the time series. From these PSs, the topography, elevation, and atmospheric phases are estimated and eliminated before the deformation phase is ultimately separated.

Firstly, the image acquired on 16 August 2019 was selected as the super master image, and the master-slave image pairs were established to generate the connection network as shown in Figure 4a. Secondly, all the slave images are co-registered on the super master image to correct the deviation caused by the incident angle and orbit position during imaging. Next, the super master and slave images are subjected to interference processing to generate differential interferogram pair sequences, and the topographic phase is eliminated by using the DEM data. Then, the stable candidate points in the time series are selected, and the amplitude dispersion value is used to represent the phase standard deviation to measure the stability of the point target on the time series. When the coherence of the point target on the time series is smaller than a fixed value, known as the amplitude dispersion index (the ratio of SAR intensity average to Standard Deviation), it can be set as a candidate point [41]. Then, Delaunay's triangulated irregular network was built between persistent scatterers. Linear deformation rate and elevation error are inverted in phase unwrapping. Data processing is greatly disturbed by atmospheric effects. Fortunately, the atmosphere is not correlated in time, only in space. According to this feature, the atmospheric phase could be removed through high-pass filtering in the time domain and low-pass filtering in the spatial domain on multi-view images. Therein, we can get the final average deformation rate and the deformation variable per phase, and finally convert the result of the Doppler coordinate system to the geographic coordinate system.

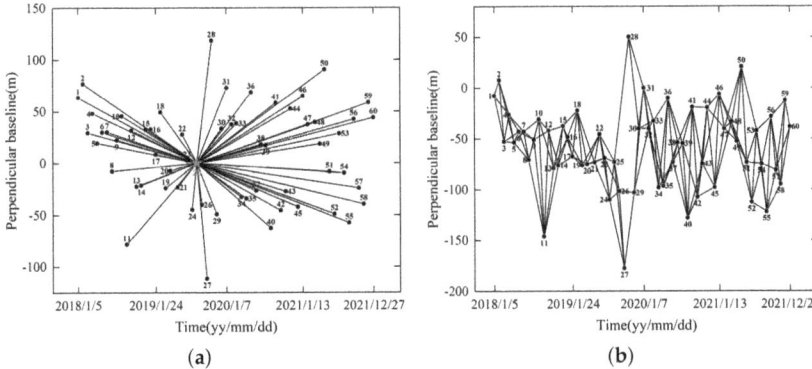

Figure 4. (a) Time−Position map of PS-InSAR. (b) Time−Position map of SBAS-InSAR. The master image is presented as a red star, the slave images are presented as black dots, and all images are marked with the serial number in Table 1. Each line represents a master-slave image pair, and the horizontal and vertical coordinates show their time and space baselines.

3.2. The Principle of SBAS-InSAR

Using a scene of the super master image, and the coherence will be weakened when the baseline becomes longer. To reduce the possibility of spatiotemporal decoherence, Berardino et al. proposed a small baseline set method that combines multiple main images to form a short spatiotemporal baseline, which ensures the coherence of the interferogram [42]. The combination of isolated data pairs with long time intervals has achieved good results for areas with fast changes in the coherence of ground objects, especially in vegetation coverage areas [43].

To begin with, the max space baseline was set as 2% of the critical baseline value, and the max time baseline was set as 90 days. The possible image connections are considered acceptable when the space and time baseline are less than the maximum thresholds. The interference pair diagram is shown in Figure 4b. Secondly, a total of 169 interferograms are generated, and the ratio of the range looks and azimuth looks is set as 4:1 in multilook processing. The Goldstein adaptive filtering method is used to remove noise, and the Delaunay MFC method is used for phase unwrapping [44–49]. The third step is orbital refinement and phase re-flattening. A Ground Control Point (GCP) file must be previously generated. Firstly, a representative image was chosen in all filtered interferograms and unwrapped images, respectively, as shown in Figure 5. Then GCPs were chosen in slant range image (Figure 5b) with reference to each phase in Figure 5a. The criteria for GCPs selection are no residual topography fringes, far away from the displacement area, and no phase jumps. Finally, 30 GCPs were selected and checked to be suitable for as many image pairs as possible. After inputting the GCP file, the phase ramp is estimated to remove the residual phase and phase ramp. Next, the deformation rate is obtained by the singular value decomposition (SVD) method, and the spatio-temporal filter is used for removing the atmospheric phase, which is the same as PS-InSAR. In the end, the displacement on the time series is calculated, and the deformation result is obtained by geocoding.

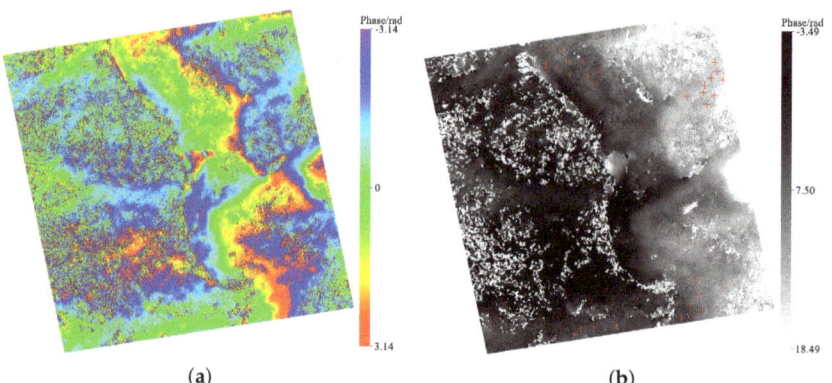

Figure 5. (**a**) Filtered interferogram, which is used for determining terrain and deformation areas. (**b**) Unwrapped interferogram with 30 GCPs. The red plus sign represents GCPs.

4. Results

4.1. Results in LOS Direction and Comparison

The results of the two time-series InSAR methods are shown in Figure 6a,b. It can be noted that there are few monitoring results in many places because of the decoherence. The ground objects in the research region are mostly bare soil, and natural characteristics like flora have low coherence, so there will be decoherence due to the long time baseline. The PS-InSAR and SBAS-InSAR select out high phase-correlation points for further analysis, which are mostly dispersed among man-made features, such as buildings and roads [50]. As a result, the raster is inconsistent with a few monitoring points. The red deformation result represents the deformation rate as positive, indicating that the objects move close to the satellite in the radar line-of-sight (LOS) direction, whereas the objects move away from the satellite in the LOS direction in the green area.

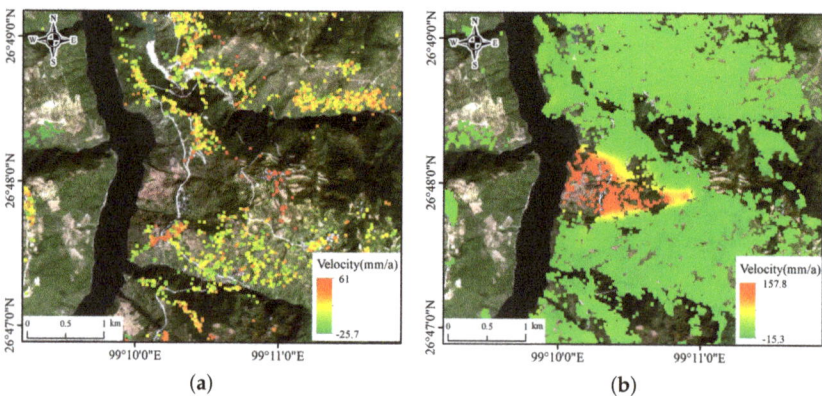

Figure 6. Monitoring results in LOS direction by PS-InSAR (**a**) and SBAS-InSAR (**b**), respectively. The annual average velocity results overlay the Sentinel-2 satellite data. The red raster shows uplifted deformation in the LOS direction, the green indicates descend.

The average deformation rate in the study area obtained by the PS-InSAR ranges from −25.7 to 61 mm/a. However, there are insufficient monitoring points on the landslide body to establish the landslide's deformation condition. The average deformation rate obtained by the SBAS-InSAR ranges from −15.3 to 157.8 mm/a, and the deformation magnitude of the landslide is much larger than that in other areas, which proves that this method is effective. Comparing the results of the two methods, the deformation trends of the two methods are generally consistent, but the SBAS-InSAR has far more monitoring points than PS-InSAR. The SBAS-InSAR results have more coherent points on the features with larger deformation, which depicts the deformation more accurately. These differences are mainly due to the different principles of the two methods combining the interferometric image pairs. The PS-InSAR uses only one main image to produce interference pairs, and when the time baseline between the main and auxiliary images becomes longer, the ground objects will change significantly in the natural vegetation coverage area, resulting in decoherence. Whereas the SBAS-InSAR uses the short baseline set criterion to generate image pairs, which greatly reduces the number of low-coherence points, and, thus, the results of SBAS-InSAR were selected to conduct further analysis of the Cheyiping landslide.

4.2. Projection of Deformation Direction

The deformation results obtained by time-series InSAR processing are along the radar line of sight (LOS). Deformation usually occurs in the direction of the steepest slope, so the deformation parallel to the direction of the maximum slope is regarded to indicate the deformation features of a landslide [51,52]. The projection method of deformation rate proposed by Colesanti et al. in 2006 is used to project the deformation from LOS to the maximum slope direction (slope) [53]. The spatial relationship between LOS direction and slope direction is shown in the following Figure 7, and the projection transformation formulas are as Formulas (1) and (2).

$$v_{slope} = \frac{1}{\cos \beta} \times v_{LOS} \quad (1)$$

$$\begin{aligned}\cos \beta = &(-\sin \alpha \times \cos \varphi) \times (-\sin \theta \times \cos \alpha_s) + \\ &(-\cos \alpha \times \cos \varphi) \times (-\sin \theta \times \sin \alpha_s) + \\ &\sin \varphi \times \cos \theta \end{aligned} \quad (2)$$

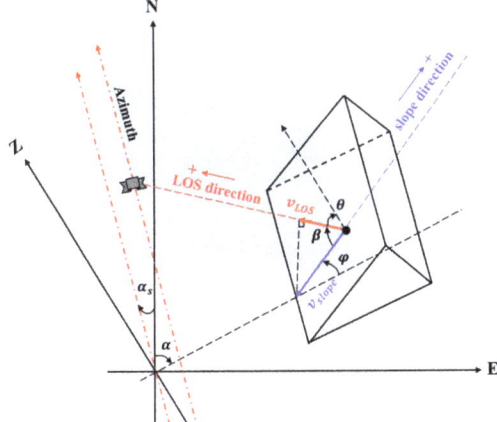

Figure 7. Spatial relationship between v_{slope} and v_{LOS} directions for a point (black dot) located on the slope. v_{slope} is the deformation rate along the slope, v_{LOS} is the deformation rate along the LOS direction. β is the angle between the v_{LOS} and v_{slope} directions, rotating from v_{slope} to LOS direction. α is the aspect angle. φ is the slope angle. θ is the angle between the vertical direction and LOS, i.e., the incidence angle with reference to flat land. α_s is the angle between the satellite azimuth and the true north direction, rotating from the north to ascending orbit direction in our study, and for the Sentinel-1A at orbit 172 is $-12°$.

Along LOS, the direction from the target to the sensor is positive, and the direction along LOS away from the sensor is negative; along the slope, the upward movement is positive, and downward movement is negative, as indicated by the red and blue plus signs in Figure 7. When the cos β is close to 0, the v_{slope} tends toward infinity. Therefore, the fixed threshold Herrera et al. proposed was used in 2013 (cos $β = ±0.3$) to avoid great anomalies in the absolute value during the conversion from v_{slope} to v_{LOS}, and v_{slope} cannot be larger than 3.33 times that of v_{LOS}. Therefore, when cos $β < −0.3$, cos $β = −0.3$; when cos $β > 0.3$, cos $β = 0.3$ [38]. The result of projecting the LOS direction result obtained by the SBAS method to the slope direction is shown in Figure 8. The positive and negative values of the deformation rate, as well as the magnitude of the value, have altered when compared to the LOS direction result in Figure 6b. It is logical that the velocity of the landslide is negative and indicates a downward movement along the slope. The dividing line of the change in slope aspect, or the location where significant deformation occurs, is the junction. Figure 8 clearly depicts the delimitation of the Cheyiping landslide (the range shown by the black solid line).

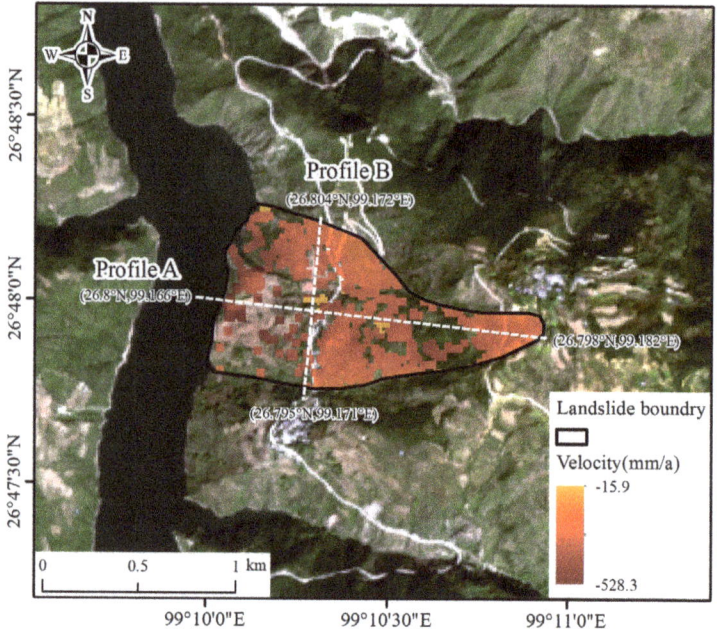

Figure 8. Deformation rate and interpretation boundary of the Cheyiping landslide. Two white dotted lines represent the lines of profiles, which are labeled with start and end coordinates.

5. Analysis and Discussion

5.1. Delimitation of the Landslide

According to the slope direction results in Figure 8, the deformation rate of the landslide varies from $−528.3 \sim −15.9$ mm/a, and the plane shape of the landslide presents an irregular triangle with a length of approximately 1500 m from east to west and a width of approximately 800 m from north to south. According to the field investigation data, the landslide covers an area of about 0.8 km^2, the thickness of the landslide body ranges from 7 to 35 m, the average thickness is about 10 m, and the volume of the landslide body is about 8 million m^3. The front edge of the landslide is bounded by the left bank of the Lancang River, a road in the south, gullies in the north, and the rear wall of the landslide which extends to Beizhiqing Village.

5.2. Time Series Change of Landslide Deformation Field

The regional distribution of the landslide surface deformation differs significantly. Figure 8 depicts cross-sections of the landslide body with the greatest deformation rate (as shown by the white dashed line), from which we retrieved the deformation rate and cumulative deformation in a partial time series. The deformation change of Profile A and Profile B is shown in Figure 9.

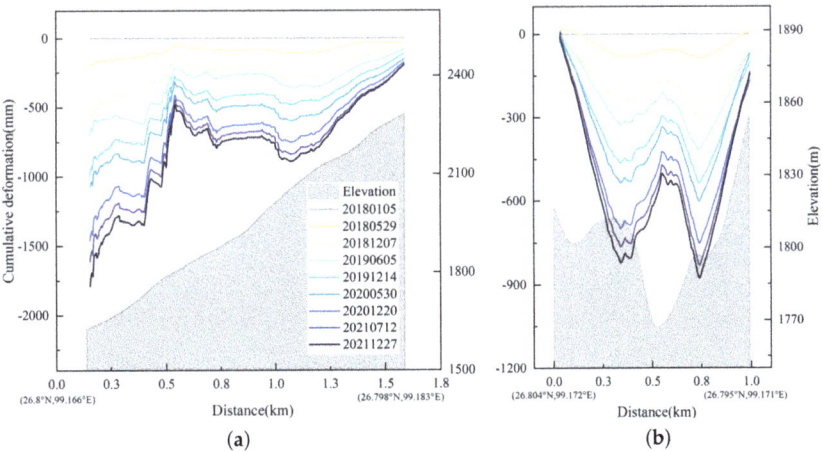

Figure 9. Time−series cumulative deformation of profiles. (**a**) Profile A. (**b**) Profile B. The solid lines represent cumulative deformation from 5 January 2018, until the date, corresponding to the left vertical axis; and the dotted lines represent the average annual deformation rate of points on the profile lines, corresponding to the right vertical axis. The grey parts represent the elevation of the profiles.

Figure 9a illustrates that the farther away from the Lancang River, the slower the landslide deforms. The sliding rate is the most extreme along the river bank, where the deformation rate surpasses −430 mm/a and the highest settlement is about 1790 mm. In Figure 9b, the landslide deformation rate increases, then decrease, and finally increases again. There are two subsidence centers on profile B located at 0.38 km and 0.8 km, and the highest subsidence rate reaches −230 mm/a, and the greatest deformation is −850 mm. It is worth noting that on the deformation curves of profiles A and B, the deformation of the middle part is smaller than that of the neighborhood. The positions of 0.57 km of profile line A and 0.6 km of profile line B are Bao–Tibet highway and Cheyiping village, and the cement floor is more stable than that of the soil. Therefore, there is an upward trend in the middle of settlement curves. Figure 10c–h shows the deformation photos. There are many cracks in the ground in the village (Figure 10c,d), on the walls of houses (Figure 10e,f), and on the roads (Figure 10g,h). In general, the landslide sinks at different rates over time. The foot of the landslide body is the most active zone of deformation, and the village and highway are relatively stable.

To further investigate the changes in landslide movement in the time dimension, we generated time series deformation of the overall landslide images presented in Figure 11. It can be seen that the deformation of the landslide developed progressively from the front edge to the trailing edge, and the sliding range on the horizontal projection surface grew. In the height direction, the magnitude of sinking is likewise increasing. It is worth noting that the front edge of the landslide along the Lancang River demonstrates deformation characteristics before the trailing edge, and the front edge drives the trailing edge to slide, demonstrating traction sliding characteristics.

Figure 10. Field survey photos. (**a**) The trailing edge has sunk. (**b**) The front edge has slid. (**c**) Cracks in the Cheyiping primary school. (**d**) Cracks in villagers' homes. (**e**) Fissures in the walls. (**f**) Cracks in houses. (**g**) Cracks in the road in the village. (**h**) Deformation of Bao–Tibet highway.

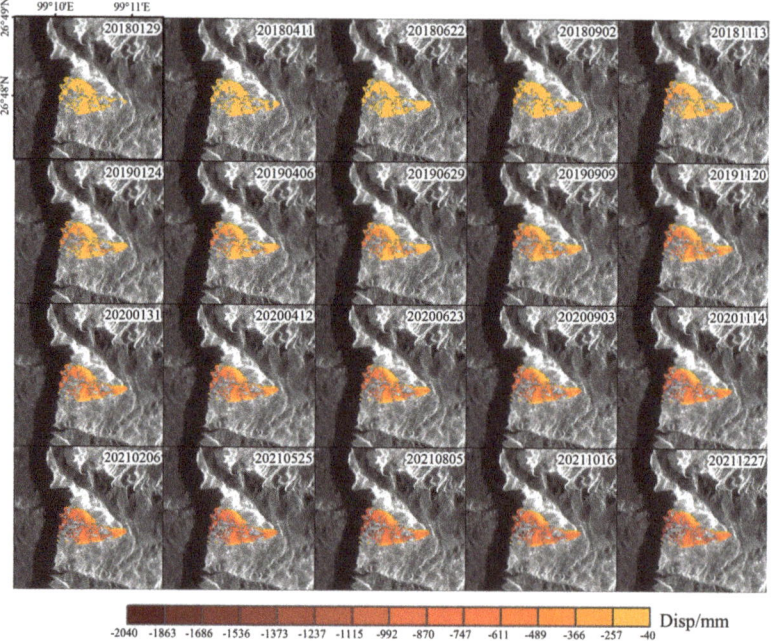

Figure 11. Time−series cumulative deformation calculated from 5 January 2018 of the landslide on each acquisition date. The base map is the SAR intensity average image, and the dates are labeled on the top right corner of every subgraph.

Based on the deformation characteristics of various portions of the landslide, five distinct areas a, b, c, d, and e were identified. Figure 12 shows manual field survey photos and monitoring results in these areas. The field investigation reveals that the trailing edge generates cracks and subsidence (Figure 12a). The deformation rate in Area a is around −140 mm/a (as shown in Figure 12f), with a cumulative deformation of −560 mm (as shown in Figure 12g). Lateral surface cracks have developed in the top and middle parts of the landslide, spreading to the north and south sides, and the vertical dislocation is visible on both sides of the crack (Figure 12b). The typical points in Area b deform at a rate of roughly −220 mm/a, with a cumulative deformation of −880 mm. However, because the Bao-Tibet Highway runs through the heart of the landslide, and it is near the village, the stability of the landslide is critical. The field photos demonstrate that the road has begun to crack, and the fissures are growing, where noticeable subsidence and dislocation are apparent. At the same time, the ground in the village was fractured, and the cracks were repaired with mortar by the villagers (Figure 12c). The deformation rate of the characteristic points in Area c is about −170 mm/a, and the accumulated settlement reaches −680 mm. Area d is located in the middle and lower part of the landslide (Figure 12d), with more cracks on the surface than the upper part, the deformation rate is around −290 mm/a, and the accumulated settlement is −1160 mm. The front edge of the landslide lies near the Lancang River, and the deformation in Area e is the most noticeable (Figure 12e). As the landslide descends, the landslide continues to crack and sink. The deformation rate in Area e is measured at a maximum of −430 mm/a, and the accumulated subsidence is −1790 mm, so the front edge is vulnerable to slide and collapse.

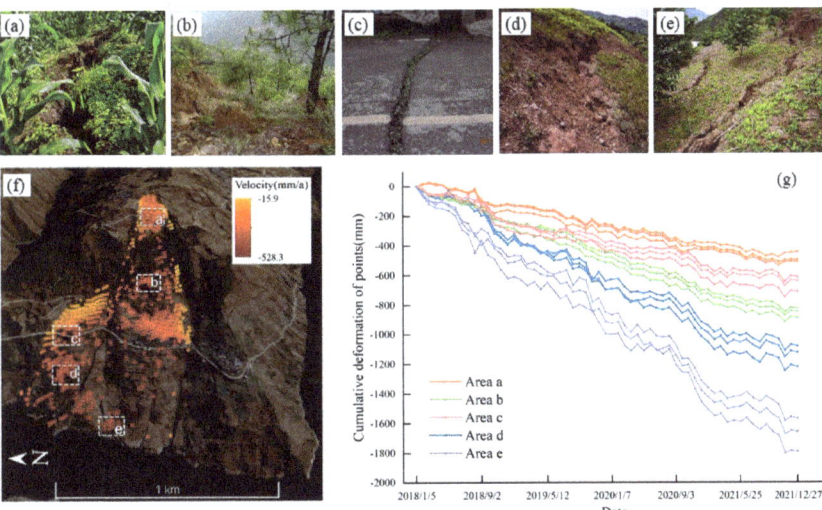

Figure 12. Field survey photos and monitoring results. (**a**) Cracks on the trailing edge (Area a in (**f**)). (**b**) Faults in the middle−upper part (Area b in (**f**)). (**c**) Fissures on the pavement in Cheyiping village (Area c in (**f**)). (**d**) Cracks in the middle and lower part (Area d in (**f**)). (**e**) Collapse on the front edge (Area e in (**f**)). (**f**) Deformation rates of regional typical points. (**g**) Accumulated settlement of three typical points in each area of (**f**).

5.3. Seasonal Movement Characteristics of Landslides

Figure 12g shows the cumulative deformation of the feature points over time, which is tentatively judged to be brought about by seasonal rainfall, given the physical setting of the study area. Specifically, the landslide deformation rate is significantly accelerated in the rainy season and slowed down in the wet season. The study area has distinct dry and wet seasons, with rainfall concentrated from May to October. According to different motion change characteristics, we plotted the cumulative deformation for selected points in areas a and b (as shown in Figure 12g) in Figure 13a and areas c, d, and e in Figure 13b. Figure 13 shows the relationship between the total deformation and the monthly average rainfall.

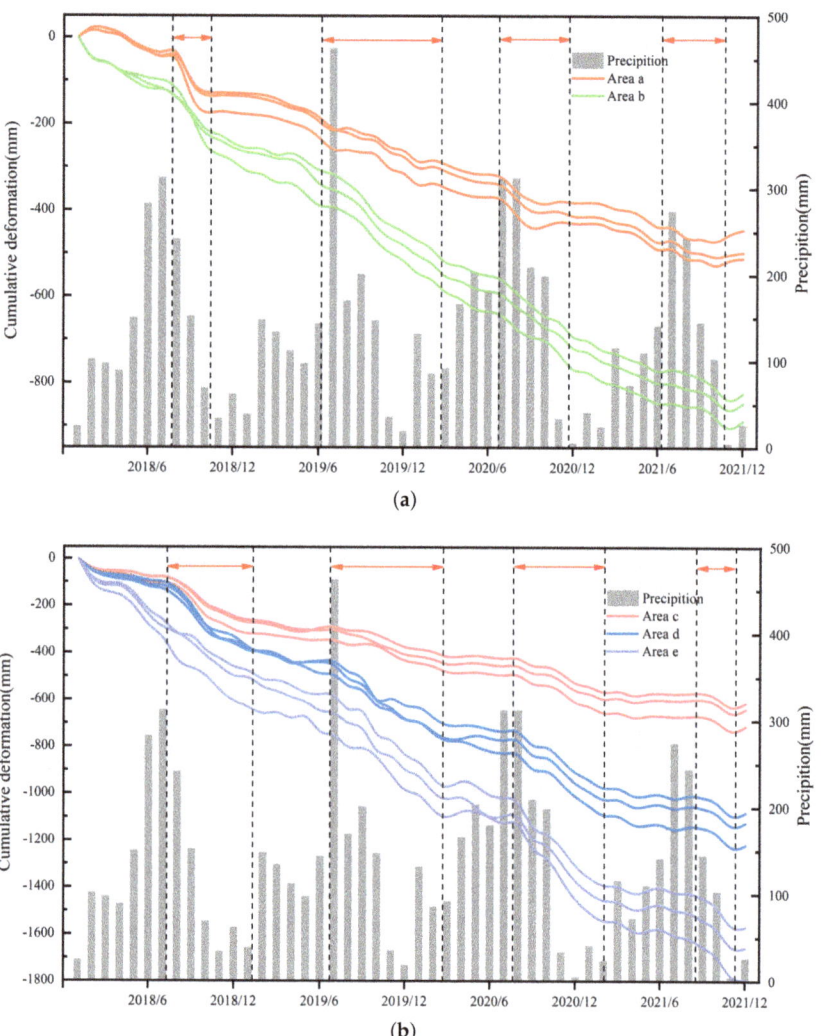

Figure 13. Time−series cumulative deformation of feature points. (**a**) The feature points in Area a, b (as shown in Figure 12f). (**b**) The feature points in Area c, d, e (as shown in Figure 12f).

It can be seen from Figure 13 that the rainfall mainly takes place from May to October, and precipitation always reaches its peak in July, with the maximum average total rainfall in July 2019 reaching 462 mm. These rainfall season months are marked as mauve blocks, and the rest are wet season months. Time series analysis reveals that the displacement is corrected to the precipitation, and the rainfall variations have more impact on the middle and lower part of the landslide body than on the top body. The cumulative deformation curves of areas a and b have a slight trend of accelerated deformation after every wet season, as shown by the periods of time in the black dashed lines and red arrows in Figure 13a. The cumulative deformation curves of areas c, d, and e show segmented changes, and there are four acceleration periods in Figure 13b. Around four rainy seasons (from July 2018 to January 2019, from July 2019 to February 2020, from August 2020 to February 2021, and from September 2021 to November 2021), the slope of the curves increased, indicating the deformation accelerated, and the curves flattened in the rest months (from February 2019 to June 2019, from March 2020 to July 2020 and from March 2021 to August 2021). The landslide has clearly changed during the last four years. The cumulative deformation of Area a, which is the slowest, grew by four times from 100 mm to 500 mm. Area e had the most dramatic deformation, reaching 1700mm by the end of 2021, more than three times that of 2018.

Analyzing and comparing the change characteristics of the deformation variable curves in these five regions, it was found that the slope of the curve becomes steeper one to two months after the first rainy month. For example, in Figure 13b, the rainy seasons started in May from 2018 to 2021, whereas the acceleration period began in July 2018, June 2019, August 2020, and August 2021 separately. Moreover, the influence of the precipitation on the landslide body often lasts for several months, as evidenced by the fact that the accelerated deformation ended in January 2019, February 2020, February 2021, and November 2021. Therefore, seasonal rainfall has a strong inducing effect on landslide deformation. The generation and disappearance of this aggravating effect will not be reflected immediately, but are usually delayed for a period of time, which has been found in many studies [54,55]. This is because it takes a period of time for rainfall to infiltrate into the landslide rock mass, so its influence on the deformation rate of the landslide has a hysteresis, which is consistent with many studies [56–59].

5.4. The Inducement of the Landslide

Whether the landslide slips or not depends on the relationship between the slope angle and the critical angle, and the critical angle is influenced by the block material composition, size, shape, and water content. Most of the non-earthquake landslides are triggered by broken structures, soil strength, intensive rainfall, and the effect of water level in many studies [60–62]. Accordingly, the causes of the Cheyiping landslide are divided into internal causes, mainly topography, geology, geotechnical properties, and external causes of water for analysis.

5.4.1. Topography and Geology

The Cheyiping landslide is located in an area of the sloping terrain, with an overall the slope direction of about 255°, a slope length of 1500 m, a terrain slope of 15° to 20°, a village side slope height of 1.5 to 4.7 m, and a slope gradient of 65° to 280°. Morphologically, it is a moderately steep and long slope, which can be classified as a loose cap rock slope according to the slope process. With the sliding of the slope body concentrated on the front edge and the middle, the movement rate of the back end is small and belongs to a traction landslide. The topography of the area has a steep slope and a large relative height difference. resulting in a large gravitational potential energy of soil on the slope, which provides the impetus for the sliding of the slope material [60]. According to regional geological data, the Cheyiping fault developed on the west side of the Bao-Tibet highway at about 350–400 m, which could cause the geological structure to fragment and change the tectonic stress field, thus, increasing the risk of landslides [56].

5.4.2. Lithology

In Lanping county, the Mesozoic strata are mainly exposed, followed by the Cenozoic and Paleozoic, and a very small amount of unidentified metamorphic rock series. The Mesozoic is almost all over the region, mainly composed of Cretaceous, Jurassic, Triassic siltstone, silty mudstone, and quartz sandstone. The Cenozoic is the sandstone, conglomerate, and calcareous siltstone of the Tertiary, and the sandy clay and sandy gravel of the Quaternary. Paleozoic strata are dominated by mudstones, sandstones, and Carboniferous bioclastic tuffs, schists, and andesites. The geology of the study area is shown in

Figure 14 (source of data: https://geocloud.cgs.gov.cn, accessed on 27 September 2022). Meanwhile, there are small amounts of basalt, andesite, and other volcanic rocks located in the eastern and western margins. The rock mass is mainly composed of layered and fractured structural soft rocks, so the weak structure affects the engineering geological properties [63].

The surface of the landslide is brownish-red and brownish-yellow clay with debris in the residual slope of the Quaternary System. The soil structure is loose, the water permeability is strong, the soil softens and collapses when it meets water, and the stability is poor; thus, excavation disturbance is prone to collapse and landslide. The underlying stratum is the purple-red and grey-green sandstone and mudstone weathering rock of the Jurassic Middle Jurassic Huakai Zuo Group (J2h), which is mainly exposed on the ridges of the north and south sides of the village and on the steeper topography of the village, with weak lithology. The rocks within the slope are strongly weathered mudstone interspersed with muddy siltstone, which is a weak structural plane due to the poor connectivity in rock and soil bodies. The dip Angle of the structural plane is similar to that of the natural slope, which forms the sliding plane.

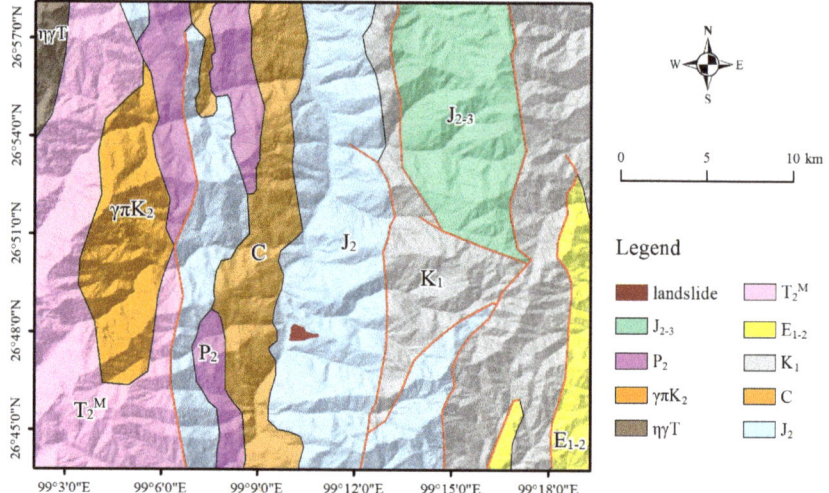

Figure 14. The geological map of the study area.

5.4.3. Influence of Seasonal Rainfall and Water Level

Water is a major cause of landslides [61]. The involvement of water removes the adsorption bond between soil particles, changes the pore water pressure, reduces the resistance to sliding, and erodes the strength of the soil. According to previous studies [64,65], continuous rainfall and rapid changes in the water level have a joint impact on the displacement rate of the landslide. The stability of the landslide decreased with the increase in rainfall intensity and the changes in the water level of the Lancang River. The combination of these two factors in the study area may be the main reason for the accelerated deformation of the Cheyiping landslide. Moreover, the unregulated discharge of water for domestic use by residents in the village and the erosion of the two gullies on the slope impair the stability of the slope.

- Seasonal rainfall. The region of the landslide is characterized by the low-latitude mountain monsoon and typical vertical distribution of the three-dimensional climate, with the highest temperature in July and the lowest temperature in January. With a clear division between wet and dry seasons, the rainfall in the study area is regular. The average annual precipitation is 1002.4 mm and the average annual rainfall is

158 days, with the rainy season from late May to mid-October, which accounts for over 90% of the annual precipitation. The monsoonal climate and seasonal precipitation concentrated in the summer provide a strong trigger for the landslide. According to the ERA5-Land reanalysis dataset, the seasonal precipitation around the Cheyiping area from 2018 to 2021 is shown in Figure 15, indicating that the amount of rainfall in the rainy season is much greater than in the wet season.

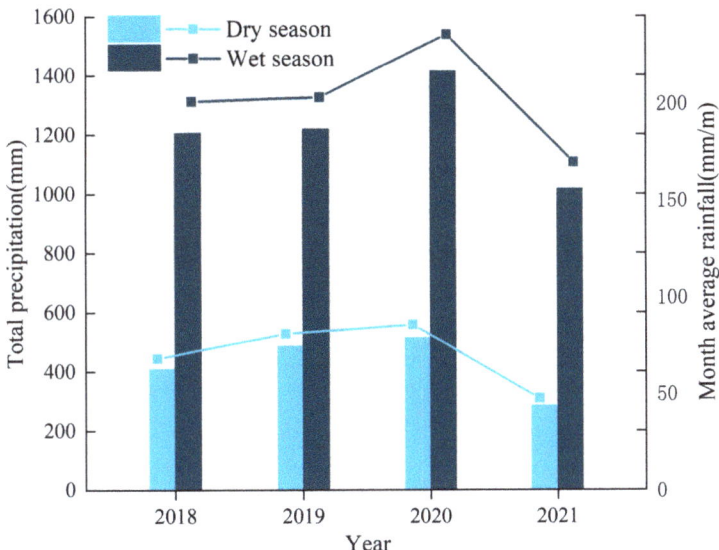

Figure 15. Seasonal precipitation. Wet season: from May to October. Dry season: from January to April, November, and December.

Persistent rainfall increases the pore pressure of the landslide, which reduces the sheer strength of the soil, the bond between the rock particles, and the friction within the landslide, resulting in a high risk of landslides [66]. Water causes expansion and contraction of geotechnical particles, which can alter the pore pressure of the landslide and seasonal rainfall makes this change frequently, whereas pore pressure changes are the main driver of landslide movement, and the larger pore pressure changes can induce landslides [67].

- Erosion and water level rise of the Lancang River. The study area is located in the high mountain area and canyon in the middle-upper reaches of the Lancang River. The Lancang River runs north to south through the mountain valley in Lanping County, with a natural drop of 127 m, an average slope of 9.8%, an average annual flow of 909 m^3, and the driest flow of 277 m^3. Moreover, the front edge of the Cheyiping landslide is adjacent to the Lancang River. The Huangdeng Hydropower Station is built at the position of 99.1197°E, 26.5597°N, which is 26 km away from the landslide, as shown in Figure 16a. The normal storage level of the reservoir is 1619 m, which started to store water in May 2018. The water level in the Cheyiping landslide section was 1557 m; however, after the impoundment, the water level rose by 62 m. By checking the width of the river surface in the radar image Figure 16b,c, it is possible to determine that the water level has significantly risen from January 2018 to January 2019.

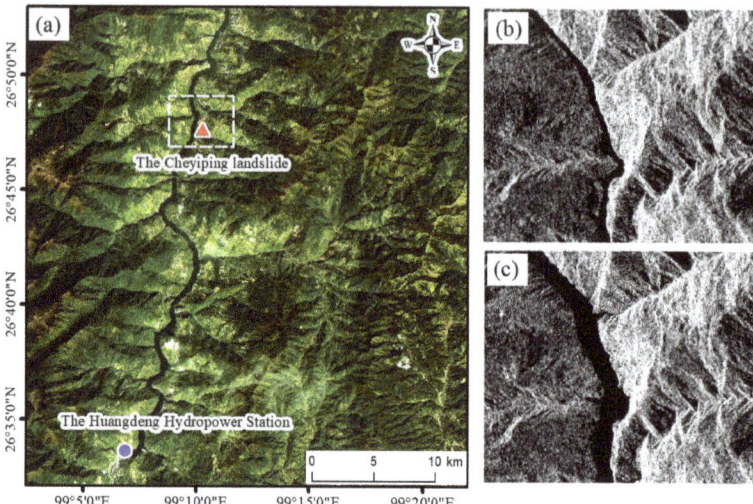

Figure 16. Variation of the Lancang River water level before and after impoundment. (**a**) The relative locations of the Cheyiping landslide and the Huangdeng Hydropower Station. (**b**) SAR image of white dotted line range in (**a**) on 5 January 2018. (**c**) SAR image of white dotted line range in (**a**) on 24 January 2019.

Changes in water level have multiple effects on the stability of landslides. The rise in water level caused by the Huangdeng Hydropower Station storage will affect the geotechnical strength of the slope, the groundwater level, and the pressure difference between the water inside and outside the slope. When the water level changes, there is a lag in the change of the groundwater level, and the pressure difference between the inside and outside of the landslide will disrupt the original equilibrium of the slope [68]. When the water level rises, the external pressure enhances the stability of the slope to a certain extent, and in this case, the accelerated deformation of the slope is typically a result of the softening impact of the water. Therefore, the deformation rate of the slope during the high water level is significantly higher than during the low water level [69].

6. Conclusions

In this study, Sentinel-1A images were collected, and the time-series deformation monitoring results of the Cheyiping landslide from 5 January 2018 to 27 December 2021 were obtained using the time-series InSAR technology. The monitoring results of the PS-InSAR and the SBAS-InSAR show the same deformation trend in most regions, while the SBAS-InSAR intensively detects the landslide with many more monitoring points. The results of the slope direction shows that the deformation rate of the landslide increases from the back edge to the front edge, the deformation rate at the foot of the slope near the Lancang River reaches approximately −430 mm/a, and the accumulated subsidence during the study period is as high as −1790 mm. The front edge of the landslide occurs first, driving the overall movement of the landslide. Based on these results, it was found that the intense concentrated seasonal rainfall accelerates the surface deformation of the slope, and the deformation velocity slows down in the dry season, meaning the landslide movement shows a periodical accelerated trend. Moreover, the water level change of the Lancang River brought by the water storage of the Huangdeng hydropower station downstream makes the landslide destabilized, and seasonal rainfall and water level changes of the Lancang River were the primary causes for the significant movement of the Cheyiping landslide.

In summary, the time-series InSAR technology is feasible for monitoring the deformation of the Cheyiping landslide. The analysis of the time-series changes of landslide deformation based on geological and geomorphological factors, seasonal rainfall, and water level changes of the Lancang River can predict the landslide movement. In the future, accurate landslide hazard warnings could be carried out by combining field survey data with remote sensing data, thus, providing protection for the life and property safety of the residents in this area.

Author Contributions: Conceptualization, L.Z. and Y.C.; methodology, L.Z., Y.C. and Y.G.; software, Y.G. and J.L.; validation, H.Z. and X.L.; formal analysis, Y.G.; investigation, Y.G. and H.Z.; writing—original draft preparation, Y.G.; writing—review and editing, L.Z.; visualization, Y.G. and Q.Z.; supervision, L.Z.; project administration, L.Z.; funding acquisition, L.Z. All authors have read and agreed to the published version of the manuscript.

Funding: This work was supported by the Joint Funds of the National Natural Science Foundation of China [grant number U2268217], the National Natural Science Foundation of China [grant number 41876326], and the Strategic Priority Research Program of the Chinese Academy of Sciences [grant numbers XDA19090135]. L. Zhang is the corresponding author of this paper (zhanglu@radi.ac.cn).

Data Availability Statement: The Sentinel-1A data is available on ASF Data Search (https://search.asf.alaska.edu), the SRTM (The Shuttle Radar Topography Mission) elevation data are available online at http://srtm.csi.cgiar.org/, precipitation data is provided by ERA5-Land reanalysis dataset (https://cds.climate.copernicus.eu), and the geological map are based on data from GeoCloud (https://geocloud.cgs.gov.cn). All links accessed on 27 September 2022.

Acknowledgments: The author would like to thank the editors and the reviewers for their suggestions and comments, which helped us improve the study greatly.

Conflicts of Interest: The authors declare no conflict of interest.

References

1. Cruden, D.M. A simple definition of a landslide. *Bull. Int. Assoc. Eng. Geol.* **1991**, *43*, 27–29. [CrossRef]
2. Dai, F.; Lee, C.; Ngai, Y. Landslide risk assessment and management: An overview. *Eng. Geol.* **2002**, *64*, 65–87. [CrossRef]
3. Froude, M.J.; Petley, D.N. Global fatal landslide occurrence from 2004 to 2016. *Nat. Hazards Earth Syst. Sci.* **2018**, *18*, 2161–2181. [CrossRef]
4. Turner, A.G.; Annamalai, H. Climate change and the South Asian summer monsoon. *Nat. Clim. Chang.* **2012**, *2*, 587–595. [CrossRef]
5. Petley, D. Global patterns of loss of life from landslides. *Geology* **2012**, *40*, 927–930. [CrossRef]
6. Kirschbaum, D.; Adler, R.; Adler, D.; Peters-Lidard, C.; Huffman, G. Global Distribution of Extreme Precipitation and High-Impact Landslides in 2010 Relative to Previous Years. *J. Hydrometeorol.* **2012**, *13*, 1536–1551. [CrossRef]
7. Huang, R. Large-Scale Landslides and Their Sliding Mechanisms in China Since the 20th Century. *Chin. J. Rock Mech. Eng.* **2007**, *26*, 433–454.
8. Runqiu, H. Some catastrophic landslides since the twentieth century in the southwest of China. *Landslides* **2009**, *6*, 69–81. [CrossRef]
9. Lin, Q.; Wang, Y. Spatial and temporal analysis of a fatal landslide inventory in China from 1950 to 2016. *Landslides* **2018**, *15*, 2357–2372. [CrossRef]
10. Iverson, R.M.; George, D.L.; Allstadt, K.; Reid, M.E.; Collins, B.D.; Vallance, J.W.; Schilling, S.P.; Godt, J.W.; Cannon, C.M.; Magirl, C.S.; et al. Landslide mobility and hazards: Implications of the 2014 Oso disaster. *Earth Planet. Sci. Lett.* **2015**, *412*, 197–208. [CrossRef]
11. Collins, B.D.; Reid, M.E. Enhanced landslide mobility by basal liquefaction: The 2014 State Route 530 (Oso), Washington, landslide. *Geol. Soc. Am. Bull.* **2020**, *132*, 451–476. [CrossRef]
12. Ma, S.; Xu, C.; Xu, X.; He, X.; Qian, H.; Jiao, Q.; Gao, W.; Yang, H.; Cui, Y.; Zhang, P.; et al. Characteristics and causes of the landslide on July 23, 2019 in Shuicheng, Guizhou Province, China. *Landslides* **2020**, *17*, 1441–1452. [CrossRef]
13. Zhao, W.; Wang, R.; Liu, X.; Ju, N.; Xie, M. Field survey of a catastrophic high-speed long-runout landslide in Jichang Town, Shuicheng County, Guizhou, China, on July 23, 2019. *Landslides* **2020**, *17*, 1415–1427. [CrossRef]
14. Tofani, V.; Raspini, F.; Catani, F.; Casagli, N. Persistent Scatterer Interferometry (PSI) Technique for Landslide Characterization and Monitoring. *Remote Sens.* **2013**, *5*, 1045–1065. [CrossRef]
15. Gili, J.; Corominas, J.; Rius, J. Using Global Positioning System techniques in landslide monitoring. *Eng. Geol.* **2000**, *55*, 167–192. [CrossRef]

16. Yin, Y.; Wang, H.; Gao, Y.; Li, X. Real-time monitoring and early warning of landslides at relocated Wushan Town, the Three Gorges Reservoir, China. *Landslides* **2010**, *7*, 339–349. [CrossRef]
17. Mora, P.; Baldi, P.; Casula, G.; Fabris, M.; Ghirotti, M.; Mazzini, E.; Pesci, A. Global Positioning Systems and digital photogrammetry for the monitoring of mass movements: Application to the Ca' di Malta landslide (northern Apennines, Italy). *Eng. Geol.* **2003**, *68*, 103–121.
18. Shi, X.; Xu, Q.; Zhang, L.; Zhao, K.; Dong, J.; Jiang, H.; Liao, M. Surface displacements of the Heifangtai terrace in Northwest China measured by X and C-band InSAR observations. *Eng. Geol.* **2019**, *259*, 105181. [CrossRef]
19. Sun, Y.J.; Zhang, D.; Shi, B.; Tong, H.J.; Wei, G.Q.; Wang, X. Distributed acquisition, characterization and process analysis of multi-field information in slopes. *Eng. Geol.* **2014**, *182*, 49–62. [CrossRef]
20. Dong, J.; Zhang, L.; Li, M.; Yu, Y.; Liao, M.; Gong, J.; Luo, H. Measuring precursory movements of the recent Xinmo landslide in Mao County, China with Sentinel-1-and ALOS-2 PALSAR-2 datasets. *Landslides* **2018**, *15*, 135–144. [CrossRef]
21. Colesanti, C.; Wasowski, J. Investigating landslides with space-borne synthetic aperture radar (SAR) interferometry. *Eng. Geol.* **2006**, *88*, 173–199.
22. Xiao, B.; Zhao, J.; Li, D.; Zhao, Z.; Xi, W.; Zhou, D. The Monitoring and Analysis of Land Subsidence in Kunming (China) Supported by Time Series InSAR. *Sustainability* **2022**, *14*, 12387 . [CrossRef]
23. Gabriel, A.K.; Goldstein, R.M.; Zebker, H.A. Method for Detecting Surface Motions and Mapping Small Terrestrial or Planetary Surface Deformations with Synthetic Aperture Radar. U.S. Patent US4975704A, 4 December 1990.
24. Gabriel, A.; Goldstein, R.; Zebker, H. Mapping Small Elevation Changes over Large Areas—Differential Radar Interferometry. *J. Geophys.-Res.-Solid Earth Planets* **1989**, *94*, 9183–9191. [CrossRef]
25. Amelung, F.; Galloway, D.; Bell, J.; Zebker, H.; Laczniak, R. Sensing the ups and downs of Las Vegas: InSAR reveals structural control of land subsidence and aquifer-system deformation. *Geology* **1999**, *27*, 483–486. .<0483:STUADO>2.3.CO;2. [CrossRef]
26. Massonnet, D.; Rossi, M.; Carmona, C.; Adragna, F.; Peltzer, G.; Feigl, K.; Rabaute, T. The Displacement Field of the Landers Earthquake Mapped by Radar Interferometry. *Nature* **1993**, *364*, 138–142. [CrossRef]
27. Kenyi, L.; Kaufmann, V. Estimation of rock glacier surface deformation using SAR interferometry data. *IEEE Trans. Geosci. Remote Sens.* **2003**, *41*, 1512–1515. [CrossRef]
28. Kimura, H.; Yamaguchi, Y. Detection of landslide areas using satellite radar interferometry. *Photogramm. Eng. Remote Sens.* **2000**, *66*, 337–344.
29. Liu, G.; Buckley, S.M.; Ding, X.; Chen, Q.; Luo, X. Estimating Spatiotemporal Ground Deformation With Improved Permanent-Scatterer Radar Interferometry. *IEEE Trans. Geosci. Remote Sens.* **2009**, *47*, 2762–2772. [CrossRef]
30. Wasowski, J.; Bovenga, F. Investigating landslides and unstable slopes with satellite Multi Temporal Interferometry: Current issues and future perspectives. *Eng. Geol.* **2014**, *174*, 103–138. [CrossRef]
31. Colesanti, C.; Ferretti, A.; Prati, C.; Rocca, F. Monitoring landslides and tectonic motions with the Permanent Scatterers Technique. *Eng. Geol.* **2003**, *68*, 3–14.
32. Ferretti, A.; Prati, C.; Rocca, F. Nonlinear subsidence rate estimation using permanent scatterers in differential SAR interferometry. *IEEE Trans. Geosci. Remote Sens.* **2000**, *38*, 2202–2212.
33. Berardino, P.; Fornaro, G.; Lanari, R.; Sansosti, E. A new algorithm for surface deformation monitoring based on small baseline differential SAR interferograms. *IEEE Trans. Geosci. Remote Sens.* **2002**, *40*, 2375–2383. [CrossRef]
34. Osmanoglu, B.; Sunar, F.; Wdowinski, S.; Cabral-Cano, E. Time series analysis of InSAR data: Methods and trends. *ISPRS J. Photogramm. Remote Sens.* **2016**, *115*, 90–102. [CrossRef]
35. Zhao, C.; Kang, Y.; Zhang, Q.; Lu, Z.; Li, B. Landslide Identification and Monitoring along the Jinsha River Catchment (Wudongde Reservoir Area), China, Using the InSAR Method. *Remote Sens.* **2018**, *10*, 993. . [CrossRef]
36. Lacroix, P.; Handwerger, A.L.; Bievre, G. Life and death of slow-moving landslides. *Nat. Rev. Earth Environ.* **2020**, *1*, 404–419. [CrossRef]
37. Hungr, O.; Leroueil, S.; Picarelli, L. The Varnes classification of landslide types, an update. *Landslides* **2014**, *11*, 167–194. [CrossRef]
38. Herrera, G.; Gutierrez, F.; Garcia-Davalillo, J.C.; Guerrero, J.; Notti, D.; Galve, J.P.; Fernandez-Merodo, J.A.; Cooksley, G. Multi-sensor advanced DInSAR monitoring of very slow landslides: The Tena Valley case study (Central Spanish Pyrenees). *Remote Sens. Environ.* **2013**, *128*, 31–43. [CrossRef]
39. Cigna, F.; Bateson, L.B.; Jordan, C.J.; Dashwood, C. Simulating SAR geometric distortions and predicting Persistent Scatterer densities for ERS-1/2 and ENVISAT C-band SAR and InSAR applications: Nationwide feasibility assessment to monitor the landmass of Great Britain with SAR imagery. *Remote Sens. Environ.* **2014**, *152*, 441–466. [CrossRef]
40. Feng, W.; Jiawei, D.; Xiaoyu, Y.I.; Zhang, G. Deformation Analysis of Woda Village Old Landslide in Jinsha River Basin Using Sbas-Insar Technology. *J. Eng. Geol.* **2020**, *28*, 384–393.
41. Ferretti, A.; Prati, C.; Rocca, F. Permanent scatterers in SAR interferometry. *IEEE Trans. Geosci. Remote Sens.* **2001**, *39*, 8–20. [CrossRef]
42. Lanari, R.; Lundgren, P.; Manzo, M.; Casu, F. Satellite radar interferometry time series analysis of surface deformation for Los Angeles, California. *Geophys. Res. Lett.* **2004**, *31*, 021294. [CrossRef]
43. Lanari, R.; Casu, F.; Manzo, M.; Zeni, G.; Berardino, P.; Manunta, M.; Pepe, A. An overview of the small BAseline subset algorithm: A DInSAR technique for surface deformation analysis. *Pure Appl. Geophys.* **2007**, *164*, 637–661.

44. Tang, H.; Wasowski, J.; Juang, C.H. Geohazards in the three Gorges Reservoir Area, China Lessons learned from decades of research. *Eng. Geol.* **2019**, *261*, 105267. [CrossRef]
45. Jia, H.; Wang, Y.; Ge, D.; Deng, Y.; Wang, R. InSAR Study of Landslides: Early Detection, Three-Dimensional, and Long-Term Surface Displacement Estimation-A Case of Xiaojiang River Basin, China. *Remote Sens.* **2022**, *14*, 1759. [CrossRef]
46. Yao, J.; Yao, X.; Liu, X. Landslide Detection and Mapping Based on SBAS-InSAR and PS-InSAR: A Case Study in Gongjue County, Tibet, China. *Remote. Sens.* **2022**, *14*, 4728. [CrossRef]
47. Soltanieh, A.; Macciotta, R. Updated Understanding of the Ripley Landslide Kinematics Using Satellite InSAR. *Geosciences* **2022**, *12*, 298. [CrossRef]
48. Jiao, R.; Wang, S.; Yang, H.; Guo, X.; Han, J.; Pei, X.; Yan, C. Comprehensive Remote Sensing Technology for Monitoring Landslide Hazards and Disaster Chain in the Xishan Mining Area of Beijing. *Remote Sens.* **2022**, *14*, 4695. . [CrossRef]
49. Mishra, V.; Jain, K. Satellite based assessment of artificial reservoir induced landslides in data scarce environment: A case study of Baglihar reservoir in India. *J. Appl. Geophys.* **2022**, *205*, 104754. [CrossRef]
50. Perissin, D.; Ferretti, A. Urban-target recognition by means of repeated spaceborne SAR images. *IEEE Trans. Geosci. Remote Sens.* **2007**, *45*, 4043–4058. [CrossRef]
51. Bejar-Pizarro, M.; Notti, D.; Mateos, R.M.; Ezquerro, P.; Centolanza, G.; Herrera, G.; Bru, G.; Sanabria, M.; Solari, L.; Duro, J.; et al. Mapping Vulnerable Urban Areas Affected by Slow-Moving Landslides Using Sentinel-1 InSAR Data. *Remote Sens.* **2017**, *9*, 876. [CrossRef]
52. Aslan, G.; Foumelis, M.; Raucoules, D.; De Michele, M.; Bernardie, S.; Cakir, Z. Landslide Mapping and Monitoring Using Persistent Scatterer Interferometry (PSI) Technique in the French Alps. *Remote Sens.* **2020**, *12*, 1305. . [CrossRef]
53. Cascini, L.; Fornaro, G.; Peduto, D. Advanced low- and full-resolution DInSAR map generation for slow-moving landslide analysis at different scales. *Eng. Geol.* **2010**, *112*, 29–42. [CrossRef]
54. Handwerger, A.L.; Huang, M.H.; Fielding, E.J.; Booth, A.M.; Burgmann, R. A shift from drought to extreme rainfall drives a stable landslide to catastrophic failure. *Sci. Rep.* **2019**, *9*, 1569. [CrossRef] [PubMed]
55. Dille, A.; Kervyn, F.; Handwerger, A.L.; d'Oreye, N.; Derauw, D.; Bibentyo, T.M.; Samsonov, S.; Malet, J.P.; Kervyn, M.; Dewitte, O. When image correlation is needed: Unravelling the complex dynamics of a slow-moving landslide in the tropics with dense radar and optical time series. *Remote Sens. Environ.* **2021**, *258*, 112402. [CrossRef]
56. Wang, Y.; Cui, X.; Che, Y.; Li, P.; Jiang, Y.; Peng, X. Automatic Identification of Slope Active Deformation Areas in the Zhouqu Region of China With DS-InSAR Results. *Front. Environ. Sci.* **2022**, *10*, 883427. [CrossRef]
57. Ma, S.; Qiu, H.; Hu, S.; Yang, D.; Liu, Z. Characteristics and geomorphology change detection analysis of the Jiangdingya landslide on July 12, 2018, China. *Landslides* **2021**, *18*, 383–396. [CrossRef]
58. Fobert, M.A.; Singhroy, V.; Spray, J.G. InSAR Monitoring of Landslide Activity in Dominica. *Remote Sens.* **2021**, *13*, 815. . [CrossRef]
59. Xue, C.; Chen, K.; Tang, H.; Liu, P. Heavy rainfall drives slow-moving landslide in Mazhe Village, Enshi to a catastrophic collapse on 21 July 2020. *Landslides* **2022**, *19*, 177–186. [CrossRef]
60. Zhu, Y.; Yao, X.; Zhou, Z.; Ren, K.; Li, L.; Yao, C.; Gu, Z. Identifying the Mechanism of Toppling Deformation by InSAR : A Case Study in Xiluodu Reservoir, Jinsha River. *Landslides* **2022**, *19*, 2311–2327. [CrossRef]
61. Medhat, N.I.; Yamamoto, M.y.; Tolomei, C.; Harbi, A.; Maouche, S. Multi-temporal InSAR analysis to monitor landslides using the small baseline subset (SBAS) approach in the Mila Basin, Algeria. *Terra Nova* **2022**, *34*, 407–423. [CrossRef]
62. Liu, Y.; Yang, H.; Wang, S.; Xu, L.; Peng, J. Monitoring and Stability Analysis of the Deformation in the Woda Landslide Area in Tibet, China by the DS-InSAR Method. *Remote Sens.* **2022**, *14*, 532. [CrossRef]
63. Ying-Wen, Y.; Wei, L.; Na, F. The Feature and Prevent-Control Policy of Geological Disaster of Lanping in Nujiang, Yunnan. *Yunnan Geol.* **2019**, *38*, 2019.
64. Li, D.; Yin, K.; Leo, C. Analysis of Baishuihe landslide influenced by the effects of reservoir water and rainfall. *Environ. Earth Sci.* **2010**, *60*, 677–687. [CrossRef]
65. Xia, M.; Ren, G.M.; Ma, X.L. Deformation and mechanism of landslide influenced by the effects of reservoir water and rainfall, Three Gorges, China. *Nat. Hazards* **2013**, *68*, 467–482. [CrossRef]
66. Handwerger, A.L.; Fielding, E.J.; Huang, M.H.; Bennett, G.L.; Liang, C.; Schulz, W.H. Widespread Initiation, Reactivation, and Acceleration of Landslides in the Northern California Coast Ranges due to Extreme Rainfall. *J. Geophys.-Res.-Earth Surf.* **2019**, *124*, 1782–1797. [CrossRef]
67. Schulz, W.H.; McKenna, J.P.; Kibler, J.D.; Biavati, G. Relations between hydrology and velocity of a continuously moving landslide-evidence of pore-pressure feedback regulating landslide motion? *Landslides* **2009**, *6*, 181–190. [CrossRef]
68. Zhao, S.; Zeng, R.; Zhang, H.; Meng, X.; Zhang, Z.; Meng, X.; Wang, H.; Zhang, Y.; Liu, J. Impact of Water Level Fluctuations on Landslide Deformation at Longyangxia Reservoir, Qinghai Province, China. *Remote Sens.* **2022**, *14*, 212. . [CrossRef]
69. Chen, M.l.; Qi, S.c.; Lv, P.f.; Yang, X.g.; Zhou, J.w. Hydraulic response and stability of a reservoir slope with landslide potential under the combined effect of rainfall and water level fluctuation. *Environ. Earth Sci.* **2021**, *80*, 25. [CrossRef]

Disclaimer/Publisher's Note: The statements, opinions and data contained in all publications are solely those of the individual author(s) and contributor(s) and not of MDPI and/or the editor(s). MDPI and/or the editor(s) disclaim responsibility for any injury to people or property resulting from any ideas, methods, instructions or products referred to in the content.

Article

Spatial Pattern and Intensity Mapping of Coseismic Landslides Triggered by the 2022 Luding Earthquake in China

Zongji Yang [1,*], Bo Pang [1,2], Wufan Dong [1,2] and Dehua Li [1,2]

[1] Institute of Mountain Hazards and Environment, Chinese Academy of Sciences, Chengdu 610041, China
[2] University of Chinese Academy of Sciences, Beijing 100049, China
* Correspondence: yzj@imde.ac.cn; Tel.: +86-139-8214-4833

Abstract: On 5 September 2022, an Mw 6.6 earthquake occurred in Luding County in China, resulting in extensive surface rupture and casualties. Sufficient study on distribution characteristics and susceptibility regionalization of the earthquake-induced disasters (especially coseismic landslides) in the region has great significance to mitigation of seismic hazards. In this study, a complete coseismic landslide inventory, including 6233 landslides with 32.4 km^2 in area, was present through multi-temporal satellite images. We explored the distribution and controlling conditions of coseismic landslides induced by the 2022 Luding event from the perspective of epicentral distance. According to the maximum value of landslide area density, the geographical location with the strongest coseismic landslide activity intensity under the influence of seismic energy, the macro-epicenter, was determined, and we found a remarkable relationship with the landslide distribution and macro-epicentral distance, that is, both the landslide area and number density associatively decreased with the increase in macro-epicentral distance. Then, a fast and effective method for coseismic landslide intensity zoning based on the obvious attenuation relationship was proposed, which could provide theoretical reference for susceptibility mapping of coseismic landslides induced by earthquakes in mountainous areas. Additionally, to quantitatively assess the impact of topographic, seismogenic and lithological factors on the spatial pattern of coseismic landslides, the relationships between the occurrences of coseismic landslides and influencing factors, i.e., elevation, slope angle, local relief, aspect, distance to fault and lithology, were examined. This study provides a fresh perspective on intensity zoning of coseismic landslides and has important guiding significance for post-earthquake reconstruction and land use in the disaster area.

Keywords: coseismic landslides; Luding earthquake; spatial distribution; micro-epicenter; macro-epicenter

Citation: Yang, Z.; Pang, B.; Dong, W.; Li, D. Spatial Pattern and Intensity Mapping of Coseismic Landslides Triggered by the 2022 Luding Earthquake in China. *Remote Sens.* **2023**, *15*, 1323. https://doi.org/10.3390/rs15051323

Academic Editor: Rachid El Hamdouni

Received: 3 January 2023
Revised: 24 February 2023
Accepted: 24 February 2023
Published: 27 February 2023

Copyright: © 2023 by the authors. Licensee MDPI, Basel, Switzerland. This article is an open access article distributed under the terms and conditions of the Creative Commons Attribution (CC BY) license (https://creativecommons.org/licenses/by/4.0/).

1. Introduction

On 5 September 2022, at 12:52 p.m. local time, an Mw6.6 earthquake struck Luding, China [1]. The Luding event's epicenter is at 29.59°N, 102.08°E with a focus depth of 16 km. This earthquake damaged a vast amount of infrastructure, resulting in 88 deaths and over 400 injuries. Simultaneously, significant disasters such as coseismic landslides and collapses were induced, seriously endangering the personal security of local residents as well as reconstruction efforts.

Coseismic landslides are a geological disaster induced by earthquakes with strong destruction [2]. Thus, analyzing the distribution of coseismic landslides, investigating the correlations between coseismic landslides and triggering factors, and assessing the vulnerability of coseismic landslides are all crucial for guiding post-disaster reconstruction and secondary disaster prevention [3–5]. The landslide inventory serves as the foundation for analyzing and evaluating the mechanism of formation and spatial distribution of coseismic landslides, and many scholars have cataloged the coseismic landslide inventories for different earthquakes, such as the 1994 Mw6.7 Northridge event, America [6]; the

1999 Mw7.6 Chi-Chi event, China [7]; the 2008 Mw7.9 Wenchuan event, China [8] and the 2013 Mw6.6 Lushan event, China [9–11].

Coseismic landslide science research has become a focused issue, and numerous studies have been carried out on coseismic landslides of different earthquake magnitudes worldwide [12–20]. The findings indicate that the spatial pattern of coseismic landslides is attributed to the ground motion mode, vibration energy, geological environment and land type [21–25]. Without the restriction of geological environment, it is generally believed that the larger the earthquake energy level is, the more landslides there are with closer epicenter distance [26,27]. Keefer [28] discovered that the spatial frequency density of the coseismic landslides caused by the 1989 California earthquake decayed exponentially with the focal fault distance. Landslide susceptibility mapping and landslide sensitivity models considering different influencing factors are conducive to understanding landslide hazard risk [29]. Su et al. [30] found that the spatial distribution of the coseismic landslides induced by the 2008 Wenchuan event in Qingchuan County was mainly determined by lithology by using the logistic regression model. Zhao et al. [31] discovered that the majority of the coseismic landslides caused by the 2008 Wenchuan event and the 2013 Lushan event were concentrated in the Longmenshan fault's hanging wall, revealing the effect of tectonic mechanism on landslide distribution. All of these studies indicate that the controlling factors make great contributions to the occurrence and distribution of coseismic landslides. Thus, a thorough understanding of the interaction between coseismic landslides and controlling factors is critical for analyzing the formation mechanism of the distribution pattern [32].

Coseismic landslides are essentially the surface deformation caused by earthquakes [33], and their spatial distribution features are often associated with release of the seismic energy. However, there is no effective reference point that can indirectly reflect the release and spread of seismic energy on the surface. The epicenter of the earthquake can hardly reflect the location of the largest release of seismic energy, because some cases have shown that the coseismic landslide distribution is not strongly interrelated with the distance from the epicenter; for example, coseismic landslides induced by the Mw6.1 Ludian event in China were not concentrated at the epicenter but 5 km away [2], and many other earthquakes have similar deviations [18,24,32,34–36]. Consequently, it is very necessary to find a benchmark observation point that can reflect the intensity of the earthquake energy on the surface.

In this study, we focused on analyzing the spatial distribution pattern of coseismic landslides with elevation, slope angle, local relief, aspect, distance to fault and lithology. The maximum value of landslide area density (LAD) was utilized to determine the geographic location with the strongest landslide activity intensity affected by the Luding earthquake, which could be used as a key parameter to evaluate the impact of earthquake energy on the spatial pattern of coseismic landslides. Then, based on the landslide number density (LND) and landslide area density (LAD) with the grading threshold, the landslide intensity zoning was divided. The spatial pattern and formation mechanism of coseismic landslides were surveyed from the perspective of macro-epicentral distance. Our study gives detailed distribution characteristics of coseismic landslides induced by the 2022 Luding event which benefit ecological restoration and disaster management in the local region. Furthermore, we provide a novel reference for susceptibility zoning of coseismic landslides.

2. Materials and Methods

2.1. Study Area

The 2022 Luding event occurred in Luding County in China (Figure 1), at the southeastern margin of the Tibetan Plateau. Affected by the Indian Ocean monsoon climate, the earthquake area is rainy in autumn and summer, providing sufficient hydrodynamic conditions for the occurrence of post-earthquake geological disasters. The bedrock adjacent to the river is constantly eroded for a long term, reducing the stiffness of the unloaded rock mass exposed to the air. In this case, the broken rock layers and weathered fracture on the valley slopes are conducive to the failure of the coseismic landslides. With regard to the plate structure, the study region is situated at the intersection of the Indian Ocean plate and the Eurasian plate, which is the junction of the Longmenshan fault, the Xianshuihe fault and the Anninghe fault. The Indian Ocean plate continues to squeeze the Tibetan Plateau at a rate of 40–50 mm/a to the Eurasian plate every year, causing the crust in this area to move toward WE at a rate of 5–15 mm/a recorded by the global positioning system (GPS) [34]. The active crustal and tectonic movements in this area lay the groundwork for earthquake susceptibility, which is also the reason for the 2008 Wenchuan earthquake. The 2022 Luding event's epicenter, sited in the south of the Xianshuihe fault zone, is located in the Hailuogou scenic area of Moxi Town, only about 110 km away from the 2013 Ms7.0 Lushan earthquake [10]. The Xianshuihe fault, located in the famous Y-shaped fault region, is a sizable left-lateral strike-slip fault with considerable activity and NNW strike. It is about 400 km long and less than 300 km away from the Longmenshan fault in the northeast [37]. The 2022 Luding event is characteristic with a sinistral strike slip earthquake, and the seismogenic fault dips westward with a strike of 160° and an inclination of 80°. The maximum slip near the epicenter is about 184 cm, and the rupture duration is about 18 s. The earthquake gave rise to wide-ranging house damage and surface failure, affecting 82 townships of 12 counties. The intensity of the earthquake is elliptically distributed with the Xianshuihe fault as the long axis.

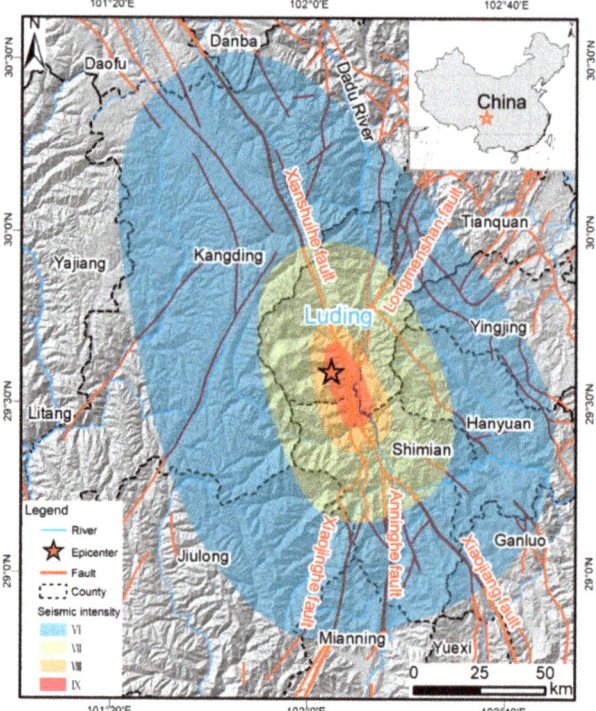

Figure 1. Location of the 2022 Luding earthquake in the southeastern Tibetan Plateau (the seismic intensity is from https://www.mem.gov.cn/xw/yjglbgzdt/202209/t20220911_422190.shtml; accessed on 15 September 2022).

2.2. Data and Methodology

It is not feasible to conduct a detailed on-site investigation for each coseismic landslide induced by the earthquake because of the rugged terrain in the region, so multi-temporal satellite images play a significant part in procuring coseismic landslide inventory data [34,38,39]. To compile a comprehensive landslide inventory database, we conducted spot observations and satellite image interpretation. The post-earthquake satellite images included GF-2 (access time: 10 September 2022; resolution: 3.2 m), GF-6 (access time: 10 September 2022; resolution: 8 m) and Beijing-3 (access time: 10 September 2022; resolution: 3 m), covering an area of about 2×10^4 km^2 (Figure 2). Coseismic landslides caused by the Luding event could be visually captured by comparing with pre-earthquake satellite images. The pre-earthquake satellite images included ZY-1 (access time: 8 July 2022; resolution: 2 m) and Sentinel-2 (access time: 29 April 2021; resolution: 10 m). We identified 6233 coseismic landslides according to the discrepancy in hue, texture, forest cover and other information of the satellite images (Figure 3). We outlined the profile of the coseismic landslide in ArcGIS platform to calculate the area of each coseismic landslide. Meanwhile, the field investigation gave great help for us to understand the coseismic landslide morphology more specifically.

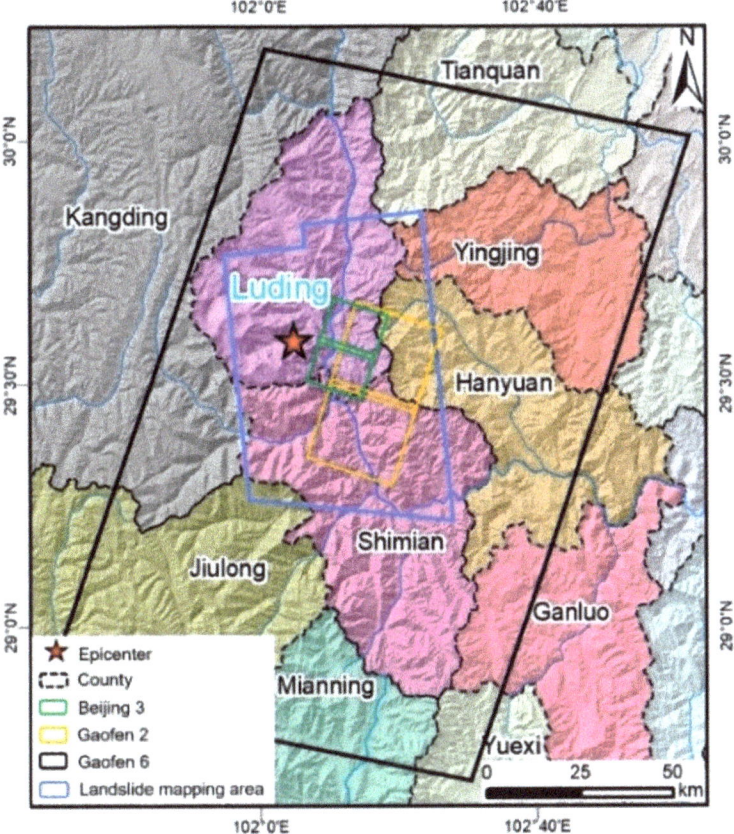

Figure 2. Coverage of satellite image.

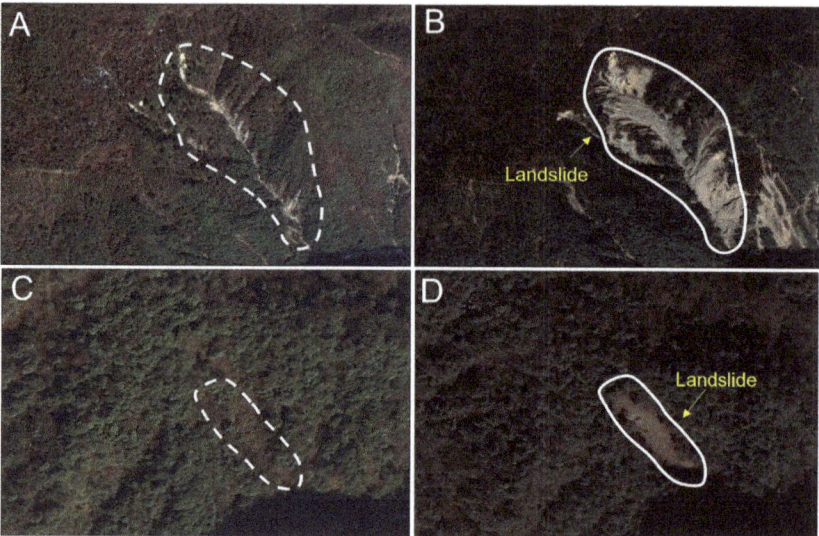

Figure 3. Coseismic landslides obtained from pre- (**A,B**) and post- (**C,D**) seismic satellite images.

In addition, to evaluate the impact of geology, seismic faults and topography on the spatial pattern of coseismic landslides, we collected elevation, slope angle, aspect, local relief, distance to seismogenic fault and lithology data. Slope angle, aspect and distance to seismogenic fault were collected from digital elevation model (DEM) data with 30 m resolution (http://www.gscloud.cn/; accessed on 15 September 2022). Geological data including lithology and faults were extracted from a geological map digitized to 1:250,000 scale. Subsequently, the spatial pattern of coseismic landslides with different factors was statistically analyzed in ArcGIS platform.

Landslide abundance is a commonly used indicator to measure the distribution scale of coseismic landslides [37,40]. We analyzed landslide area density (LAD) and landslide number density (LND) of the coseismic landslide inventory through the grid-based maps produced by small squares of 1 km in length and width with an area of 1 km^2 (LAD refers to the total area of coseismic landslides per km^2; LND refers to the total number of coseismic landslides per km^2).

3. Results

3.1. Landslide Inventory

The energy aroused by the Mw6.6 Luding event is dozens of times smaller than that of the 2008 Wenchuan event [41], so the type of landslide differs from that of the Wenchuan event dominated by a large landslide. The spot survey reveals that the type of coseismic landslide is mainly shallow landslide including natural slopes and cut slopes, manifested as mountain peeling. Affected by seismic amplification effect along the slope and shear vibration, coseismic landslides are mainly developed in the steep and gentle slope break section of watershed, ridge and mountainside, mainly including soil collapses, strongly weathered bedrocks (mainly granite in lithology) and rockfalls (Figure 4).

Figure 4. Typical coseismic landslide types induced by the Luding event. (**A**) collapse flow; (**B–F**) soil landslides.

The coseismic landslides triggered by the 2022 Luding event are primarily concentrated along the Moxi–Wanggangping section and distributed along both sides of the seismogenic fault. The coseismic landslides are most concentrated about 10 km to the south of Moxi Town, which is also a severe disaster area of coseismic landslides. However, there are relatively few coseismic landslides at the epicenter (Figure 5A,B). Landslide densities (LAD and LND) are mainly located south of the epicenter and are asymmetrically distributed along the fault. Over a 3545 km² affected region, the 2022 Luding event caused 6233 coseismic landslides at a minimum. In accordance with the correlations between the area affected by coseismic landslides and earthquake magnitude, most events are located at the lower side of the envelope (dashed and solid lines) [26,27]. The 2022 Luding earthquake follows the criteria as well (Figure 6A). For the landslide number and total area, the Luding event is as close to the trend line as the previous earthquakes and is located below the fitting line, demonstrating the coseismic landslides are more numerous and larger in area than the earthquakes with same magnitude (Figure 6B,C) [33,42]. With regard to the coseismic landslide frequency density p, Figure 6D compares the distribution frequency of the landslide area with other earthquakes near the fault: the 2008 Wenchuan event (Mw 7.9), the 2013 Lushan event (Mw 6.6) and the 2017 Jiuzhaigou event (Mw 6.5) [18]. The 2022 Luding earthquake also fits the inverse gamma distribution, i.e., $\log(p) = -1.56 \times \log(A) + 3.0$. For the size distribution, we divided the coseismic landslides into five scales in Table 1.

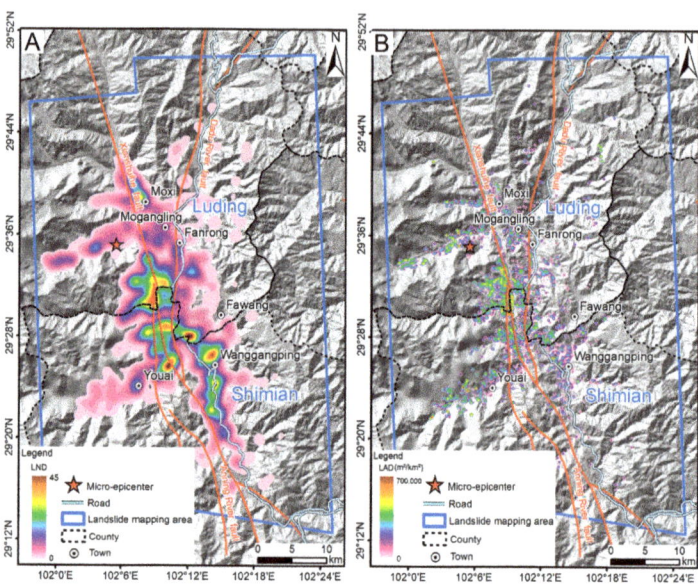

Figure 5. Regional distribution of the coseismic landslides caused by the 2022 Luding event. (**A**) LND; (**B**) LAD.

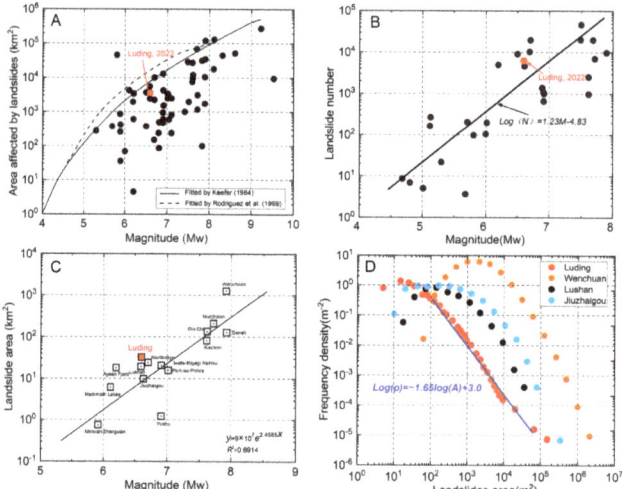

Figure 6. Comparison of landslide inventory caused by the 2022 Luding event with other events. (**A**) the earthquake magnitude and the area affected by coseismic landslides (the black circle represents other earthquake cases from [26,27]); (**B**) the earthquake magnitude and the number of landslides (other cases are referred from [33,42]); (**C**) the total area of landslides and earthquake magnitude; (**D**) the correlations of landslide area frequency density for the 2022 Luding event, the 2008 Wenchuan event, the 2013 Lushan event and the 2017 Jiuzhaigou event. The base map of (**C**,**D**) are referred from [18].

Table 1. Distribution characteristics of coseismic landslides in different scales.

Classification	Landslide Area/m^2	Landslide Number	Total Ratio %
I	areas < 1000	2337	37.50
II	1000 ≤ areas < 5000	2409	38.64
III	5000 ≤ areas < 10,000	720	11.56
IV	10,000 ≤ areas < 50,000	688	11.04
V	areas ≥ 50,000	79	1.26
Total		6233	100

3.2. Spatial Pattern of Coseismic Landslides with Epicentral Distance

The epicentral distance for a coseismic landslide is considered as the distance from the coseismic landslide point to the seismogenic epicenter [2,34]. The spatial distribution of coseismic landslides is significantly impacted by the epicentral distance as well [28,43]. However, the initial rupture site of the Xianshuihe fault which is regarded as the micro-epicenter of the 2022 Luding event is not the most intensive zone of the coseismic landslides (Figure 7A). The pertinence between the distribution of the coseismic landslide and the micro-epicentral distance is not correlative; the closer to the micro-epicenter of the earthquake, the lower occurrence probability of the coseismic landslides is, manifesting that the initial rupture point of the seismogenic fault cannot generate the energy that can trigger the occurrence of large-scale coseismic landslides. Since landslide concentration can assess earthquake damage to the ground [44], we set the geographical location at the maximum landslide area density as the macro-epicenter (located in 29.5°N, 102.15°E). Figure 7B clearly demonstrates that coseismic landslides are concentrated near the macro-epicenter, and the coseismic landslide number decreases inch by inch with the extension of the macro-epicentral distance. Notably, coseismic landslides with large area (red circle) appear sporadically far away from the macro-epicenter on the Xianshuihe fault's hanging wall, indicating that the spatial distribution of coseismic landslides caused by the 2022 Luding event is not only driven by the magnitude of seismic energy, but may be related to other influencing factors as well, such as terrain and stratum [43].

In order to quantitatively obtain the relationships between the spatial pattern of coseismic landslides and the epicentral distance, we compared the correlations between LAD and LND with the micro- and macro-epicenter, respectively, where the distance from the epicenter is the Euclidean distance [3]. The LAD and LND of coseismic landslides have no obvious correlation with micro-epicenter distance (Figure 8A,B), but are quantitatively related with macro-epicenter distance, i.e., $y = 216881 \times x^{(-0.69)}$ with $R^2 = 0.956$ for LAD and $y = 0.011x^2 - 0.89x + 18.78$ with $R^2 = 0.791$ for LND (Figure 8C,D). Therefore, the macro-epicentral distance, as a metric, can better indicate the degree of harm on the surface in the process of seismic energy diffusion when exploring the spatial pattern of coseismic landslides triggered by earthquakes compared with the micro-epicentral distance. Emphatically, the reason why two different functions were used is that we wanted to obtain a best goodness-of-fit of each LAD and LND with satisfying R^2 to ensure that the subsequent quantitative analysis had a smaller deviation.

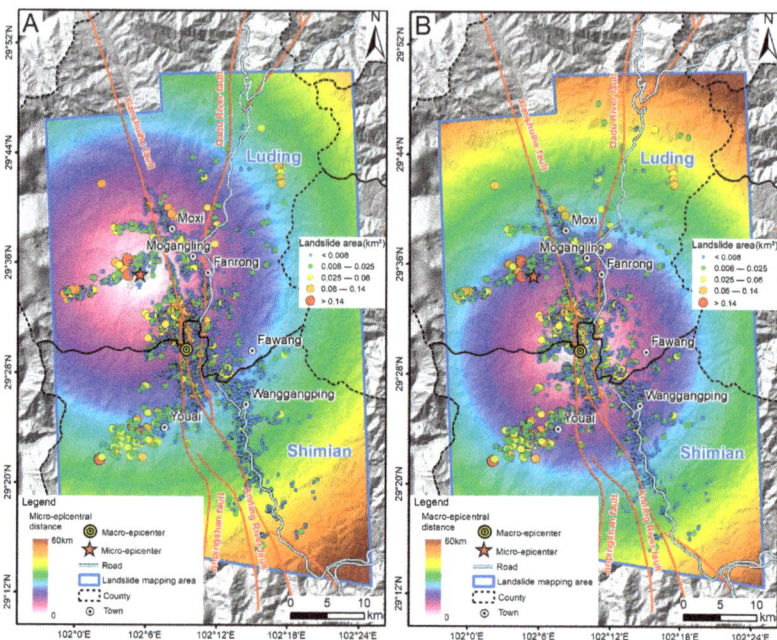

Figure 7. Coseismic landslide distribution with epicentral distance. (**A**) micro-epicenter; (**B**) macro-epicenter.

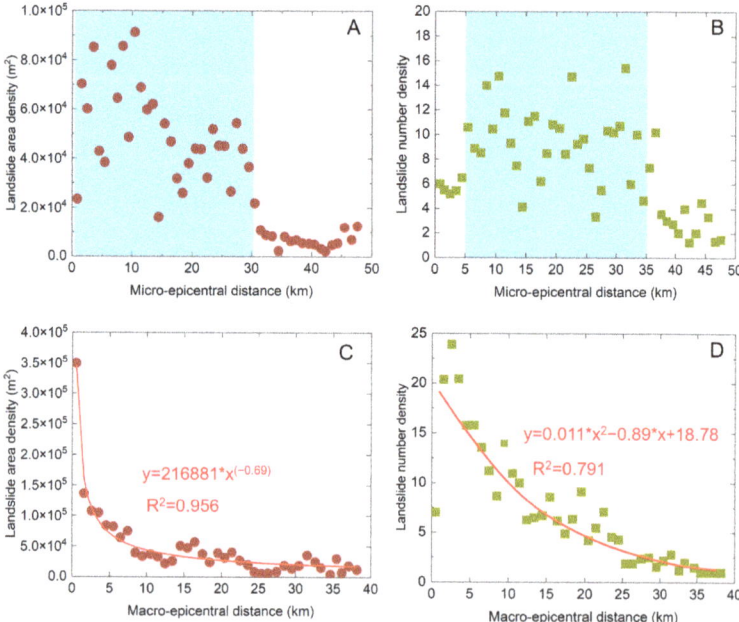

Figure 8. Correlations between epicentral distance and landslide abundance (LAD and LND). (**A**,**B**) micro-epicenter; (**C**,**D**) macro-epicenter.

The spatial distribution of coseismic landslides takes on a clear gradient reduction tendency with the macro-epicentral distance in Figure 8. Based on this, considering the landslide abundance, we proposed a fast and effective landslide intensity zoning method. The partition thresholds were calculated by the fitting function of LAD and LND with macro-epicenter distance in Table 2. Figure 9 is the landslide intensity map, depicting the spatial pattern of coseismic landslides in high-, mid- and low-prone area. According to statistics, the landslide number induced by the 2022 Luding earthquake in the high-, mid- and low-prone areas is 3829, 2164 and 240, respectively, with areas for 18.78 km^2, 12.30 km^2 and 1.34 km^2, respectively.

Table 2. Zoning value for landslide intensity.

Intensity Level	LAD (m^2/km^2)	LND	Macro-Epicentral Distance (km)
high-prone	50,000	10	11.5
mid-prone	25,000	5	22.9
low-prone	18,000	1	39.0

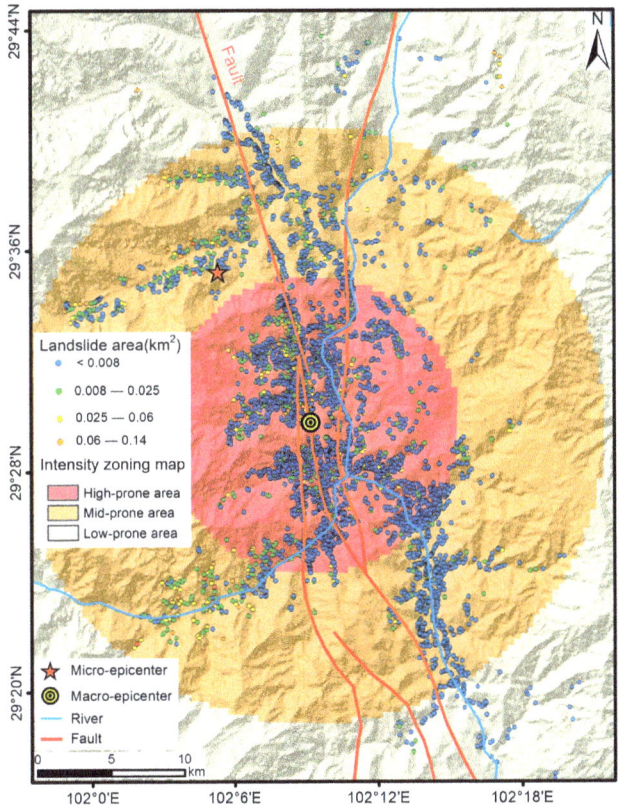

Figure 9. Landslide intensity map.

3.3. Controlling Factors of Coseismic Landslide Distribution

Earlier research has found the nonuniformity in the coseismic landslide spatial pattern [20,45]. In this part, we aim to analyze the related influencing factors that lead to the phenomenon. Six related factors were taken into account to thoroughly understand the impact of controlling factors on the spatial pattern of coseismic landslides.

3.3.1. Topographic Factors

Elevation is a crucial topographic feature that affects the occurrence of coseismic landslides [24]. The spatial pattern of coseismic landslides caused by the 2022 Luding event with elevation was statistically analyzed based on DEM data (Figure 10A). For the landslide abundance, the LAD and LND are mainly concentrated in the range of 0–10 km from the macro epicenter and 10–35 km from the micro epicenter, and the larger values of the LAD and LND correspond to the elevation of 1400–1800 m (Figure 10B,C). For the individual landslide, the regions with the elevation ranking from 1000 to 2300 m are more prevalent for coseismic landslides (Figure 10D), with 5518 in total, accounting for 88.5% of the total. This prone area is a concentrated area of human activities (housing construction, mining, road construction and water conservancy projects), manifesting that these activities have a significant effect on the susceptibility of coseismic landslides. After the elevation exceeds 1500 m, the landslide number decreases gradually as the elevation rises. With the increasing elevation, the landslide area expands inch by inch, and the relationship is approximately Log (y) = 0.52x + 2.35 (where y is landslide area, x is elevation of landslide).

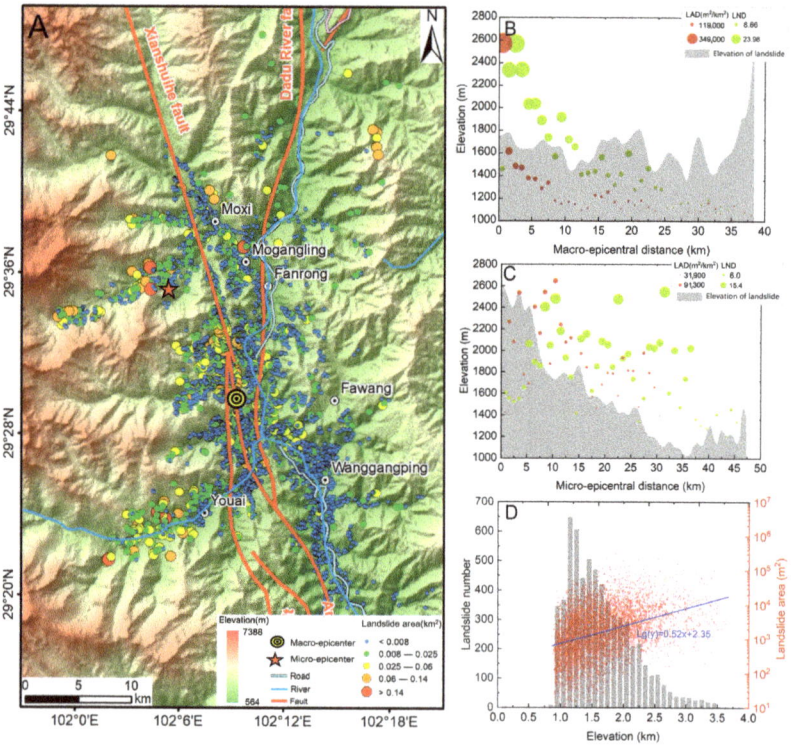

Figure 10. (**A**) coseismic landslide distribution with elevation; (**B**,**C**) elevation of landslide with macro- and micro-epicentral distance; (**D**) individual landslide distribution with elevation.

As is known, the slope angle has a massive effect on the distribution pattern of coseismic landslides. The shear stress of the rock mass along the slope increases with the increasing angle, in which case slope failure occurs in steep places even without earthquake. Numerous studies have found that most landslides are concentrated around 20–50° [8,46]. Figure 11A–C shows that the most intensive landslide abundance (LAD and LND) occurs in the range of 30–40°, and the epicentral distance has little influence on the angle. Throughout the whole affected region, most of the landslide cluster is in the range of 30 to 40°, totaling 4211, accounting for 67.6% of the total (Figure 11D). The landslide number shows a Gaussian distribution with the slope angle, reaching a peak at 35°.

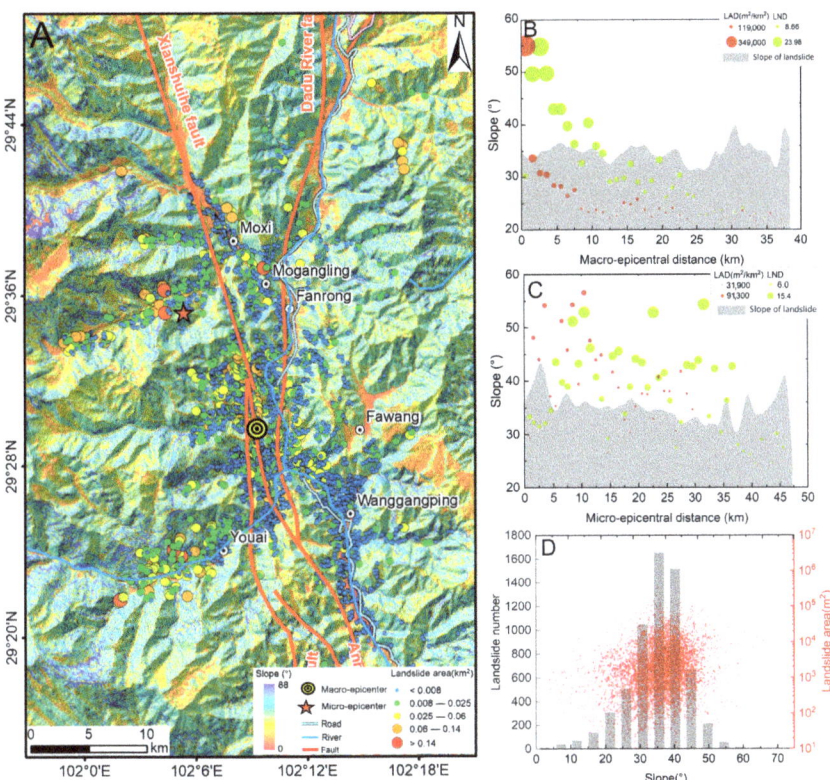

Figure 11. (**A**) coseismic landslide distribution with slope angle; (**B**,**C**) slope angle with macro- and micro-epicentral distance; (**D**) individual landslide distribution with slope angle.

Local relief reflects the surface distortion and is also a quantitative indicator of the gravity potential energy in the region. Figure 12A depicts the coseismic landslide distribution with the local relief as the background (the local relief map is extracted in GIS platform based on 5 × 5 km window). As shown in Figure 12B,C, the most intensive LAD and LND are primarily gathered at the elevation difference ranking from 1400 to 1800 m. In addition, 90.7% of the landslides occurred in the elevation of 1200–2000 m, 5654 in total (Figure 12D).

During seismic wave propagation, the development of coseismic landslides would be impacted by the aspect of slope [40,45,47]. In addition, the influence of climate on slopes with different aspect is also not consistent, resulting in different sensitivity to the instability of slopes of different aspect. For example, the slope on the windward side is more prone to runoff due to rain erosion, and these unstable slopes are more likely to be triggered by earthquakes [13]. The spatial pattern of landslides in various slope aspects is shown in Figure 13A. Statistical analysis indicates that the aspect distribution presents primarily S-E predominance, consistent with the Xianshuihe fault's strike and the travelling direction of seismic waves, which can be explained by the stronger amplification effect on the slopes that are back to the seismic wave's propagation direction (Figure 13B,C) [48]. The coseismic landslides in N, NE, E, SE, S, SW, W and NW are 292, 740, 1081, 1245, 908, 737, 684 and 546, respectively, accounting for 4.7%, 11.9%, 17.3%, 20.0%, 14.6%, 11.8%, 11.0% and 8.7% of total, respectively.

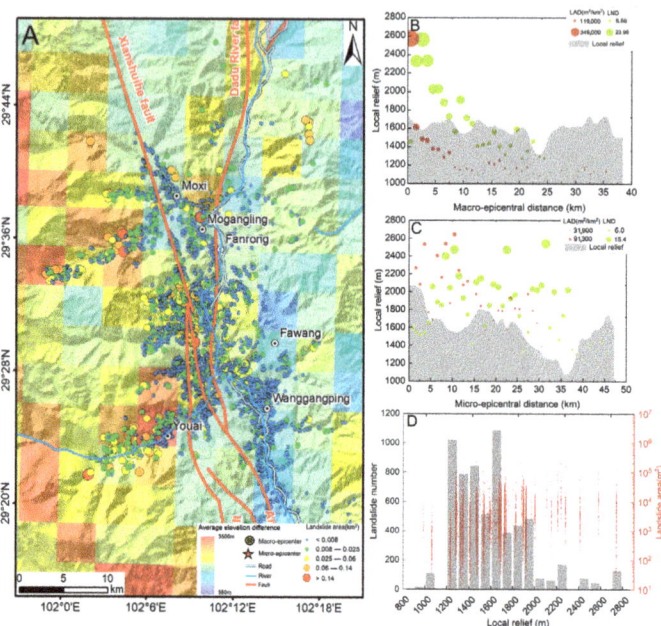

Figure 12. (**A**) coseismic landslide distribution with local relief; (**B**,**C**) local relief with macro- and micro-epicentral distance; (**D**) individual landslide distribution with local relief.

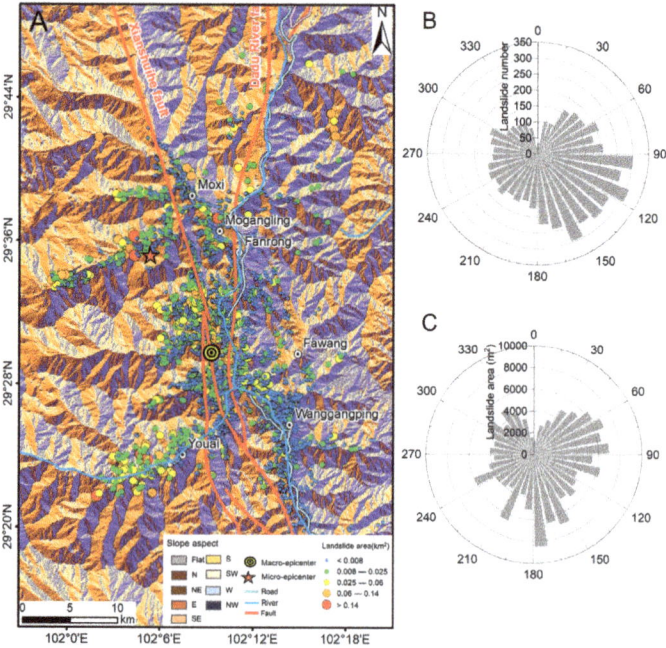

Figure 13. (**A**) coseismic landslide distribution with slope aspect; (**B**) landslide number and aspect; (**C**) landslide area and aspect.

3.3.2. Seismogenic Factor

The distribution of coseismic landslides is predominantly controlled by the seismogenic fault, confirmed in other cases [28,37]. In general, coseismic landslides occur on both sides of the seismogenic fault, and the landslide number exponentially decreases as the distance to fault increases [49]. Figure 14A shows the landslide distribution pattern with different distance to fault. The landslide distributed in 0–5 km, 5–10 km, >10 km counts 4652, 1242 and 339, respectively, occupy 74.7%, 19.9 and 5.4% of total, respectively, and follows an exponential distribution $y = 1613 \times e^{(-x/4.2)} - 17.5$ (Figure 14B). Furthermore, the area of the coseismic landslides increases with the increase in the distance to fault, following a relationship of approximately $Log(y) = 0.04x + 3.05$.

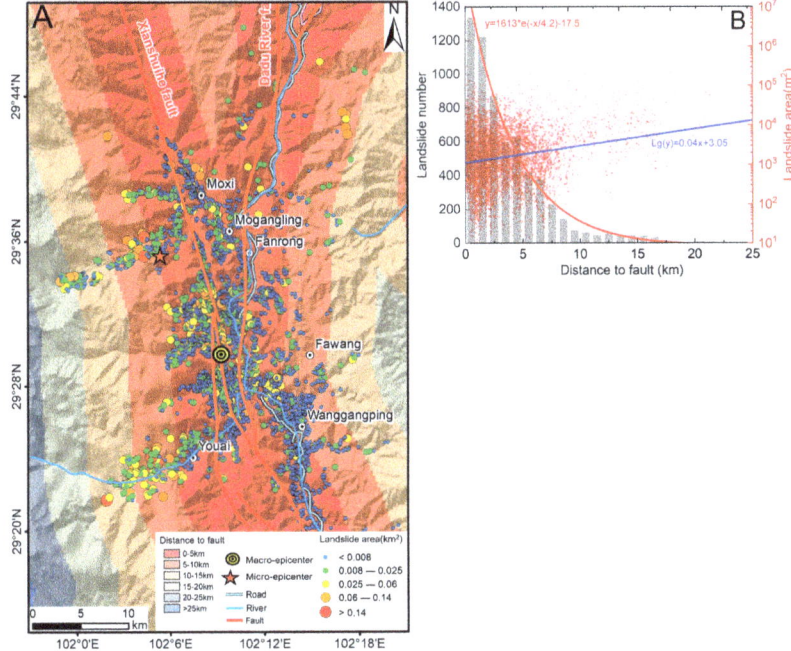

Figure 14. (**A**) coseismic landslide distribution with distance to fault; (**B**) correlations between landslide distribution and distance to fault.

3.3.3. Geological Factor

Figure 15A shows the landslide spatial distribution pattern related to different stratums. The potential impact areas of landslides have complex controlling lithologies, mainly including sedimentary rocks and intrusive rocks. Statistically, there are 2261, 1358, 956, 745, 369, 227 and 177 landslides occurring in granite, quartz diorite, tuff sandstone, metasandstone, quartz sandstone, carbonatite and ultrabasic rock, accounting for 36.3%, 21.9%, 15.3%, 12.0%, 5.9% and 2.3% of the total, respectively (Figure 15B). Granite is the main factor affecting the distribution of coseismic landslides. This "weakening effect" may be that the granite rock mass has very developed fissure joints due to the long-term tectonic activity in this area, leading to the decline of rock mass stability [40,50].

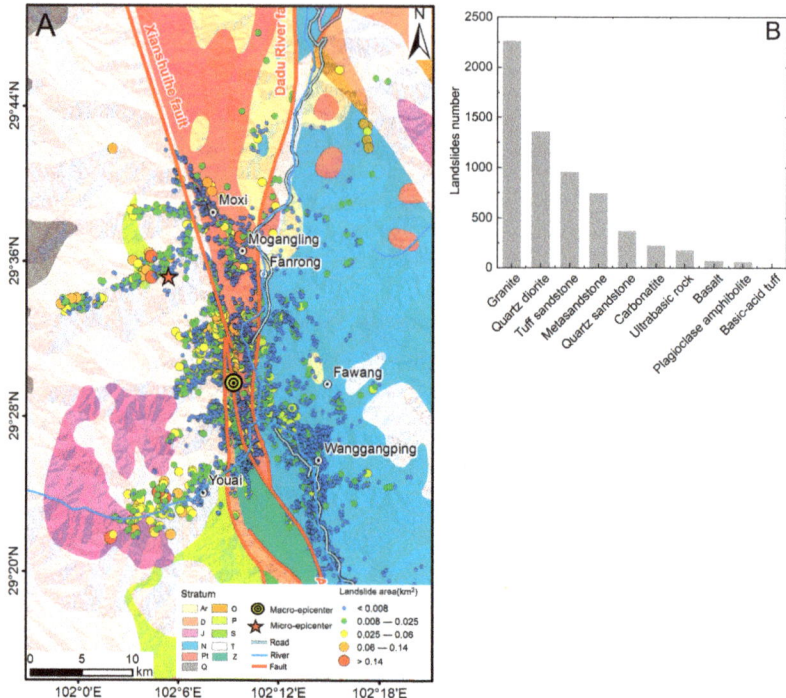

Figure 15. (**A**) coseismic landslide distribution with lithology; (**B**) correlations between landslide number and lithology.

4. Discussion

4.1. Landslide Intensity Mapping

Typically, seismic events cause varying extents of damage to the site. The Environmental Macroseismic Scale (EMS-98) can evaluate the level of ground damage of earthquakes, mainly determined by the damage to objects or buildings and the feelings of people in the epicentral area [51]. However, the EMS-98 is limited in a sparsely populated mountain area. Subsequently, the Environmental Seismic Intensity Scale (ESI-07) was exclusively developed to evaluate the impact of earthquakes in mountain areas on the natural environment [52]. The ESI-07 defines earthquake damage by considering the occurrence and area distribution of earthquake environmental effect (EEE), including surface fault, geological uplift and settlement, landslide, rockfall, liquefaction, surface subsidence and tsunami [53]. Gosar [54] determined the seismic intensity map of the 1998 Mw 5.6 Krn Mountains earthquake by investigating the spatial pattern of 78 rockfalls triggered by the earthquake, indicating that the seismic damage can be reflected by the spatial pattern of coseismic landslides/rockfalls when the geo-disasters caused by the seismic event are dominated by slope movements. The intensity isoseism can be determined by the distribution probability of landslides/rockfalls of different sizes. However, it is not easy to gauge the intensity isoseism and coseismic landslide regional intensity when there are mixed numerous coseismic landslides/rockfalls with various size. The 2022 Luding event also conforms to this characteristic.

The intensity zoning map of coseismic landslides developed in this paper reflected the concentration of coseismic landslides from the distribution abundance, which could therefore avoid some uncertainty caused by non-uniform landslide distribution with variable size using ESI-07. The threshold values of different partition levels are determined according to the specific distribution of coseismic landslides induced by earthquakes with different magnitudes. Emphatically, the landslide intensity map may not accurately reflect the macro earthquake intensity, but provide a suggestion. The landslide intensity map is the manifestation of the joint control of topographic conditions, seismogenic faults and stratum lithology, which is helpful for people to better comprehend the damage caused by seismic events on the spatial scale. The intensity mapping of coseismic landslides is not only applicable to the 2022 Luding event, but is also worth exploring in other earthquakes in the future. The zoning method based on macro-epicentral distance has better guiding significance for post-earthquake landslide prevention, rapid evaluation of seismic intensity and land-use planning.

4.2. Tectonic Genesis for the Discrepancy of Landslide Distribution

Many earthquakes have occurred on the Xianshuihe fault in history due to abundant tectonic activities (Figure 16). According to the record, the GPS horizontal displacement velocity in the seismogenic fault's hanging wall is significantly larger than that in the footwall and the direction of velocity is nearly parallel, which contributes to a sinistral strike-slip earthquake for the 2022 Luding event. The spatial pattern of coseismic landslides is profoundly affected by the fault slip mode. Coseismic landslides caused by strike-slip earthquakes, particularly deep landslides, are often localized within 5 km of the seismogenic fault [55]. The coseismic landslide spatial pattern of the Luding event also conforms to this rule. The majority of coseismic landslides towards SE also reveal that there is a strong correlation between the direction of seismic waves and the distribution of coseismic landslides. Additionally, the preponderance of the coseismic landslides localized in the hanging wall implies that the Xianshuihe fault's hanging wall exhibits more robust vibrational characteristics than the footwall wall [1].

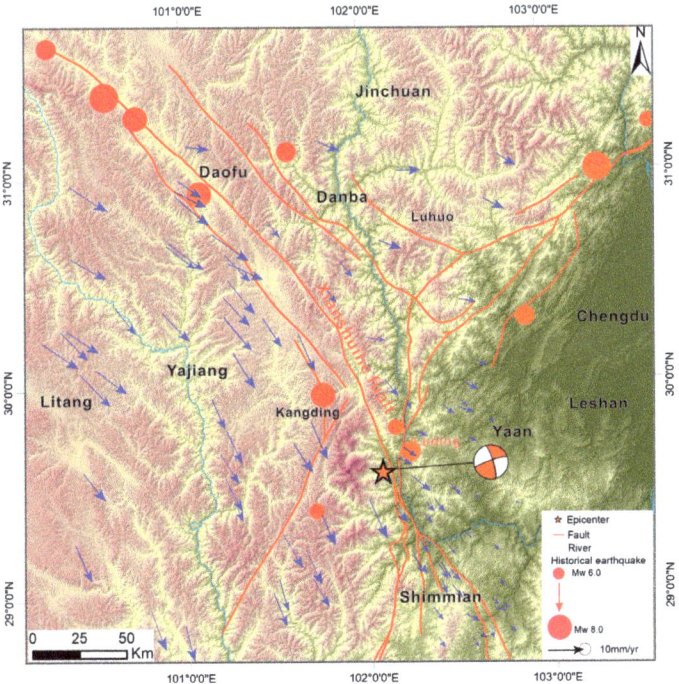

Figure 16. Regional map showing the velocity field of GNSS horizontal motion before the 2022 Luding M6.8 earthquake (from http://data.earthquake.cn; accessed on 20 October 2022) and the historical earthquakes (>Mw6) since 1900 (from USGS.gov I Science for a changing world; accessed on 20 October 2022).

The location of the micro-epicenter, projection point from source to surface, is determined by inversion from nearby stations based on seismic waves released by the initial rupture of the fault. Seen in Figure 17A, the distance between the fault and the micro-epicenter is a particular amount related to dip angle of the fault, slip angle and focal depth. The accumulated stress in the process of plate compression is released suddenly after an earthquake, and the fault plane releases seismic energy onto the surrounding area (Figure 17B). During energy transmission, the seismic energy attenuates along the path [56], resulting that the micro-epicenter is not the place with the largest surface energy, which explains why the coseismic landslide spatial pattern is more closely related to the macro-epicentral distance, and the macro-epicenter has more control over the occurrence and spatial pattern of coseismic landslides than the micro-epicenter. In fact, the specific position of the macro-epicenter depends vastly on the rupture direction of the seismogenic fault during an earthquake. On account of the southward rupture of the seismogenic fault during the Luding event [57], the macro-epicenter is located on the south side of the micro-epicenter. This phenomenon is also confirmed in the 2008 Wenchuan earthquake because the controlled area of coseismic landslides induced by the Wenchuan earthquake is just on the northward rupture of the seismogenic fault, rather than the micro-epicenter [24]. The impact of tectonics on spatial pattern of coseismic landslides induced by the 2022 Luding event is emphasized, which differs from the combination of topography and tectonics proposed by Zhao et al. [1]. Thus, we propose that more focus should be placed on the macro-epicentral distance rather than the micro-epicentral distance in the future study of the spatial characteristics of coseismic landslides controlled by epicentral distance.

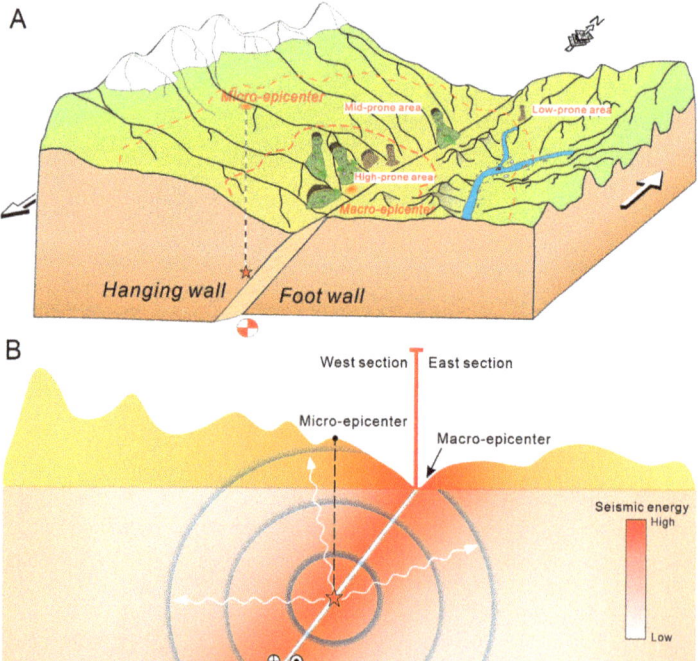

Figure 17. Schematic Diagram of the 2022 Luding Mw6.6 earthquake. (**A**) three-dimensional focal mechanism; (**B**) propagation process of earthquake energy.

4.3. Limitations

This study aims to compile a thorough coseismic landslide inventory for the 2022 Luding event and analyze the impact of potential controlling factors on the distribution pattern of coseismic landslides. However, there are still a few drawbacks in landslide mapping and corresponding analysis.

For landslide mapping, we extracted 6233 landslides totaling 32.4 km² in size, covering the area 50 km away from the epicenter. However, some small landslides may not be effectively identified due to the inadequate resolution of satellite images and lush vegetation, resulting in a modest undercount of landslides compared to the actual situation. Despite the fact that the data for the landslide inventory is overestimated, the present data of coseismic landslides covered the whole meizoseismal region and will not change the assessment results.

In addition, the mismatch between the DEM data (resolution: 30 m × 30 m) and the geological map at 1:250,000 scale may lead to deviation in results, but this can be avoided because it is not our main research purpose. The weathering of granite can also be considered in the spatial pattern of coseismic landslides to obtain more comprehensive outcomes, which requires more detailed geological and lithological mapping at a scale larger than 1:250,000 (such as 1:25,000, 1:10,000 or 1:5000). These further studies are able to add to our awareness of the relationship between geology and coseismic landslides and additional details of granite weathering grade maps as a predisposing factor [58], which contributes to our comprehension of landslide distribution for further landslide risk and hazard assessment.

With regard to the analysis of epicentral distance, we took the place with the maximum value of LAD in the study region as the macro-epicenter of the earthquake. However, whether this location is the projection point on the surface where the maximum energy is released when the fault breaks remains to be debated. Surely, the macro-epicenter, the location where the surface is most affected by the earthquake, is related to the release of earthquake stress. Moreover, we solely explored the correlation between epicentral distance and spatial pattern of coseismic landslides from the macroscopic phenomenon on the surface, without considering the intrinsic influence of seismic physical parameters on landslides, such as seismic attenuation acceleration (α), ground motion period (T), seismic vibration duration (t), etc. because these not only involve the research content of the earthquake itself, but also involve the relationship between seismic physical parameters and landslide material characteristic parameters. If the seismic physical parameters and the coseismic landslide physical parameters are studied together, there will be many complex functional relationships, and no satisfactory solution can be obtained.

5. Conclusions

In order to clarify the spatial pattern characteristics of coseismic landslides caused by the 2022 Luding event, we provided a complete landslide inventory containing 6233 coseismic landslides through remote sensing interpretation and field investigation. The associations between the spatial pattern of coseismic landslides and six potential controlling factors encompassing elevation, slope angle, slope aspect, local relief, distance to the seismogenic fault and lithology were analyzed. We found that mostly coseismic landslides are primarily concentrated on the slopes at elevation from 1000 to 2300 m with slope of 30–40°, an E–S aspect and local relief from 1200 to 2000 m. The main coseismic landslide occurred in granite, accounting for the largest proportion (36.3%). Within 5 km from the fault, there is an intensive concentration of coseismic landslides, clustered along both sides of the fault. The seismogenic fault and focal mechanism play an important role in the spatial pattern of coseismic landslides in this earthquake.

Through the maximum value of LAD of coseismic landslides, the position of the macro-epicenter is established. The LAD and LND of coseismic landslides exhibit a fairly satisfactory function relationship with the macro-epicentral distance (compared with the micro epicenter) as follows: $y = 216{,}881 \times x^{(-0.69)}$ for LAD with $R^2 = 0.956$; $y = 0.011x^2 - 0.89x + 18.78$ for LND with $R^2 = 0.791$. Then, the intensity distribution map of coseismic landslides was proposed. The intensity distribution of landslides can reveal the dissipation process of seismic energy propagation and provide information on the damage of the earthquake to mountain areas. In addition, we also revealed the reason why the spatial pattern of coseismic landslides deviated from the micro-epicenter in the 2022 Luding earthquake from the perspective of tectonic activities, assisting us in better comprehending the distribution mechanism of earthquake-induced landslides.

Author Contributions: Author Contributions: Conceptualization, Z.Y.; methodology, Z.Y.; software, B.P.; validation, B.P., W.D. and D.L.; formal analysis, B.P.; investigation, Z.Y.; resources, Z.Y.; data curation, B.P.; writing—original draft preparation, B.P.; writing—review and editing, B.P., W.D. and D.L.; visualization, B.P.; supervision, Z.Y.; project administration, Z.Y.; funding acquisition, Z.Y. All authors have read and agreed to the published version of the manuscript.

Funding: This work was financially supported by the Projects of Western Light in Chinese Academy of Sciences (Grant Nos. E1R2090) and the National Natural Science Foundation of China (Grant Nos. U22A20565).

Data Availability Statement: The datasets used and analyzed during the current study are available from the corresponding authors upon reasonable request.

Acknowledgments: The authors are grateful to the editors and the anonymous reviewers for their extensive and profound comments and suggestions, which substantially improved the quality of the paper. The authors are thankful for the data support from "China Earthquake Networks Center, National Earthquake Data Center. (http://data.earthquake.cn; accessed on 20 October 2022)". We also thank Zhang Jianqiang, Hu Kaiheng, Chen Huayong, Zhao Bo, Zhu Lei, Zhang Weifeng, Liu Qiao and Zou Qiang for their support in collecting the data and interpretation.

Conflicts of Interest: The authors declare no conflict of interest.

References

1. Zhao, B.; Hu, K.-H.; Yang, Z.-J.; Liu, Q.; Zou, Q.; Chen, H.-Y.; Zhang, B.; Zhang, W.-F.; Zhu, L.; Su, L.-J. Geomorphic and tectonic controls of landslides induced by the 2022 Luding earthquake. *J. Mt. Sci.* **2022**, *19*, 3323–3345. [CrossRef]
2. Zou, Y.; Qi, S.W.; Guo, S.F.; Zheng, B.W.; Zhan, Z.F.; He, N.W.; Huang, X.L.; Hou, X.K.; Liu, H.Y. Factors controlling the spatial distribution of coseismic landslides triggered by the Mw 6.1 Ludian earthquake in China. *Eng. Geol.* **2022**, *296*, 106477. [CrossRef]
3. Karakas, G.; Nefeslioglu, H.A.; Kocaman, S.; Buyukdemircioglu, M.; Yurur, T.; Gokceoglu, C. Derivation of earthquake-induced landslide distribution using aerial photogrammetry: The January 24, 2020, Elazig (Turkey) earthquake. *Landslides* **2021**, *18*, 2193–2209. [CrossRef]
4. Miles, S.B.; Keefer, D.K. Evaluation of CAMEL—Comprehensive areal model of earthquake-induced landslides. *Eng. Geol.* **2009**, *104*, 1–15. [CrossRef]
5. Wu, W.; Xu, C.; Wang, X.; Tian, Y.; Deng, F. Landslides Triggered by the 3 August 2014 Ludian (China) Mw 6.2 Earthquake: An Updated Inventory and Analysis of Their Spatial Distribution. *J. Earth Sci.* **2020**, *31*, 853–866. [CrossRef]
6. Budimir, M.E.A.; Atkinson, P.M.; Lewis, H.G. Seismically induced landslide hazard and exposure modelling in Southern California based on the 1994 Northridge, California earthquake event. *Landslides* **2015**, *12*, 895–910. [CrossRef]
7. Lee, Y.T.; Turcotte, D.L.; Rundle, J.B.; Chen, C.C. Aftershock Statistics of the 1999 Chi-Chi, Taiwan Earthquake and the Concept of Omori Times. *Pure Appl. Geophys.* **2013**, *170*, 221–228. [CrossRef]
8. Chigira, M.; Wu, X.; Inokuchi, T.; Wang, G. Landslides induced by the 2008 Wenchuan earthquake, Sichuan, China. *Geomorphology* **2010**, *118*, 225–238. [CrossRef]
9. Cui, P.; Zhang, J.Q.; Yang, Z.J.; Chen, X.Q.; You, Y.; Li, Y. Activity and distribution of geohazards induced by the Lushan earthquake, April 20, 2013. *Nat. Hazards* **2014**, *73*, 711–726. [CrossRef]
10. Wang, G. Comparison of the landslides triggered by the 2013 Lushan earthquake with those triggered by the strong 2008 Wenchuan earthquake in areas with high seismic intensities. *Bull. Eng. Geol. Environ.* **2015**, *74*, 77–89. [CrossRef]
11. Xu, C.; Xu, X.; Shyu, J.B.H. Database and spatial distribution of landslides triggered by the Lushan, China Mw 6.6 earthquake of 20 April 2013. *Geomorphology* **2015**, *248*, 77–92. [CrossRef]
12. Chen, C.-W.; Sato, M.; Yamada, R.; Iida, T.; Matsuda, M.; Chen, H. Modeling of earthquake-induced landslide distributions based on the active fault parameters. *Eng. Geol.* **2022**, *303*, 106640. [CrossRef]
13. Chang, M.; Cui, P.; Xu, L.; Zhou, Y. The spatial distribution characteristics of coseismic landslides triggered by the Ms7.0 Lushan earthquake and Ms7.0 Jiuzhaigou earthquake in southwest China. *Environ. Sci. Pollut. Res.* **2021**, *28*, 20549–20569. [CrossRef] [PubMed]
14. Zhuang, J.; Peng, J.; Xu, C.; Li, Z.; Densmore, A.; Milledge, D.; Iqbal, J.; Cui, Y. Distribution and characteristics of loess landslides triggered by the 1920 Haiyuan Earthquake, Northwest of China. *Geomorphology* **2018**, *314*, 1–12. [CrossRef]
15. Tian, Y.; Xu, C.; Ma, S.; Xu, X.; Wang, S.; Zhang, H. Inventory and Spatial Distribution of Landslides Triggered by the 8th August 2017 MW 6.5 Jiuzhaigou Earthquake, China. *J. Earth Sci.* **2019**, *30*, 206–217. [CrossRef]

16. Has, B.; Noro, T.; Maruyama, K.; Nakamura, A.; Ogawa, K.; Onoda, S. Characteristics of earthquake-induced landslides in a heavy snowfall region—Landslides triggered by the northern Nagano prefecture earthquake, March 12, 2011, Japan. *Landslides* **2012**, *9*, 539–546. [CrossRef]
17. Guo, C.-w.; Huang, Y.-d.; Yao, L.-k.; Alradi, H. Size and spatial distribution of landslides induced by the 2015 Gorkha earthquake in the Bhote Koshi river watershed. *J. Mt. Sci.* **2017**, *14*, 1938–1950. [CrossRef]
18. Fan, X.; Scaringi, G.; Xu, Q.; Zhan, W.; Dai, L.; Li, Y.; Pei, X.; Yang, Q.; Huang, R. Coseismic landslides triggered by the 8th August 2017 Ms 7.0 Jiuzhaigou earthquake (Sichuan, China): Factors controlling their spatial distribution and implications for the seismogenic blind fault identification. *Landslides* **2018**, *15*, 967–983. [CrossRef]
19. Chen, X.; Liu, C.; Wang, M. A method for quick assessment of earthquake-triggered landslide hazards: A case study of the Mw6.1 2014 Ludian, China earthquake. *Bull. Eng. Geol. Environ.* **2019**, *78*, 2449–2458. [CrossRef]
20. Collins, B.D.; Kayen, R.; Tanaka, Y. Spatial distribution of landslides triggered from the 2007 Niigata Chuetsu–Oki Japan Earthquake. *Eng. Geol.* **2012**, *127*, 14–26. [CrossRef]
21. Meunier, P.; Hovius, N.; Haines, A.J. Regional patterns of earthquake-triggered landslides and their relation to ground motion. *Geophys. Res. Lett.* **2007**, *34*, 1–5. [CrossRef]
22. Meunier, P.; Hovius, N.; Haines, J.A. Topographic site effects and the location of earthquake induced landslides. *Earth Planet. Sci. Lett.* **2008**, *275*, 221–232. [CrossRef]
23. Lee, C.-T.; Huang, C.-C.; Lee, J.-F.; Pan, K.-L.; Lin, M.-L.; Dong, J.-J. Statistical approach to earthquake-induced landslide susceptibility. *Eng. Geol.* **2008**, *100*, 43–58. [CrossRef]
24. Xu, C.; Xu, X. Statistical analysis of landslides caused by the Mw 6.9 Yushu, China, earthquake of April 14, 2010. *Nat. Hazards* **2014**, *72*, 871–893. [CrossRef]
25. Zhong, X.M.; Xu, X.W.; Chen, W.K.; Liang, Y.X.; Sun, Q.Y. Characteristics of loess landslides triggered by the 1927 Mw8.0 earthquake that occurred in Gulang County, Gansu Province, China. *Front. Environ. Sci.* **2022**, *10*, 1–19. [CrossRef]
26. Keefer, D.K. Landslides caused by earthquakes. *Geol. Soc. Am. Bull.* **1984**, *95*, 406–421. [CrossRef]
27. Rodrıguez, C.; Bommer, J.; Chandler, R. Earthquake-induced landslides: 1980–1997. *Soil Dyn. Earthq. Eng.* **1999**, *18*, 325–346. [CrossRef]
28. Keefer, D.K. Statistical analysis of an earthquake-induced landslide distribution—The 1989 Loma Prieta, California event. *Eng. Geol.* **2000**, *58*, 231–249. [CrossRef]
29. Huang, F.M.; Cao, Z.S.; Jiang, S.H.; Zhou, C.B.; Huang, J.S.; Guo, Z.Z. Landslide susceptibility prediction based on a semi-supervised multiple-layer perceptron model. *Landslides* **2020**, *17*, 2919–2930. [CrossRef]
30. Su, F.H.; Cui, P.; Zhang, J.Q.; Xiang, L.Z. Susceptibility assessment of landslides caused by the wenchuan earthquake using a logistic regression model. *J. Mt. Sci.* **2010**, *7*, 234–245. [CrossRef]
31. Zhao, B.; Li, W.L.; Su, L.J.; Wang, Y.S.; Wu, H.C. Insights into the Landslides Triggered by the 2022 Lushan Ms 6.1 Earthquake: Spatial Distribution and Controls. *Remote Sens.* **2022**, *14*, 4365. [CrossRef]
32. Zhao, B.; Liao, H.J.; Su, L.J. Landslides triggered by the 2018 Lombok earthquake sequence, Indonesia. *Catena* **2021**, *207*, 105676. [CrossRef]
33. He, X.L.; Xu, C. Spatial distribution and tectonic significance of the landslides triggered by the 2021 Ms6.4 Yangbi earthquake, Yunnan, China. *Front. Earth Sci.* **2022**, *10*, 1–17. [CrossRef]
34. Dai, F.C.; Xu, C.; Yao, X.; Xu, L.; Tu, X.B.; Gong, Q.M. Spatial distribution of landslides triggered by the 2008 Ms 8.0 Wenchuan earthquake, China. *J. Asian Earth Sci.* **2011**, *40*, 883–895. [CrossRef]
35. Lu, J.Y.; Li, W.L.; Zhan, W.W.; Tie, Y.B. Distribution and Mobility of Coseismic Landslides Triggered by the 2018 Hokkaido Earthquake in Japan. *Remote Sens.* **2022**, *14*, 3957. [CrossRef]
36. Zhang, S.; Li, R.; Wang, F.W.; Iio, A. Characteristics of landslides triggered by the 2018 Hokkaido Eastern Iburi earthquake, Northern Japan. *Landslides* **2019**, *16*, 1691–1708. [CrossRef]
37. Gorum, T.; Fan, X.M.; van Westen, C.J.; Huang, R.Q.; Xu, Q.; Tang, C.; Wang, G.H. Distribution pattern of earthquake-induced landslides triggered by the 12 May 2008 Wenchuan earthquake. *Geomorphology* **2011**, *133*, 152–167. [CrossRef]
38. Hungr, O.; Leroueil, S.; Picarelli, L. The Varnes classification of landslide types, an update. *Landslides* **2014**, *11*, 167–194. [CrossRef]
39. Martino, S.; Bozzano, F.; Caporossi, P.; D'angiò, D.; Della Seta, M.; Esposito, C.; Fantini, A.; Fiorucci, M.; Giannini, L.; Iannucci, R. Impact of landslides on transportation routes during the 2016–2017 Central Italy seismic sequence. *Landslides* **2019**, *16*, 1221–1241. [CrossRef]
40. Shao, X.Y.; Ma, S.Y.; Xu, C. Distribution and characteristics of shallow landslides triggered by the 2018 Mw 7.5 Palu earthquake, Indonesia. *Landslides* **2022**, *20*, 1–19. [CrossRef]
41. Tang, R.; Fan, X.; Scaringi, G.; Xu, Q.; van Westen, C.J.; Ren, J.; Havenith, H.-B. Distinctive controls on the distribution of river-damming and non-damming landslides induced by the 2008 Wenchuan earthquake. *Bull. Eng. Geol. Environ.* **2019**, *78*, 4075–4093. [CrossRef]
42. Keefer, D.K. Investigating landslides caused by earthquakes—A historical review. *Surv. Geophys.* **2002**, *23*, 473–510. [CrossRef]
43. Xu, C.; Xu, X.; Yao, X.; Dai, F. Three (nearly) complete inventories of landslides triggered by the May 12, 2008 Wenchuan Mw 7.9 earthquake of China and their spatial distribution statistical analysis. *Landslides* **2014**, *11*, 441–461. [CrossRef]
44. Xu, C.; Xu, X.W.; Zhou, B.G.; Yu, G.H. Revisions of the M 8.0 Wenchuan earthquake seismic intensity map based on co-seismic landslide abundance. *Nat. Hazards* **2013**, *69*, 1459–1476. [CrossRef]

45. Sato, H.P.; Harp, E.L. Interpretation of earthquake-induced landslides triggered by the 12 May 2008, M7.9 Wenchuan earthquake in the Beichuan area, Sichuan Province, China using satellite imagery and Google Earth. *Landslides* **2009**, *6*, 153–159. [CrossRef]
46. Yin, Y.; Wang, F.; Sun, P. Landslide hazards triggered by the 2008 Wenchuan earthquake, Sichuan, China. *Landslides* **2009**, *6*, 139–152. [CrossRef]
47. Havenith, H.B.; Vanini, M.; Jongmans, D.; Faccioli, E. Initiation of earthquake-induced slope failure: Influence of topographical and other site specific amplification effects. *J. Seismol.* **2003**, *7*, 397–412. [CrossRef]
48. Celebi, M. Topographical and geological amplifications determined from strong-motion and aftershock records of the 3 March 1985 Chile earthquake. *Bull. Seismol. Soc. Amer.* **1987**, *77*, 1147–1167. [CrossRef]
49. Qi, S.; Xu, Q.; Lan, H.; Zhang, B.; Liu, J. Spatial distribution analysis of landslides triggered by 2008.5. 12 Wenchuan Earthquake, China. *Eng. Geol.* **2010**, *116*, 95–108. [CrossRef]
50. Osmundsen, P.T.; Henderson, I.; Lauknes, T.R.; Larsen, Y.; Redfield, T.F.; Dehls, J. Active normal fault control on landscape and rock-slope failure in northern Norway. *Geology* **2009**, *37*, 135–138. [CrossRef]
51. Rossi, A.; Tertulliani, A.; Azzaro, R.; Graziani, L.; Rovida, A.; Maramai, A.; Pessina, V.; Hailemikael, S.; Buffarini, G.; Bernardini, F.; et al. The 2016-2017 earthquake sequence in Central Italy: Macroseismic survey and damage scenario through the EMS-98 intensity assessment. *Bull. Earthq. Eng.* **2019**, *17*, 2407–2431. [CrossRef]
52. Michetti, A.; Esposito, E.; Guerrieri, L.; Porfido, S.; Serva, L.; Tatevossian, R.; Vittori, E.; Audemard, F.; Azuma, T.; Clague, J. Environmental seismic intensity scale-ESI 2007. *Mem. Descr. Carta Geol. D'Ital* **2007**, *74*, 7–23.
53. Serva, L.; Vittori, E.; Comerci, V.; Esposito, E.; Guerrieri, L.; Michetti, A.M.; Mohammadioun, B.; Mohammadioun, G.C.; Porfido, S.; Tatevossian, R.E. Earthquake Hazard and the Environmental Seismic Intensity (ESI) Scale. *Pure Appl. Geophys.* **2016**, *173*, 1479–1515. [CrossRef]
54. Gosar, A. Application of Environmental Seismic Intensity scale (ESI 2007) to Krn Mountains 1998 M-w=5.6 earthquake (NW Slovenia) with emphasis on rockfalls. *Nat. Hazards Earth Syst. Sci.* **2012**, *12*, 1659–1670. [CrossRef]
55. Chen, C.W.; Iida, T.; Yamada, R. Effects of active fault types on earthquake-induced deep-seated landslides: A study of historical cases in Japan. *Geomorphology* **2017**, *295*, 680–689. [CrossRef]
56. Zuccaro, G.; De Gregorio, D.; Titirla, M.; Modano, M.; Rosati, L. On the simulation of the seimic energy transmission mechanisms. *Ing. Sismica* **2018**, *35*, 109–130. [CrossRef]
57. Sun, D.; Yang, T.; Cao, N.; Qin, L.; Hu, X.; Wei, M.; Meng, M. Characteristics and Prevention of Coseismic Geohazard Induced by Luding Ms 6.8 Earthquake, Sichuan, China. *Earth Sci. Front.* **2022**, *1*, 1–18.
58. Borrelli, L.; Coniglio, S.; Critelli, S.; La Barbera, A.; Gulla, G. Weathering grade in granitoid rocks: The San Giovanni in Fiore area (Calabria, Italy). *J. Maps* **2016**, *12*, 260–275. [CrossRef]

Disclaimer/Publisher's Note: The statements, opinions and data contained in all publications are solely those of the individual author(s) and contributor(s) and not of MDPI and/or the editor(s). MDPI and/or the editor(s) disclaim responsibility for any injury to people or property resulting from any ideas, methods, instructions or products referred to in the content.

Article

Characterization and Analysis of Landslide Evolution in Intramountain Areas in Loja (Ecuador) Using RPAS Photogrammetric Products

Belizario A. Zárate [1,*], Rachid El Hamdouni [2] and Tomás Fernández del Castillo [3,4]

1 Department Civil Engineering, Private Technical University of Loja, Loja AP. 1101608, Ecuador
2 Department of Civil Engineering, ETSICCP, University of Granada, Campus Fuentenueva s/n, 18071 Granada, Spain; rachidej@ugr.es
3 Department of Cartographic, Geodetic and Photogrammetric Engineering, University of Jaén, Campus de las Lagunillas s/n, 23071 Jaén, Spain; tfernan@ujaen.es
4 Centre for Advanced Studies in Earth Sciences, Energy and Environmental, University of Jaén, Campus de las Lagunillas s/n, 23071 Jaén, Spain
* Correspondence: bazarate@utpl.edu.ec

Abstract: This case study focuses on the area of El Plateado near the city of Loja, Ecuador, where landslides with a high impact on infrastructures require monitoring and control. The main objectives of this work are the characterization of the landslide and the monitoring of its kinematics. Four flights were conducted using a remotely piloted aerial vehicle (RPAS) to capture aerial images that were processed with SfM techniques to generate digital elevation models (DEMs) and orthoimages of high resolution (0.05 m) and sufficient accuracy (below 0.05 m) for subsequent analyses. Thus, the DEM of differences (DoD) and profiles are obtained, but a morphometric analysis is conducted to quantitatively characterize the landslide's elements and study its evolution. Parameters such as slope, aspect, topographic position index (TPI), terrain roughness index (TRI), and topographic wetness index (TWI) are analyzed. The results show a higher slope and roughness for scarps compared to stable areas and other elements. From TPI, slope break lines have been extracted, which allow the identification of landslide features such as scarps and toe tip. The landslide shows important changes in the landslide body surface, the retraction of the main scarp, and advances of the foot. A general decrease in average slope and TRI and an increase in TWI are also observed due to the landslide evolution and stabilization. The presence of fissures and the infiltration of rainfall water in the unsaturated soil layers, which consist of high-plasticity clays and silts, contribute to the instability. Thus, the study provides insights into the measurement accuracy, identification and characterization of landslide elements, morphometric analysis, landslide evolution, and the relationship with geotechnical factors that contribute to a better understanding of landslides. A higher frequency of the RPAS surveys and quality of geotechnical and meteorological data are required to improve the instability analysis together with a major automation of the GIS procedures.

Keywords: landslide characterization; evolution; RPAS; DEM; slope; aspect; TPI; TRI; TWI; Loja-Ecuador

Citation: Zárate, B.A.; El Hamdouni, R.; Fernández del Castillo, T. Characterization and Analysis of Landslide Evolution in Intramountain Areas in Loja (Ecuador) Using RPAS Photogrammetric Products. *Remote Sens.* **2023**, *15*, 3860. https://doi.org/10.3390/rs15153860

Academic Editor: Domenico Calcaterra

Received: 8 June 2023
Revised: 27 July 2023
Accepted: 31 July 2023
Published: 3 August 2023

Copyright: © 2023 by the authors. Licensee MDPI, Basel, Switzerland. This article is an open access article distributed under the terms and conditions of the Creative Commons Attribution (CC BY) license (https://creativecommons.org/licenses/by/4.0/).

1. Introduction

The study of landslides is of vital importance to understand and mitigate the risks associated with natural phenomena as well as to promote the sustainable development of affected areas. Understanding the characterization and analysis of the evolution of landslides plays a fundamental role in the Sustainable Development Goals, because it allows an adequate management of natural resources, more efficient territorial planning, and protection of the communities that inhabit areas prone to landslides [1–4]. Understanding the characterization and analysis of landslide evolution can significantly contribute to the Sustainable Development Goals in several ways:

Disaster risk management: Understanding the characterization of landslides and their evolution makes it possible to identify landslide-prone areas and assess the associated risk. This facilitates the implementation of adequate prevention and mitigation measures to reduce the exposure of communities to landslides [5,6]. By minimizing the risks of natural disasters, both people and the natural environment are protected, contributing to long-term sustainability.

Sustainable urban planning: Analysis of the evolution of landslides helps to understand how human activities can contribute to their occurrence. This is especially relevant in the context of urban planning, where the growth of cities can increase the pressure on slopes and unstable terrain [7,8]. By taking this information into account, regulations and development guidelines can be established to avoid construction in areas of high landslide risk, thus promoting more sustainable urban planning.

Ecosystem conservation: Landslides can have a significant impact on ecosystems, altering soils, vegetation, and local hydrology. Understanding the characterization of landslides and their evolution makes it possible to identify the factors that contribute to their appearance, such as deforestation or soil degradation [9]. By taking steps to preserve and restore natural ecosystems, you can strengthen the resilience of vulnerable areas to landslides and promote environmental sustainability.

Natural resource management: Landslides can impact the availability and quality of natural resources such as water and fertile soil. Through the characterization and analysis of landslides, it is possible to understand how these processes affect natural resources and take measures for their sustainable management. For example, soil and water conservation practices can be implemented to reduce erosion and prevent landslides, thus ensuring the availability of essential resources for future generations. To study natural hazards, specifically landslides, it is necessary to have techniques and procedures that provide information on terrain evolution with sufficient spatial and temporal resolution [10–15] to determine geomorphic changes. Currently, Global Navigation Satellite Systems (GNSS) [16–20] and Terrestrial Laser Scanning (TLS) are commonly used, which provide high-density point clouds and high-quality digital elevation models [21]. However, these techniques require significant processing time and are costly. In this context, the use of remotely piloted aerial systems (RPAS), also known as drones, represents a low-cost alternative [22–24] that enables the acquisition of high-resolution aerial imagery for mapping and monitoring small-scale areas [5,25–27].

The use of RPAS in the study of landslides has allowed for the evaluation of their kinematic behavior and temporal evolution [5,25,26,28–30]. This is achieved through the acquisition of photogrammetric products derived from precise processing, enabling measurements that can even detect small-scale terrain changes. Structure from Motion (SfM) algorithms are employed for images orientation [31–34] providing accuracies of about 0.10 m from which photogrammetric products such as DEMs and orthoimages are obtained. The former are usually compared by means of DEM of differences (DoDs) and the latter are used for interpretation and digital image correlation (DIC) [25–27,35–45]. The accuracy of photogrammetric products largely depends on the number and distribution of ground control points (GCPs) and checkpoints (CHK) measured using GNSS techniques [16,25,38,46,47], which enable model orientation [17]. Additionally, geodetic measurements [45] using GNSS techniques provide high-precision coordinate estimation, making them valuable for monitoring surfaces undergoing both slow and rapid deformations at different scales [48].

The geomorphometric approach [49] is employed to analyze the morphology of the terrain based on high-resolution digital elevation models (DEMs) derived from LiDAR systems or through RPAS photogrammetry, which allows for the detection of subtle changes in the topography. This information can be used to generate maps of slope, aspect, curvature and roughness, among others, which help to characterize landslides, identify their elements, and understand the kinematics and dynamics of the slope [50–54].

Recent findings regarding landslides in Ecuador have significant implications for the understanding and management of this natural phenomenon. These discoveries allowed for a greater understanding of the determinant and triggering factors of landslides, such as topography, geology, seismic activity, and weather conditions. In addition, they have made it possible to identify areas prone to this type of event [55,56], which contributes to better urban and rural planning, as well as the implementation of preventive and adequate mitigation measures. These findings have also improved the ability to monitor land changes in landslides areas based on classical approaches such as photointerpretation, DoDs calculation and point extraction and measurement [57], which is essential to alert communities at risk and take timely measures to ensure the safety of the population. However, other approaches are required such as DIC for a better measurement of displacements; object-based analysis (OBIA) for features identification; and morphometric analysis, the approach implemented in this work, for landslide and terrain characterization. In general, scientific advances in this field are strengthening Ecuador's resilience against landslides and require a solid foundation for the development of effective natural risk management strategies.

Thus, the main objectives of this study are: (a) the characterization of the landslide, identifying its elements and describing its morphology; and (b) the monitoring of its kinematics. For it, we have analyzed not only the direct photogrammetric products such as DEM and orthoimages but also topographic or morphometric parameters and their changes. The detailed and systematic analysis of these parameters is the main contribution of this study, which has allowed precise landslide characterization and monitoring. Previous works have focused on some parameters such as slope [52,53], aspect [53], curvature [51–54], roughness [50,53] or TWI [50], but none have examined all of them and even less with a multitemporal approach that allows landslide monitoring.

Thus, multitemporal RPAS flights were captured in a landslide in the El Plateado sector of the city of Loja, Ecuador. One of the main advantages of RPAS is their capability to fly at altitudes below 100 m, allowing for the integration of new sensors on aerial platforms. RPAS can capture images from various angles, provide flexibility in conducting work at different scales, offer cost-effective solutions, and deliver high-quality results [15–18]. RPAS technology has been successfully used for the cartography and monitoring of areas spanning few square kilometers.

The scheduled flights allowed for the acquisition of images, which once processed, produced orthoimages and DEMs. For this purpose, the photogrammetric image blocks were oriented using SfM techniques with the support of ground control points whose positions were determined using differential GNSS. The interpretation of the orthoimages has allowed the observation of morphological features and changes in land cover and elements affected by the landslide. Moreover, from DEMs, longitudinal profiles, DEMs of differences (DoDs), and especially detailed maps of morphometric parameters derived from DEMs such as slope, aspect, TPI, TRI, and TWI were obtained. These models have enabled the detection of morphologies that contribute to the characterization of landslides in the study area, and furthermore, due to their multitemporal nature, the analysis of their temporal evolution.

2. Materials and Methods

2.1. Study Area

The study area is situated in the El Plateado sector on the western side of the city of Loja, located in southern Ecuador, along the bypass road (Figure 1). Geologically, the study area is part of the Trigal Formation (Miocene), which is primarily exposed in the western portion of the basin (Figure 1). It predominantly consists of a homogeneous, finely laminated brown clay with occasional gypsum veins. Additionally, it comprises coarse-grained sandstones with thin layers of conglomerate and minor occurrences of limonite. The sandstones exhibit horizontal stratification with crossbedding planes.

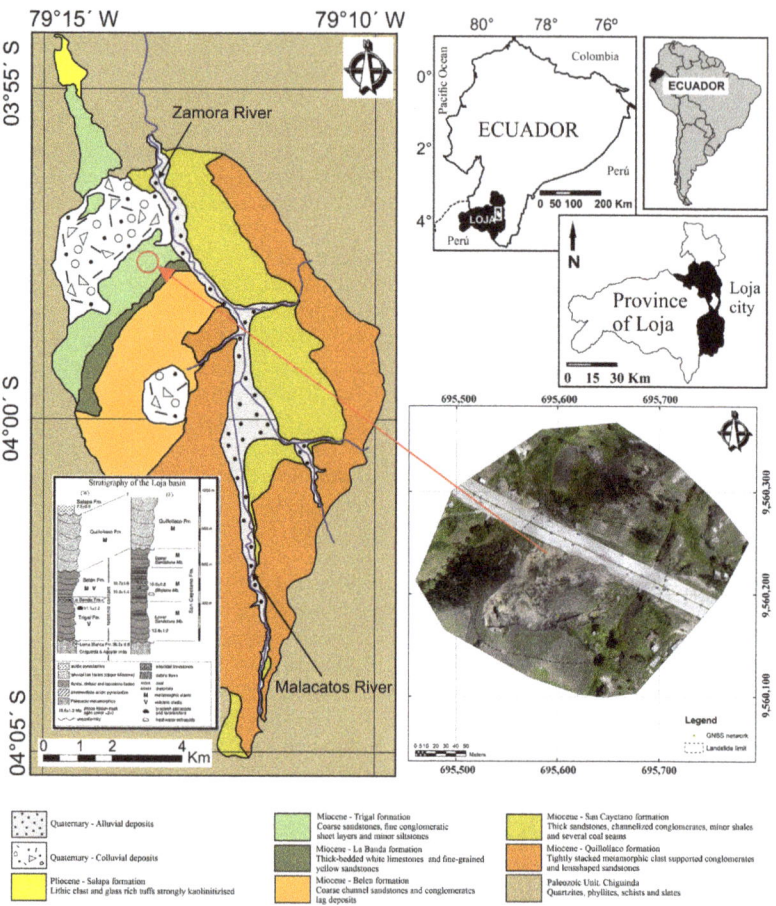

Figure 1. Location and geological framework of the El Plateado study area. Adapted from (Zárate et al., 2021 [57]).

The activity of the landslide is evident through the deposition of material at the low part of the hillslope as well as the presence of cracks, lobes, and the formation of main, lateral, and secondary scarps resulting from the ongoing movement and material detachment (Figure 2b–d). The progressive development of the main scarp and the right (southern) flank of the landslide caused the collapse of a residential structure (Figure 2a) without any reported loss of human life.

Figure 2. (**a**) Collapse of houses due to slope movement activity; (**b**) Presence of cracks in the slope body; (**c**) View of the main scarp on the southern flank of the slope; (**d**) Accumulation zone along the roadway.

2.2. Materials (RPAS and GNSS)

The RPAS used in this study consisted of a DJI Phantom 2 vehicle (Figure 3a) with the following specifications: a maximum horizontal range of 1000 m, maximum horizontal speed of 12 m/s, ascent speed of 6 m/s, descent speed of 2 m/s, net weight including the battery of 1 kg, horizontal displacement accuracy of 2.50 m, vertical displacement accuracy of 0.80 m, operating angle and temperature range from −10 to 50 °C. It was equipped with a Zenmuse H3-3D gimbal made of aluminum alloy, which maintained the camera's position fixed in three axes using an Inertial Measurement Unit (IMU). The gimbal weighed 22 g (excluding the camera) and was compatible with GoPro and MAPIR cameras. The power for the gimbal was supplied by a DJI Phantom 2 intelligent battery.

For autonomous flight of the RPAS, the DJI 2.4 GHz Datalink system (Figure 3b) was used, enabling communication between the ground base and the aerial system through bidirectional data communication modules. This system allowed the flight plan to be loaded onto the RPAS via Bluetooth for subsequent execution. Flight planning and execution were carried out using the DJI Ground Station Version 1.4.63 application. Images were captured using a GoPro Silver Edition 3+ camera with a resolution of 10 Mp. The image capture interval was set to 2 s.

The ground control point (GCP) coordinates were measured using the differential GNSS technique with a Trimble R6 GNSS receiver (Figure 3c).

Figure 3. (**a**) Phantom 2 equipped with a ventral camera and gimbal; (**b**) DJI 2.4 GHz Datalink system for ground-to-air connection; (**c**) Dual-frequency GNSS receiver used for control network point measurements; (**d**) Plastic markers used as ground control points (GCPs).

2.3. Methods

The methodology employed can be summarized as follows:

(1) Data capture of images from Remotely Piloted Aircraft Systems (RPASs) and ground control points (GCPs) measurement using differential GNSS; software used Trimble Business Center (TBC) [58];
(2) Processing of photogrammetric blocks and generation of DEMs and orthophotographs; software used: Agisoft Photoscan V 1.4.5 [29,35,59,60];
(3) Generation of DoDs, profiles, and derivative models by a Geographic Information System (GIS); software used: QGIS V 3.30.0–ArcGIS 10.2.2 [61–63];
(4) Mapping, morphometric analysis, and evolutionary assessment; software used: QGIS V 3.30.0, SAGA GIS V 9.0.1 [64];
(5) Geotechnical characterization of materials.

2.3.1. Data Capture: Images and GCPs

The investigation of the slope movement in the El Plateado sector was conducted from 24 January 2017 to 12 March 2020, during which 4 RPAS campaigns were carried out for image capture. Table 1 displays the dates and characteristics of the flights.

Table 1. Details of RPAS missions in the study area.

RPAS Mission	Control 1	Control 2	Control 3	Control 4
Date	24 January 2017	9 June 2017	8 June 2018	12 March 2020
RPAS drone	Phantom 2			
Camera	GoPro Silver Edition 3+			
Flight height (m)	84.2	87	87.6	89
Area (km^2)	0.265	0.285	0.258	0.251
Resolution (cm/pix)	4.20	4.33	4.10	4.00
Number of images	327	489	335	356
Longitudinal overlap (%)	70	70	70	70
Transversal overlap (%)	70	70	70	70
Number of GCP	9	9	9	9
X error (m)	0.022	0.033	0.036	0.028
Y error (m)	0.028	0.024	0.039	0.021
XY error (m)	0.036	0.041	0.053	0.035
Z error (m)	0.031	0.034	0.023	0.031
Number of CHK	6	6	7	6
X error (m)	0.011	0.023	0.036	0.014
Y error (m)	0.031	0.024	0.019	0.021
XY error (m)	0.033	0.033	0.041	0.025
Z error (m)	0.025	0.041	0.028	0.026

To provide photogrammetric support and orientation for the models, an in situ network of control and checkpoints was implemented. These points consisted of simple concrete markers with a diameter of 0.15 m and a depth of 0.30 m, with a steel rod of 12 mm diameter anchored at their center. Plastic markers measuring 1.00 m × 1.00 m were placed on the control points to ensure proper model orientation during the processing phase with the software (Figure 3d). Metal rings were installed at the center and ends of the markers, allowing for precise centering over the network points and fixation to the ground using metal hooks.

The coordinates of the control points were measured using differential GNSS technique employing a Trimble R6 GNSS receiver (Figure 3d) with an occupation time of 10 min at each point of the network. The post-processing of GNSS data used the data from the LJEC GNSS reference station of the Military Geographic Institute (IGM) belonging to the SIRGAS network and the Trimble Business Centre software version 2.6. The coordinates were oriented in the UTM WGS 84 zone 17 South coordinate system. The number of control and checkpoints is indicated in Table 1.

2.3.2. Photogrammetric Processing and Generation of Products

The aerial images obtained in the four RPAS campaigns were processed using Agisoft PhotoScan Professional software version 1.4.5. The accuracies achieved in the orientation process are shown in Table 1, and they are expressed as the root mean square error (RMSE).

The horizontal errors, both in the GCPs and the CHKs, ranged from 0.035 to 0.053 m, while the vertical errors ranged from 0.023 to 0.041 m, all of which were consistently below the recommended threshold of 0.10 m [40].

Subsequently, photogrammetric and SfM techniques were applied in the software to orient the photogrammetric blocks. Point clouds and dense point clouds were generated for the four processed flights, which were then filtered using a tool to remove outliers and noise. On average, the dense point clouds encompassed 11,238,000 points for the four flights. The result of this process was the corresponding digital elevation models (DEMs), which were exported in TIFF format for analysis in a GIS, with a resolution of 0.05 m.

Finally, orthophotos were also produced at a resolution of 0.05 m with the aforementioned accuracy being below the pixel size. The orthophotos can be used to delineate landslides in relation to the surrounding environment as well as to identify features such as scarps, cracks, etc., through photointerpretation.

2.3.3. Generation of DoDs, Profiles and Derivative Models

First, the DEM of differences (DoD) is obtained using raster calculators in the GIS. For the surface monitoring of landslides, at least two data acquisition periods are required [65]. In the case of multitemporal analysis [25,27,57], the DEMs are compared in pairs. DoDs allow for the observation of vertical changes rather than vertical displacements. In fact, in many cases, what is observed are horizontal displacements and advances of the mass, which result in modifications of the DEMs [24,26,57]. DoDs can have negative values when the surface is lower in the later date and positive values when the surface is higher in the later date. Thus, excluding changes due to vegetation growth or decline, construction activities, and human interventions on the terrain (excavations, flattening, or fillings), natural changes in the surface of the terrain can occur due to surface descents (negative DoDs) in scarps and head areas, or horizontal displacements that result in excavation (negative DoDs) or accumulation of material (positive DoDs) in different areas of the landslide (head, body or foot).

Second, to analyze the topographic parameters in the study area, profile lines were established, as shown in Figure 4. Profiles were generated based on the DEMs using Profiles tool in ArcGIS software, with a horizontal sampling interval of 0.50 m. The placement of the profiles considered soil displacements observed during field inspections and surface features. Profile A is longitudinal with a direction of N72°E, Profiles B and C are oblique with a direction of N33°E and N30°E, respectively, where B is more centered and C is more displaced to the foot area. The profiles provide a clear visualization of the topography and microtopography of the slope and the landslide, including all characteristic features such as scarps, head area, main body, foot, toes, and even cracks. By comparing the profiles, the changes in topography over time, the movement (downward and forward) of the mass, and its eventual retreat can be observed.

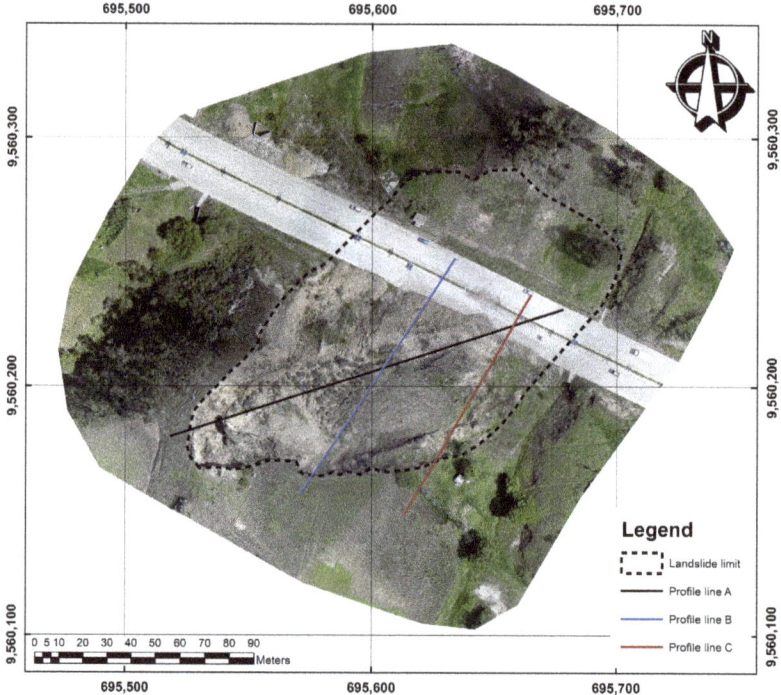

Figure 4. Placement of profiles in the study area for the analysis of topographic factors.

Lastly, in the third step, various derived models from the DEM have been obtained. These models consist of a series of topographic parameters that characterize the morphology of the slope and extract terrain features, focusing specifically on the main landslide but also on other instability processes observed in the study area. The derived models include slope angle, slope aspect, topographic position index (TPI), terrain roughness index (TRI), and topographic wetness index (TWI) for each analyzed date.

Slope angle or simply slope is the most relevant factor in determining slope stability [52,53,66–70]. Alterations in slope can increase shear stresses within the soil mass due to gravitational forces, leading to slope failure. Depending on the slope gradient, slow or rapid surface movements can occur.

Slope aspect or orientation, defined as the direction of the terrain's inclination in each cell, influences the physical properties of soils primarily through the effects of rainfall, wind, and solar exposure [53,71–73].

The topographic position index (TPI) enables the description of morphological aspects of the terrain by determining and segmenting the hillslope [49,53,54,74–76]. It is calculated for the i-th pixel of the DEM elevation h_i, where u_i represents the standard deviation of the DEM pixels and, σ_i is the standard deviation of the DEM pixels within that same range. The calculation is given by Equation (1):

$$TPI = \frac{(h_i - u_i)}{\sigma_i} \quad (1)$$

The terrain roughness index (TRI) [50,54,77] is a measure that quantifies the variability of terrain height in a given area. It can be expressed using Equation (2):

$$TRI = \frac{\sigma_z}{Z_0} \quad (2)$$

where σ_z is the standard deviation of the terrain heights and Z_0 is the mean height of the terrain surface. The result of the equation is a dimensionless number.

Finally, the topographic wetness index (TWI) [50,69,78] has been considered, which is related to the slope of a terrain and is used to identify areas where moisture or water accumulates. It is represented by Equation (3):

$$TWI = ln\left(\frac{a}{tan\beta}\right) \quad (3)$$

where a is the drained area for a specific cell, and $tan\beta$ is the slope of the analyzed cell.

For the determination of slope and aspect, ArcGIS V 10.2.2 software was used with the employment of Slope and Aspect tools, while QGIS V 3.30.0 and SAGA (System for Automated Geoscientific Analyses) V 9.0.1 software were used for TPI, TRI, and TWI [69].

2.3.4. Mapping, Morphometric Analysis, and Evolution Assessment

From DEMs and derivative models, terrain and hillslope forms in general and landslide features in particular can be identified and extracted. Slope and roughness (TRI) allow for the identification of scarps (steep slopes) and the body or foot of the landslide. However, it is the curvature or in this case the topographic position index (TPI) [51–54,79–81] that enables the identification of slope break lines and consequently delineates different parts of the movement. Specifically, the upper and lower break lines of scarps can be extracted. Additionally, other elements such as the foot of the landslides and especially its toe can also be detected [53].

The procedure involved obtaining the TPI map and symbolizing it with different classification and palette schemes to find the most appropriate thresholds for detecting break lines. The thresholds were ultimately set at +0.05 for the upper break line of scarps and −0.05 for the lower break line. Once the thresholds were selected, the skeletonization tool (SAGA tools) was applied to trace the lines (vectors) in shapefile format. The classified and color-symbolized rasters as well as the vectors from different dates can be used to observe the evolution of the scarps (for instance, retreat of main and lateral scarps and advancement of secondary scarps). Displacement measurements can be made using the measuring tool in GIS. Other resolutions different to the reference one at 0.05 m (e.g., 1 m) were also tested to better detect other features such as the toe, where the slope break is not as abrupt and thus more visible over a longer profile.

Finally, after identifying the main landslide elements and other features through photointerpretation, observation of the different derivative models, and the mentioned automatic extraction methods, a zoning of the study area was performed in order to analyze the characteristic morphometric parameters that define them. On one hand, landslide features and on the other hand, vegetation and constructions (roads, and buildings) were digitized. The latter were digitized to create masks that excluded the areas occupied by these elements from the analysis of characteristic landslide and terrain elements. In the case of vegetation, trees, shrubs, and undifferentiated areas of trees and shrubs were differentiated. Additionally, as support for photointerpretation, the alternative vegetation index GLI (Green Leaf Index) [82–84], which works with the native digital values of the red, green and blue (RGB) bands, was calculated.

Once the vector layers of landslide elements, unstable areas, vegetation, and constructions are obtained, they are rasterized. Subsequently, the raster calculator is applied to obtain a new raster excluding the vegetation and construction areas using Equation (4):

$$Raster = raster\ elements \times (raster\ vegetation - 1) \times (raster\ constructions - 1) \quad (4)$$

The elements identified and differentiated in the main landslide are as follows: main scarp, head, lateral flank and scarps, secondary and counterslope scarps, body, secondary body or lobe, and foot. Scarps, bodies, and foots of other minor instabilities in the study area were also detected.

Finally, the analysis of morphometric parameters (slope, aspect, TPI, TRI, and TWI) for the different elements was performed using zonal analysis tools (zonal statistics and zonal histogram) in QGIS.

2.3.5. Geotechnical Characterization of Materials

Laboratory tests were carried out on different soil samples (3 soil samples at the head, 3 samples at the body and 3 samples at the foot of the slope) all obtained at a depth of 3 m with an open pit. Using the Center-pivot backhoe loader 450, a Soil Moisture SM300 Kit was used to measure soil moisture (w%) at the time of extracting the samples for the respective analyses. The following tests were carried out in the laboratory, and the respective standards used for their execution are indicated:

- Water Content of Soil (ASTM D4643-17) [85];
- Liquid Limit, Plastic Limit, and Plasticity Index of Soils (ASTM D4318-17e1) [86];
- Particle-Size Analysis of Soils (ASTM D422-63) [87];
- Unified Soil Classification System (ASTM D2487-17e1) [88];
- Direct Shear Test of Soils Under Consolidated Drained Conditions (ASTM D3080) [89]

3. Results

3.1. Orthoimages and DEMs

The orthoimages obtained in the study area for the different flights are shown in Figure 5. Visual analysis of these orthoimages allowed for the identification of the main scarp, the head, the lateral flank and scarps, the secondary and counterslope scarps, the body, the foot, and other features. Among these, tension cracks in the head, body, and foot of the landslide were noteworthy both in longitudinal and transverse arrangement. At the crown level, the presence of induced cracks led to the failure of crown material, resulting in multiple scarps. The crown of the main scarp maintained a semicircular shape during the three observation periods. The landslide body could be clearly identified, as well as the accumulation zone at the foot and the advancement of material on one of the road lanes, as shown in Figure 5a–d. The main scarp was continuously eroded due to the material falling onto the slope body; however, its retreat was not very relevant, as will be discussed later.

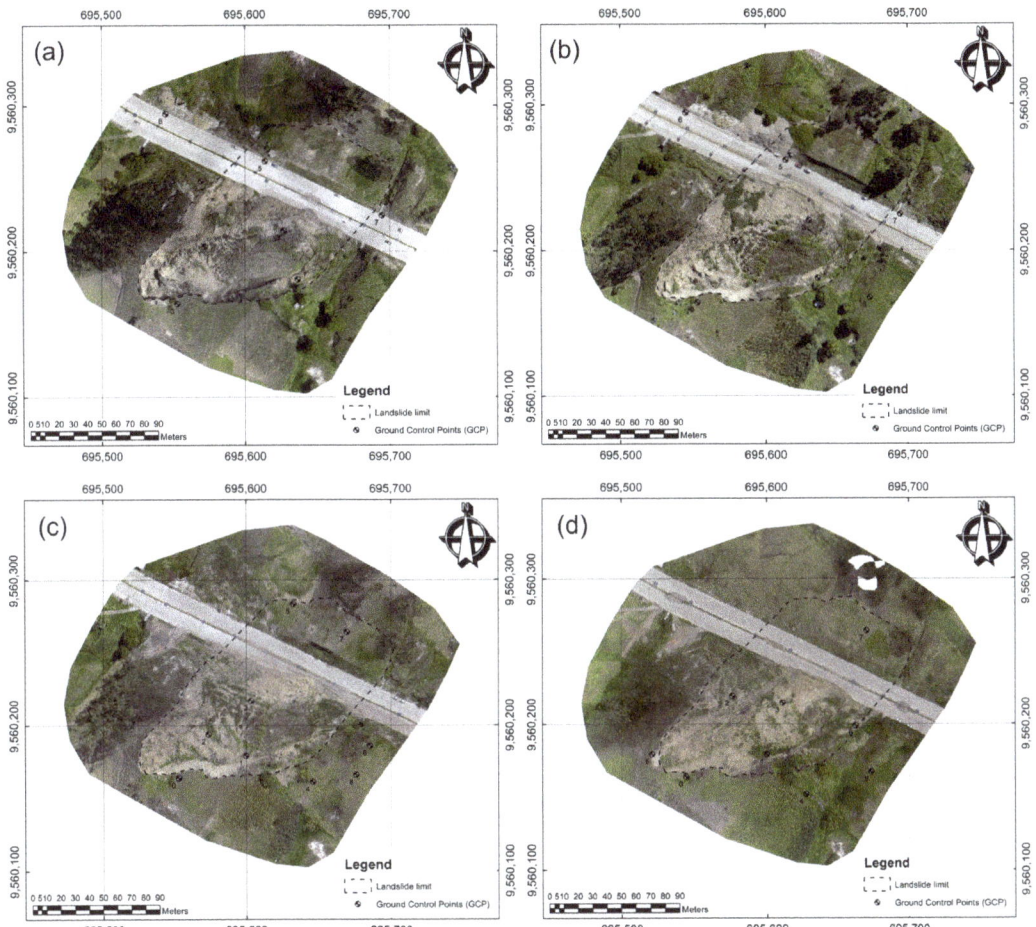

Figure 5. Orthoimages obtained from the study area, where superficial changes can be observed in the four-control dates. The dotted line shows the landslide limits: (**a**) 24 January 2017, (**b**) 9 June 2017, (**c**) 8 June 2018, and (**d**) 12 March 2020.

In Figure 6, the digital elevation models (DEMs) generated from the dense point cloud for the four-monitoring dates are displayed.

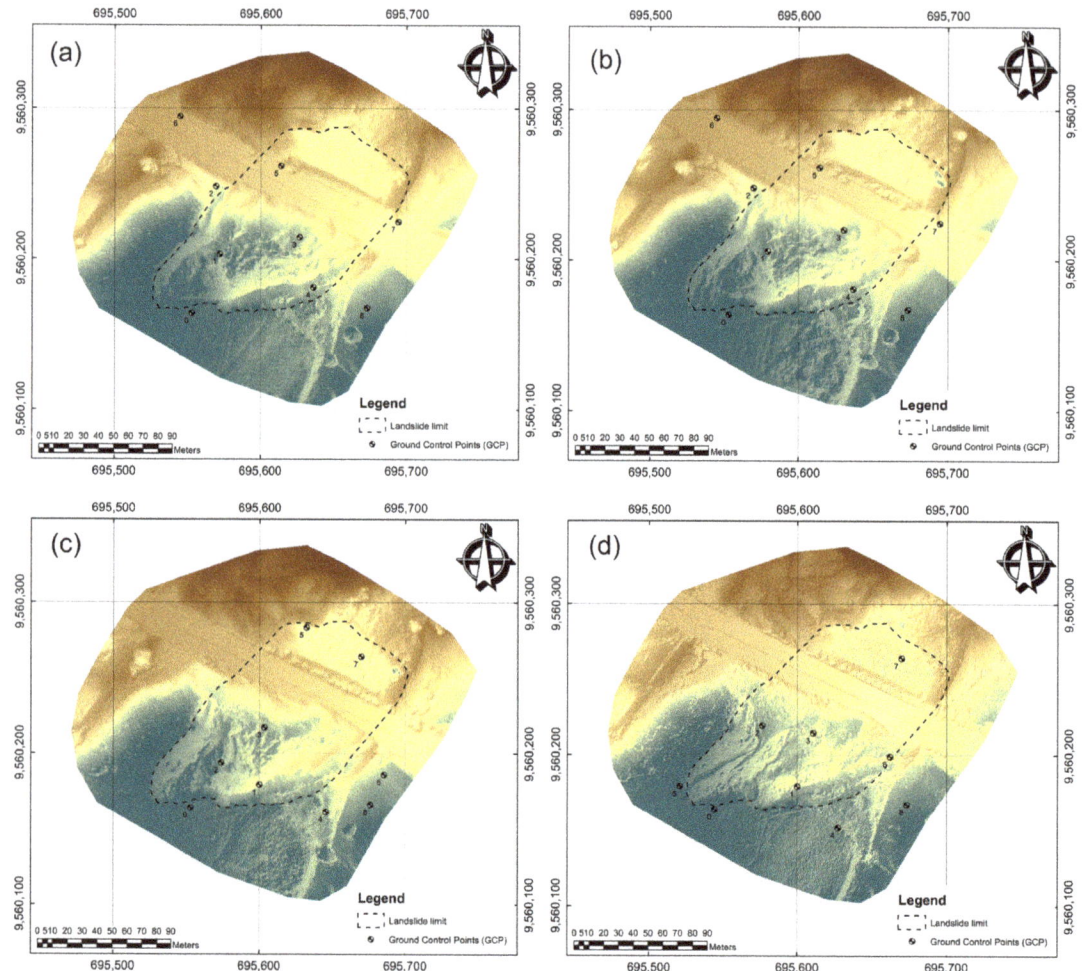

Figure 6. DEMs obtained from the study area for the four monitoring dates. The dotted line represents the landslide limit as well as the locations of the GCPs: (**a**) 24 January 2017, (**b**) 9 June 2017, (**c**) 8 June 2018, and (**d**) 12 March 2020.

3.2. DEM of Differences (DoDs)

Figure 7 presents the DoDs of the El Plateado sector. The color palette was adjusted to visualize subtle movements in the DoDs. The color palette represents positive values (red) and negative values (green), with the former indicating an increase in terrain elevation and the latter indicating a decrease in terrain elevation.

The DoDs were generated from two study periods. The first period corresponds to January 2017 to June 2017; the second one covers from June 2017 to June 2018; and finally, the third one covers from June 2018 to March 2020. In the first period (Figure 7a), it is clearly visible that there are areas of terrain surface descent in the head and the upper part of the landslide body. Similarly, in the landslide body, there are areas of terrain surface descent related to secondary scarps and ascent due to the accumulation of material from the main scarp and secondary scarps. At the foot, the terrain surface ascent due to material accumulation or advancement is notable. As mentioned initially, points cloud filtering was performed to remove vegetation, although there were areas where filtering was not applied to preserve the topographic surface, as seen in the southern zone of the slope where positive values (light brown color) representing a surface elevation are observed, corresponding to vegetation growth (corn crops).

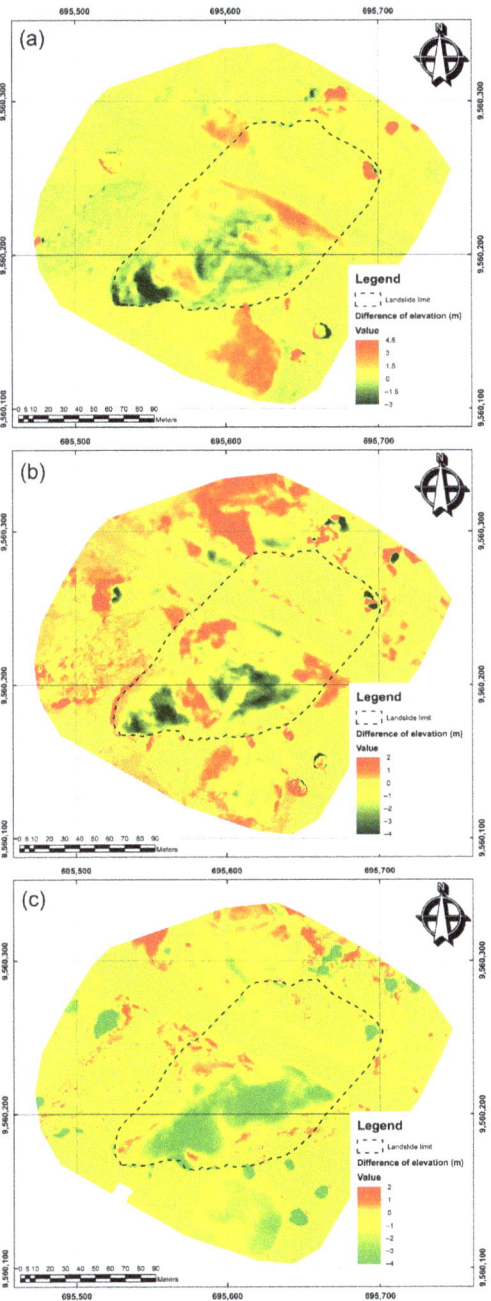

Figure 7. DEM of difference (DoD): (**a**) January 2017 to June 2017; (**b**) June 2017 to June 2018; (**c**) June 2018 to March 2020.

Figure 7b corresponds to the DoDs for the second period (June 2017 to June 2018). In both the head and the body of the landslide, areas of surface descent or ascent are identified, slightly displaced downhill compared to the previous period, indicating the progressive downslope movement of the slope, mainly in the ENE direction. In the foot, surface descent is observed, which is possibly related to the removal of material from the road. There is also clear evidence of instability under the road, with a material descent indicating the formation of a scarp.

In Figure 7c, a generalized descent is observed throughout the landslide body, which can be attributed to the anthropogenic action of material removal for slope stabilization carried out in early 2020.

3.3. Profiles

The results of the profiles (Figure 8) obtained from the DEMs allowed for the recognition of changes in the landslide body and foot corresponding mainly to horizontal displacements, taking the profiles obtained from the first DEM (24 January 2017) as reference. Profile line A, which corresponds to a longitudinal profile in the main direction of landslide progress from the crown at the WSW toward the foot at the ENE, clearly reveals different parts of the landslide: the main scarp and the head; followed by the body with a counterslope scarp in the upper part and secondary scarps in the lower part; and finally the foot that reaches the road. This same profile is observed in the subsequent dates (9 June 2017; and 8 June 2018), although it is displaced downslope and has less pronounced scarps due to the evolution of the landslide. These features allow for estimating greater mass descents and advances in the head and upper part of the body (total descent of about 6–8 m and advances of around 20–25 m; 3–4 m and 10–12 m in each period); while at the foot, the mass advances about 5–6 m, which may be underestimated due to material removal in the road area. However, the profile from the last date (12 March 2020) is different, showing a much more uniform slope from the crown area to the foot. Thus, what is observed is a flattening of the slope with material removal and the disappearance of the typical morphology, both in the scarp, in the body and even in the foot, where nevertheless a steeper slope is present in the road embankment.

In the case of profile line B, which represents an oblique profile from the right flank-scarp to the road, the main scarp is clearly observed in the first date, which is followed by the head with a slight counterslope toward the scarp. Further down the body, there is a convex shape, which then leads to a steeper slope toward the lower part of the body and the foot at the road. This shape remains consistent with time, although the surface gradually descends in the second and third dates. In the last date, there was smoothing of the slope shape from the scarp to the road, although a certain slope was still observable. We hardly observed any significant mass advancement in this profile due to its nearly transverse direction compared to the main movement direction.

Profile line C, parallel to the previous one but shifted toward the foot area, shows on the first date a more gentle right flank with almost no main scarp. Slightly below, the body exhibits a secondary scarp, and finally, there is a slightly steeper slope in the foot and road area. In the following two dates, there was a descent of about 3–5 m and a horizontal advancement of about 6–8 m, as expected in the foot area. However, this advancement could be attenuated by material removal from the road. In the last date, the change was minor and mainly corresponded to a surface descent due to slope repair works.

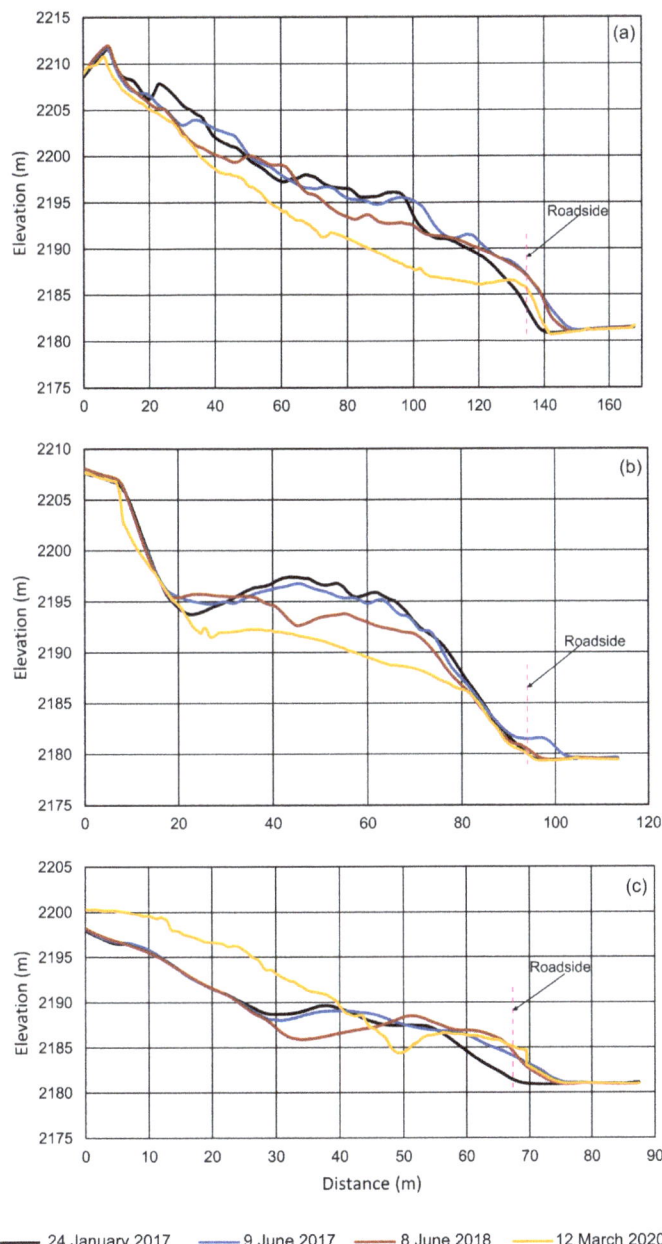

Figure 8. Details of profiles obtained from DEMs. The dashed line represents the lateral edge of the road as a reference point for the corresponding analysis: (**a**) profile A; (**b**) profile B; and (**c**) profile C.

3.4. Topographic Parameter Maps

Before describing the topographic or morphometric parameters, the map of landslide elements and unstable zones (Figure 9) mentioned in Section 2.3.4 is presented. This map identifies the different types of scarps (main, lateral, secondary, counterslopes), the head, body, and foot of the main landslide, as well as other unstable zones present in the study area.

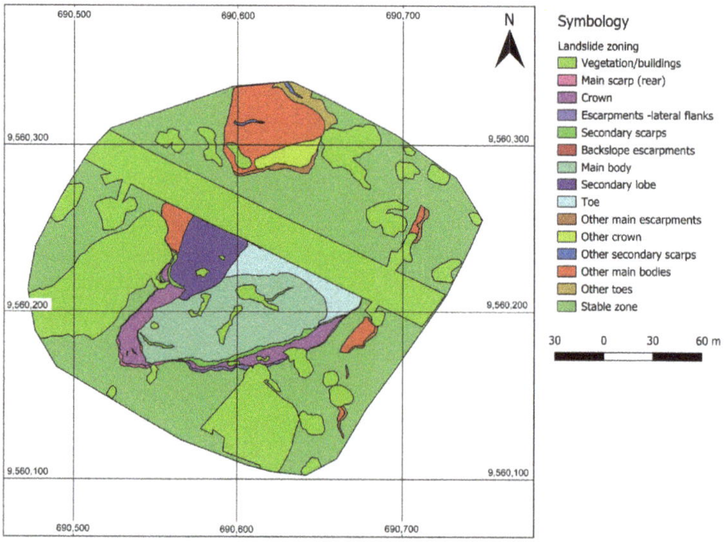

Figure 9. Zoning of the landslide where the main features and elements obtained through photointerpretation and digitization are identified.

This zoning also serves for the analysis of the morphometric parameters presented in the following subsections.

3.4.1. Distribution and Evolution of Slope

Slope is the most important topographic parameter that describes the behavior of the terrain surface along time caused by slope kinematics [90]. Slope maps of the different monitoring dates are shown in Figure 10, and these allow for analyzing visually the temporal changes in slope distribution.

In the maps shown in Figure 10, areas with low slopes (<10°) can be observed on the road and the embankment below it as well as in the high and stable zone above the landslide crown. Areas with steep slopes (above 30°) can be seen in vegetated areas, especially trees and bushes, which need to be disregarded. Focusing on the landslide area, steep slopes are clearly observed in the different scarps (main, lateral, and secondary), while the slopes in the body are lower, alternating between flat areas with steeper slopes corresponding to secondary scarps. Slightly higher slopes are also observed in the foot area.

The slopes for the landslide elements for the first date are presented in Table 2. It can be observed that the scarp areas have average slopes greater than 40° with the main scarp reaching nearly 54°. The head has a slope of 23.25°, the main body has a slope of 19.84°, and the foot has a slope of 27.76°. The overall slope of the landslide was 24.15° compared to the stable zone with a slope of 15.61°. Other unstable zones have average slopes of 22.59°, with the slopes of their scarps slightly lower than those of the main landslide scarps.

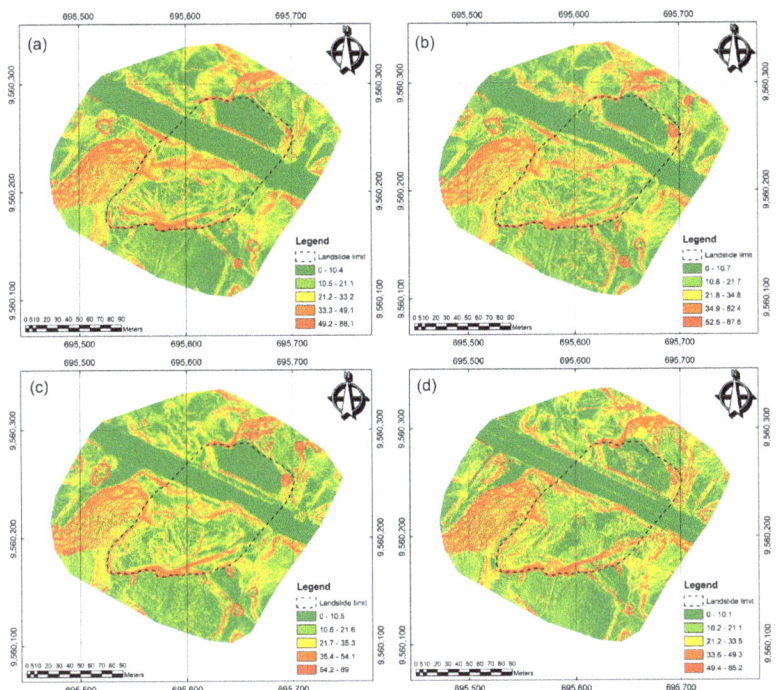

Figure 10. Slope maps of the study area for the four monitoring dates: (**a**) 24 January 2017; (**b**) 9 June 2017; (**c**) 8 June 2018; (**d**) 12 March 2020.

Table 2. Slope statistics for landslide elements.

Element	Average	Mode	Min.	Max.	Range	Std-D.	C.Var.
Main landslide							
Main scarp	53.53	20.61	0.15	85.63	85.48	16.14	0.30
Head	23.25	20.48	0.00	75.60	75.60	12.39	0.53
Lateral scarps-flanks	40.97	19.12	0.56	83.23	82.67	16.08	0.39
Secondary scarps	43.65	25.46	0.30	82.28	81.98	13.67	0.31
Counterslope scarps	48.02	12.93	2.80	74.53	71.73	15.09	0.31
Body	19.84	7.10	0.00	77.69	77.69	10.60	0.53
Secondary body	26.75	25.46	0.07	81.11	81.04	12.49	0.47
Foot	27.76	25.46	0.07	70.36	70.29	12.50	0.45
Other landslides							
Main scarps	40.95	32.01	0.00	83.05	83.05	13.60	0.33
Heads	18.83	12.93	0.00	74.74	74.74	10.41	0.55
Secondary scarps	35.99	14.64	1.05	65.63	64.59	13.54	0.38
Bodies	20.21	12.93	0.00	75.27	75.27	11.97	0.59
Foots	29.98	30.99	0.25	63.58	63.33	9.34	0.31
Stable area							
Stable area	15.61	7.10	0.00	84.30	84.30	10.28	0.66

In Figure 11, the distribution of slopes in some of the different landslide elements is shown.

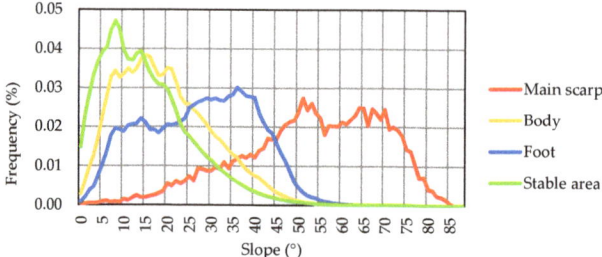

Figure 11. The distribution of the slope in landslide elements. The main scarp shows slope values greater than 50°. The landslide body has a slope range between 5° and 25°; the foot of the landslide has its highest slope is in the range between 25° and 45°; and the stable zone has a slope range is between 0° and 20°.

The evolution of the slope can be observed through their frequency distribution in each profile (Figure 12). In longitudinal profile A, taking the first date (24 January 2017) as a reference with an average slope of 18.11°, a gradual decrease in slope is observed, becoming 17.75° in the second date, 16.72° in the third date, and more significantly in the fourth date, decreasing to 13.30°. In profile B, oblique to the direction of the landslide and with an initial average slope of 21.79°, the trend is a progressive decrease, reaching 21.40° in the second date, 20.40° in the third date, and 18.10° in the fourth date. Meanwhile, profile C, which is located more marginally than the others, starts with an initial average slope of 17.43° but exhibits different and more irregular behavior. In the second date, it increases to 20.24°, after which it decreases to 18.70° in the third date and increases again to 21.42° in the fourth date.

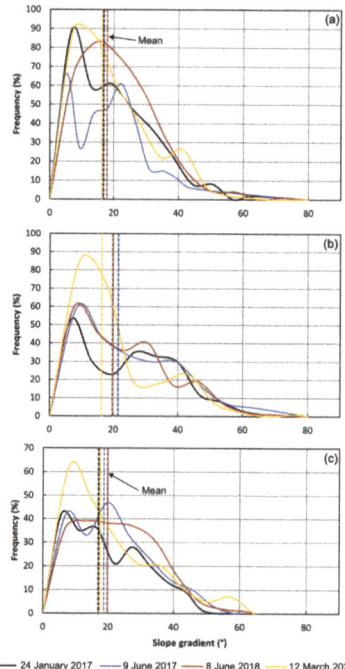

Figure 12. Temporal evolution of slope in profiles: (**a**) Profile A; (**b**) Profile B; (**c**) Profile C. Dashed lines represent the mean slope values for each monitoring date and corresponding profile. Each colored line corresponds to the monitoring date.

The frequency of slopes below 20° increases in each profile. In profile A, it starts at 60%, slightly decreases in the second date to 58%, then increases significantly in the third date (65%) and particularly in the fourth date (76%). In profile B, a similar pattern is observed with an initial percentage of 46%, which slightly increases in the second date and further increases in the third (52%) and fourth (64%) dates. Profile C starts at 66%, decreases in the second date to 56%, increases in the third date to 68%, and decreases again in the fourth date to 60%.

3.4.2. Distribution and Evolution of Aspect

Figure 13 displays the aspect maps of the study area for the four monitoring dates.

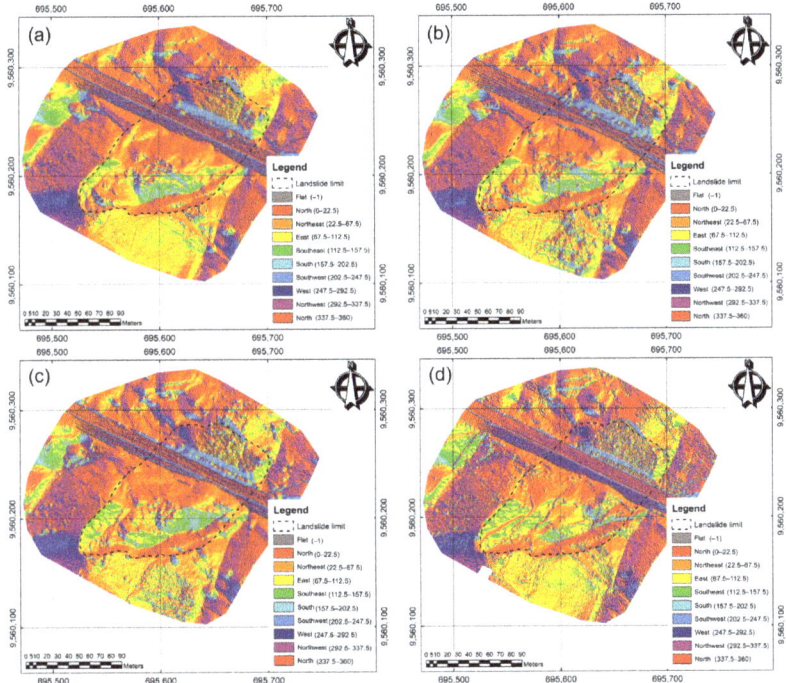

Figure 13. Aspect maps in the study area for the four monitoring periods: (**a**) 24 January 2017; (**b**) 9 June 2017; (**c**) 8 June 2018; (**d**) 12 March 2020.

In the maps shown in Figure 13, a set of slope units alternately oriented toward the NW and SE can be observed, the landslide slope being oriented toward the NE but with a wide range between the N and SE directions. There are also some localized areas in the landslide with slopes facing the SW. This general arrangement remains consistent in all analyzed dates, although the counterslope areas (between the S and W) generally appear downslope in the other dates, except for the last date where they practically disappear.

In the element analysis for the first date presented in Table 3 and Figure 14, it can be observed that the main scarp has an average orientation toward the NE although with several relative maxima toward the N and E. The secondary scarps have a similar average orientation to the NE, while the lateral scarps face the E, with two maxima toward the SE and N, corresponding to the left and right flanks, respectively. Finally, the counterslope scarps are oriented toward W. The body has an orientation in a wide range between the N in the lower part and the S in the upper part. The orientation of the foot is highly concentrated toward the NNE. The other unstable areas exhibit variable orientations, while the stable zone shows a relatively insignificant average with maxima between the N and E or even the NW.

Table 3. Aspect statistics for landslide elements.

Element	Average	Mode	Min	Max	Range	Std-D.	C.Var.
Main landslide							
Main scarp	68	0	0.00	359.99	359.99	139.43	1.05
Head	90	90	0.00	359.98	360.98	104.06	0.91
Lateral scarps-flanks	89	0	0.00	359.99	359.99	130.09	1.02
Secondary scarps	64	0	0.00	359.98	359.98	154.46	0.94
Counterslope scarps	271	270	1.28	358.73	357.45	29.81	0.11
Body	91	90	0.00	359.97	360.97	99.17	0.82
Secondary body	46	0	0.00	359.98	359.98	124.71	1.18
Foot	31	0	0.00	359.97	359.97	120.77	1.40
Other landslides							
Scarps	313	0	0.00	359.99	360.99	107.93	0.40
Heads	280	0	0.00	359.97	360.97	121.74	0.52
Secondary scarps	110	180	0.00	358.79	358.79	74.40	0.68
Bodies	267	0	0.00	359.98	360.98	142.33	0.76
Foots	32	0	0.00	359.97	359.97	101.22	1.56
Stable area							
Stable area	124	0	0.00	359.98	359.98	123.87	0.72

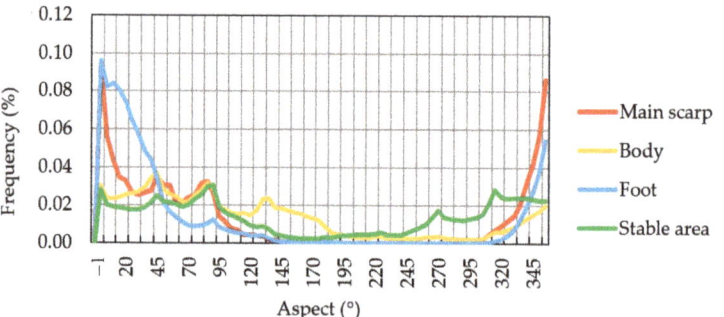

Figure 14. The distribution of aspects in landslide elements. The graph shows a greater distribution of frequencies of all elements between 0° and 145° and from 270° to 359°. However, the main scarp presents a maximum at 0° (N) and other relative maximums at 45° (NE) and 90° (E). Body orientations extended in the ranges of 0–145° and 330–359° (NW-SE); foot orientation is concentrated in the ranges of 0–45° and 345–359° (N-NE); and stable area orientations are distributed in the ranges of 0–90° (NE) and 270–359° (NW).

The evolution of orientation in profile A (Figure 15a) starts in the first date with a predominant orientation toward the N and NE (scarp, body and foot), although there is also a relative maximum toward the south (upper part of the body and counterslope scarps). From there, the majority orientation gradually shifts more toward the NE, reducing the orientations toward the S and W.

In profile B (Figure 15b), a similar pattern is observed, with an absolute maximum orientation toward the N and NE in the first date, along with a relative maximum toward the south. In the following two dates, no significant changes are observed although the orientations toward the south decrease. In the last date, this trend continued, but the absolute maximum clearly shifted toward the east–northeast.

Finally, in profile C (Figure 15c), the orientation is distributed between N and E in the first date, remaining similar in the second date. In the third date, a relative maximum appears toward the SE, and in the fourth date, the maximum shifts toward the NE.

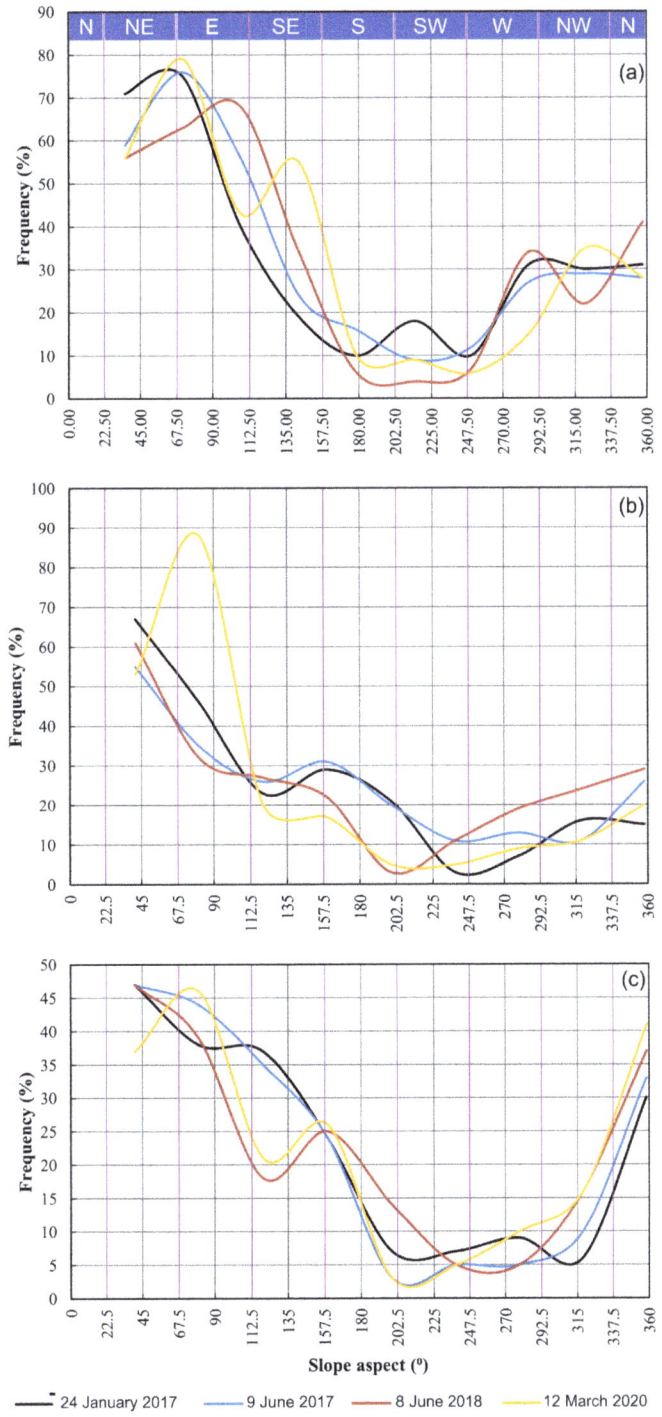

Figure 15. Temporal evolution of aspect in profiles: (**a**) Profile A; (**b**) Profile B; (**c**) Profile C.

3.4.3. Distribution and Evolution of the Topographic Position Index (TPI)

Figure 16 displays the maps of the topographic position index (TPI) at a 1 m resolution for the four considered dates.

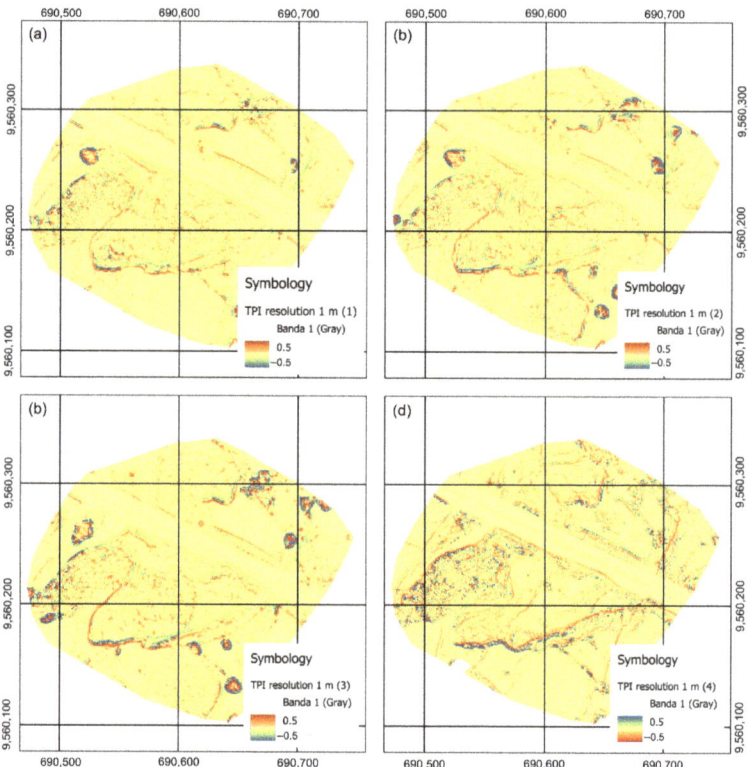

Figure 16. Topographic position index (TPI) maps at 1 m resolution in the study area for the four monitoring periods: (**a**) 24 January 2017; (**b**) 9 June 2017; (**c**) 8 June 2018; (**d**) 12 March 2020.

In these maps, there is a general predominance of TPI values close to zero throughout the area, with concentrated sectors that are elongated in a certain direction, alternating between positive and negative values. These sectors correspond to areas with clear slope breaks, which correspond to the upper and lower boundaries of the main scarp as well as the lateral and secondary scarps.

In the element analysis of the TPI at 0.05 m resolution for the first date, as shown in Table 4 and Figure 17, all elements and zones show average values close to zero. The differences occur in the range and, especially, the standard deviation, which is higher in the scarps, particularly in the main scarp (0.13), and lower in the body and foot of the landslide (0.02) as well as in the stable zone (0.01), which shows the least variability among all the analyzed zones.

Meanwhile, Figure 18 displays the results of applying the skeletonization tool on the four-measurement dates for TPI at 1 m resolution, showing the corresponding extracted lines. The lines representing the upper boundary (blue) and lower boundary (red) of the scarps are clearly visible.

Table 4. TPI statistics for the landslide elements.

Element	Average	Mode	Minimum	Maximum	Range	Std-D.	C.Var.
			Main landslide				
Main scarp	0.00	−0.02	−1.04	0.88	1.92	0.13	−103.11
Head	0.00	0.00	−0.35	0.36	0.70	0.02	−11.94
Lateral scarps-flanks	0.01	0.00	−0.63	0.40	1.03	0.06	9.27
Secondary scarps	0.00	0.00	−0.52	0.56	1.09	0.05	40.75
Counterslope scarps	0.00	−0.03	−0.26	0.31	0.57	0.08	−55.03
Body	0.00	0.00	−0.28	0.35	0.63	0.02	−28.22
Secondary body	0.00	0.00	−0.40	0.39	0.79	0.02	−23.97
Foot	0.00	0.00	−0.19	0.21	0.40	0.02	−217.18
			Other landslides				
Scarps	0.00	0.00	−0.54	0.81	1.35	0.05	−188.40
Heads	0.00	0.00	−0.17	0.43	0.60	0.02	−8.23
Secondary scarps	0.00	0.00	−0.22	0.17	0.39	0.04	58.95
Bodies	0.00	0.00	−0.25	0.24	0.49	0.02	−17.44
Foots	0.00	0.00	−0.20	0.20	0.39	0.02	−12.25
			Stable area				
Stable area	0.00	0.00	−0.68	0.53	1.21	0.01	108.33

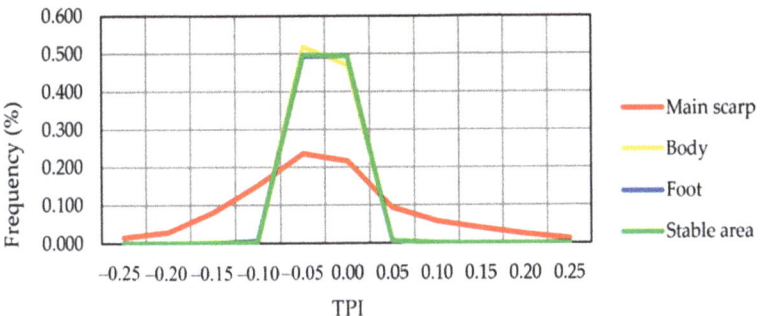

Figure 17. Distribution of TPI at 0.05 m resolution in landslide elements. It can be seen that all the elements present a maximum near 0, but the body, foot and stable area present a steeper peak, while the main scarp has a smoother peak, that is, higher absolute values typical of zones with slope break. The distribution of the TPI of the foot (red line) and the stable zone (green line) are coincident.

Throughout the different analyzed dates, excluding areas with vegetation changes, changes in the position of the upper and lower slope break or edge lines of the scarps can be observed, particularly the upper lines of the main scarp and the lateral scarps, which show a retreat toward the upper part of the slope, as will be discussed later. Meanwhile, the lower lines of these scarps exhibit less variation, and the lines of other scarps are more discontinuous and irregularly distributed throughout the landslide area. Regarding the foot area, the line defined by the tip is identified by the lower break or edge line of the TPI at 1 m resolution. In this case, advancements of the lines in the downslope direction can be observed, occupying the road area in the second and third dates, with a retreat in the fourth date.

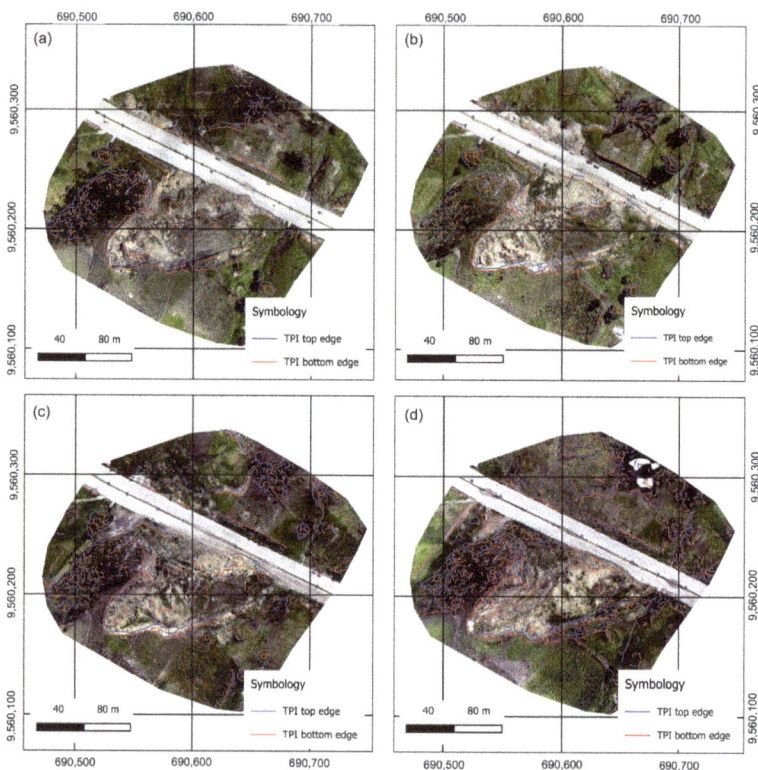

Figure 18. Slope break lines obtained from TPI at 1 m resolution superimposed over the corresponding orthoimages for the four monitoring dates: (**a**) 24 January 2017; (**b**) 9 June 2017; (**c**) 8 June 2018; (**d**) 12 March 2020. The upper lines are drawn in blue and the lower ones in red.

3.4.4. Distribution and Evolution of the Terrain Roughness Index (TRI)

In the maps of Figure 19, areas of high roughness can be observed, particularly in relation to the scarps, especially the main scarp and the lateral scarps, but also in the secondary scarps and the foot of the main landslide, while the landslide body exhibits low roughness. Areas of relatively high roughness are also observed in the scarps of other minor landslides, while the stable zone presents the lowest roughness if areas with vegetation are excluded.

The roughness values for the different landslide elements are shown in Table 5 and Figure 20. It can be observed that the scarps have an average roughness of 0.14, which is slightly higher in the main scarp (0.22). Meanwhile, the head area shows an average roughness value of 0.06, which is slightly higher than the body (0.05) and lower than the foot (0.07). The other instability areas exhibit generally lower but comparable values, which are always higher than the stable zone (0.04).

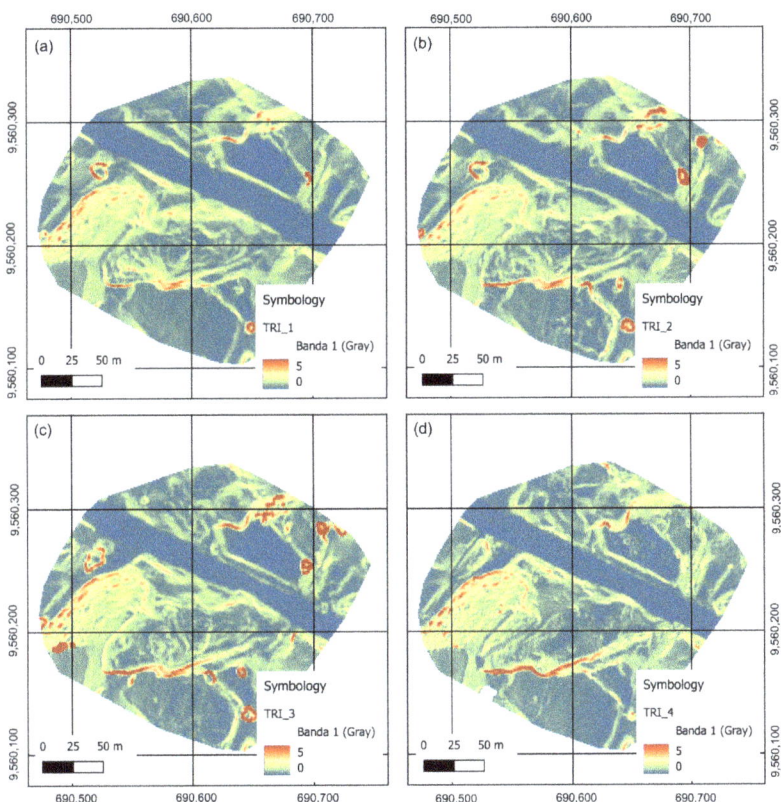

Figure 19. Terrain roughness index (TRI) maps in the study area for the four monitoring periods: (**a**) 24 January 2017; (**b**) 9 June 2017; (**c**) 8 June 2018; (**d**) 12 March 2020.

Table 5. TRI statistics for the landslide elements.

	Average	Mode	Min.	Max.	Range	Std-D.	C.Var.
			Main landslide				
Main scarp	0.22	0.05	0.00	1.60	1.59	0.16	0.71
Head	0.06	0.05	0.00	0.52	0.52	0.04	0.65
Lateral scarps-flanks	0.13	0.04	0.00	1.03	1.03	0.09	0.68
Secondary scarps	0.13	0.06	0.00	0.90	0.90	0.07	0.55
Counterslope scarps	0.16	0.03	0.01	0.44	0.43	0.09	0.52
Body	0.05	0.02	0.00	0.56	0.56	0.03	0.61
Secondary body	0.07	0.03	0.00	0.78	0.78	0.04	0.58
Foot	0.07	0.06	0.00	0.34	0.34	0.04	0.53
			Other landslides				
Scarps	0.12	0.08	0.00	1.00	1.00	0.08	0.65
Heads	0.04	0.03	0.00	0.45	0.45	0.03	0.64
Secondary scarps	0.10	0.03	0.00	0.27	0.27	0.05	0.50
Bodies	0.05	0.03	0.00	0.47	0.47	0.03	0.69
Foots	0.07	0.07	0.00	0.24	0.24	0.03	0.37
			Stable area				
Stable area	0.04	0.02	0.00	1.22	1.22	0.03	0.76

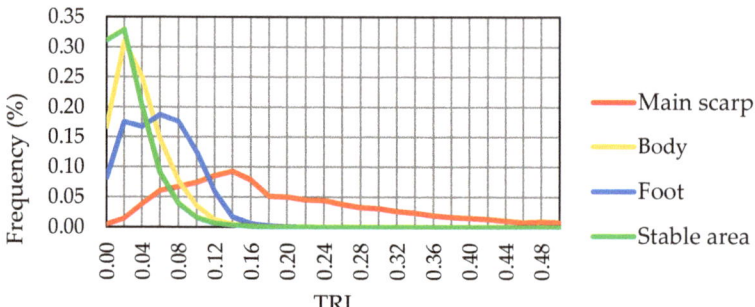

Figure 20. Distribution of TRI in landslide elements. The stable area and landslide body present a distribution with low roughness indicative of a terrain with little topographic variability. Meanwhile, the foot but especially the main scarp present higher roughness: that is, a greater topographic variability.

The evolution of roughness observed in the maps is relatively smooth at the overall landslide level and is mainly concentrated in the scarps, where roughness increases in the second and third dates. Changes in the position of high-roughness areas can be observed in the body and foot areas due to landslide displacement. However, in the fourth date, a general decrease in roughness can be observed in the body and foot. Some changes can also be seen in other unstable zones, especially the one occurring under the road, over the entire analyzed period.

3.4.5. Distribution and Evolution of the Topographic Wetness Index (TWI)

Figure 21 shows the distribution of TWI for each monitoring date, where higher positive TWI values are represented in shades of blue. These areas indicate a high potential for water accumulation or surface water runoff, primarily from rainfall, thus representing in some way the drainage network of the slope. Negative values correspond to areas where water accumulation is not possible, generally corresponding to the higher parts of the slope.

Thus, it can be observed that in the upper part of the landslide, near the main scarp and lateral scarps, the values are low, gradually increasing in the head and along the body where a drainage network formed within the landslide. Two drainage lines stand out on both sides of the body, between it and the scarps, which developed from the cracks generated by friction and displacement of the mass. Additionally, it is important to note how the different networks converge in the foot zone and the road, where the highest index values are reached. The remaining sectors of the study area outside the movement show a better organized drainage network except in the areas of crops and vegetation, which are more irregular. The drainage configuration changed throughout the analyzed dates, with a tendency to accumulate higher index values in the foot zone around the road from the first to the third date, while in the last date, the area shows a less hierarchical and more irregular drainage structure.

These observations are corroborated by the results of the element analysis shown in Table 6 and Figure 22. Thus, the scarps have low values of the TPI index, especially the main scarp, which even presents negative values (-0.69). The remaining parts of the landslide show increasing values from the head (1.74), body (1.88), and foot, where the highest values are reached (2.22). The other unstable zones also show increasing values from the scarps (1.35) to the foot (2.07). Moreover, the stable zone presents even higher average values (2.38).

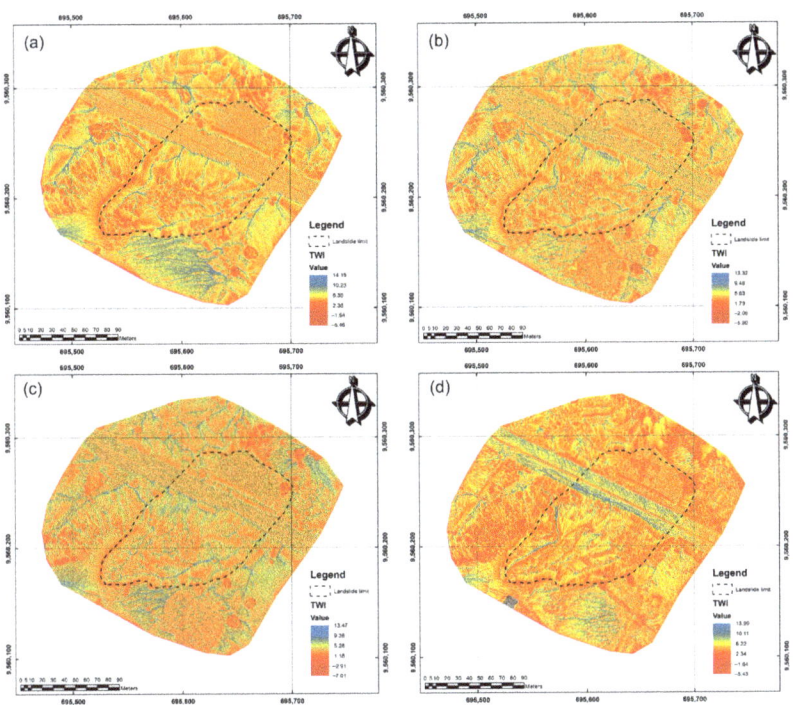

Figure 21. Topographic wetness index (TWI) maps in the study area for the four monitoring periods: (**a**) 24 January 2017; (**b**) 9 June 2017; (**c**) 8 June 2018; (**d**) 12 March 2020.

Table 6. TWI statistics for the landslide elements.

	Average	Mode	Min.	Max.	Range	Std-D.	C.Var.
			Main landslide				
Main scarp	−0.69	−2.65	−5.25	7.95	13.19	1.28	−1.85
Head	1.74	2.83	−4.55	12.18	16.72	1.51	0.87
Lateral scarps-flanks	1.02	−1.03	−4.73	8.43	13.16	1.55	1.52
Secondary scarps	1.04	−2.76	−5.34	7.11	12.44	1.39	1.33
Counterslope scarps	−0.24	−4.15	−4.15	5.11	9.26	1.44	−6.00
Body	1.88	2.83	−4.46	13.57	18.03	1.59	0.85
Secondary body	2.06	0.59	−4.98	13.74	18.72	1.53	0.74
Foot	2.22	2.83	−3.56	11.50	15.05	1.21	0.54
			Other landslides				
Scarps	1.35	−1.34	−5.24	7.31	12.55	1.32	0.98
Heads	2.01	0.75	−4.72	11.71	16.44	1.39	0.70
Secondary scarps	0.77	2.68	−3.40	6.78	10.18	1.04	1.35
Bodies	2.35	2.83	−4.17	14.15	18.32	1.56	0.67
Foots	2.07	0.62	−3.86	8.54	12.41	1.23	0.59
			Stable area				
Stable area	2.38	2.83	−5.02	15.26	20.28	1.66	0.70

In Figure 23, the temporal changes of TWI in each profile are shown. In the case of longitudinal profile A (Figure 23a), it can be observed that the mean TWI values gradually increase from 1.38 (24 January 2017) to 1.49 (9 June 2017), 1.60 (8 June 2018), and 1.81 (12 March 2022), mainly in the foot area, as previously pointed out. In profile B (Figure 23b), which is oblique to the landslide, the mean TWI values are 1.99, 1.66, 1.79, and 2.56 for the considered date, showing an increasing trend of TWI in general. In profile C (Figure 23c), which occupies a more marginal position toward the foot area, a decrease in mean TWI values can be observed, from 2.24 to 2.12, 1.92, and finally to 1.60. The discussion will further analyze the variations of TWI and its relationship with the behavior of the slope movement.

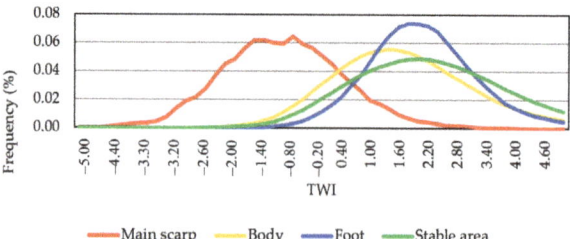

Figure 22. Distribution of TWI in landslide elements. The main scarp present TWI values significantly lower (negative average) than the body, foot and stable area of the slope (positive average).

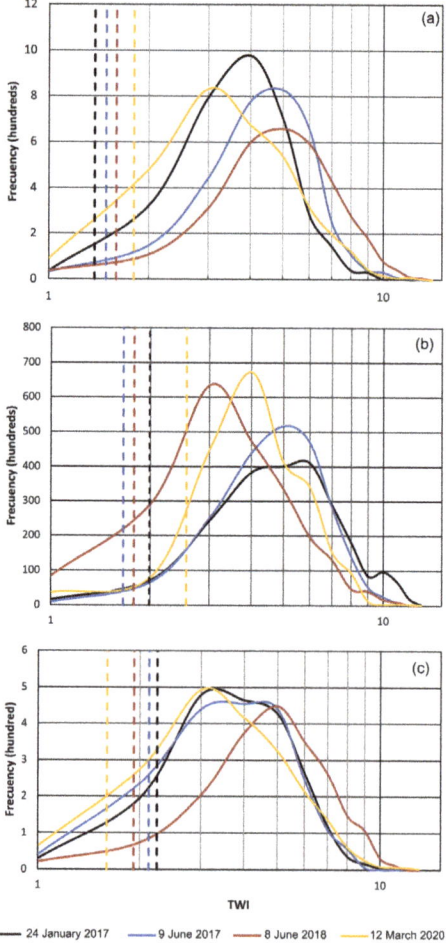

Figure 23. Temporal evolution of TWI in profiles: (**a**) Profile A; (**b**) Profile B; (**c**) Profile C. Dashed lines represent the mean TWI values in each monitoring date and corresponding profile. Each colored line corresponds to the monitoring date.

3.5. Geotechnical Characterization of the Affected Materials

Table 7 shows the results of the tests conducted on soil samples. At the head, soils with high plasticity clay (CH) have been determined, while at the body and foot levels, inorganic silts (MH-OH) are present. LL represents the liquid limit, LP represents the plastic limit, SUCS is the system classification of soil, φ represents the friction angle, and C represents cohesion.

Table 7. Summary of laboratory test results conducted on soil samples obtained from the slope.

Ubication	Sampling	w (%) +	LL	w (%) *	LP	SUCS	φ	C (kg/cm^2)
	1	28.55	59.7	27.6	5.86	CH		
Crown	2	28.00	59.4	27.3	6.04	CH	27°	1.65
	3	28.26	60.1	27.4	6.17	CH		
	1	34.22	67.9	33.8	38.3	MH-OH		
Body	2	34.12	67.0	33.8	37.6	MH-OH	22°	0.37
	3	34.31	66.3	33.6	37.4	MH-OH		
	1	40.51	86.2	39.9	37.6	MH-OH		
Foot	2	41.02	84.6	39.9	37.1	MH-OH	11°	0.34
	3	40.74	86.0	39.1	38.7	MH-OH		

+ Laboratory determined moisture content. * Moisture content measured in the field with Soil Moisture SM300 Kit.

4. Discussion

4.1. Accuracy and Uncertainty of DEMs and Orthoimages

Considering the root mean square errors (RMSE) at the checkpoint (CHK) locations shown in Table 1, it can be observed that the horizontal errors (XY) range from 0.025 to 0.041 m through the different flights, which is even lower than those found in GCPs (0.035–0.053), and they are always below the 0.05 m resolution of the orthoimages and DEMs. Both these values and those calculated for the control points are similar to the errors obtained by other authors in RPAS surveys under comparable conditions [25,26,30,34,38,57,91,92]. Therefore, the uncertainty for horizontal measurements is established at 0.05 m.

Regarding vertical RMSE (Z), the obtained values vary from 0.025 to 0.041 m with an average of 0.030 m, which is similar to the average error obtained at GCPs (0.023–0.034 m). According to previous studies [26,27,67], the uncertainty of the DEMs is estimated to be two to three times the value of these errors, which amounts to approximately 0.10 m. Meanwhile, the vertical uncertainty of the DoDs, also known as the minimum level of detection (minLoD), is estimated as [23,26,46,93–96]:

$$Uncert_{YEAR\ 1-YEAR\ 2} = \left(Uncert^2_{YEAR\ 1} + Uncert^2_{YEAR\ 2}\right)^{0.5} \quad (5)$$

Thus, based on the overall uncertainty of DEMs, the uncertainty of DoDs can be stated in 0.15 m.

In both cases, the displacements of metric order, both horizontal and vertical, caused by the landslide in the study area far exceed these uncertainty thresholds, suggesting that the models and orthophotos have more than sufficient quality for this study.

4.2. Detection of Elements through Photointerpretation and Semiautomatic Extraction

The detection and mapping of terrain features, specifically those related to the landslide, have been carried out through both photointerpretation and semiautomatic extraction using orthoimages, DEMs, and derived models.

Photointerpretation of the orthoimages has allowed for the identification, delineation and digitization of several features and elements of the main landslide, such as different types of scarps (main, lateral, secondary and counterslope), crown and head, body and foot. Other areas of instability and their corresponding elements as well as areas with vegetation and structures have also been identified and mapped to exclude them from subsequent analyses. The identification of features and elements has followed the classic guidelines of photointerpretation primarily involving the analysis of color (RGB images), including shadows, and texture analysis.

However, this identification and mapping of elements has also relied on the observation of DEMs, especially the derived models. There are many works based on topographic parameters carried out at different resolutions, from models corresponding to aerial LiDAR [51,53] to models obtained with RPAS images [52–54]. Thus, the slope model, for instance, provides a good approximation for identifying different types of scarps in steep areas (greater than $30°-45°$), but it is also useful for identifying other landslide elements such as the foot, with slope contrasting to the body or the stable zone where the slope is gentler.

Meanwhile, aspect or orientation allows us to observe some features such as scarps or lobes; thus, the scarps in some cases interrupt the general orientation of the slope, such as the right lateral scarp that has a N orientation within a slope exposed to the ESE. Within the landslide body, the lobed surface of a flow-type movement can be seen oriented to the N on the left part and SE on the right part, suggesting an ENE axis as the direction of advance of the landslide. In this area, counterslope scarps (WNW) are also very visible.

The TPI makes it possible to clearly delimit scarps and other elements such as the toe or tip of the foot where slope breaks occur. Thus, the skeletonization of the zones with high absolute values (positive and negative) of the index leads to quite clearly identifying the lines corresponding to the upper and lower edge of the scarps, respectively. This index and the curvature, which are usually closely related, have been used in previous works for detecting landslide scarps [51,53,54,81]. In scarps, the slope breaks are very pronounced, so they are well detected in the very high-resolution models (0.05 m) as well as in the high-resolution models (1 m), although they are logically more precisely in the former. However, other elements such as the foot and more specifically the toe, in which the break is less pronounced, are better extracted with models of 1 m resolution.

The roughness also allows the identification of scarps in a very similar way to the slope, but it does not provide any significant improvement with respect to it, so it is not analyzed in detail in this work. Finally, the TWI makes it possible to identify the hillslope areas where there is a great potential to accumulate or circulate water mainly from rain and runoff. Thus, the drainage network of the hillslope is somehow represented, and its higher or lower development and hierarchical order are analyzed. It can be seen that in the upper part of the landslide, the values are low and increase in the body until reaching higher values (water accumulation) at the foot next to the road. The formation of drainage channels on the flanks can also be seen. The relationship of this index with the soil and its incidence on the stability of the landslide are discussed later.

4.3. Morphometric Analysis

The morphometric analysis makes it possible to characterize in a quantitative way the different landslide elements and distinguish them from the stable area of the hillslopes based on the statistical values obtained for each of the parameters considered.

Thus, the scarps are one of the elements that are best characterized as they present high slopes and roughness. Slope present average values greater than $40°$, and even $50°$ in the main scarp, which allows this parameter to be used for detecting areas that can eventually be mapped as scarps. Meanwhile, the average slope of the landslide is about $24°$, more than $8°$ higher than that of the stable zone, which is consistent with the fact that the slope, together with roughness, is a determining factor in stability and susceptibility analyses both probabilistic and deterministic [97]. Furthermore, this slope is characteristic of slide-flow type movements such as the one studied [98]. For the remaining elements, the main body barely reaches an average slope of $20°$, the head of $23°$, and the foot of almost $28°$.

The aspect logically does not show characteristic values as the slope angle, but it does allow distinguishing elements of the landslide regarding the overall hillslope. Thus, in a sequence of hillslopes oriented to the NW and SE, the landslide shows an orientation with a greater N component (NE in the body and NE in the foot). There are also areas with an orientation contrary to the general one, corresponding to counterslope scarps.

The TPI average values are very uniform across all considered elements; in this case, the statistical measure that marks the differences is the standard deviation, which is significantly lower in the stable zone but increases noticeably in the landslide area, especially in the scarp but also in the foot. In the scarps, although the mean value is always close to 0 due to the compensation of positive curvature (upper limit) and negative curvature (lower zone), the absolute values and consequently the standard deviation are higher than in the stable zone, where they are always close to 0.

The TRI exhibits a distribution similar to the slope, with the highest values found in the scarp zones (0.14), particularly on the main scarp (0.22). Meanwhile, the head area shows an average roughness value of 0.06, which is slightly higher than the body (0.05) and lower than the foot (0.07). The other instability zones also present higher values in the scarps compared to the other elements, although these are slightly lower. In both cases, the average roughness values are always higher than in the stable zone (0.04).

Finally, the TWI within the landslide area shows increasing values from the head to the foot where the highest values are reached, indicating water concentration in the foot and road area, which can promote instability in this part and the overall movement. However, it is interesting to note that the mean TWI value is lower in the sliding zone and other unstable areas compared to the stable zone, which may be attributed to the disorganization of the hydrographic network within the landslides areas.

4.4. Landslide Evolution and Kinematics

The analysis of the landslide evolution has been carried out using various techniques: photointerpretation, observation of DEM of differences and derived model maps, topographic and parameter profiles, and the break lines extracted from TPI. Based on these analyses and the obtained results, areas of depletion and accumulation of material can be identified in the head and foot, respectively, resulting from the landslide kinematics. In summary, the following observations can be made regarding landslide evolution:

- The landslide had already developed before the start of the monitoring campaigns with RPAS flights (24 January 2017), as clearly seen in the maps, profiles, and lines extracted from the TPI maps, showing a well-formed main scarp and various secondary scarps along the mass body. The body appeared individualized with a lobed shape and an ENE direction axis, as indicated by the aspect map discussed previously, ending in a foot that progressed in a direction with a greater N component due to the presence of the road. The TWI allows for the observation of two drainage channels formed close to the two-lateral flanks and a more irregular and less hierarchical drainage network within the landslide. The average initial slope in the landslide area was 24°, calculated from the total area, and 18–22° in the two most significant profiles (longitudinal and oblique-centered). This can be explained by the fact that the profiles include part of the terrain on the crown and the road beneath the toe. The frequency of slopes less than 20° was 60% in profile A and 46% in profile B.
- In the next two dates (flights on 9 June 2017 and 8 June 2018), an advancement of the landslide mass could be observed, which can be estimated from the secondary and counterslope scarps observed in the DoDs and derived models (slopes, aspect, TPI, and TRI) but especially in the topographic profiles. In the longitudinal profile (A), mass descents of approximately 6–8 m and advancements of around 20–25 m were estimated in the head area and upper part of the body, which were distributed almost equally in each period. These results lead to an average velocity approximately twice as high in the first period (descent of about 1 m/month and advancement of about 4–5 m/month) compared to the second period (0.5 m/month and 2–2.5 m/month, respectively). In the foot and toe, the mass advanced about 5–6 m, mostly occurring in the first period with an estimated velocity of around 1 m/month, although it may be underestimated due

to material removal in the road area. The other profiles of a transversal nature allow for estimating a descent of about 3–5 m and an advancement of about 6–8 m in the toe area, which were also higher in the first period. The slope gradually decreased in both dates in profiles A and B, as a result of the landslide advancement and material evacuation, while profile C shows a more irregular evolution due to its marginal position in the landslide. Meanwhile, the percentages of slopes exceeding 20% increase up to 65% in profile A and up to 52% in profile B.

- In the last date (12 March 2020), a much flatter shape of the slope was clearly observed, with material removal from the body and foot, although the slope above the road was still present. The secondary scarps and counterslopes disappeared, as observed in the aspect models, and the overall morphology of the body was smoothed, with a descent of the surface that became progressively greater toward the lower part of the body and the foot, reaching about 5 m compared to the third date and about 8 m compared to the first date. The slope angle clearly decreased compared to the previous dates, especially in the longitudinal profile (from 17° to 13°). The percentage of slopes less than 20° increased to 76% in profile A and 64% in profile B.
- Details about the evolution of the landslide, such as the formation of scarps and the development of the foot, can also be deduced from the models and profiles. In this case, the most appropriate analysis is the comparison of break lines extracted from the TPI, both the 0.05 m resolution for the scarps and the 1 m resolution for the toe (specifically the tip). Thus, the retreat of the main and lateral scarps of about 2–3 m can be observed (Figure 24a), irregularly distributed (more in some sectors and less in others) between the first date (24 January 2017) and the third date (8 June 2018). Regarding the toe, advancements of about 5 m were observed between the first and third dates (Figure 24b). This generally coincides with what is observed in the profiles.

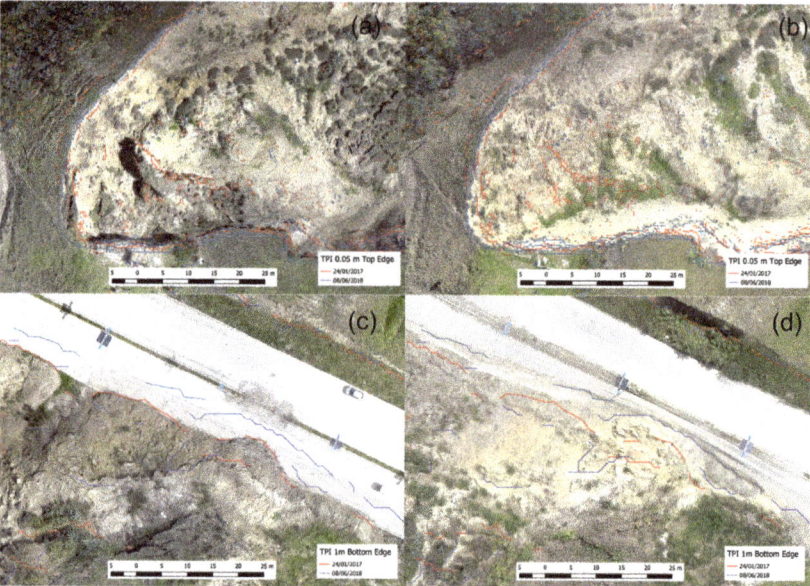

Figure 24. Details of the evolution of the landslide based on slope break lines: retreat of the main scarp with top edges extracted from TPI 0.05 m over the image of 24 January 2017 (**a**) and the image of 8 June 2018 (**b**); (**c**) advancement of the toe with bottom edges extracted from TPI 1 m over the image of 24 January 2017 and the image of 8 June 2018 (**d**).

Therefore, the use of DEM with centimetric resolutions allows the detailed representation of the topographic changes caused by landslides, so it is feasible to detect changes in some topographic parameters such as the slope, aspect, TRI and TWI.

4.5. Relationship between TWI and Soil Characteristics

Landslides occur when, in addition to the slope intrinsic factors or instability determinants (such as lithology, slope, morphology, and vegetation cover), other triggering external factors come into play, including rainfall [25,99–102], earthquakes [70], anthropogenic factors [80], etc. However, in this case, rainfall is the main factor that affects landslide processes, as established in the region surrounding the study area, where rainfall thresholds triggering landslides have been determined [103]. A superficial exploration of the terrain reveals the presence of well-identified cracks and jumps at the head (Figure 1b,c).

Analyzing the TWI maps, it is clear that blue shades represent areas of water accumulation or circulation, which are concentrated in the lower part of the body and the foot of the landslide, particularly over the road. This is confirmed by morphometric analysis. This accumulation in the foot causes it to tend to flow, contributing to the instability of the overall movement, which acquires a complex typology with a greater component of slide in the head and earth/mud flow in the foot [104,105].

4.6. Relationship between the Landslide and Geotechnical Data

The data in Table 6 establish that the landslide head is composed of high-plasticity clays, while the body and foot are made of high-plasticity silts. During the rainfall events, the superficial runoff water with the infiltration causes the unsaturated soil layers to decrease in resistance to cutting, causing instability. In this case, the presence of fissures concentrates the flow paths, which increases the infiltration of rainfall water and therefore causes alterations in the interstitial pressures that change the properties of the soil.

When there are clayey soils with high-plasticity indices (>29), the swelling and contraction processes can contribute to the opening and closing of cracks, significantly affecting them [73]. Likewise, the presence of a more permeable layer underlying unsaturated soils can create a capillary barrier effect, cause the storage of water at the bottom of the cracks and, in the presence of finer soils, reach critical saturation conditions, in particular to the variation of soil pressures that affect stability. The accumulation of water and saturation of soil, with the consequent loss of shear strength due to an increase in interstitial pressure, occurs to a greater extent in the lower part of the landslide and the foot, which causes it to flow.

5. Conclusions

The use of remote sensing techniques such as remote piloted aircraft systems (RPAS) has enabled the capture of high-resolution images and the generation of photogrammetric products that can be employed for various analyses and applications. From these images and the corresponding GCPs surveyed with GNSS techniques, DEMs and orthoimages of high precision and resolution can be obtained for detailed landslide analysis. If several surveys are available, multitemporal and evolutionary analysis can be addressed.

The case study of this work is a landslide in the area of El Plateado near the city of Loja in Ecuador. Landslides in Ecuador are a widespread hazard with a high impact on infrastructures; therefore, their characterization and monitoring are mandatory and urgent. Thus, the main objectives of this study are: (1) to characterize the landslide, identify its elements and describe its morphology; and (2) to monitor its kinematics.

Regarding the methodology developed to achieve these objectives, the following observations can be made:

- Four RPAS flights were conducted for capturing images, which were processed with SfM techniques to generate digital elevation models (DEMs) and orthoimages with a resolution of 0.05 m. The horizontal uncertainty estimated was under resolution (0.05 m), while the vertical uncertainty was 0.10 m for DEMs and 0.15 m for DoDs.

Since the features observed and the displacements measured in the landslide are at least an order of magnitude higher than the uncertainties, the quality of images is more than sufficient for this study.

- We analyzed not only direct photogrammetric products (DEMs and orthoimages) but also topographic or morphometric parameters such as slope, aspect, topographic position index (TPI), terrain roughness index (TRI) and topographic wetness index (TWI), which were determined using GIS tools. The systematic and detailed analysis of these parameters and the obtained results can be considered as the main contribution of this paper.
- Thus, the detection and mapping of landslide features have been carried out by means of photointerpretation and identification from DEM, profiles, and derived models. Morphometric analysis with GIS areal tools has allowed the characterization of these features and landslide elements to be used in their automatic identification in other areas.
- Meanwhile, multitemporal analysis by the calculation of DoDs, visual comparison of orthoimages and DEM derivative maps and profiles has allowed the study and monitoring of landslide evolution.
- The integration of the previous analysis, especially the TWI parameter maps, with geotechnical data and soil properties leads to establishing the role of rainfalls as a triggering factor.
- The main results obtained are as follows:
- Several scarps (main, lateral and secondary) were identified, and the landslide body and foot were differentiated from the stable area. Moreover, lines of slope break such as those at the upper and lower part of scarps or at the tip (toe) of the landslide foot were extracted from TPI analysis at different resolutions (0.05 and 1 m).
- Scarps present the highest values of slope (up 40°) and TRI, which are followed by the foot and the body, and finally the stable area (8° lower than the whole landslide area). A general decrease in average slope and TRI was observed due to the displacement of the landslide mass and, in particular, due to the works of stabilization in the last period.
- TWI within the landslide area shows increasing values from the head toward the foot where the highest values are reached, indicating water concentration. Moreover, its value is lower in the landslide area than in the stable zone, which may be due to the disorganization of the hydrographic network within the landslide. An increase in TWI is also observed, especially in the landslide foot, which can accumulate a certain amount of water.
- Horizontal displacements of 20–25 m and terrain descents of 6–8 m have been measured in the upper part (head and body), while in the foot, the mass advances about 5–6 m in the more active period (24 January 2017–8 June 2018). These displacements coincide with those observed with the comparison of break lines at the foot area. This comparison also allows us to observe a retraction of 2–3 m in the main scarp.
- The integration of the previous analysis, especially the TWI maps, with geotechnical data and soil properties leads to establishing the role of rainfalls as a triggering factor. Thus, rainfall produces water infiltration favored by the presence of cracks on the terrain surface and therefore the flow accumulation of runoff water in the foot area. This fact together with the geotechnical conditions, such as the presence of high-plasticity clays in the landslide head and high-plasticity silts in the foot, leads to soil saturation, an increase in pore pressure and then a loss of soil strength and slope instability. The limitations of this study are related first to the temporal resolution of images captured with RPAS and the scarcity and even the lack of geotechnical and meteorological data, which does not allow real landslide monitoring. Meanwhile, another important limitation deals with the low automation of the procedure both in the feature extraction and morphometric analysis.

Future work should endeavor to overcome these limitations: both the capture and processing of larger amounts of different data and the automation of the different procedures.

Thus, a higher frequency of RPAS surveys would allow better landslide monitoring and deeper analysis of its kinematics and dynamics through the study of relationships with geotechnical parameters. In this sense, more and more accurate geotechnical and meteorological data would support this analysis. Regarding automation, GIS models, and especially machine learning methods (ML) for feature extraction and characterization, would be an interesting development for this approach. Moreover, techniques of digital image correlation (DIC) could also support kinematic analysis.

Author Contributions: Conceptualization: B.A.Z., R.E.H. and T.F.d.C.; Data curation and formal analysis: B.A.Z.; Funding acquisition and resources: B.A.Z.; Investigation: B.A.Z., R.E.H. and T.F.d.C.; Methodology: R.E.H. and T.F.d.C.; Supervision: R.E.H., T.F.d.C.; Validation: B.A.Z., R.E.H. and T.F.d.C.; Visualization: B.A.Z.; Writing—original draft: B.A.Z.; Writing—review and editing: R.E.H. and T.F.d.C. All authors have read and agreed to the published version of the manuscript.

Funding: This study has the support of the Private Technical University of Loja through the internal research project PROY_GMIC_1285.

Data Availability Statement: The data presented in this study are available on request from the corresponding author.

Acknowledgments: We thank Camila Alvarado and María Paula Peña for field assistance and the Private Technical University of Loja through the Department of Civil Engineering and its Highway Engineering research group for its logistical and laboratory support; the Photogrammetric and Topometric Systems Research Group (TEP-213 of the Andalusian Plan of Research, Development and Innovation, PAIDI) and Environmental Research: Geological Risks and Ground Engineering Research Group (RNM-121 PAIDI) for the realization of this study.

Conflicts of Interest: The authors declare no conflict of interest. The funders had no role in the design of the study; in the collection, analyses, or interpretation of data; in the writing of the manuscript; or in the decision to publish the results.

References

1. Sassa, K. Monthly Publication of Landslides: Journal of International Consortium on Landslides (ICL). *Landslides* **2018**, *15*, 1–3. [CrossRef]
2. Sassa, K. Participants in the Fourth World Landslide Forum and Call for ICL Members, Supporters, and Associates. *Landslides* **2017**, *14*, 1839–1842. [CrossRef]
3. Konagai, K. More than Just Technology for Landslide Disaster Mitigation: Signatories to The Kyoto Landslide Commitment 2020—No. 1. *Landslides* **2021**, *18*, 513–520. [CrossRef]
4. Yavuz, M.; Koutalakis, P.; Diaconu, D.C.; Gkiatas, G.; Zaimes, G.N.; Tufekcioglu, M.; Marinescu, M. Identification of Streamside Landslides with the Use of Unmanned Aerial Vehicles (UAVs) in Greece, Romania, and Turkey. *Remote Sens.* **2023**, *15*, 1006. [CrossRef]
5. Casagli, N.; Frodella, W.; Morelli, S.; Tofani, V.; Ciampalini, A.; Intrieri, E.; Raspini, F.; Rossi, G.; Tanteri, L.; Lu, P. Spaceborne, UAV and Ground-Based Remote Sensing Techniques for Landslide Mapping, Monitoring and Early Warning. *Geoenviron. Disasters* **2017**, *4*, 9. [CrossRef]
6. Assilzadeh, H.; Levy, J.K.; Wang, X. Landslide Catastrophes and Disaster Risk Reduction: A GIS Framework for Landslide Prevention and Management. *Remote Sens.* **2010**, *2*, 2259–2273. [CrossRef]
7. Casagli, N.; Tofani, V.; Morelli, S.; Frodella, W.; Ciampalini, A.; Raspini, F.; Intrieri, E. Remote Sensing Techniques in Landslide Mapping and Monitoring, Keynote Lecture. In *Workshop on World Landslide Forum*; Springer: Cham, Switzerland, 2017; pp. 259–266. [CrossRef]
8. Lollino, G.; Manconi, A.; Guzzetti, F.; Culshaw, M.; Bobrowsky, P.; Luino, F. *Engineering Geology for Society and Territory. Volume 5: Urban Geology, Sustainable Planning and Landscape Exploitation*; Springer: Cham, Switzerland, 2015; pp. 1–1400. [CrossRef]
9. Mateos, R.M.; López-Vinielles, J.; Poyiadji, E.; Tsagkas, D.; Sheehy, M.; Hadjicharalambous, K.; Liscák, P.; Podolski, L.; Laskowicz, I.; Iadanza, C.; et al. Integration of Landslide Hazard into Urban Planning across Europe. *Landsc. Urban Plan.* **2020**, *196*, 103740. [CrossRef]
10. Giordan, D.; Manconi, A.; Remondino, F.; Nex, F. Use of Unmanned Aerial Vehicles in Monitoring Application and Management of Natural Hazards. *Geomat. Nat. Hazards Risk* **2017**, *8*, 1–4. [CrossRef]
11. Palenzuela, J.A.; Jiménez-Perálvarez, J.D.; El Hamdouni, R.; Alameda-Hernández, P.; Chacón, J.; Irigaray, C. Integration of LiDAR Data for the Assessment of Activity in Diachronic Landslides: A Case Study in the Betic Cordillera (Spain). *Landslides* **2016**, *13*, 629–642. [CrossRef]

12. Jiménez-Perálvarez, J.D.; El Hamdouni, R.; Palenzuela, J.A.; Irigaray, C.; Chacón, J. Landslide-Hazard Mapping through Multi-Technique Activity Assessment: An Example from the Betic Cordillera (Southern Spain). *Landslides* **2017**, *14*, 1975–1991. [CrossRef]
13. Kalantar, B.; Ueda, N.; Saeidi, V.; Ahmadi, K.; Halin, A.A.; Shabani, F. Landslide Susceptibility Mapping: Machine and Ensemble Learning Based on Remote Sensing Big Data. *Remote Sens.* **2020**, *12*, 1737. [CrossRef]
14. Karagianni, A.; Lazos, I.; Chatzipetros, A. *Remote Sensing Techniques in Disaster Management: Amynteon Mine Landslides, Greece*; Springer International Publishing: Cham, Switzerland, 2019; ISBN 9783030053291.
15. Liu, P.; Wei, Y.; Wang, Q.; Chen, Y.; Xie, J. Research on Post-Earthquake Landslide Extraction Algorithm Based on Improved U-Net Model. *Remote Sens.* **2020**, *12*, 894. [CrossRef]
16. Buša, J.; Rusnák, M.; Kušnirák, D.; Greif, V.; Bednarik, M.; Putiška, R.; Dostál, I.; Sládek, J.; Rusnáková, D. Urban Landslide Monitoring by Combined Use of Multiple Methodologies—A Case Study on Sv. Anton Town, Slovakia. *Phys. Geogr.* **2020**, *41*, 169–194. [CrossRef]
17. Cina, A.; Piras, M.; Bendea, H.I. Monitoring of Landslides with Mass Market GPS: An Alternative Low Cost Solution. *Int. Arch. Photogramm. Remote Sens. Spat. Inf. Sci. ISPRS Arch.* **2013**, *40*, 131–137. [CrossRef]
18. Hastaoglu, K.O.; Sanli, D.U. Accuracy of GPS Rapid Static Positioning: Application to Koyulhisar Landslide, Central Turkey. *Surv. Rev.* **2011**, *43*, 226–240. [CrossRef]
19. Wang, G.Q. Millimeter-Accuracy GPS Landslide Monitoring Using Precise Point Positioning with Single Receiver Phase Ambiguity (PPP-SRPA) Resolution: A Case Study in Puerto Rico. *J. Geod. Sci.* **2013**, *3*, 22–31. [CrossRef]
20. Zárate, B. Monitoreo de Movimientos de Ladera En El Sector de San Pedro de Vilcabamba Mediante Procedimientos GPS. *Maskana* **2011**, *2*, 17–25. [CrossRef]
21. Isè, L. Application of a Terrestrial Laser Scanner (TLS) to the Study of the Sé. *Remote Sens.* **2010**, *2*, 2785–2802. [CrossRef]
22. Cook, K.L. An Evaluation of the Effectiveness of Low-Cost UAVs and Structure from Motion for Geomorphic Change Detection. *Geomorphology* **2017**, *278*, 195–208. [CrossRef]
23. Giordan, D.; Manconi, A.; Tannant, D.D.; Allasia, P. UAV: Low-Cost Remote Sensing for High-Resolution Investigation of Landslides. In Proceedings of the 2015 IEEE International Geoscience and Remote Sensing Symposium (IGARSS), Milan, Italy, 26–31 July 2015; pp. 5344–5347. [CrossRef]
24. Kršák, B.; Blišťan, P.; Pauliková, A.; Puškárová, P.; Kovanič, L.; Palková, J.; Zelizňaková, V. Use of Low-Cost UAV Photogrammetry to Analyze the Accuracy of a Digital Elevation Model in a Case Study. *Meas. J. Int. Meas. Confed.* **2016**, *91*, 276–287. [CrossRef]
25. Fernández, T.; Pérez, J.L.; Cardenal, F.J.; López, A.; Gómez, J.M.; Colomo, C.; Delgadoa, J.; Sánchez, M. Use of a Light UAV and Photogrammetric Techniques to Study the Evolution of a Landslide in Jaén (Southern Spain). *Int. Arch. Photogramm. Remote Sens. Spat. Inf. Sci. ISPRS Arch.* **2015**, *40*, 241–248. [CrossRef]
26. Fernández, T.; Pérez, J.L.; Cardenal, J.; Gómez, J.M.; Colomo, C.; Delgado, J. Analysis of Landslide Evolution Affecting Olive Groves Using UAV and Photogrammetric Techniques. *Remote Sens.* **2016**, *8*, 837. [CrossRef]
27. Cardenal, J.; Fernández, T.; Pérez-García, J.L.; Gómez-López, J.M. Measurement of Road Surface Deformation Using Images Captured from UAVs. *Remote Sens.* **2019**, *11*, 1507. [CrossRef]
28. Cui, Y.; Cheng, D.; Choi, C.E.; Jin, W.; Lei, Y.; Kargel, J.S. The Cost of Rapid and Haphazard Urbanization: Lessons Learned from the Freetown Landslide Disaster. *Landslides* **2019**, *16*, 1167–1176. [CrossRef]
29. Mokhtar, M.R.M.; Matori, A.N.; Yusof, K.W.; Embong, A.M.; Jamaludin, M.I. Assessing UAV Landslide Mapping Using Unmanned Aerial Vehicle (UAV) for Landslide Mapping Activity. *Appl. Mech. Mater.* **2014**, *567*, 669–674. [CrossRef]
30. Niethammer, U.; Rothmund, S.; Schwaderer, U.; Zeman, J.; Joswig, M. Open Source Image-Processing Tools for Low-Cost Uav-Based Landslide Investigations. *ISPRS-Int. Arch. Photogramm. Remote Sens. Spat. Inf. Sci.* **2012**, *38*, 161–166. [CrossRef]
31. Haas, F.; Hilger, L.; Neugirg, F.; Umstädter, K.; Breitung, C.; Fischer, P.; Hilger, P.; Heckmann, T.; Dusik, J.; Kaiser, A.; et al. Quantification and Analysis of Geomorphic Processes on a Recultivated Iron Ore Mine on the Italian Island of Elba Using Long-Term Ground-Based Lidar and Photogrammetric SfM Data by a UAV. *Nat. Hazards Earth Syst. Sci.* **2016**, *16*, 1269–1288. [CrossRef]
32. Thomas, A.F.; Frazier, A.E.; Mathews, A.J.; Cordova, C.E. Impacts of Abrupt Terrain Changes and Grass Cover on Vertical Accuracy of UAS-SfM Derived Elevation Models. *Pap. Appl. Geogr.* **2020**, *6*, 336–351. [CrossRef]
33. Warrick, J.A.; Ritchie, A.C.; Schmidt, K.M.; Reid, M.E.; Logan, J. Characterizing the Catastrophic 2017 Mud Creek Landslide, California, Using Repeat Structure-from-Motion (SfM) Photogrammetry. *Landslides* **2019**, *16*, 1201–1219. [CrossRef]
34. Eltner, A.; Kaiser, A.; Castillo, C.; Rock, G.; Neugirg, F.; Abellán, A. Image-Based Surface Reconstruction in Geomorphometry-Merits, Limits and Developments. *Earth Surf. Dyn.* **2016**, *4*, 359–389. [CrossRef]
35. Immerzeel, W.W.; Kraaijenbrink, P.D.; Shea, J.M.; Shrestha, A.B.; Pellicciotti, F.; Bierkens, M.F.; de Jong, S.M. High-Resolution Monitoring OfHimalayan Glacier Dynamics Using Unmanned Aerial Vehicles. *Remote Sens. Environ.* **2014**, *150*, 93–103. [CrossRef]
36. Hu, S.; Qiu, H.; Wang, X.; Gao, Y.; Wang, N.; Wu, J.; Yang, D.; Cao, M. Acquiring High-Resolution Topography and Performing Spatial Analysis of Loess Landslides by Using Low-Cost UAVs. *Landslides* **2018**, *15*, 593–612. [CrossRef]
37. Huang, H.; Long, J.; Lin, H.; Zhang, L.; Yi, W.; Lei, B. Unmanned Aerial Vehicle Based Remote Sensing Method for Monitoring a Steep Mountainous Slope in the Three Gorges Reservoir, China. *Earth Sci. Inform.* **2017**, *10*, 287–301. [CrossRef]
38. Lindner, G.; Schraml, K.; Mansberger, R.; Hübl, J. UAV Monitoring and Documentation of a Large Landslide. *Appl. Geomat.* **2016**, *8*, 1–11. [CrossRef]

39. Mateos, R.M.; Azañón, J.M.; Roldán, F.J.; Notti, D.; Pérez-Peña, V.; Galve, J.P.; Pérez-García, J.L.; Colomo, C.M.; Gómez-López, J.M.; Montserrat, O.; et al. The Combined Use of PSInSAR and UAV Photogrammetry Techniques for the Analysis of the Kinematics of a Coastal Landslide Affecting an Urban Area (SE Spain). *Landslides* **2017**, *14*, 743–754. [CrossRef]
40. Peppa, M.V.; Mills, J.P.; Moore, P.; Miller, P.E.; Chambers, J.E. Accuracy Assessment of a Uav-Based Landslide Monitoring System. *Int. Arch. Photogramm. Remote Sens. Spat. Inf. Sci. ISPRS Arch.* **2016**, *41*, 895–902. [CrossRef]
41. Peternel, T.; Kumelj, Š.; Oštir, K.; Komac, M. Monitoring the Potoška Planina Landslide (NW Slovenia) Using UAV Photogrammetry and Tachymetric Measurements. *Landslides* **2017**, *14*, 395–406. [CrossRef]
42. Ruzgiene, B.; Berteška, T.; Gečyte, S.; Jakubauskiene, E.; Aksamitauskas, V.Č. The Surface Modelling Based on UAV Photogrammetry and Qualitative Estimation. *Meas. J. Int. Meas. Confed.* **2016**, *73*, 276–287. [CrossRef]
43. Niethammer, U.; James, M.; Rothmund, S.; Travelletti, J.; Joswig, M. UAV-Based Remote Sensing of the Super-Sauze Landslide: Evaluation and Results. *Eng. Geol.* **2011**, *128*, 2–11. [CrossRef]
44. Agüera-Vega, F.; Carvajal-Ramírez, F.; Martínez-Carricondo, P. Assessment of Photogrammetric Mapping Accuracy Based on Variation Ground Control Points Number Using Unmanned Aerial Vehicle. *Meas. J. Int. Meas. Confed.* **2017**, *98*, 221–227. [CrossRef]
45. Al-Rawabdeh, A.; Moussa, A.; Foroutan, M.; El-Sheimy, N.; Habib, A. Time Series UAV Image-Based Point Clouds for Landslide Progression Evaluation Applications. *Sensors* **2017**, *17*, 2378. [CrossRef] [PubMed]
46. Fernández, T.; Pérez, J.L.; Colomo, C.; Cardenal, J.; Delgado, J.; Palenzuela, J.A.; Irigaray, C.; Chacón, J. Assessment of the Evolution of a Landslide Using Digital Photogrammetry and LiDAR Techniques in the Alpujarras Region (Granada, Southeastern Spain). *Geosciences* **2017**, *7*, 32. [CrossRef]
47. Lucieer, A.; de Jong, S.M.; Turner, D. Mapping Landslide Displacements Using Structure from Motion (SfM) and Image Correlation of Multi-Temporal UAV Photography. *Prog. Phys. Geogr.* **2014**, *38*, 97–116. [CrossRef]
48. Pesci, A.; Teza, G.; Casula, G.; Fabris, M.; Bonforte, A. Remote Sensing and Geodetic Measurements for Volcanic Slope Monitoring: Surface Variations Measured at Northern Flank of La Fossa Cone (Vulcano Island, Italy). *Remote Sens.* **2013**, *5*, 2238–2256. [CrossRef]
49. Franklin, S.E. Interpretation and Use of Geomorphometry in Remote Sensing: A Guide and Review of Integrated Applications. *Int. J. Remote Sens.* **2020**, *41*, 7700–7733. [CrossRef]
50. Rózycka, M.; Migoń, P.; Michniewicz, A. Topographic Wetness Index and Terrain Ruggedness Index in Geomorphic Characterisation of Landslide Terrains, on Examples from the Sudetes, SW Poland. *Z. Geomorphol.* **2017**, *61*, 61–80. [CrossRef]
51. Tarolli, P.; Sofia, G.; Dalla Fontana, G. Geomorphic Features Extraction from High-Resolution Topography: Landslide Crowns and Bank Erosion. *Nat. Hazards* **2012**, *61*, 65–83. [CrossRef]
52. Peppa, M.V.; Mills, J.P.; Moore, P.; Miller, P.E.; Chambers, J.E. Brief Communication: 3D landslide motion from cross correlation of UAV-derived morphological attributes. *Nat. Hazards Earth Syst. Sci.* **2017**, *17*, 2143–2150. [CrossRef]
53. Chudý, F.; Slámová, M.; Tomaštík, J.; Prokešová, R.; Mokroš, M. Identification of Micro-Scale Landforms of Landslides Using Precise Digital Elevation Models. *Geosciences* **2019**, *9*, 117. [CrossRef]
54. Mauri, L.; Straffelini, E.; Cucchiaro, S.; Tarolli, P. UAV-SFM 4D Mapping of Landslides Activated in a Steep Terraced Agricultural Area. *J. Agric. Eng.* **2021**, *52*. [CrossRef]
55. Soto, J.; Palenzuela, J.A.; Galve, J.P.; Luque, J.A.; Azañón, J.M.; Tamay, J.; Irigaray, C. Estimation of Empirical Rainfall Thresholds for Landslide Triggering Using Partial Duration Series and Their Relation with Climatic Cycles. An Application in Southern Ecuador. *Bull. Eng. Geol. Environ.* **2017**, *78*, 1971–1987. [CrossRef]
56. Bravo-López, E.; Fernández Del Castillo, T.; Sellers, C.; Delgado-García, J. Analysis of Conditioning Factors in Cuenca, Ecuador, for Landslide Susceptibility Maps Generation Employing Machine Learning Methods. *Land* **2023**, *12*, 1135. [CrossRef]
57. Zárate, B.; El Hamdouni, R.; Fernández, T. GNSS and RPAS Integration Techniques for Studying Landslide Dynamics: Application to the Areas of Victoria and Colinas Lojanas, (Loja, Ecuador). *Remote Sens.* **2021**, *13*, 3496. [CrossRef]
58. McColl, S.T.; Holdsworth, C.N.; Fuller, I.C.; Todd, M.; Williams, F. Disproportionate and Chronic Sediment Delivery from a Fluvially Controlled, Deep-Seated Landslide in Aotearoa New Zealand. *Earth Surf. Process. Landf.* **2022**, *47*, 1972–1988. [CrossRef]
59. Ajayi, O.G.; Salubi, A.A.; Angbas, A.F.; Odigure, M.G. Generation of Accurate Digital Elevation Models from UAV Acquired Low Percentage Overlapping Images. *Int. J. Remote Sens.* **2017**, *38*, 3113–3134. [CrossRef]
60. Tempa, K.; Peljor, K.; Wangdi, S.; Ghalley, R.; Jamtsho, K.; Ghalley, S.; Pradhan, P. UAV Technique to Localize Landslide Susceptibility and Mitigation Proposal: A Case of Rinchending Goenpa Landslide in Bhutan. *Nat. Hazards Res.* **2021**, *1*, 171–186. [CrossRef]
61. An, K.; Kim, S.; Chae, T.; Park, D. Developing an Accessible Landslide Susceptibility Model Using Open-Source Resources. *Sustainability* **2018**, *10*, 293. [CrossRef]
62. Liashenko, D.; Belenok, V.; Spitsa, R.; Pavlyuk, D.; Boiko, O. Landslide GIS-Modelling with QGIS Software. *XIV Int. Sci. Conf. Monit. Geol. Process. Ecol. Cond. Environ.* **2020**, *2020*, 1–5. [CrossRef]
63. Sansare, D.A.; Mhaske, S.Y. Natural Hazard Assessment and Mapping Using Remote Sensing and QGIS Tools for Mumbai City, India. *Nat. Hazards* **2020**, *100*, 1117–1136. [CrossRef]
64. Lemenkova, P. Sentinel-2 for High Resolution Mapping of Slope-Based Vegetation Indices Using Machine Learning by SAGA GIS. *Transylv. Rev. Syst. Ecol. Res.* **2020**, *22*, 17–34. [CrossRef]

65. Samodra, G.; Ramadhan, M.F.; Sartohadi, J.; Setiawan, M.A.; Christanto, N.; Sukmawijaya, A. Characterization of Displacement and Internal Structure of Landslides from Multitemporal UAV and ERT Imaging. *Landslides* **2020**, *17*, 2455–2468. [CrossRef]
66. Shubina, D.D.; Fomenko, I.K.; Gorobtsov, D.N. The Specifities of Landslides Danger Assessment Accepted in Eurocode. *Procedia Eng.* **2017**, *189*, 51–58. [CrossRef]
67. Liu, C.; Li, W.; Wu, H.; Lu, P.; Sang, K.; Sun, W.; Chen, W.; Hong, Y.; Li, R. Susceptibility Evaluation and Mapping of China's Landslides Based on Multi-Source Data. *Nat. Hazards* **2013**, *69*, 1477–1495. [CrossRef]
68. Piras, M.; Taddia, G.; Forno, M.G.; Gattiglio, M.; Aicardi, I.; Dabove, P.; Russo, S.L.; Lingua, A. Detailed Geological Mapping in Mountain Areas Using an Unmanned Aerial Vehicle: Application to the Rodoretto Valley, NW Italian Alps. *Geomat. Nat. Hazards Risk* **2017**, *8*, 137–149. [CrossRef]
69. Yang, D.; Qiu, H.; Hu, S.; Pei, Y.; Wang, X.; Du, C.; Long, Y.; Cao, M. Influence of Successive Landslides on Topographic Changes Revealed by Multitemporal High-Resolution UAS-Based DEM. *Catena* **2021**, *202*, 105229. [CrossRef]
70. Zeng, T.; Ghulam, A.; Yang, W.N.; Grzovic, M.; Maimaitiyiming, M. Estimating the Contribution of Loose Deposits to Potential Landslides over Wenchuan Earthquake Zone, China. *IEEE J. Sel. Top. Appl. Earth Obs. Remote Sens.* **2015**, *8*, 750–762. [CrossRef]
71. Vorpahl, P.; Elsenbeer, H.; Märker, M.; Schröder, B. How Can Statistical Models Help to Determine Driving Factors of Landslides? *Ecol. Modell.* **2012**, *239*, 27–39. [CrossRef]
72. Conforti, M.; Pascale, S.; Robustelli, G.; Sdao, F. Evaluation of Prediction Capability of the Artificial Neural Networks for Mapping Landslide Susceptibility in the Turbolo River Catchment (Northern Calabria, Italy). *Catena* **2014**, *113*, 236–250. [CrossRef]
73. Zhang, J.; Zhu, W.; Cheng, Y.; Li, Z. Landslide Detection in the Linzhi-ya'an Section along the Sichuan-Tibet Railway Based on Insar and Hot Spot Analysis Methods. *Remote Sens.* **2021**, *13*, 3566. [CrossRef]
74. Ghosh, A.; Bera, B. Landform Classification and Geomorphological Mapping of the Chota Nagpur Plateau, India. *Quat. Sci. Adv.* **2023**, *10*, 100082. [CrossRef]
75. Grabowski, D.; Laskowicz, I.; Ma, A.; Rubinkiewicz, J. Geomorphology Geoenvironmental Conditioning of Landsliding in River Valleys of Lowland Regions and Its Significance in Landslide Susceptibility Assessment: A Case Study in the Lower Vistula Valley, Northern Poland. *Geomorphology* **2022**, *419*, 108490. [CrossRef]
76. Roy, L.; Das, S. The Egyptian Journal of Remote Sensing and Space Sciences GIS-Based Landform and LULC Classifications in the Sub-Himalayan Kaljani Basin: Special Reference to 2016 Flood. *Egypt. J. Remote Sens. Sp. Sci.* **2021**, *24*, 755–767. [CrossRef]
77. Dall'Asta, E.; Forlani, G.; Roncella, R.; Santise, M.; Diotri, F.; Morra di Cella, U. Unmanned Aerial Systems and DSM Matching for Rock Glacier Monitoring. *ISPRS J. Photogramm. Remote Sens.* **2017**, *127*, 102–114. [CrossRef]
78. Hong, H.; Naghibi, S.A.; Pourghasemi, H.R.; Pradhan, B. GIS-Based Landslide Spatial Modeling in Ganzhou City, China. *Arab. J. Geosci.* **2016**, *9*, 1–26. [CrossRef]
79. Zeybek, M.; Şanlıoğlu, İ. Point Cloud Filtering on UAV Based Point Cloud. *Meas. J. Int. Meas. Confed.* **2019**, *133*, 99–111. [CrossRef]
80. Borkowski, A. Advancing Culture of Living with Landslides. *Adv. Cult. Living Landslides* **2017**. [CrossRef]
81. Mauri, L.; Straffelini, E.; Cucchiaro, S.; Tarolli, P. RPAS-SfM 4D Mapping of Shallow Landslides Activated in a Steep Terraced Vineyard. In Proceedings of the EGU General Assembly Conference Abstracts, Virtual, 19–30 April 2021.
82. Viña, A.; Gitelson, A.A.; Nguy-robertson, A.L.; Peng, Y. Remote Sensing of Environment Comparison of Different Vegetation Indices for the Remote Assessment of Green Leaf Area Index of Crops. *Remote Sens. Environ.* **2011**, *115*, 3468–3478. [CrossRef]
83. Scheidl, C.; Heiser, M.; Kamper, S.; Thaler, T.; Klebinder, K.; Nagl, F.; Lechner, V.; Markart, G.; Rammer, W.; Seidl, R. The Influence of Climate Change and Canopy Disturbances on Landslide Susceptibility in Headwater Catchments. *Sci. Total Environ.* **2020**, *742*, 140588. [CrossRef]
84. Salleh, M.R.M.; Ishak, N.I.; Razak, K.A.; Rahman, M.Z.A.; Asmadi, M.A.; Ismail, Z.; Khanan, M.F.A. Geospatial Approach for Landslide Activity Assessment and Mapping Based on Vegetation Anomalies. *Int. Arch. Photogramm. Remote Sens. Spat. Inf. Sci.-ISPRS Arch.* **2018**, *42*, 201–215. [CrossRef]
85. ASTM International D4643; Standard Test Method for Determination of Water (Moisture) Content of Soil by the Microwave Oven Heating. American Society for Testing and Materials: West Conshohocken, PA, USA, 2000.
86. ASTM International D4318-17; Standard Test Methods for Liquid Limit, Plastic Limit, and Plasticity Index of Soils. American Society for Testing and Materials: West Conshohocken, PA, USA, 2018.
87. ASTM International D422-63; Particle Size Analysis of Soils. American Society for Testing and Materials: West Conshohocken, PA, USA, 2007.
88. ASTM International D2487-0; Standard Practice for Classification of Soils for Engineering Purposes (Unified Soil Classification System) 1. American Society for Testing and Materials: West Conshohocken, PA, USA, 2017.
89. West Conshohocken, PA, USA; Standard Test Method for Direct Shear Test of Soils under Consolidated Drained Conditions. American Society for Testing and Materials: West Conshohocken, PA, USA, 2011.
90. Chalkias, C.; Ferentinou, M.; Polykretis, C. GIS-Based Landslide Susceptibility Mapping on the Peloponnese Peninsula, Greece. *Geosciences* **2014**, *4*, 176–190. [CrossRef]
91. Turner, D.; Lucieer, A.; de Jong, S.M. Time Series Analysis of Landslide Dynamics Using an Unmanned Aerial Vehicle (UAV). *Remote Sens.* **2015**, *7*, 1736–1757. [CrossRef]
92. Carvajal, F.; Agüera, F.; Pérez, M. Surveying a Landslide in a Road Embankment Using Unmanned Aerial Vehicle Photogrammetry. *ISPRS-Int. Arch. Photogramm. Remote Sens. Spat. Inf. Sci.* **2012**, *38*, 201–206. [CrossRef]

93. Bossi, G.; Cavalli, M.; Crema, S.; Frigerio, S.; Quan Luna, B.; Mantovani, M.; Marcato, G.; Schenato, L.; Pasuto, A. Multi-temporal LiDAR-DTMs as a tool for modelling a complex landslide: A case study in the Rotolon catchment (eastern Italian Alps). *Nat. Hazards Earth Syst. Sci.* **2015**, *15*, 715–722. [CrossRef]
94. Brasington, J.; Rumsby, B.T.; McVey, R.A. Monitoring and modelling morphological change in a braided gravel-bed river using high resolution GPS-based survey. *Earth Surf. Proc. Land.* **2000**, *25*, 973–990. [CrossRef]
95. Wheaton, J.M.; Brasington, J.; Darby, S.E.; Sear, D.A. Accounting for uncertainty in DEMs from repeat topographic surveys: Improved sediment budgets. *Earth Surf. Proc. Land.* **2010**, *35*, 136–156. [CrossRef]
96. Prokešová, R.; Kardoš, M.; ved'ová, A. Landslide Dynamics from High-Resolution Aerial Photographs: A Case Study from the Western Carpathians, Slovakia. *Geomorphology* **2010**, *115*, 90–101. [CrossRef]
97. Van Westen, C.; van Asch, T.; Soeters, R. Landslide hazard and risk zonation—Why is it still so difficult? *Bull. Eng. Geol. Environ.* **2006**, *65*, 167–184. [CrossRef]
98. Fernández, T.; Pérez-García, J.L.; Gómez-López, J.M.; Cardenal, J.; Moya, F.; Delgado, J. Multitemporal Landslide Inventory and Activity Analysis by Means of Aerial Photogrammetry and LiDAR Techniques in an Area of Southern Spain. *Remote Sens.* **2021**, *13*, 2110. [CrossRef]
99. Baldi, P.; Cenni, N.; Fabris, M.; Zanutta, A. Kinematics of a Landslide Derived from Archival Photogrammetry and GPS Data. *Geomorphology* **2008**, *102*, 435–444. [CrossRef]
100. Galeandro, A.; Simunek, J.; Simeone, V. Simulating Infiltration Processes into Fractured and Swelling Soils as Triggering Factors of Landslides. *Landslide Sci. Pract. Spat. Anal. Model.* **2013**, *3*, 135–141.
101. Ram, A.R.; Brook, M.S.; Cronin, S.J. Engineering Geomorphological Investigation of the Kasavu Landslide, Viti Levu, Fiji. *Landslides* **2019**, *16*, 1341–1351. [CrossRef]
102. Guzzetti, F. Landslide Hazard Assessment and Risk Evaluation: Limits and Prospectives. In Proceedings of the 4th EGS Plinius Conference, Mallorca, Spain, 14–16 October 2003; pp. 1–4.
103. Palenzuela, J.A.; Soto, J.; Irigaray, C. Characteristics of Rainfall Events Triggering Landslides in Two Climatologically Dierent Areas: Southern Ecuador and Southern Spain. *Hydrology* **2020**, *7*, 807. [CrossRef]
104. Marino, P.; Peres, D.J.; Cancelliere, A.; Greco, R.; Bogaard, T.A. Soil Moisture Information Can Improve Shallow Landslide Forecasting Using the Hydrometeorological Threshold Approach. *Landslides* **2020**, *17*, 2041–2054. [CrossRef]
105. Mirus, B.B.; Morphew, M.D.; Smith, J.B. Developing Hydro-Meteorological Thresholds for Shallow Landslide Initiation and Early Warning. *Water* **2018**, *10*, 1274. [CrossRef]

Disclaimer/Publisher's Note: The statements, opinions and data contained in all publications are solely those of the individual author(s) and contributor(s) and not of MDPI and/or the editor(s). MDPI and/or the editor(s) disclaim responsibility for any injury to people or property resulting from any ideas, methods, instructions or products referred to in the content.

Article

Landslide Susceptibility Analysis on the Vicinity of Bogotá-Villavicencio Road (Eastern Cordillera of the Colombian Andes)

María Camila Herrera-Coy [1], Laura Paola Calderón [2], Iván Leonardo Herrera-Pérez [1,3], Paul Esteban Bravo-López [1,4], Christian Conoscenti [2], Jorge Delgado [1], Mario Sánchez-Gómez [5,6] and Tomás Fernández [1,6,*]

1. Department of Cartographic, Geodetic and Photogrammetric Engineering, University of Jaén, 23071 Jaén, Spain; mchc0009@red.ujaen.es (M.C.H.-C.); ilhp0001@red.ujaen.es (I.L.H.-P.); pebl0001@red.ujaen.es (P.E.B.-L.); jdelgado@ujaen.es (J.D.)
2. Department of Earth and Marine Sciences (DiSTeM), University of Palermo, 90123 Palermo, Italy; laurapaola.calderoncucunuba@unipa.it (L.P.C.); christian.conoscenti@unipa.it (C.C.)
3. Department of Geographic and Environmental Engineering, University of Applied and Environmental Sciences (U.D.C.A.), Bogotá 111166, Colombia
4. Institute for Studies of Sectional Regime of Ecuador (IERSE), University of Azuay, Cuenca 010107, Ecuador
5. Department of Geology, University of Jaén, 23071 Jaén, Spain; msgomez@ujaen.es
6. Natural Hazards Lab of the Centre for Advanced Studies in Earth Sciences, Energy and Environment (CEACTEMA), University of Jaén, 23071 Jaén, Spain
* Correspondence: tfernan@ujaen.es; Tel.: +34-53-212843

Citation: Herrera-Coy, M.C.; Calderón, L.P.; Herrera-Pérez, I.L.; Bravo-López, P.E.; Conoscenti, C.; Delgado, J.; Sánchez-Gómez, M.; Fernández, T. Landslide Susceptibility Analysis on the Vicinity of Bogotá-Villavicencio Road (Eastern Cordillera of the Colombian Andes). Remote Sens. 2023, 15, 3870. https://doi.org/10.3390/rs15153870

Academic Editor: Domenico Calcaterra

Received: 11 June 2023
Revised: 24 July 2023
Accepted: 31 July 2023
Published: 4 August 2023

Copyright: © 2023 by the authors. Licensee MDPI, Basel, Switzerland. This article is an open access article distributed under the terms and conditions of the Creative Commons Attribution (CC BY) license (https://creativecommons.org/licenses/by/4.0/).

Abstract: Landslide occurrence in Colombia is very frequent due to its geographical location in the Andean mountain range, with a very pronounced orography, a significant geological complexity and an outstanding climatic variability. More specifically, the study area around the Bogotá-Villavicencio road in the central sector of the Eastern Cordillera is one of the regions with the highest concentration of phenomena, which makes its study a priority. An inventory and detailed analysis of 2506 landslides has been carried out, in which five basic typologies have been differentiated: avalanches, debris flows, slides, earth flows and creeping areas. Debris avalanches and debris flows occur mainly in metamorphic materials (phyllites, schists and quartz-sandstones), areas with sparse vegetation, steep slopes and lower sections of hillslopes; meanwhile, slides, earth flows and creep occur in Cretaceous lutites, crop/grass lands, medium and low slopes and lower-middle sections of the hillslopes. Based on this analysis, landslide susceptibility models have been made for the different typologies and with different methods (matrix, discriminant analysis, random forest and neural networks) and input factors. The results are generally quite good, with average AUC-ROC values above 0.7–0.8, and the machine learning methods are the most appropriate, especially random forest, with a selected number of factors (between 6 and 8). The degree of fit (DF) usually shows relative errors lower than 5% and success higher than 90%. Finally, an integrated landslide susceptibility map (LSM) has been made for shallower and deeper types of movements. All the LSM show a clear zonation as a consequence of the geological control of the susceptibility.

Keywords: landslide; susceptibility analysis; modelling; Bogotá-Villavicencio road; Eastern Cordillera; Colombian Andes

1. Introduction

Landslides are considered one of the most important natural hazards worldwide, causing thousands of casualties and costs amounting to billions of euros each year [1–3]. Compared to other risk phenomena such as earthquakes or floods, the effect of landslides is more diffuse and continuous in space and time, so their impact can be underestimated according to some evaluations [2]. Nevertheless, they cause significant damage to infrastructure, properties and the environment, as well as the interruption of socioeconomic activity [1,4].

In Colombia, the occurrence of natural hazard phenomena, including landslides, earthquakes, and volcanic eruptions, is very frequent due to its geographical location in the Andean mountain range, with a steep orography, a great geological complexity and a significant climatic variability [5–8]. In fact, according to [9], it is one of the most prominent countries in global databases such as the Disaster database (EM-DAT [10]), the Disaster Inventory System (DesInventar [11]), the Global Landslide Catalog (GLC [12]) and the Global Fatal Landslide database (GFLD [13]). Specifically, DesInventar [11] reports 10,559 incidents and 7400 deaths for Colombia.

Thus, landslides in Colombia represent almost half of all natural catastrophic events, far exceeding disasters caused by floods, earthquakes and volcanic eruptions [14], with an average of 47 landslides and 59 deaths each year from 1993 to 2004. According to information available in the Mass Movements Information System (SIMMA) of the Colombian Geological Service (SGC [15]), 135,632 mass movements have been reported in the country since 1900. Due to this, 31,631 people have lost their lives, and 68,792 families have been affected. Combining different national and international databases, García-Delgado et al. [8] collected a total of 2351 fatal landslides that caused almost 40,000 deaths, with some of them in historical times (prior to 1912) and the majority in modern times (1912–2020), with an upward trend in the last 20 years. In another work, Aristizábal and Sánchez [6] compiled about 30,730 landslides that caused 34,198 fatalities and economic losses of more than 600 million dollars in the period of 1900–2018.

According to SIMMA [15], among the most affected regions in absolute terms are the departments of Cundinamarca, Boyacá and Norte de Santander, located in the Eastern Cordillera at the central and northern part of the country, as well as Cauca in the Colombian Massif in the south. Other smaller departments, such as Caldas, Risaralda or Quindío in the Central Cordillera or Atlántico in the Sierra de Santa Marta, also present a considerable density in relative terms [15]. These data coincide with the compilation of García-Delgado et al. [8], in which the highest densities of landslides occur in the departments of Tolima, Caldas, Risaralda, Bogotá, Quindío, Cauca and Cundinamarca. Specifically, the departments of Cundinamarca and Meta in the central sector of the Eastern Cordillera are exposed to medium and high probabilities of occurrences of catastrophic phenomena, particularly landslides, caused, among other reasons, by a high anthropic intervention on the slopes with the consequent deterioration of the hydrographic basins and their stability conditions. Thus, the Subdirectorate of Geoambiental Engineering of Ingeominas (now Subdirectorate of Geo-Hazards of the SGC) prioritized six regions with a higher concentration of phenomena: the Guavio river basin, the area around the Bogotá-Villavicencio road, the eastern slope of the Negro River, the Sumapaz river basin, the middle basin of the Bogotá River and the municipality of San Cayetano [16]. This work focuses on the vicinity of the road from Bogotá to Villavicencio, specifically on its section towards the Orinoco river basin.

One of the most effective measures for risk prevention and mitigation is the evaluation of both the hazard of the phenomenon and of the exposure and vulnerability of the elements at risk [17]. In Colombia, some studies have been carried out that evaluate risk assessment and reduction [7,18–20], but there are many more that evaluate hazard or susceptibility. For hazard, there are numerous deterministic and probabilistic methods, the latter of which are generally the most applied for extensive areas due to the lack of precise and exhaustive data in such areas. Within the deterministic methods, different hydrological models [21,22] or stability analysis methods such as infinite slope [23], r.slope.stability of GRASS [24], FOSM [25–27], SLIP [26,27], PEM point estimates [25] or deformation analysis [28] have been applied. These models have been used in high-impact landslides such as the Mocoa debris flow in the southern part of the country [21,29–31], hillslopes around Medellín [22] and San Eduardo in Boyacá [28] or watersheds such as La Arenosa and La Liboriana [24–27] in the department of Antioquia in the Western Cordillera.

Meanwhile, within probabilistic methods, most are extensively applied in the estimation of spatial probability or susceptibility, which are based on the statistical analysis of correlation between determinant factors and landslides, according to the susceptibility definition established by Brabb [32]. To develop susceptibility models using statistical techniques, numerous methods are available, which, according to Reichenbach et al. [33], can be grouped into models based on indices [34–36] and bivariate statistics [37]; multicriteria evaluation [34]; multivariate statistics [4,38,39]; machine learning, ML [40] and artificial neural networks (ANN) [41,42]. The last two groups, sometimes with the ANN integrated in the more general group of ML, have advanced regarding classical statistical methods due to their greater versatility and better performance in nonlinear systems such as models developed from factors of different nature [43]. These methods allow the integration of a great number of factors that are not usually analyzed and selected since the algorithms directly perform the fit of the models [33]. In our opinion, this lack of factor control and selection leads to a loss of knowledge in the elaboration of the models and sometimes to an overfitting of them.

Different studies have been conducted in the Colombian Andes, using all types of methods from index-based methods such as frequency ratio or weight of evidence [44–48]; statistical methods such as logistic regression [31,46,48–51]; machine learning methods such as random forest [46], graded boosted regression trees, GBRT [46] or multivariate adaptive regression [31] and simple [46,52] or convolutional neural networks [53]. The methods have been applied to different areas such as Capitanejo in the NE of the country [52], Medellín and the department of Antioquia to the NW [47–49], Caldas [44], Cauca [45,50], Boyacá [52], Bogotá and Cundinamarca [46,51] and Mocoa in the south [31]. Most of these studies, not only in Colombia but also throughout the world, do not take into account the landslide typology, which produces less precision and noise in the models.

Finally, it should be mentioned that those methods for determining rainfall thresholds and, where appropriate, establishing early warning systems, exist throughout the whole country [54] or in different departments such as Bolívar, Antioquia and Caldas to the northwest [50,55–58] or in Bogotá [59]. Regarding precipitation, the influence of deep convective systems [60] and the impact of climate change on the generation of landslides and other risks [61] have also been evaluated in these previous works.

The main objective of this work is to present a detailed inventory and susceptibility models of the study area. The inventory will allow the understanding of the different processes that occur in this mountainous area and their characteristics, while the subsequent factor analysis by typologies will allow the factor selection and the determination of the conditions under which they originate, as a previous step to modeling landslide susceptibility. Thus, an important limitation of most current studies, such as the lack of knowledge of the different landslide typologies, factors and conditions, can be overcome. From this knowledge, susceptibility models for the different typologies have been developed using several methods and introducing an increasing number of factors, which will allow a control of the models' behavior in relation to overfitting and noise. Examples of the main groups of methods that have been used are the matrix method (index), linear discriminant analysis (multivariate statistics), random forest (machine learning) and a perceptron ANN (neural networks). The results have been compared in order to provide consistent results in determining the landslide hazard in the region and ensuring robust models that can be applied in other areas. Finally, once the susceptibility models from the different typologies are obtained and selected, they have been integrated into susceptibility maps for shallower and deeper processes in order to assess the hazard in the study area.

2. Materials and Methods

2.1. Study Area

The study area is located in the central sector of the Eastern Cordillera of the Colombian Andes (Figure 1a). This is a mountain chain that extends for about 1000 km from southern Colombia (Colombian Massif, where the cordillera divides into its three main branches, Western, Central and Eastern) to near the border with Venezuela in the north (Sierra Nevada del Cocuy), where the highest altitudes of around 5400 m are reached. To the west of the Cordillera lies the Cundinamarca Plateau, where the capital of Bogotá (Figure 1b) and many other towns are located; from there, it descends to the Magdalena River, which flows between the Eastern and Central Cordilleras; to the east, the basins of the Orinoco and Amazon rivers extend. In this central sector of the Eastern Cordillera, altitudes range from a few hundred meters in the river basins to 4000 m at the summits. The Eastern Cordillera of Colombia is an intracontinental mountain belt 100 to 200 km wide [62] with a SW-NE trend. The materials correspond mainly to marine deposits but also transitional to continental, ranging from the Cretaceous to Paleocene in age [63,64]. On both sides of the Cordillera, in the Magdalena River valley and the Amazon and Orinoco basins, there are Tertiary sedimentary or volcanoclastic deposits. Above all of them appear Quaternary materials: alluvial, colluvial and paludal fillings. The structure consists of thrusts and folds, which, in some cases, bring to the surface metamorphic materials of the Paleozoic substrate (Ordovician and Devonian) without reaching the Proterozoic crystalline basement.

More specifically, the study area has an extension of approximately 746 km^2 on the vicinity of the road between Bogotá and Villavicencio (Figure 1b,c), a city located 75 km southeast of the country capital (120 km by this road, also known as "Vía al Llano" or Route 40). The area extends through the municipalities of Cáqueza, Fosca, Quetame and Guayabetal in the Oriente province of the department of Cundinamarca and the municipality of Villavicencio in the department of Meta. The municipalities' total number of inhabitants and percentage of rural population are Cáqueza with 15,594 inhabitants and 58%; Fosca, 5578 inhabitants and 75%; Quetame with 4929 inhabitants and 77%; Guayabetal with 5809 inhabitants and 70% and Villavicencio with 451,212 inhabitants, of which only the 7% are rural population [65]. The road between Bogotá and Villavicencio connects the entire area together other minor roads.

The elevations range from 600 m in Villavicencio up to 3500 m in the mountain range to the east of Quetame. The average annual rainfall varies from 500 mm in the western sector to over 3000 mm in the Villavicencio sector. The slopes are generally quite steep, with almost 64% in the range of 20–45°. Hydrographically, it corresponds to the Negro river basin, a tributary of the Guayuriba River that in turn flows into the Orinoco river basin.

From a geological point of view, materials from the Paleozoic substrate and Cretaceous sedimentary series [66] outcrop in this area, with the former in the lower part of the basin and the latter in the higher part (Figure 1d). Within the Paleozoic, there are two sets of materials: first, metamorphic rocks of low grade, phyllites, schists and quartzites of Ordovician age; over them, discordantly, there are quartz sandstones and shales of Devonian-Carboniferous age. In the Cretaceous series, there is a small outcrop of conglomerates and transitional environment sands at the base of the series, which pass to lutites of a marine environment, which are predominant in the area. These series end in Paleogene, and then Miocene sedimentary and volcanoclastic deposits fill the basins formed within the Cordillera and at the east over the Paleozoic and Precambrian basement. The structure is of thrusts and folds with NNE-SSW main direction, which allow the outcrop of the underlying Paleozoic formations to the Cretaceous series in the lower part of the area. On top of all these sets, there are Quaternary materials consisting of terraces, colluvial deposits and current alluvial deposits in the riverbeds.

In the municipalities of the area, landslide activity is very high, according to SIMMA [15]: Cáqueza has had 169 incidences; Quetame, 22 incidences; Guayabetal, 9 incidences and Villavicencio, 94 incidences.

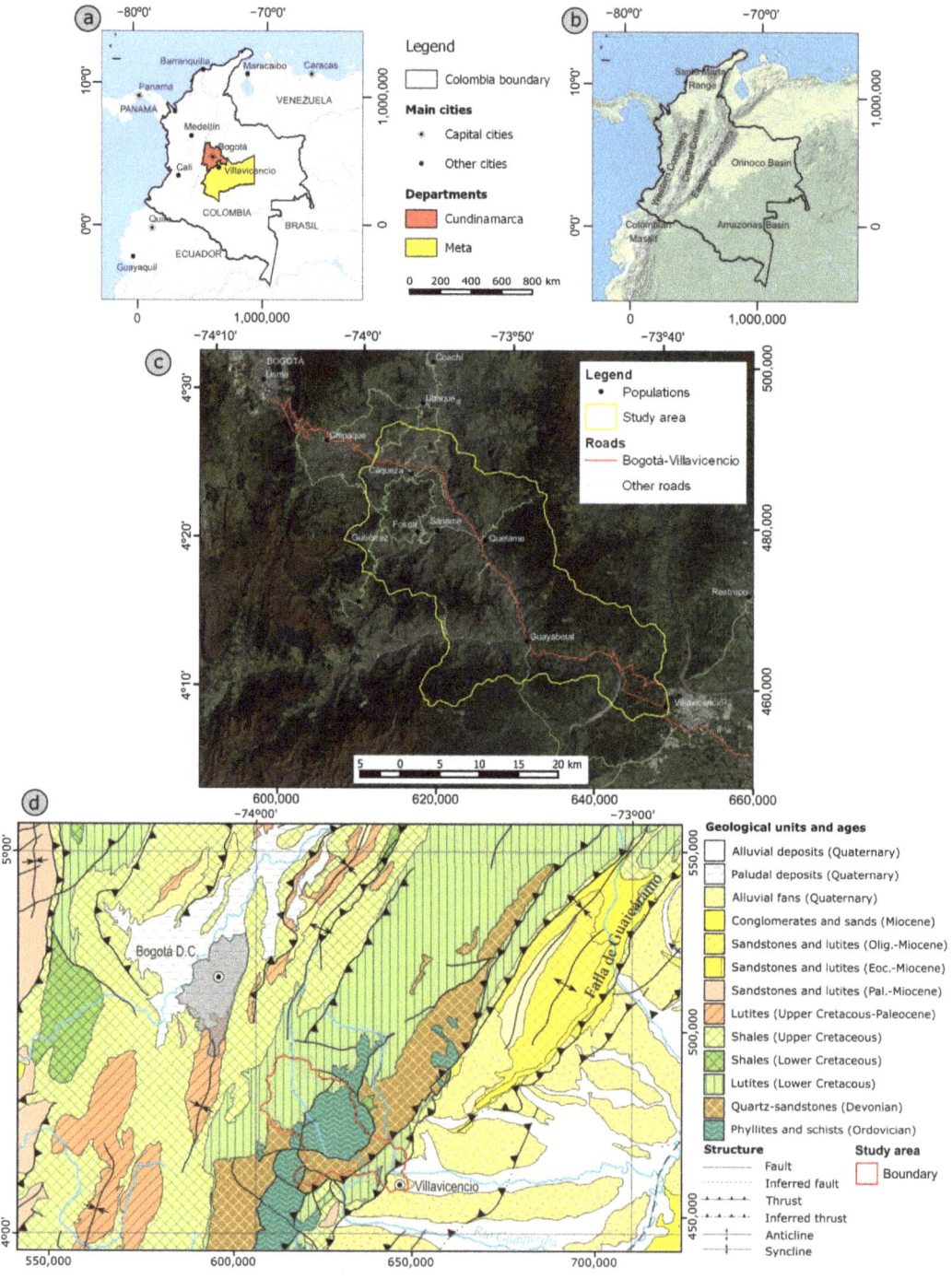

Figure 1. Location and geology of study area: (**a**) Location of Cundinamarca and Meta Departments in Colombia (own elaboration on AutoNavi Base Maps); (**b**) Colombian Andes and Eastern Cordillera

(own elaboration on Esri Physical Map); (**c**): Study area and main populations (own elaboration on Google Satellite); (**d**) Geological setting adapted from the Geological Map of Colombia [64]. Coordinates are in WGS84 (lat/long at left and top margins) and in WGS84-UTM 18 (projected, in right and down margins).

2.2. Materials

The sources of information and software used are listed in Table 1. For the inventory, background images from Google Maps—Google Earth (GE-GM) and Bing Maps (BM) were used, corresponding to Airbus (Pleiades), Maxar and Copernicus (Sentinel-2). In GE, images of different dates and resolutions could be observed.

Regarding the factor layers, the digital elevation model (DEM) obtained by InSAR from ALOS PALSAR images of 2011 [67] was used, with a spatial resolution of 12.5 m, downloaded from the Alaska Satellite Facility [68]. The geology comes from the Geological Atlas of Colombia [69], available as vector information, from which the lithological units have been extracted. A Sentinel image from 2020 [70] was used to calculate the NDVI index and obtain the classification of land cover. Finally, the map of average precipitation available in IDEAM [71] was used.

Table 1. Sources of information and software used in this study.

Information	Resources	Software
Digital elevation model (12.5 m resolution)	JAXA/METI ALOS PALSAR, 2011 [67,68]	Google Earth 7.3.6.9345 [72]
Background images	GM, GE, BM (Airbus, Maxar, Copernicus)	QGIS versión 3.18.3 [73]
Geology: Geological Atlas of Colombia	Layer files (shp): Servicio Geológico Colombiano, 2015 [69]	SAGA versión: 7.9.1 [74]
Sentinel-2 image	Copernicus, 2020 [70]	Rstudio 2022.02.2 [75]
Precipitation in Colombia	Raster files (tif): IDEAM, 2015 [71]	

Regarding software, Google Earth Pro 7.3.6.9345 [72] was used for image visualization, and QGIS 3.18.3 [73] and SAGA GIS 7.9.1 [74] for data processing and analysis. Additionally, R studio 2022.02.2 statistical software [75] was used for generating multivariate statistical and machine learning susceptibility models.

2.3. Methodology

The methodology is summarized in the flowchart of Figure 2. It includes first the elaboration of a detailed landslide inventory in the study area. Second, it includes the analysis of landslide determinant factors by typologies in order to the factor selection and understanding the conditions of their occurrence. Third, it includes the elaboration of susceptibility models (LSM) for each landslide typology using different methods, as well as their validation, and finally, it includes the integration of LSM in synthesis maps.

2.3.1. Landslide Inventory

The landslide inventory was carried out using photointerpretation and digitization from the GM-GE and BM background images. In addition, the database of the Colombian Mass Movements Information System, SIMMA [15], was used as support. The digitization of the identified movements was carried out by connecting to these images through WMS from the open-source software QGIS, although the photointerpretation was helped by the pseudo-3D views of the images in GE (Airbus/Pleiades, Maxar and Copernicus/Sentinel-2).

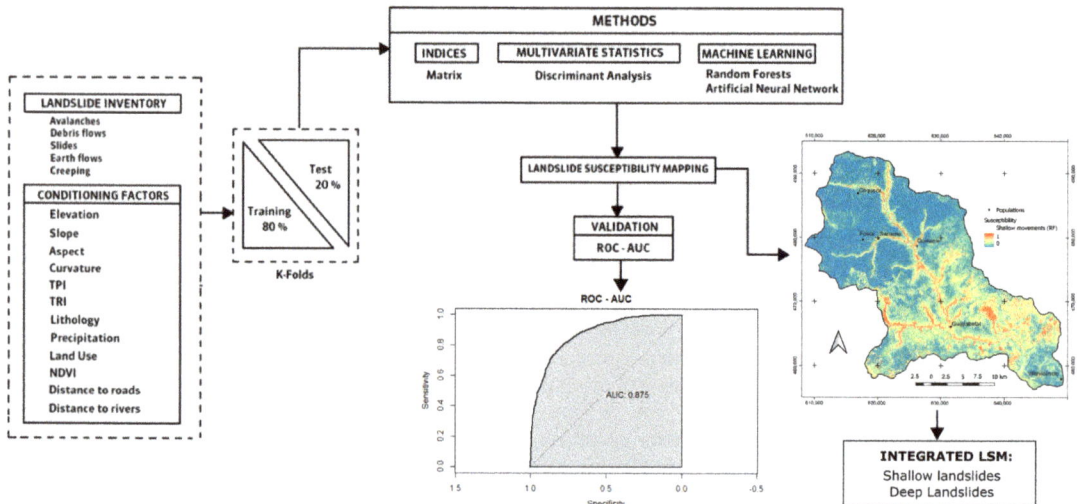

Figure 2. Flowchart followed in this study.

Once digitized, a database was created, which, according to Chacon et al., 2006 [76] and Guzzetti et al., 2012 [77], should include the spatial location, temporal dating and thematic attributes of the landslides. Specifically, this database or inventory includes as attributes the typology [78] and the activity [79], which were supported by the multi-temporal GE images, and their area, calculated with the attribute calculation tools of QGIS. From this database, an analysis has been carried out that will allow us to know the frequency and total extension, the activity and the average area of each landslide typology.

2.3.2. Analysis of Determinant Factors

For the analysis of determinant factors and the elaboration of LSM with GIS, it is necessary to have the factor layers. In this case, factor layers and maps have been obtained from different official geographic information sources in Colombia (Table 2). Figure 3 shows the factor maps both the quantitative (DEM derivatives, precipitation, NDVI and distance to roads and rivers) and the qualitative (lithology and land cover).

Table 2. Factors used in the analysis and their corresponding sources of information.

Factor	Origin
Elevation	
Slope	Derived from DEM
Aspect	of 12.5 m resolution from JAXA-ALOS
Curvature	PALSAR [67,68]
Topographical Position Index (TPI)	
Terrain Roughness Index (TRI)	
Lithology	Geological Atlas of Colombia, SGC [69]
Precipitation	Raster files (tif): IDEAM, 2015 [71]
Land Cover	Sentinel-2 image, Copernicus 2020 [70]
Normalized Difference Vegetation Index (NDVI)	
Distance to roads	Roads digitized on GE-GM image
Distance to rivers	Rivers digitized on GE-GM image

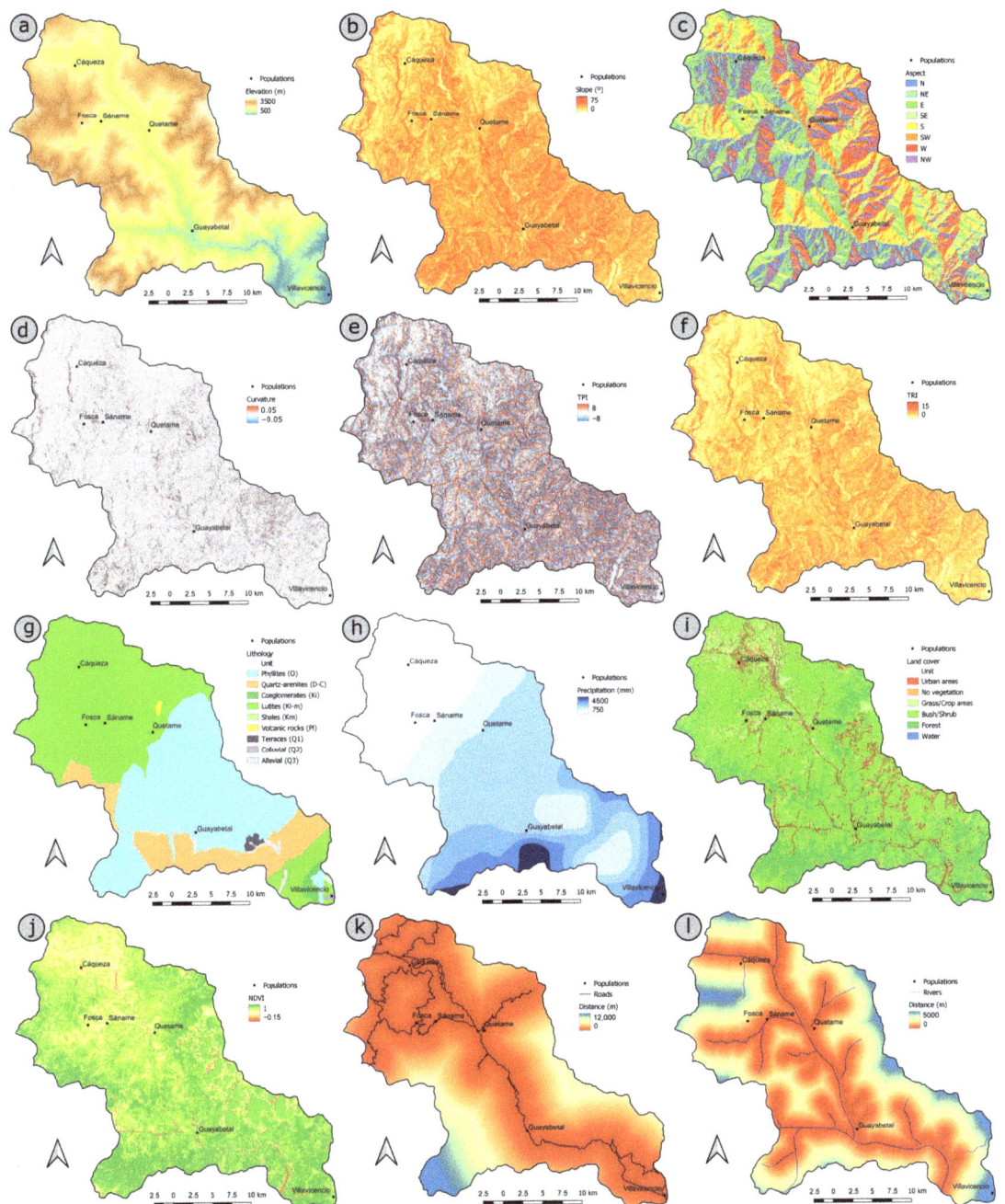

Figure 3. Factor maps considered in the study area; (**a**): Elevation; (**b**): Slope; (**c**): Aspect; (**d**): Curvature; (**e**): TPI; (**f**): TRI; (**g**): Lithology; (**h**): Precipitation; (**i**): Land Cover; (**j**): NDVI; (**k**): Distance to roads; (**l**): Distance to rivers.

From the digital elevation model (DEM), derivative models such as slope, aspect or orientation, terrain curvature, topographic position index (TPI) and terrain roughness index (TRI) have been obtained using QGIS analysis functions. Additionally, lithological units extracted from the Geologic Atlas of Colombia [69] were used, which were rasterized; then, a quantitative value was assigned to each unit based on material resistance [80], as shown in Table 3 (lower to hard rocks and higher to soft rocks). The NDVI and land use were obtained from a Sentinel-2 image using the corresponding formula [81] and supervised classification (maximum probability), respectively. In the case of land cover, a similar scheme to lithology was followed, assigning a value to each unit based on the vegetation cover or other considerations [82], also shown in Table 3. Precipitation data were obtained directly from a raster layer of precipitation intervals [71]. Finally, the distance to rivers and roads was obtained through vector digitization on the GM/GE background image and subsequent distance calculation using the corresponding QGIS function.

Table 3. Values assigned to lithological and land cover units. In lithological units, the values are assigned to each unit based on material resistance (lower to hard rocks and higher to soft rocks). In land cover units, values are assigned to each unit based mainly on the vegetation cover.

Lithology		Land Cover	
Unit	Value	Unit	Value
Phyllites-Schists	0.3	Urban	0.6
Quartzarenites	0.4	Scarce vegetation	0.8
Conglomerates	0.8	Grass-Crops	0.5
Lutites	1.0	Bush-Shrubs	0.4
Shales	0.9	Forest	0.2
Volcanic	0.2	Water	0
Terraces	0.5		
Alluvial fans	0.6		
Alluvial deposits	0.7		

The factorial analysis consisted of cross-tabulating the maps of determinant factors and the landslide inventory, both global and/or by typologies. Then, the Kolmogorov–Smirnov (K–S) coefficient was calculated to compare the distributions of factors in areas affected and not affected by movements, thus estimating the correlation between factors and landslides. At the same time, an analysis was conducted among the factors themselves by determining the Pearson linear correlation coefficient in order to estimate the collinearity between them. These analyses allowed for the selection of factors involved in the models and the identification of conditions for the occurrence of the different landslide typologies. The analysis has been made taking into account the landslide activity, although in most typologies, the differences are not significant, so only the general results are shown in the next section.

2.3.3. Susceptibility Models

Susceptibility models have been developed using different methodologies [33]: matrix method (indices), lineal discriminant analysis, LDA (classical multivariate statistics), random forest, RF (machine learning) and a simple artificial neural network (ANN) (Figure 4).

In the matrix method, the procedure starts by combining the raster layers of factors in order to obtain the units of unique condition. For qualitative factors, discrete values are used (Table 2), and for quantitative factors, continuous values are classified in intervals. Next, a cross-tabulation is performed between the raster layer of unique condition and the binary raster layers of presence/absence of landslides by typologies. From this table, the percentage area of each combination of factors affected by landslides can be obtained [35,36,83,84]. This is the susceptibility index, which is obtained by the expression:

$$I_i = \frac{z_i}{s_i} \times 100 \qquad (1)$$

where I_i is the susceptibility index; z_i and s_i are, respectively, the area of each factor combination affected by landslides and the total area of the combination.

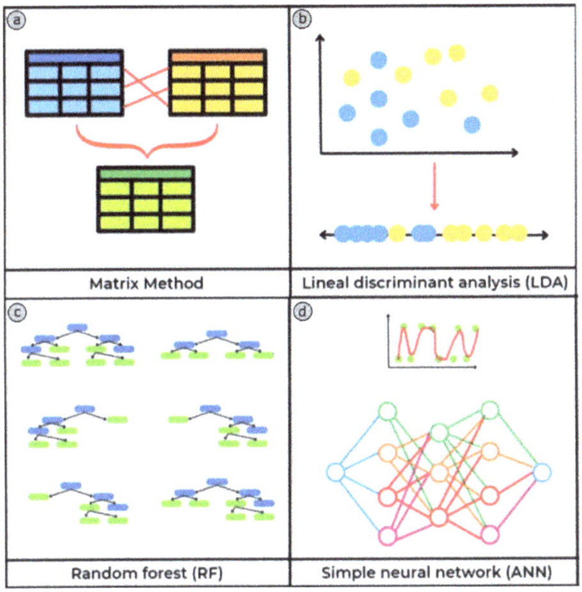

Figure 4. LSM methods: (**a**) Matrix; (**b**) Linear discriminant analysis; (**c**) Random forest; (**d**) Artificial neural network.

This index is used as a classification template, and thus, the susceptibility map is obtained by reclassifying the unique condition raster layer with it [35,36,83,84].

Meanwhile, the remaining methods follow a different scheme that starts with obtaining a random sampling of points. Based on previous works [85–89], a total of 5000 points have been obtained with GIS tools (QGIS) in the stable zone (absence of landslides) and 5000 points in each of the differentiated landslide typologies (presence). A table was then created by extracting the values of the different factors' layers at each point or pixel. Therefore, continuous (for quantitative factors) or discrete (for qualitative factors) values are used. The integrated tables of landslides' presence/absence were introduced in R studio statistical software [75] with the aim of obtaining susceptibility models and maps (LSM).

Linear discriminant analysis is one of the multivariate statistical models most used in slope instability or landslide susceptibility. These models assume that the factors that caused landslides in a given area are the same ones that will cause landslides in the future. The general linear models take the form [4,90]:

$$L = B_0 + B_1 X_1 + B_2 X_2 + B_3 X_3 + \ldots + B_n X_n + \varepsilon \qquad (2)$$

where L is the presence/absence or area percentage of landslides in each mapping unit; X's are input predictor variables or factors in each mapping unit; B's are coefficients estimated from the data through statistical techniques and ε represents the model error.

In discriminant analysis, the probability or susceptibility of landslides in a given area (a pixel in our case) is calculated by adjusting the linear discriminant function to data inputs and then minimizing the model error [38,39]. These data inputs are, on one hand, the presence/absence of landslide in the area and, on the other hand, the values of determinant factors considered in the same area (pixels in our case).

Random Forest (RF) is a nonlinear supervised method used for data classification and regression. It is considered an ensemble method consisting of a combination of deep decision trees so that each tree depends on values taken from a vector sampled randomly for growing [40,46,91,92]. Each decision tree grows splitting the input data (in our case, factors values) recursively so that each division contains more or less homogeneous states of the target variable (in our case, landslide susceptibility) [93]. In RF, each tree is trained on a subset of the data set and returns a result (in our case, landslide presence/absence). Therefore, the result of each decision tree is considered a vote, and thus, the final result is the one with the most votes or, in our case, the highest probability of landslide occurrence [91,94]. Some relevant RF characteristics are its predictive accuracy, low tendency to overfitting, relatively low computational cost and its ability to work with high dimensional data [40,92,95].

An artificial neural network (ANN) is a set of interconnected nodes or neurons useful for modeling problems with a complex relationship between analysis factors, so it is ideal for dynamic and nonlinear phenomena such as landslide occurrence [41,42]. The ANN architecture consists of a set of inputs (determinant factors); a set of intermediate layers (hidden layers) that perform the processing and an output layer with the prediction result [96]. Neural networks generally refer to supervised classification algorithms, which compare a given output with a predicted output, adapting the necessary parameters based on this comparison [97]. There are several neural network algorithms such as convolutional neural networks (CNNs) or recurrent neural networks (RNNs); however, one of the most widely used is multilayer perceptron (MLP), which has been applied in several studies [41,42,98,99]. A perceptron is an individual neuron that allows us to classify a set of inputs into one or two categories by means of a step function, which returns 1 if the weighted sum of inputs exceeds a threshold or otherwise returns 0 [96]:

$$z = b + \sum w_i x_i$$
$$y = \begin{cases} 1 \; if \; z \geq 0 \\ 0 \; if \; z < 0 \end{cases} \qquad (3)$$

where y is the label or output variable (to predict); x_i is the feature or input variable; w_i and b are the weights and the bias, both parameters that the model has to learn during the training process.

Another important feature is the activation function, which allows an ANN to work with nonlinear problems [96] and can be of linear, sigmoid or logistic types, hyperbolic tangent or rectified linear unit (ReLU).

The MLP algorithm consists of a set of perceptrons organized in layers connected by synapses that are assigned a weight. Connection weights, hidden layers and the output layer were initialized and then updated using the backpropagation algorithm [42]. In our case, the MLP implemented had only one hidden layer with 3 or 4 neurons.

In practice, modeling using discriminant analysis methods [4,38,39,90] and random forest [40,46,91,92,94] involved partitioning the sample into training (80%) and validation (20%) sub-samples, giving a 80/20 ratio as some works recommend [100], although other proportions have been tested, such as 70/30 and 60/40. Next, a k-folds procedure (5 folds) was applied with the training sub-sample in order to fit the models while avoiding skewed partitions [101], in turn with an 80/20 ratio in training/testing. This procedure allowed us to obtain the corresponding susceptibility models and maps (LSM) and their validation through the area under the Receiver Operating Characteristic curve (AUC-ROC) [102]. In the case of random forest, model refinement methods are applied based on hyperparameter control, mainly number of trees (ntree, nodesize and tuneGrid) [40,91,95].

In the neural network models [41,42,46,52,95–99,103–107], the same landslide presence/absence samples as in the previous methods were used, from which a partition as also made into training (80%) and validation (20%) sub-samples. With the training sample, a one-layer ANN with 3 or 4 hidden neurons was adjusted using the rprop algorithm for backpropagation [95]. The R script returned the corresponding LSM, and the AUC-ROC value was estimated with the validation sample.

Every method described allows the elaboration of LSM for each typology independently, but from the results of factor analysis, synthesis maps have been made, grouping the types of landslides that present similar conditions. In this way, an integrated LSM was obtained for avalanches and debris flows and another for slides, earth flows and creeping processes.

2.3.4. Models Validation

A key aspect in the use of all these methods and models is validation, which enables them to be used as predictive models for estimating hazard and proposing prevention and mitigation measures. The susceptibility models developed can be validated through random, spatial and temporal partitioning of the inventory [108]. In this study, a random partition validation was mainly used, based on obtaining training and validation samples [33]. As mentioned before, in LDA, RF and ANN methods, a sample of 2000 points (20%) was used to validate the models fitted with the training sample (80%). However, to validate the results of the matrix method, which is based on a different scheme, a new random sampling of 2000 points (1000 in each of the landslide typologies and 1000 in the stable area) was performed in QGIS. In all cases, the susceptibility values were extracted from LSM by means of QGIS tools. The value tables were imported into R studio [75], where the theoretical values (susceptibility) were compared with the actual values (presence/absence) to calculate the AUC-ROC values.

AUC-ROC values were calculated from the ROC curves, which were built representing some values derived from the confusion matrix [109] such as True Positive Rate (TPR) or sensitivity in Y axis, versus the False Positive Rate (FPR) or 1—specificity in X-axis for different thresholds of the predicted values (for instance, for intervals of 0.1). The expressions for *TPR* and *FPR* are:

$$TPR \ (sensitivity) = \frac{TP}{TP+FN}$$
$$FPR = \frac{FP}{FP+TN} = 1 - specificity \quad (4)$$

where: *TP* are the true positives; *TN* are the true negatives; *FP* are the false negatives and *FN* are the false negatives.

In addition to the validation with the AUC-ROC values, derived from the confusion matrix, another independent validation method has been applied. The degree of fit (DF), calculated in previous works from slope units [39] or landslide polygon areas [35,36,84] has been adapted to random point samples. Thus, the LSM obtained with the procedure described before were classified into five levels by means of the quantile method. Then, the additional point sample obtained for validation of the LSM of the matrix method (1000 points in stable area and 1000 points in each landslide typology) were used to extract the susceptibility levels in them (very low, low, moderate, high, very high). The DF of each susceptibility level was calculated as:

$$DF_i = \frac{pl_i/pt_i}{\sum pl_i/pt_i} \quad (5)$$

where pl_i is the number of points in landslide areas in each susceptibility level and pt_i is the total number of points in each susceptibility level (approximately 200 points). Sum of pl_i is 1000, and sum of pt_i is 2000.

The sum of DF in very low to low susceptibility levels was considered the relative error of LSM, while the sum of DF in high to very high levels was the relative success [36].

Finally, a temporal validation was conducted, obtaining the training samples from the landslides catalogued as latent and relict (4000 points in each typology), while the validation samples were extracted from the active landslides (1000 points in each typology). Additionally, from the 5000-point sample obtained for the previous validation strategy, 4000 points were added to complete the training sample, and 1000 points were added to the validation sample, giving again a training/validation ratio of 80/20. The next steps are the same as in the previous strategy, thus calculating the AUC-ROC values. This validation has been applied only for LDA, RF and ANN methods.

3. Results
3.1. Landslide Inventory

The landslide inventory of Villavicencio-Bogotá (Figure 5) shows a total of 2506 landslides, representing 8.13% of the study area (Table 4). Five basic typologies have been differentiated according to Varnes [77] and Hungr et al. [110]: avalanches or collapses; debris flows; slides and earth flows, often as complex movements but classified according to the dominant type and soil creep, with a reduced speed and slow activity over time.

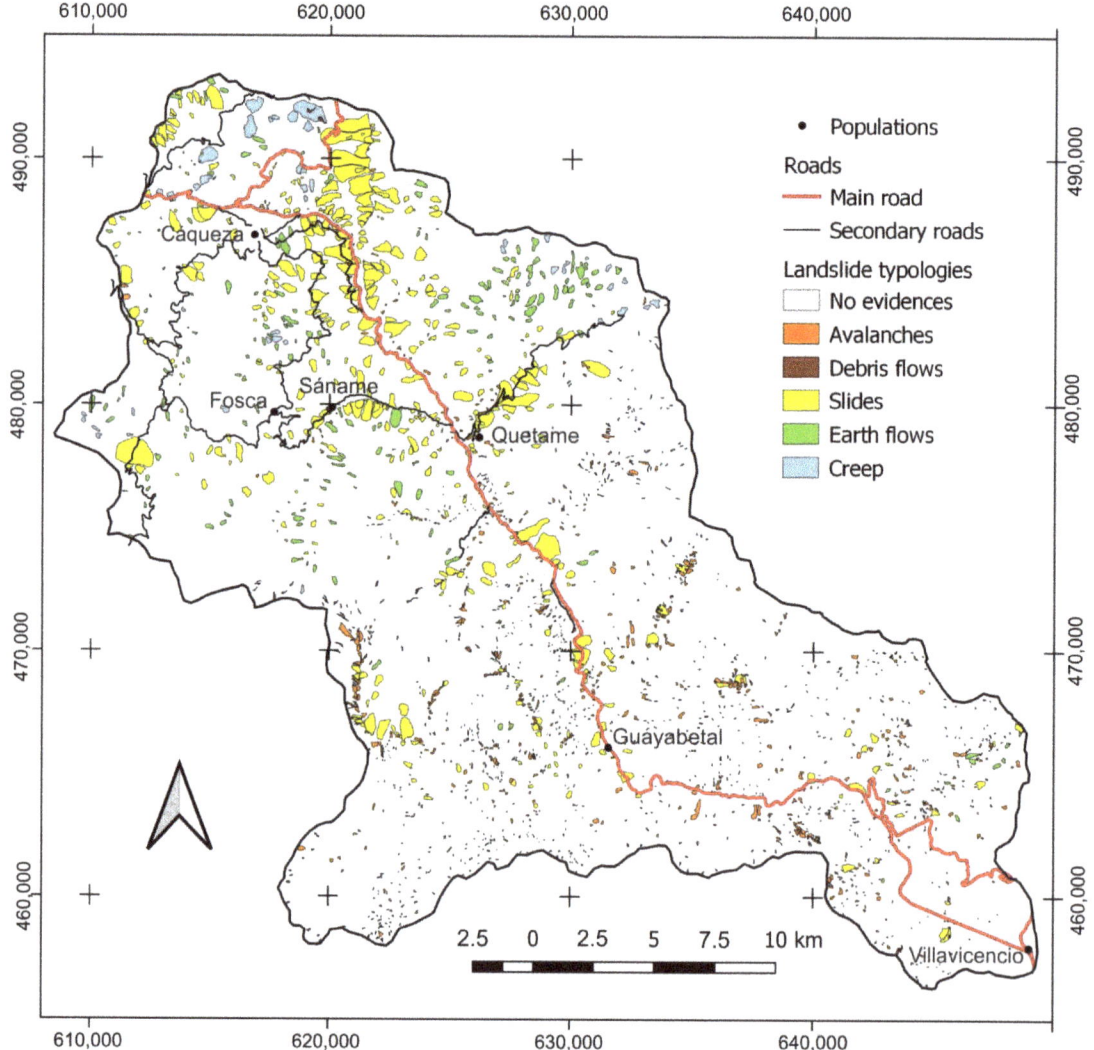

Figure 5. Landslide inventory in the study area. Five landslide typologies are distinguished.

Table 4. Analysis of landslide inventory. Areas are in m².

Typologies	Number		Total Area		Area Ind. (m²)	Active		Latent		Relict	
	No	%	A (m²)	%		No	%	No	%	No	%
Avalanches	979	39	7.95	13	8123	760	78	199	20	20	2
Debris flows	866	35	2.29	4	2649	594	69	271	31	1	0
Slides	437	17	39.00	64	89,261	59	14	122	28	256	59
Earth flows	179	7	7.47	12	41,747	4	2	70	39	105	59
Creep	45	2	3.99	7	88,803	6	13	36	80	3	7
All landslides	2506	-	60.71	8.13 [1]	24,231	1423	57	698	28	385	15

[1] This percentage corresponds to the area occupied by all landslides regarding the total study area.

The most predominant movements are avalanches (39%) and debris flows (35%), followed by slides (17%), earth flows (7%) and creep processes (2%). However, in terms of area, landslides occupy the largest extent (64%), with the percentages of avalanches and debris flows decreasing to 13% and 4%, respectively and the percentages of earth flows and creep increasing to 12% and 7%, respectively. These data are consistent with the fact that the average size (area) of slides and creeping processes is almost 90,000 m², whereas that of earth flows is approximately 42,000 m² and avalanches and debris flows have average areas of only 8100 and 2650 m², respectively.

From the activity perspective, most landslides are catalogued as active (57%), while remaining are considered latent (28%) and relict (15%). However, this distribution is different by type (Table 3), with avalanches and debris flows being predominantly active (78% and 69%, respectively), whereas slides and earth flows are mostly relict (59%) and creeping processes are latent (80%).

3.2. Analysis of Determinant Factors

The factors used in this study (shown in Figure 3) were analyzed; thus, their distribution in the area and their correlation with landslides are shown in Table 5 and Figure 6.

Table 5. Distribution of factor classes and cross-correlation with the landslides (all the landslides and for landslide typologies). The distribution of factor classes is expressed in % of area of every class respect to the total area of the study area. The density of landslides is expressed in % of landslide area for each class respect to the total area of this class. The correlation is expressed as the Kolmogorov-Smirnov (K–S) coefficient. The predominant classes of each factor, the classes with the highest density of landslides and K–S coefficients considered as significant are shown in bold.

Factors	Classes	All Landslides	Avalanches	Debris Flows	Slides	Earth Flows	Creep
			Elevation (m)				
500–1000	5.90%	4.00%	1.90%	0.11%	1.61%	0.37%	0.00%
1000–1500	15.53%	11.51%	**3.76%**	0.72%	6.73%	0.30%	0.00%
1500–1800	16.08%	**15.40%**	1.61%	0.50%	**12.05%**	0.88%	0.37%
1800–2000	11.86%	10.35%	0.97%	0.41%	7.55%	1.07%	0.34%
2000–2400	**23.10%**	8.85%	0.80%	0.46%	4.58%	**2.30%**	**1.29%**
2400–2600	9.75%	6.98%	0.60%	0.48%	3.17%	1.75%	0.44%
2600–2800	6.83%	5.85%	0.62%	0.64%	1.75%	1.87%	0.97%
2800–3600	10.96%	2.87%	0.29%	**0.91%**	0.32%	0.73%	0.62%
K–S		**0.18**	**0.32**	0.14	**0.29**	**0.25**	**0.34**
			Slope (°)				
0–5	1.94%	6.96%	0.62%	0.29%	3.46%	0.76%	**1.83%**
5–10	6.38%	7.53%	0.58%	0.15%	4.00%	1.26%	**1.53%**
10–20	22.69%	9.02%	0.79%	0.21%	5.12%	**1.73%**	1.17%

Table 5. Cont.

Factors	Classes	All Landslides	Avalanches	Debris Flows	Slides	Earth Flows	Creep
20–30	**31.08%**	9.91%	1.22%	0.42%	**6.34%**	1.46%	0.47%
30–45	**32.62%**	8.89%	1.87%	0.86%	5.43%	0.64%	0.09%
45–90	5.29%	9.93%	**3.24%**	**1.37%**	5.08%	0.23%	0.00%
K–S		0.03	**0.19**	**0.27**	0.05	**0.20**	**0.38**
Aspect							
N	11.25%	7.72%	1.34%	0.42%	4.77%	0.98%	0.21%
NE	11.88%	9.11%	1.49%	0.58%	5.71%	0.92%	0.40%
E	12.13%	10.37%	1.50%	0.80%	6.40%	1.08%	0.58%
SE	14.15%	9.47%	1.51%	0.83%	5.41%	0.99%	0.73%
S	13.23%	9.07%	1.52%	0.51%	4.33%	1.47%	**1.24%**
SW	13.17%	10.11%	1.50%	0.44%	5.85%	1.55%	0.77%
W	12.36%	9.37%	1.20%	0.37%	6.46%	1.06%	0.28%
NW	11.82%	7.90%	1.02%	0.37%	5.06%	1.21%	0.24%
K–S		0.04	0.05	0.14	0.06	0.08	**0.24**
Curvature							
−1–−0.02	6.26%	10.45%	**3.20%**	0.98%	5.30%	0.77%	0.21%
−0.02–−0.01	18.06%	10.27%	1.62%	0.60%	6.10%	1.34%	0.62%
−0.01–0.01	51.22%	9.25%	1.16%	0.47%	5.65%	1.29%	0.68%
0.1–0.2	18.28%	8.04%	1.09%	0.47%	4.97%	0.99%	0.52%
0.02–1	6.19%	7.30%	1.67%	0.76%	4.17%	0.51%	0.19%
K–S		0.04	0.11	0.07	0.03	0.06	0.06
TPI							
−100–−6	16.94%	12.29%	**3.24%**	0.82%	6.65%	1.22%	0.37%
−6–−2.5	16.52%	11.41%	1.47%	0.57%	6.87%	**1.70%**	**0.80%**
−2.5–0	16.47%	9.75%	1.01%	0.47%	5.98%	1.47%	**0.81%**
0–2.5	16.02%	8.58%	0.87%	0.44%	5.28%	1.25%	0.75%
2.5–6	16.29%	7.29%	0.88%	0.46%	4.58%	0.90%	0.47%
6–100	17.76%	5.83%	0.84%	0.51%	3.69%	0.49%	0.29%
K–S		0.12	**0.24**	0.09	0.10	**0.15**	0.12
TRI							
0–2	16.02%	7.93%	0.58%	0.17%	4.31%	1.36%	**1.52%**
2–3	16.68%	9.36%	0.87%	0.23%	5.40%	**1.83%**	1.03%
3–4	18.12%	9.96%	1.14%	0.36%	6.34%	1.57%	0.54%
4–6	17.04%	9.72%	1.45%	0.57%	6.34%	1.10%	0.27%
5–6	13.65%	8.90%	1.74%	0.76%	5.59%	0.70%	0.10%
>6	18.50%	8.99%	**2.49%**	**1.15%**	4.94%	0.38%	0.03%
K–S		0.03	**0.19**	**0.27**	0.06	**0.19**	**0.38**
Lithology							
Phylites-Schists	**38.43%**	5.77%	1.85%	0.89%	2.90%	0.14%	0.00%
Quartzarenites	13.62%	5.62%	**2.84%**	**1.27%**	1.28%	0.22%	0.01%
Shales	0.00%	0.00%	0.00%	0.00%	0.00%	0.00%	0.00%
Lutites	**45.48%**	**13.32%**	0.56%	0.06%	**9.07%**	**2.37%**	**1.26%**
Conglomerates	0.89%	4.22%	0.50%	0.20%	3.53%	0.00%	0.00%
Volcanic	0.13%	4.77%	0.85%	0.00%	1.92%	2.00%	0.00%
Alluvial fans	0.09%	0.00%	0.00%	0.00%	0.00%	0.00%	0.00%
Alluvial deposit	0.93%	4.41%	1.88%	0.00%	2.53%	0.00%	0.00%
Terraces	0.42%	6.69%	2.86%	0.44%	3.39%	0.00%	0.00%
K–S		**0.23**	**0.28**	**0.42**	**0.31**	**0.48**	**0.55**

Table 5. Cont.

Factors	Classes	All Landslides	Avalanches	Debris Flows	Slides	Earth Flows	Creep
		Precipitation (mm)					
500–1000	2.43%	**20.56%**	0.38%	0.06%	**15.44%**	1.74%	**2.93%**
1000–1500	12.31%	9.80%	0.29%	0.01%	7.58%	1.63%	0.29%
1500–2000	7.74%	9.10%	1.51%	0.60%	3.58%	**3.07%**	0.35%
2000–2500	**32.25%**	7.91%	**2.07%**	0.85%	4.45%	0.38%	0.17%
2500–3000	14.37%	3.47%	**1.88%**	0.52%	0.90%	0.17%	0.00%
3000–4000	13.50%	3.40%	1.39%	0.86%	0.95%	0.20%	0.00%
4000–5000	17.40%	2.64%	1.12%	**1.12%**	0.40%	0.00%	0.00%
K–S		0.21	0.22	0.27	0.33	0.42	0.59
		Land cover					
Urban	5.00%	18.68%	7.06%	1.09%	9.58%	0.53%	0.43%
No vegetation	1.94%	17.16%	**5.70%**	1.12%	8.46%	0.58%	**1.30%**
Grass	9.83%	16.15%	1.70%	0.50%	**11.41%**	1.21%	1.33%
Bush-Shrubs	62.01%	10.20%	1.07%	0.48%	6.27%	**1.68%**	0.71%
Forest	21.03%	2.40%	0.39%	0.33%	1.38%	0.27%	0.03%
Water	0.20%	15.44%	8.45%	0.58%	6.32%	0.00%	0.09%
K–S		0.17	0.29	0.09	0.17	0.17	0.20
		NDVI					
−0.5–0.1	5.08%	4.40%	1.81%	**1.09%**	1.48%	0.01%	0.02%
0.1–0.25	19.68%	2.40%	0.39%	0.34%	1.38%	0.26%	0.02%
0.25–0.4	**58.66%**	10.06%	1.11%	0.49%	6.14%	**1.64%**	0.69%
0.4–0.6	11.07%	**16.61%**	2.47%	0.60%	**11.23%**	1.13%	**1.19%**
0.6–1	5.52%	13.24%	**5.35%**	1.25%	5.51%	0.46%	0.66%
K–S		0.19	0.25	0.08	0.20	0.21	0.24
		Distance to roads (m)					
0–100	4.91%	**17.53%**	1.84%	0.02%	**13.88%**	1.17%	0.62%
100–250	6.49%	15.93%	1.31%	0.11%	12.38%	1.34%	0.79%
250–500	9.08%	13.50%	1.12%	0.22%	10.57%	0.83%	0.76%
500–1000	14.66%	11.81%	0.92%	0.30%	8.28%	1.07%	1.24%
>1000	64.85%	6.61%	1.50%	**0.73%**	2.79%	1.21%	0.38%
K–S		0.20	0.05	0.22	0.34	0.03	0.23
		Distance to rivers (m)					
0–100	3.95%	17.43%	**6.33%**	0.15%	10.57%	0.23%	0.15%
100–250	5.85%	17.44%	2.86%	0.46%	**13.45%**	0.45%	0.21%
250–500	9.00%	15.25%	1.93%	0.67%	11.55%	0.98%	0.11%
500–1000	16.32%	11.54%	1.31%	0.52%	7.98%	1.56%	0.18%
>1000	64.88%	6.47%	0.89%	0.57%	3.00%	**1.21%**	0.80%
K–S		0.21	0.24	0.04	0.31	0.08	0.21

The factors and their correlation with landslides are described below:

Elevation shows a wide range between 500 and 3600 m, although the majority (68%) is well distributed between 1500 and 2800 m. Landslides have their highest density in the range of 1500 to 1800 m. By typologies, avalanches are mainly concentrated in the range of 1000 to 1500 m, debris flows above 2800 m, slides between 1500 and 1800 m, earth flows between 2400 and 2600 m and creeping processes between 2000 and 2400 m. They show a significant correlation in practically all the typologies, especially in avalanches and creeping processes (more than 0.3).

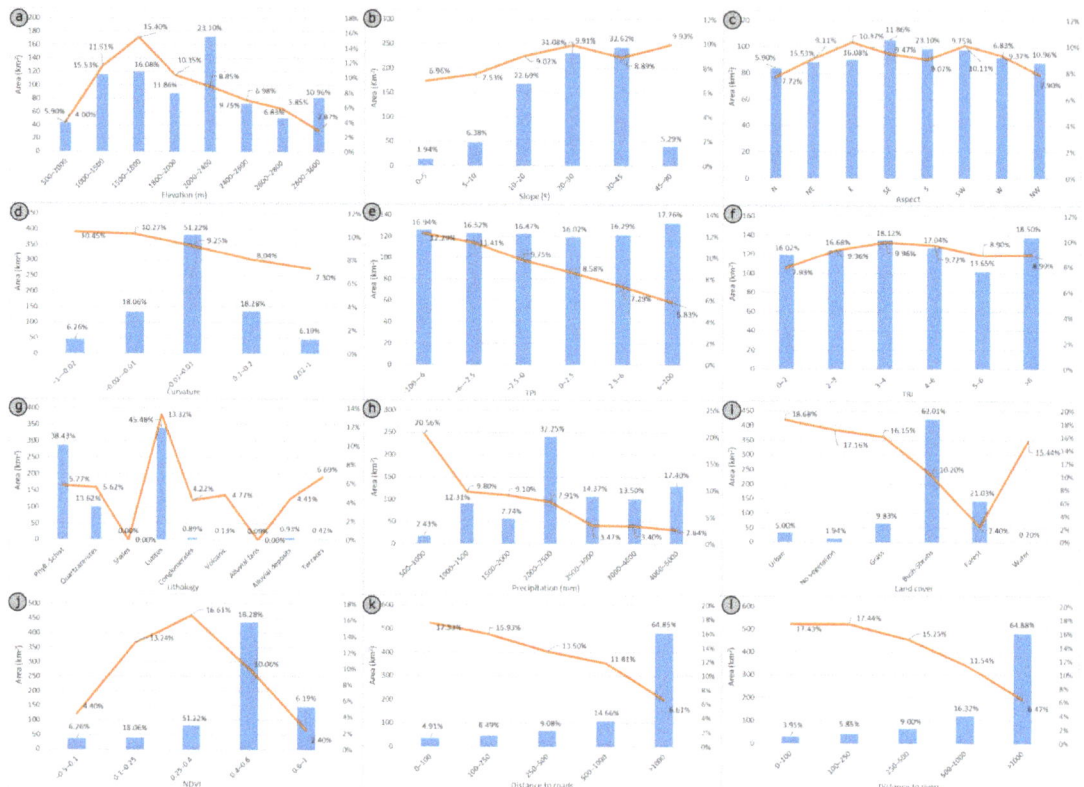

Figure 6. Distribution of classes and landslide density by class in each of the factors considered: (**a**): Elevation; (**b**): Slope; (**c**): Aspect; (**d**): Curvature; (**e**): TPI; (**f**): TRI; (**g**): Lithology; (**h**): Precipitation; (**i**): Land Cover; (**j**): NDVI; (**k**): Distance to roads; (**l**): Distance to rivers. Percentage area of the different classes and intervals are shown as histogram bars in blue; Landslide density in each class is shown as line diagram in red.

Slope is distributed practically throughout the total range from 0 to 90°, but more than 80% is between 10 and 45°. The landslide density is equally well distributed in the different ranges (between 7 and 10%), but by typologies, the distribution is different. Thus, avalanches and debris flows have a higher density as the slope increases, while slides reach their highest density in the range of 20 to 30°, earth flows between 10 and 20° and creeping processes between 0 and 10°. Practically all typologies, except slides, show a significant correlation.

Aspect appears to be fairly well distributed in all main orientations (11–14%). Meanwhile, the density is equally similar for all the movements (8–10%) and even in the analysis by typologies. Only the creeping processes have a higher density on south-facing slopes, so the correlations are generally low in all typologies, except for these processes.

Curvature presents a normal distribution concentrated around values close to 0. However, the distribution of landslides is higher in negative values (concave shapes) than in positive values (convex shapes). This is more evident in some typologies such as avalanches and slides, while in creeping processes, the higher density occurs in values close to 0. Correlations are not significant in any case.

Topographic Position Index (TPI): The classification allows for a balanced distribution (16–17%) of all intervals. However, the landslides' overall density is higher in negative values (10–12%) compared to positive values (6–8%). This asymmetry is very clear in avalanches, which concentrate in the lower sections of the hillslopes with negative index, and not so much in slides, earth flows and creeping processes that move towards the lower-middle and middle sections of the hillslopes. The correlation is significant in avalanches, landslides and creeping processes.

Terrain Roughness Index (TRI): As in the previous case, the classification produces a fairly uniform distribution of the different classes (14–18%). In this case, the landslide density is relatively uniform in the different classes, although the analysis by types does show important differences. Thus, avalanches and debris flows mainly concentrate in high roughness classes, while landslides and especially creeping processes do so in low roughness classes. The correlation is significant in all types, except for slides.

Lithology: Of the nine differentiated classes, three of them, such as lutites (45%), phyllites and schists (38%) and quartz sandstones (14%), have a significant extension, while the remaining classes barely reach 1%, including quaternary materials. Taking this into account, a higher density is observed in Cretaceous lutites in all the landslide types; but especially, a clear difference is observed between the distribution of avalanches and debris flows with higher density in phyllites and quartz sandstones compared to slides, earth flows and creeping processes, with higher density in Cretaceous lutites and conglomerates. Avalanches and slides also involve quaternary materials (alluvial and colluvial). Meanwhile, correlations are significant in all types and even for the landslides as a whole.

Precipitation has a wide range from 500 to nearly 5000 mm of average annual precipitation, with the interval of 2000 to 2500 m having the greatest extension (32%). Unlike that which might be expected, the landslide density decreases in the intervals of higher precipitation for all the movements, but certain differences are observed between the different typologies. Thus, avalanches and debris flows have a higher density in areas of medium or high precipitation, while slides, earth flows and creep processes have a higher density in areas of lower precipitation. In all cases, correlations are significant.

Land cover: Of the six differentiated classes, there is a predominance of shrub areas (62%) over-forested areas (21%), grasslands and crops (10%), areas with scarce vegetation (2%), urban areas (5%) and water (0.2%). Excluding water and urban areas, which are occasionally affected by movements, the highest landslide density occurs in areas with scarce vegetation and grass/crop areas (19% and 17%, respectively) compared to shrub areas (10%) and forested areas (2.4%). There are certain differences by typologies, as avalanches and debris flows have a higher density in areas with scarce vegetation, while slides, earth flows and creep processes occur mainly in grass/crop areas and even shrub areas. In all cases, forest areas produce the lowest densities. The certain incidence of avalanches and slides in urban or water areas is noteworthy. The correlations are also significant in all cases.

NDVI: The most extended class is the 0.4–0.6 with almost 60%, followed by the 0.6–1 with almost 20% and the 0.25–0.4 with 10%. In terms of landslide density, the highest is achieved in the middle values of the index, decreasing towards the low and high values. By typologies, avalanches and debris flows present higher densities in the low values than in the high ones, except in the very low values generally associated with urban areas; meanwhile, slides, earth flows and creeping processes have the highest densities in the middle values of the index. The correlations are significant in practically all cases.

The distance to roads logically shows an increasing distribution of the area as the distance increases, with almost 65% of the surface area being more than 1 km from the main roads. The landslide density, however, is higher at shorter distances than at longer ones. By typologies, avalanches and earth flows barely present a relationship with roads, while in debris flows and creeping processes, the density increases with distance; only in slides is there an increase in density at shorter distances. The correlations are only significant in slides, debris flows and creeping processes.

Distance to rivers: As in the case of roads, it shows an increasing distribution of the area with distance, and almost 65% of the surface is more than 1 km away from the riverbeds. Similarly, the landslide density increases at shorter distances compared to longer ones, although this does not happen in the same way for all typologies. Thus, avalanches and slides have higher density at shorter distances, while creep processes have higher density at longer distances, and debris flows and earth flows do not present a clear relationship. Correlations are only significant in avalanches, slides, and creeping processes.

Meanwhile, the results of multicollinearity analysis among the factors are shown in Table 6. As can be seen, strong correlations are only found between slope and TRI (0.97), and land use and NDVI (0.75); moderate correlations appear between TPI and curvature (0.65) and lithology and precipitation (0.59). Finally, weak correlations appear between elevation and distance to roads and distance to rivers and between lithology and slope and roughness.

Table 6. Correlation coefficients (Pearson) between factors. Strong and moderate correlations are shown in bold.

Factors	Elevation	Slope	Aspect	Curvat.	TPI	TRI	Lithol.	Precip.	Land C.	NDVI	D. Roads	D. Rivers
Elevation	1.000											
Slope	0.014	1.000										
Aspect	0.001	0.022	1.000									
Curvature	0.034	0.005	0.000	1.000								
TPI	0.090	0.009	0.000	**0.654**	1.000							
TRI	0.010	**0.971**	0.023	0.003	0.006	1.000						
Lithology	0.006	0.355	0.045	0.003	0.007	0.326	1.000					
Precipitation	0.248	0.231	0.029	0.000	0.001	0.210	**0.585**	1.000				
Land cover	0.114	0.073	0.140	0.000	0.004	0.058	0.022	0.215	1.000			
NDVI	0.107	0.041	0.175	0.005	0.007	0.026	0.043	0.150	**0.754**	1.000		
D.Roads	0.380	0.179	0.026	0.003	0.013	0.169	0.470	0.401	0.044	0.014	1.000	
D.Rivers	0.471	0.086	0.012	0.011	0.036	0.071	0.029	0.119	0.021	0.006	0.185	1.000

Based on these analyses, a factor selection has been made for the elaboration of susceptibility models and maps (Table 7). At the first level, all 12 factors used were considered. At the second level, those factors that showed a clear collinearity (strong and moderate correlation) were discarded, retaining those eight factors considered as independent. At the third level, only those non-collinear factors that showed significant correlation with the different landslide typologies were considered. Finally, the four factors that show total independence between them but correlation with most typologies were maintained: elevation, slope, TPI and lithology.

Table 7. Selected factors for susceptibility models: 1: All factors; 2: Non-collinear factors; 3: Factors with significant correlation with the different landslide typologies; 4: Independent factors between them but correlated with most typologies: Elevation, slope, TPI and lithology factors.

Factors	Elevation	Slope	Aspect	Curvat.	TPI	TRI	Lithol.	Precip.	Land C.	NDVI	D. Roads	D. Rivers
Avalanches	1234	1234	12	1	1234	1	1234	1	123	1	12	123
Debris flows	1234	1234	123	1	124	1	1234	1	123	1	12	12
Slides	1234	124	12	1	124	1	1234	1	123	1	123	123
Earth flows	1234	1234	12	1	1234	1	1234	1	123	1	12	12
Creep	1234	1234	123	1	124	1	1234	1	123	1	12	12

In summary, the conditions under which landslides occur preferentially are lithology of lutites; anthropic land use, areas with scarce vegetation and grass-crop lands, with NDVI between 0.2 and 0.4; elevation range between 1500 and 2000 m; precipitation between 500 and 1000 mm and the lower-concave section of the hillslopes.

If the analysis is considered by typologies:

- Avalanches show a higher density in Paleozoic quartz sandstones and phyllites, areas with scarce vegetation and NDVI between 0.1 and 0.25, altitudes between 1000 and 1500 m, slopes greater than 30°, the lower-concave sections of the hillslopes, areas with high roughness and areas near streams.
- Debris flows occur mainly in phyllites and quartz sandstones in areas with scarce vegetation, elevations above 2800 m, slopes greater than 30°, areas facing the east and southeast and areas with high roughness.
- Slides occur more frequently in Cretaceous lutites and grass-crop areas with NDVI between 0.25 and 0.4, elevations between 1500 and 1800 m, the middle-lower sections of the hillslopes and areas near streams and roads.
- Earth flows are concentrated mainly in lutites in areas with shrub vegetation with NDVI between 0.4 and 0.6, elevations between 2400 and 2800 m, slopes between 10 and 20° and the middle-lower sections of the hillslopes with low roughness.
- Creeping processes occur in lutites and grass-crop areas with NDVI between 0.25 and 0.4, elevations between 2000 and 2400 m, slopes of 0 to 10° and areas facing south with low roughness.

As can be observed, there is a certain similarity between the conditions for the occurrence of avalanches and debris flows on the one hand, and slides, earth flows and creep processes on the other hand. Thus, the first group is mainly associated with the lithology of phyllites, schists and quartz sandstones, areas with scarce vegetation, slopes greater than 30° and lower sections of the hillslopes with high roughness. Meanwhile, the second group is associated with Cretaceous lutites, grass-crop areas with NDVI between 0.25 and 0.4 and slopes of 0 to 20° in middle-lower sections of hillslopes with low roughness.

3.3. Susceptibility Models and Validation

The results of the susceptibility models and maps (LSM) performed with training samples of 8000 points (80%), using various methods and sets of factors, are shown in Figure 7. Meanwhile, Table 8 shows the results of the AUC-ROC obtained with sample validation (20%), Table 9 the DF for the same validation and Table 10 the AUC-ROC values obtained with temporal validation.

It can be observed that in general, for all landslide typologies, techniques and numbers of input factors, the models were well fitted, with the AUC-ROC for the validation samples always above 0.70. Regarding typologies, the creeping processes presented AUC-ROC values generally higher than 0.90; the avalanches and earth flows also reached quite high values (average above 0.84), while slides and debris flows had the lowest values, although still high (average above 0.80).

Regarding methods, the matrix method provides very high fits in general (average close to 0.90), followed by RF (0.88), ANN (0.84) and finally LDA (0.82). Generally, starting from the models obtained with the four basic factors (elevation, slope, TPI and lithology), all statistical and machine learning methods underwent an improvement when one or two factors were introduced that had some correlation with the landslides and low collinearity (reaching AUC-ROC from approximately 0.80 to 0.84). Then, they moderated their growth when non-collinear factors that did not show a high correlation with landslides (AUC-ROC up to 0.86) were introduced and even stabilized when all factors were introduced, including those that showed collinearity (0.87). The matrix method also improved when factors correlated with landslides were introduced (AUC-ROC from 0.80 to 0.90), stabilized with non-correlated factors (0.90) and increased again when all factors, even those showing collinearity, were considered (0.98).

An analysis carried out with other training/validation ratios (70/30 and 60/40) showed similar results. Thus, creeping processes presented the maximum AUC-ROC values (average of 0.93 in both cases), and debris flows presented the minimum values (0.80–0.81). Meanwhile, excluding matrix methods in which these ratios are not considered, RF presented the highest AUC-ROC values (0.87–0.88) and LDA the lowest (0.82).

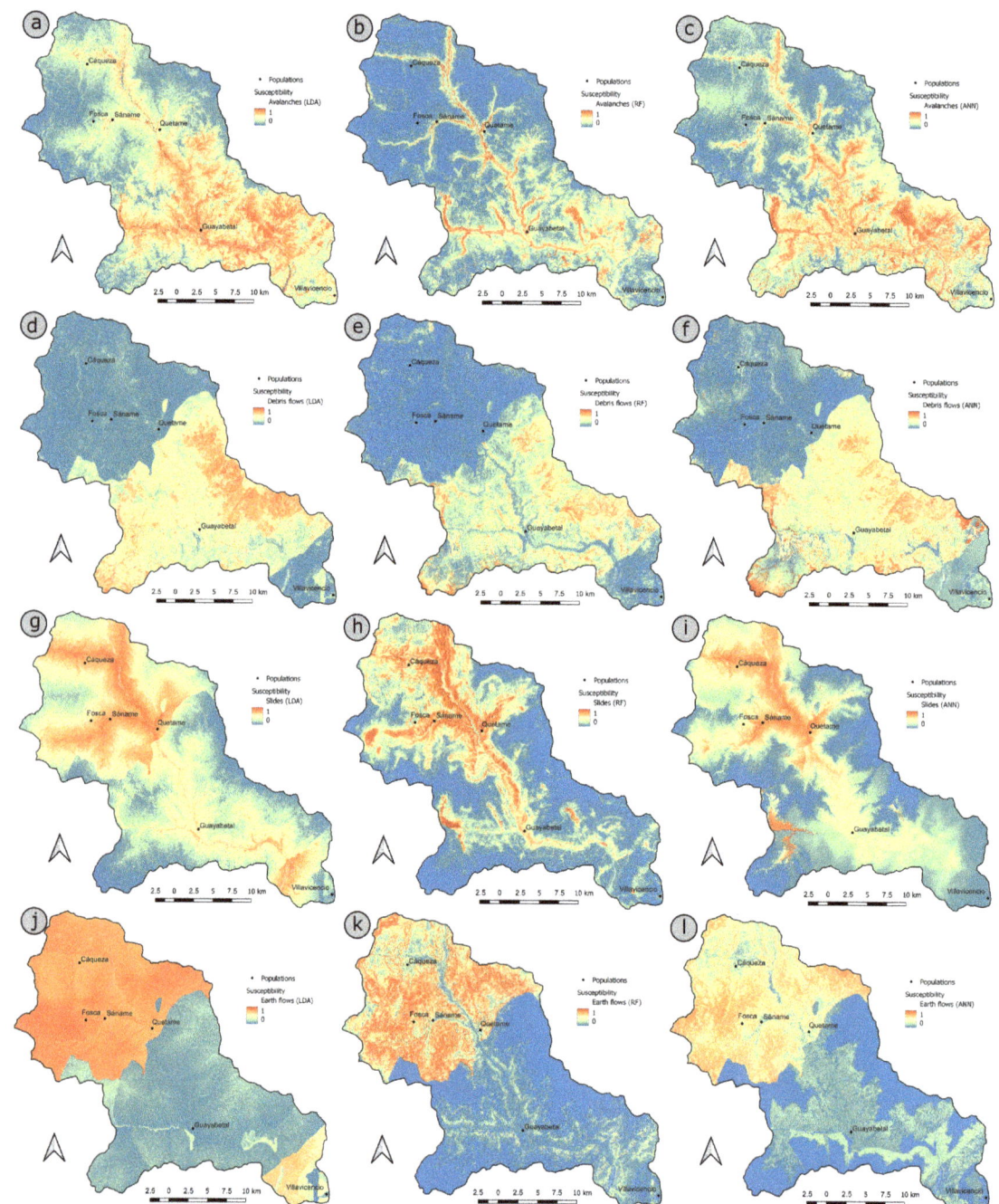

Figure 7. Cont.

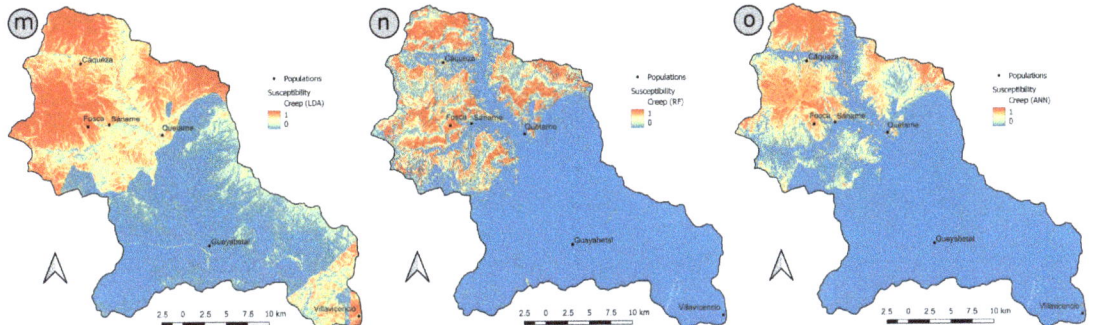

Figure 7. Landslide susceptibility models: Avalanches (**a**): LDA; (**b**): RF; (**c**): ANN; Debris flows: (**d**): LDA; (**e**): RF; (**f**): ANN; Slides: (**g**): LDA; (**h**): RF; (**i**): ANN; Earth flows: (**j**): LDA; (**k**): RF; (**l**): ANN; Creep: (**m**): LDA; (**n**): RF; (**o**): ANN. Color scale from blue-green (lower levels of susceptibility) to orange-red (higher levels of susceptibility.

Table 8. AUC-ROC values of LSM in sample validation for the different typologies, methods and number of factors.

Methods	N Factors	Avalanches	Debris fl.	Slides	Earth Flows	Creep	All mov.
Matrix	4 f	0.804	0.739	0.835	0.842	0.884	0.825
	5–6 f	0.908	0.897	0.881	0.911	0.942	0.908
	8 f	0.914	0.861	0.904	0.924	0.952	0.910
	12 f	0.977	0.978	0.977	0.984	0.987	0.982
LDA	4 f	0.790	0.750	0.733	0.805	0.875	0.791
	5–6 f	0.832	0.781	0.780	0.805	0.901	0.820
	8 f	0.834	0.791	0.782	0.808	0.919	0.827
	12 f	0.848	0.794	0.807	0.815	0.932	0.839
RF	4 f	0.773	0.727	0.776	0.817	0.908	0.800
	5–6 f	0.874	0.870	0.874	0.830	0.916	0.873
	8 f	0.894	0.882	0.891	0.915	0.984	0.913
	12 f	0.885	0.873	0.897	0.939	0.988	0.916
ANN	4 f	0.800	0.715	0.802	0.810	0.867	0.796
	5–6 f	0.857	0.806	0.814	0.822	0.909	0.842
	8 f	0.857	0.799	0.818	0.833	0.934	0.848
	12 f	0.844	0.819	0.841	0.877	0.940	0.845
Average		0.856	0.818	0.838	0.862	0.927	0.860
Methods	Matrix	0.901	0.869	0.899	0.915	0.941	0.906
	LDA	0.826	0.779	0.776	0.808	0.907	0.819
	RF	0.857	0.838	0.860	0.875	0.949	0.876
	ANN	0.840	0.785	0.819	0.844	0.903	0.838
N. Factors	4 f	0.792	0.733	0.787	0.821	0.884	0.803
	5–6 f	0.868	0.839	0.837	0.842	0.917	0.861
	8 f	0.875	0.833	0.849	0.870	0.947	0.875
	12 f	0.889	0.866	0.881	0.904	0.969	0.902

Table 9. Degree of Fit of lower/higher susceptibility levels for the different typologies, methods and number of factors.

Methods	N Factors	Avalanches	Debris fl.	Slides	Earth Flows	Creep	All mov.
Matrix	4 f	5/79	11/58	6/88	3/95	2/98	5/84
	5–6 f	2/96	3/96	1/92	1/99	1/98	2/96
	8 f	3/95	5/87	1/94	1/97	1/97	2/94
	12 f	2/97	1/99	1/99	1/99	1/99	1/99
LDA	4 f	5/82	4/72	9/78	7/93	1/99	5/85
	5–6 f	6/86	4/81	8/84	7/93	1/99	5/89
	8 f	5/88	4/81	7/84	6/93	1/99	4/89
	12 f	4/90	4/81	5/85	6/93	1/99	4/90
RF	4 f	4/82	5/74	4/87	1/95	1/99	3/88
	5–6 f	2/93	2/90	1/98	1/97	1/99	1/95
	8 f	1/96	1/95	1/98	1/99	1/99	1/98
	12 f	2/95	1/94	1/97	1/99	0/100	1/97
ANN	4 f	4/82	4/79	5/87	2/93	1/99	3/87
	5–6 f	4/89	3/86	3/92	1/94	1/99	3/92
	8 f	4/89	4/86	4/88	1/94	1/99	3/91
	12 f	4/90	4/88	4/89	1/95	1/99	3/92
Average		4/89	4/84	4/90	3/96	1/99	3/92
Methods	Matrix	3/92	5/85	2/93	1/98	1/97	2/93
	LDA	5/86	4/79	7/83	6/93	1/99	3/89
	RF	2/91	2/88	1/95	1/98	1/99	2/94
	ANN	4/86	4/84	4/89	1/94	1/99	3/91
N. Factors	4 f	5/81	6/71	6/85	4/94	1/98	4/86
	5–6 f	4/91	3/88	3/91	3/96	1/99	3/93
	8 f	3/92	4/87	3/91	2/96	1/99	2/93
	12 f	3/94	2/91	2/94	2/98	1/99	2/95

Table 10. AUC-ROC values of LSM in temporal validation for the different typologies, methods and number of factors.

Methods	N Factors	Avalanches	Debris fl.	Slides	Earth Flows	Creep	All mov.
LDA	4 f	0.803	0.794	0.724	0.781	0.857	0.792
	5–6 f	0.845	0.800	0.716	0.790	0.894	0.809
	8 f	0.845	0.811	0.724	0.784	0.900	0.813
	12 f	0.848	0.786	0.779	0.786	0.921	0.824
RF	4 f	0.745	0.699	0.687	0.755	0.868	0.751
	5–6 f	0.789	0.795	0.735	0.787	0.865	0.794
	8 f	0.794	0.724	0.743	0.806	0.913	0.796
	12 f	0.832	0.748	0.790	0.829	0.923	0.824
ANN	4 f	0.801	0.770	0.705	0.768	0.831	0.775
	5–6 f	0.819	0.793	0.726	0.764	0.847	0.790
	8 f	0.834	0.793	0.734	0.795	0.869	0.805
	12 f	0.926	0.785	0.785	0.808	0.926	0.846
Average		0.823	0.775	0.737	0.788	0.885	0.802
Methods	LDA	0.835	0.798	0.736	0.785	0.893	0.809
	RF	0.790	0.742	0.739	0.794	0.892	0.791
	ANN	0.845	0.785	0.737	0.784	0.868	0.804
N. Factors	4 f	0.783	0.754	0.705	0.768	0.852	0.773
	5–6 f	0.818	0.796	0.726	0.780	0.869	0.798
	8 f	0.824	0.776	0.734	0.795	0.894	0.805
	12 f	0.869	0.773	0.785	0.808	0.923	0.831

The validation made with degree of fit showed that the error/success ratio was very suitable in most cases, with an average of 3/92, with the creeping processes being those that presented the best ratio (1/99) and the debris flows the worst (4/84). By methods, the matrix method and RF presented slightly better average ratios (2/93 and 2/94) than ANN (3/91) and LDA (3/89). Regarding the number of factors included in the models, the aver-age ratio of the models with four factors was the worst (4/86), while the remaining ones reached 2–3/93–95.

The temporal validation shows acceptable results in general, with AUC-ROC values being mostly higher than 0.7 but between 0 and 12 points lower than in the sample validation. The average is 0.80, which is 4.5 points lower than in the sample validation. By typologies, the best results are obtained also in creeping processes (0.89) and the worst in slides (0.74), in which the AUC-ROC values decrease about eight points. By methods, all of them (LDA, RF and ANN) present similar AUC-ROC average values, about 0.80. Regarding the number of factors involved, the models present increasing AUC-ROC values, from 0.77 with 4 factors to 0.83 with 12 factors.

Finally, the integrated LSM of shallower processes (avalanches and debris flows) and deeper processes (slides, earth flows and creep), modeled with the random forest method and eight factors, are shown in Figure 8. In this case, the AUC-ROC values using sample validation and an 80/20 ratio are 0.88 and 0.83, respectively.

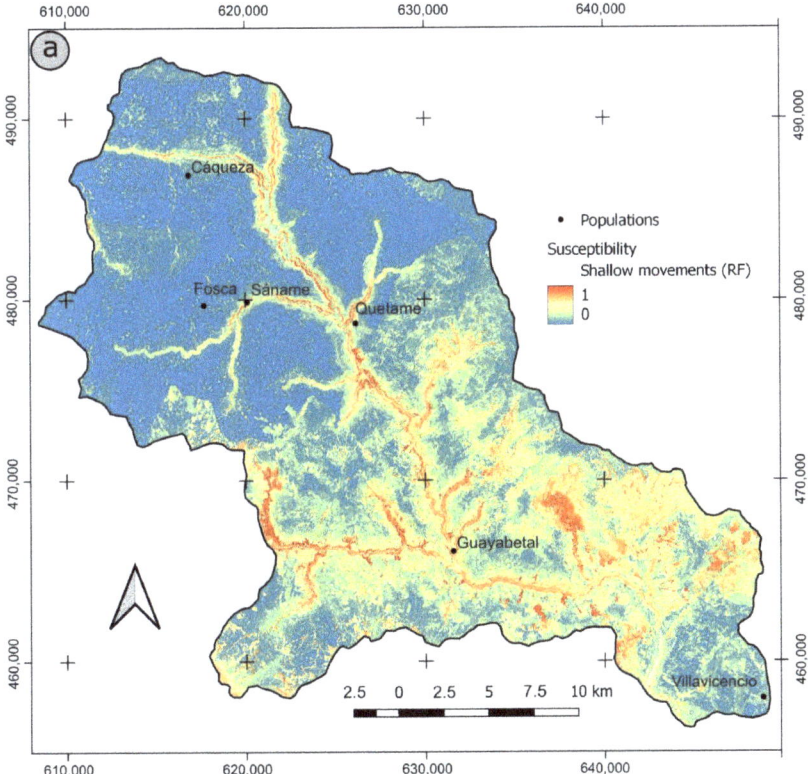

Figure 8. *Cont.*

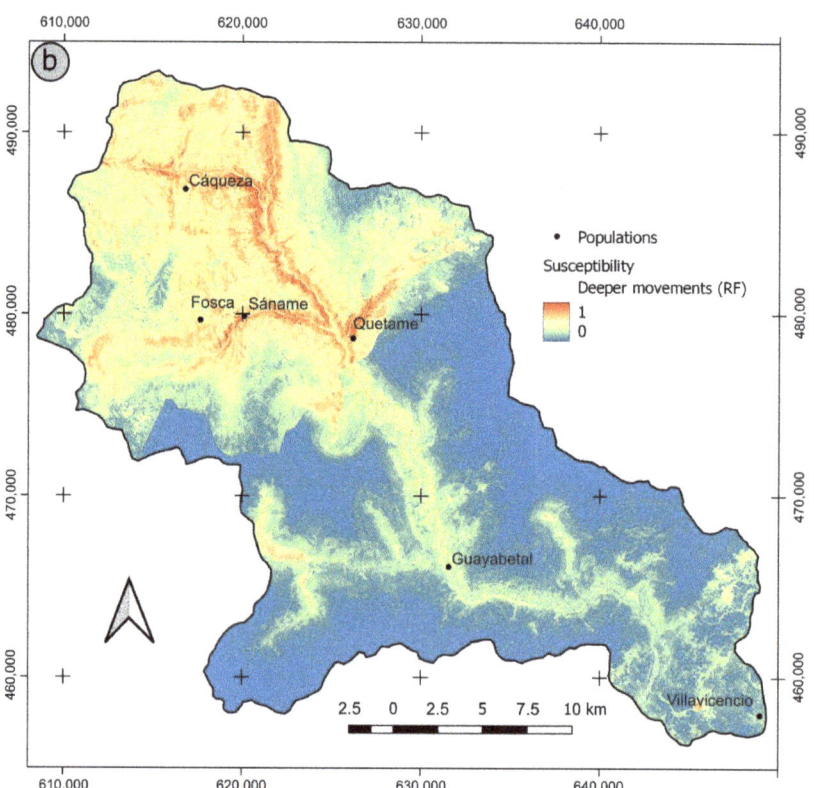

Figure 8. Integrated LSM: (**a**) Shallow movements; (**b**) Deeper movements. Color scale from blue-green (lower levels of susceptibility) to orange-red (higher levels of susceptibility).

4. Discussion

This study was conducted in an area of the central sector of the Eastern Andes mountain range in Colombia, which is characterized by intense landslide activity due to particular geological, topographical and climatic conditions. The geology of the area consists of Cretaceous sedimentary series, mostly composed of lutites and sandstones, structured by thrust faults and folds with a SSW-NNE direction, through which materials from the Paleozoic substrate with a certain degree of metamorphism outcrop [63,64]. The predominant elevation in the area ranges from 1000 to 3000 m, with relatively steep slopes (modal range between 20 and 30°), generally above 1000–1500 mm of precipitation, and land cover made up mainly of shrub-bush, grass-crop and forest areas.

4.1. Lanslides Inventory

A predominance of slide and earth flow type movements was observed when the total area or extension of landslides were considered, although avalanches, debris flows and creeping processes were also present. The individual areas were larger in slides and soil creep (almost 90,000 m^2), whereas that of earth flows was approximately 42,000 m^2 and only 8100 and 2650 m^2, respectively, for avalanches and debris flows. Meanwhile, slides and earth flows occasionally had a complex character, although in this work, they were considered the dominant process, which generally depends on their greater or lesser evolution, respectively. On the other hand, the creeping processes corresponded to undifferentiated flows with slow movement in general. Avalanches, which in other terminologies may be called collapses [36,111], are frequently in transition with both rock falls and debris slides-debris flows; however, considering the morphology and slopes observed (steep but not sub-vertical) and the materials in which they originate (phyllites, schists and quartzites often superficially weathered), they are classified as debris avalanches. Nevertheless, these avalanches can evolve into debris flows if the hillslope morphology allows it.

The inventory and differentiated typologies generally coincided with other studies of the Eastern Cordillera of the Colombian Andes. Thus, in the study by Calderón et al. [46], translational, rotational and wedge slides were differentiated in the vicinity of the Bogotá-Villavicencio highway. In the study by Valencia and Martínez-Graña [52], debris-flow, debris slides and rock falls were also inventoried in the Capitanejo area (Santander), further north in the Eastern Cordillera. Garcia-Delgado et al. [28] studied deep landslides and gravitational processes in San Eduardo, which is also to the North. Finally, Pradhan et al. [51] and Ramos-Cañon et al. [59] catalogued rock falls, avalanches, rotational and translational landslides, earth-mud flows and debris flows. In other parts of the country, such as the Western and Central Cordilleras, shallow slides, falls, debris flows and mud-flows have been identified [5,24–27,29,45,50,55]; in the southern Colombian Massif, specifically in Mocoa, debris flows, debris avalanches and shallow slides have been found [8,21,30,31].

Regarding the estimated activity based on photointerpretation, smaller movements such as avalanches, debris flows and small slides show higher activity than larger slides, earth flows and creeping processes, which have lower activity. This generally agrees with what happens in other regions of the world where these types of analyses have been addressed [111]. Activity generally depends on precipitation, which is abundant in the region due to the influence of deep convective systems [60] and which will probably have an even greater impact in the coming years [61,112]. This influence of precipitation has been analyzed by applying hydrological models [21,22,29,39] or determining rainfall thresholds [54–59,113]. The influence of other phenomena as triggering factors, such as earthquakes [8,114], active faults [115] or deforestation [116], has been also considered.

4.2. Analysis of Determinant Factors

Regarding factor analysis, the number of factors to be used in the models can be very high, especially in machine learning models [33], where it is common not to perform factor selection and allow the algorithms to fit the models. In this work, 12 variables have been used, which more or less coincide with those used in previous studies on susceptibility modelling, both globally [35,36,38,40–42,90,97,103,104,106,107] and in Colombian Andes [31,44–53]. Among them, the ones derived from the DEM stand out, which are related to the spatial distribution of important parameters such as slope, morphology, soil moisture or flow direction [51,117,118]; those related to geology and the geomechanical characteristics of materials [119] or those related to land use and land cover [81].

Despite what was said above, some authors recommend performing a certain factor selection to optimize the predictive capacity of the models and the performance of computational processes [120], generally based on the multicollinearity between factors or through methods of dimensionality reduction, which allows the selection of the most determinant factors and discarding other ones [121]. In addition, this analysis allows us to determine the conditions of the different landslide typologies [36,74,122]. The analysis carried out in this work shows that the factors that mainly condition the landslides' generation are elevation and lithology; although in the differentiated analysis by typologies, slope must be considered, and in some cases, TPI, aspect, land cover, distance to roads and distance to rivers should be considered as well. Some factors such as TRI, precipitation and NDVI also show some correlation with different landslides typologies but at the same time are strongly correlated to slope, TPI, lithology and land cover, respectively. These factors are similar to those found in previous works in other areas of the Colombian Andes, where a factor selection has been made. Thus, in Calderón et al.'s work [46], landslides were found to be related to the distance to faults, profile curvature, flow length, accumulated flow and land use. Salazar et al. [44] considered elevation, slope, planar curvature, landform shape, distance to faults, geological units, distance to rivers and distance to roads.

Correa et al. [50] considered slope, flow length, TWI, convergence index and soil types. Goyes-Peñafiel et al. [45] considered slope, curvature, TWI, landform shape, geological units and land use.

The conditions in which landslides occur preferably are the lithology of lutites, which is usually lower-resistance material and therefore more prone to landslides; areas with scarce vegetation and grass-crop lands, with NDVI between 0.2 and 0.4, that is, areas where there is a scarce or null vegetation cover protecting the soil from erosion and weathering processes; elevation between 1500 and 2000 m, where various geological, soil and morphological conditions favourable to instability are concentrated; rainfall between 500 and 1000 mm, which has no justification but is a consequence of the correlation of this factor with lithology and is therefore discarded in the analysis and, finally, lower-concave sections of the hillslopes, where hydrological and erosive phenomena promote the landslide generation. These conditions are similar to those found in other parts of the world and specifically in the Colombian Andes. Thus, in Salazar et al. [44] and Valencia and Martínez-Graña [52], landslides were associated with steep slopes (above 25–30°), southern orientation, concave morphologies, high roughness and a certain proximity to rivers but not so much to roads. However, they differed in altitude intervals and especially in the most affected lithology, which in the case of Valencia and Martínez-Graña [52] in the Eastern Cordillera were lutites or shales similar to those in the study area; and in Salazar et al. [44] in the Western Cordillera, with a different geological environment, they were volcanic rocks. Meanwhile, in Grima et al.'s work [116], landslides occurred in a proportion of six times more in deforested areas than in forested areas, while in Renza et al.'s work [53], no clear relationship was found between NDVI and other vegetation indices and landslides.

One of the interesting aspects of this work is the typology-based factor analysis, which has allowed us to observe differences in the conditions of occurrence of the different landslide typologies. From the results, a dichotomy was observed in the conditions in which landslides occur, with a first group of shallower landslides corresponding to avalanches and debris flows and a second group of larger and usually deeper landsides corresponding to slides, earth flows and creep processes. These conditions are shown in the Figure 9 for every determinant factor. Lithology appears as a crucial factor that in turn influences other factors such as slope, elevation or land cover, among others.

Thus, since lithology is the most determinant factor, a higher resolution of the geological map is needed, which allows greater precision in the identification of the conditions in which landslides and their corresponding typologies originate. The same can be stated about other factors such as the land cover, NDVI and even DEM derivatives.

Despite this, it can be observed that avalanches and debris flows present similar occurrence conditions that are different from the remaining typologies. As previously noted, there can be a certain transition between avalanches and debris flows, such that the former may eventually evolve into the latter if the morphological conditions allow for it. Thus, the conditions that lead to a higher landslide density are mainly the lithology of Paleozoic quartz sandstones and phyllites, which are rocks more resistant a priori, enabling the formation of quite steep slopes where these processes originate [36]. Therefore, the slopes on which they occur are generally higher than 30° and even up to 45°, with a high terrain roughness. However, avalanches occur in the lower-concave sections of the hillslopes and close to the channels, which is not the case with debris flows, which extend to higher sections, as the elevation analysis shows (1000–1500 m and over 2800 m, respectively). Furthermore, both types are associated with areas of low vegetation cover and low NDVI values (0.10–0.25).

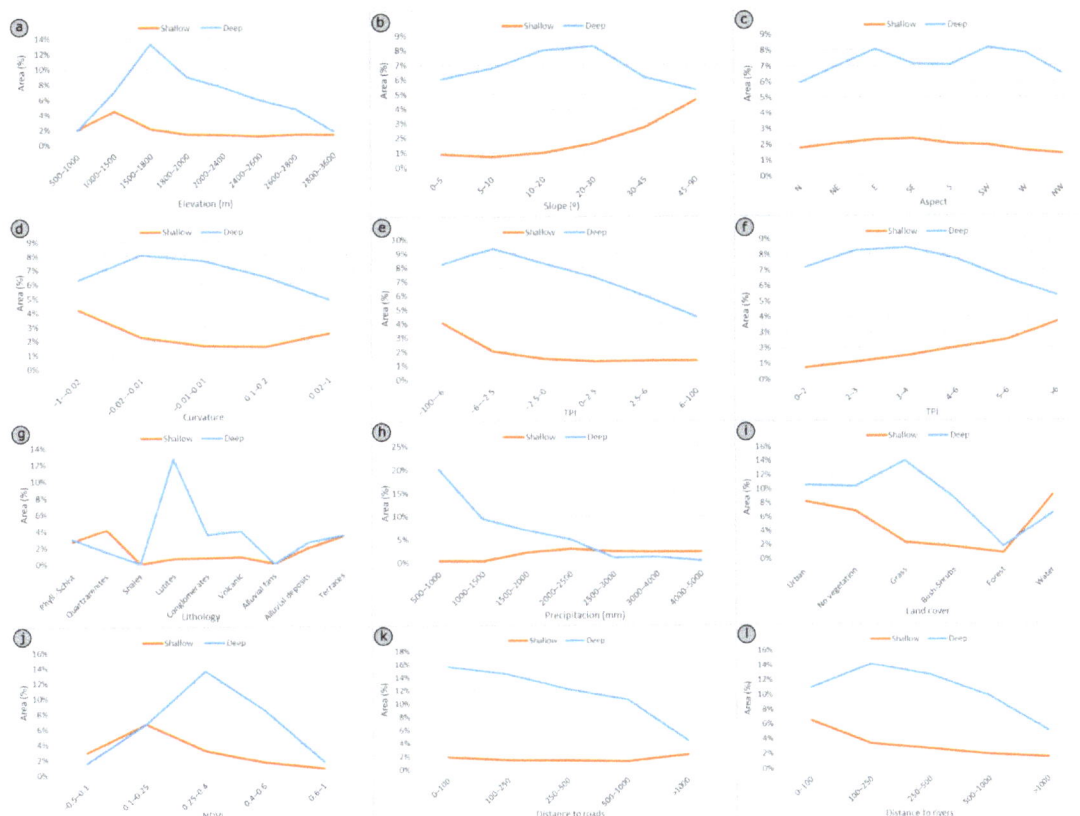

Figure 9. Conditions in which the shallower (orange line) and deeper landslides (blue line) occur. X axis: Classes of the different factors; Y axis: % of area occupied for landslides in each class. (**a**): Elevation; (**b**): Slope; (**c**): Aspect; (**d**): Curvature; (**e**): TPI; (**f**): TRI; (**g**): Lithology; (**h**): Precipitation; (**i**): Land Cover; (**j**): NDVI; (**k**): Distance to roads; (**l**): Distance to rivers.

Meanwhile, slides, earth flows and creeping processes occur more frequently in Cretaceous lutites, which are rocks with less resistance than the previous ones. Hence, landslides can occur even at lower slopes than with the previous types. Specifically, slides occur at a wide range of slopes but more frequently between 20 and 30°, with earth flows between 10 and 20° and creeping processes on slopes below 10°. Additionally, they often occur in the middle and lower sections of the slopes, with low roughness. Regarding land cover and NDVI, they are more concentrated in grass-crop and shrub areas, with medium NDVI values (0.25–0.6). However, the elevations range from lower for slides to medium-high for earth flows and creeping processes.

4.3. Susceptibility Models and Validation

Regarding susceptibility models and maps (LSM), four methods have been applied according to the groups established in Reichenbach et al. [33]: matrix method (index), discriminant analysis (multivariate statistics), random forest and an artificial neural network (machine learning). In general, all methods showed good results, with AUC-ROC values above 0.70 for the different typologies and number of factors considered in the models (Table 8, Figure 10a). For typologies, creeping processes generally had AUC-ROC values above 0.90, which shows that these processes are associated with very specific conditions that make the models very fitted. In fact, in this case, there are methods such as RF that have higher ROC AUC values (0.95) than even the matrix method (0.94). Next are avalanches and earth flows with also quite high values (average around 0.86), reaching maximum values in the matrix method (0.90–0.91). Finally, slides and debris flows present the lowest AUC-ROC average values (0.84 and 0.82, respectively), since the conditions for the landslide occurrence are not so clearly defined in these typologies, reaching maximum values also in the matrix method (0.88–0.90).

Figure 10. AUC-ROC values: (**a**): By typologies; (**b**): By methods; By number of factors (4 to 12) in the different methods (**c**): Matrix; (**d**): LDA; (**e**): RF; (**f**): ANN.

Linking to the above, the comparison by methods shows that the matrix method provides very high fits in general (average close to 0.90), followed by RF (0.88), ANN (0.84), and finally LDA (about 0.82) (Figure 8b). As can be seen, the matrix method elaborated from unique condition units provides very good fits even with randomly selected testing samples. This agrees with the results obtained in previous works, where other validation techniques such as temporal validation (performed with inventories elaborated after the one used in the model) produce degree of fit [39] of 5–10% in the classes of very low to low susceptibility (errors) and 70–80% in the classes of high to very high susceptibility (success) [35,36,83]. Despite that, matrix and other index-based or bivariate statistical methods have some limitations related to the simplification of the conditioning factors (especially when they are classified as in matrix approach) and the assumption of conditional independence between them [123]. Thus, they fit very well to the specific conditions of an area when inventories are exhaustive, but their performance is reduced when they are transferred to other areas [124] or when the starting inventories are less exhaustive.

The LDA models showed relatively lower fits, although they can be considered acceptable, as they reached average values of AUC-ROC around 0.80 for most landslide typologies and even 0.90 for creeping processes. This is corroborated by previous studies where 70–80% of the slopes were correctly classified in high susceptibility classes [4,39]. However, multivariate statistical methods maintain some of the limitations of the bivariate methods, such as the simplification and dependence between factors [123]. Moreover, these limitations cannot be overcome due to the linearity of the discriminant functions compared to the greater versatility and better performance of machine learning methods in non-linear systems such as susceptibility models developed from factors of different nature [105,125].

The RF method produced excellent results, with AUC values between 0.84 and 0.88 in most landslide typologies and 0.95 in creeping processes, which agrees with other studies where values close to and above 0.90 were achieved [40,92,126–128]. In the studies carried out Colombian Andes, Calderón et al. [46] applied RF starting from 14 similar factors, with very good results (area under the success rate curve, A_{SRC}, of 93%).

Finally, the application of a perceptron-type ANN with a single hidden layer and three–four neurons provided fits that were somewhat lower to RF but higher to LDA. Thus, except for the lower AUC-ROC value for debris flows (0.79), the remaining typologies had values between 0.82 and 0.84, and even the creeping processes reached a value of 0.90. These values are consistent with the results obtained by numerous authors who have applied neural networks of different types, from MLP to convolutional ones, passed through Radial Basis Funtion (RBF) networks and others. Focusing on perceptron-type networks, Pradham and Lee [41] obtained an AUC of 0.91–0.94; Tien Bui et al. [42], 0.92; Bravo et al. [95], 0.76; Zare et al. [98], 0.88; Pham et al. [99], 0.87; Park et al. [104], 0.81 and Aslam et al. [107], 0.87. In the Colombian Andes, Calderón et al. [46], in an area that encompasses that used in this article, applied an MLP-type ANN with 14 factors (input neurons) and two hidden layers of 16 neurons, which provided an A_{SRC} of 0.88. When factor selection was applied (five factors), the A_{SRC} became 0.86, and after applying factor reduction by means of PCA, the A_{SRC} decreased to 0.81. Meanwhile, Valencia and Martínez-Graña [52], further north in the Santander department, applied a perceptron-type ANN with backpropagation algorithm, starting from 14 factors and a layer of 20 neurons, which provides a very high AUC-ROC value (0.988).

Regarding the number of factors, although all methods show a better fit as the number of factors used in the models increases, not all do so in the same way, which allows us to extract considerations about the behavior of these models and the opportunity to make or not make factor selection (Figure 10c–f). Thus, starting from models with four basic factors common to all typologies and independent of each other (elevation, slope, TPI and lithology), statistical methods (LDA) and machine learning (RF and ANN) undergo a significant increase when introducing one or two factors that have a certain correlation with landslides and low collinearity. Then, the AUC-ROC value increases from approximately 0.80 to 0.845, mainly in RF. However, this growth is attenuated (0.86) when up to eight non-collinear factors are included, although some of them did not show a correlation with landslides, and it practically stabilizes when all factors are introduced, including those that show collinearity (0.87). This behavior, in which the value of the AUC-ROC on the validation sample stabilizes, may be evidence of noise when introducing factors that do not show an influence on landslides and of overfitting or overtraining [129] when introducing redundant factors. In this sense, it is important to mention that the strong correlation between factors also increases the probability of overfitting [130]. This shows the interest of the factor analysis that allows for factor selection.

These observations are partially corroborated by the behavior observed in the matrix method, which does not aim to develop a statistical or learning model but rather fit the susceptibility zoning to the data on factors and movements in a specific area. Thus, the value of AUC increases as correlated factors with landslides are introduced (from 0.80 to 0.90), which is consistent with what was observed in the previous methods, indicating the importance of including as many determinant factors of landslides as possible. Then, the AUC-ROC value stabilizes when introducing factors uncorrelated to landslides (0.90), which confirms that these factors do not improve the model and may introduce noise. However, unlike the other methods, the AUC-ROC value again increases when all factors are introduced, even those that show collinearity (0.98), which could be a clear indication of overfitting in this method when a large number of factors are introduced. It should be noted that, despite using a random validation sample, this validation sample is not entirely independent of the total mobilized surface used to develop the susceptibility map, since the matrix method uses the entire landslide area to fit the model.

Regarding the other methods and strategies of validation, the analysis carried out with other training/validation ratios showed similar results for AUC-ROC values, which ensured that the ratio used (80/20) is valid for this case and in general [100]. Meanwhile, the degree of fit showed an error/success ratio that was very suitable in most cases. Relative errors were lower than 5% in practically every case, and relative successes reached values usually higher than 90% and even 95%. Moreover, the results by typologies, methods and number of factors involved were in agreement with those obtained with the AUC-ROC. Finally, the temporal validation showed acceptable results in general, with AUC-ROC values mostly higher than 0.7. However, these values were between 0 and 12 points lower than in the sample validation, being an average of 4.5 points lower. By typologies, methods and number of factors, similar tendencies were observed, with the models of creeping processes and the models with a greater number of factors presenting higher AUC-ROC values. However, there were some differences such as the lowest values observed in slides that probably are related to the different size and, thus, the conditions of the active slides (smaller, in areas of higher slopes and lower sections of hillslopes) and non-active (larger, in areas of moderate slopes and lower-middle sections of the hillslopes). The other typologies presented similar conditions between active and non-active landslides, so the AUC-ROC values obtained in temporal validation were higher. The other difference was the similar behavior between the methods analyzed (LDA, RF and ANN since this validation was not applied for the matrix approach). In this case, the global reduction of AUC-ROC values made the values of different methods less distinguishable.

In summary, we can conclude about the importance of choosing the more adequate method as well as the factor selection for LSM. Regarding the method, it seems that machine learning methods, especially random forest, show better performance than statistical methods due to their greater flexibility and non-linear adjustment. Regarding the matrix method, it has less statistical basis and produces significant overfitting, especially when there is no factor selection. Factor analysis and selection appear to be a recommended procedure to avoid overfitting and noise, even in machine learning methods where the importance of factor selection is not as apparent.

Finally, regarding the distribution of landslide susceptibility, a clear zoning can be observed in the maps of the different typologies whatever the method and the number of factors that have been used. Thus, the LSM corresponding to avalanches and debris flows show greater susceptibility in the south-eastern part (lower basin), which is also clearly observed in the integrated map of these typologies; meanwhile, the LSM of slides, earth flows and creep processes have a higher susceptibility in the north-western part (upper basin), as can also be observed in the integrated map. This zoning is the consequence of geological control, as mentioned before in the discussion of factor analysis, since metamorphic rocks outcrop in the lower basin through fold and thrust structures; these rocks are generally more coherent and resistant, developing steep slopes where avalanches and debris flows occur (affecting mainly the superficial layers of weathered rocks). Meanwhile, the Cretaceous shales outcrop predominantly in the upper part of the basin, and these sedimentary rocks, usually less resistant, are more susceptible to slides, earth flows and creeping processes. Logically, this general zoning is discriminated by the remaining factors, such as slopes, TPI, land cover and distance to roads in those maps that show better results (e.g., maps obtained with RF). Moreover, given the importance of the lithological factor, a higher resolution of the geological maps would have allowed a better discrimination of susceptibility.

5. Conclusions

The present study has allowed the characterization of landslides in a sector of almost 750 km² in the Eastern Cordillera of the Colombian Andes, by means of the elaboration of inventories using photointerpretation, GIS factors analysis and landslide susceptibility maps (LSM). A total of 2506 landslides were inventoried, occupying approximately 8% of the study area, including avalanches, debris flows, slides, earth flows and creeping processes. Debris flows (39%) and avalanches (35%) were the most abundant in number, while landslides (64%) occupied the largest area due to their larger individual size. Avalanches and debris flows were predominantly active, while most slides and earth flows were relict and creeping processes were latent.

The factors analysis showed that elevation, lithology and land cover were the factors that most influenced the generation of landslides. However, in the differentiated analysis by typologies, slope and, in some cases, TPI, the aspect, distance to roads and distance to rivers had to also be considered. Some factors, such as TRI, precipitation and NDVI, also showed some correlation with different landslide typologies but were strongly related to slope, TPI, lithology and land cover, respectively. This analysis also allowed us to observe differences in the conditions of occurrence of the different typologies. Thus, avalanches and debris flows had similar conditions of occurrence, which were the lithology of quartz-sandstone and phyllites from the Paleozoic era (more resistant rocks) and areas with scarce vegetation and low NDVI values (0.10–0.25); slopes higher than 30° and lower sections of the hillslopes with high roughness. Meanwhile, slides, earth flows and creeping processes occurred mainly in Cretaceous lutites and grass/crop areas; morphologically, they occurred on a wide range of slopes, generally lower than 30° in the middle and lower sections of the hillslopes with low roughness.

Regarding LSM, different types of methods were tested to provide consistent results in determining landslide hazards in the region, including index-based methods (matrix), multivariate statistical methods (discriminant analysis, LDA), machine learning methods (random forest, RF) and a perceptron-type neural network (ANN). In general, all methods produced good results, with the AUC-ROC values always above 0.7 and obtained with the validation sample (20%). For landslide typologies, the best fits occur in cases where the conditions are more specific, such as creeping processes (0.90) and debris flows and avalanches (0.84), and the worst fits occur where they are not so specific, such as slides and debris flows (0.82–0.80). By methods, although the matrix method provides very high fits (average close to 0.90), there is a certain tendency toward overfitting, especially when no factor selection is addressed. LDA offers relatively lower adjustments (average around 0.82), while RF and ANN present very good fits in general (average around 0.88 and 0.84, respectively). In all these methods, starting from four common and non-collinear factors with a high correlation with all typologies (elevation, slope, TPI and lithology), an increase in AUC-ROC occurs when one or two additional factors specific for each typology are introduced. Then, they moderate their increase when non-collinear factors with lower correlation with the movements are included (introducing noise), and especially when collinear factors are also considered (producing overfitting). But whatever the method used, the LSM maps show a zoning as a consequence of geological control, which produces higher susceptibility to avalanches and debris flow in the lower part of the basin, with more resistant rocks, while a higher susceptibility to slides, earth flows and creep occurs in the upper part of the basin where less resistant rocks appear.

The importance of the choice of susceptibility modeling method and factor selection is also checked in order to avoid noise and overfitting. This ensures the development of robust and coherent models that can be in neighboring areas of this region. Thus, the use of RF or other machine learning methods is recommended due to their greater versatility and better behavior in nonlinear systems such as LSM. Likewise, a certain factor analysis is also recommended that allows a better understanding of the conditions in which landslides occur and also an improvement in the factor selection. In this case, models with six or eight factors are considered to provide the most reliable results, and thus, synthesis maps have been prepared for shallower or deeper movements, with AUC-ROC values of 0.87 and 0.83, respectively.

The main limitations of this work are related to the quality and resolution of input data, especially the determinant factors. Thus, although the resolution of DEM can be considered as sufficient, its precision as a DEM derived from PalSAR could be limited. Nevertheless, the most limiting aspect is the resolution of thematic factors, especially the geological map as the main determinant factor; in addition, a better quality and temporal signification of the land cover, even with derived land use, and vegetation indices is also required. Meanwhile, another limiting aspect is the estimation of landslide activity that can allow not only a refinement of susceptibility maps but also the elaboration of hazard maps.

Thus, for future improvements and work, the introduction of new determinant factors, especially the improvement of those used in this work, is proposed, both derived from the DEM and thematic maps, which allow the models to be refined. This would also allow the analysis of the influence of spatial and thematic resolution on the models. Thus, robust models are developed, which can be applied to other neighboring areas. Automated methods for factor selection and complex models for susceptibility maps can be tested, but always subject to control by the analysts. Finally, the estimation of temporal probability based on the realization of multi-temporal inventories and the introduction of triggering factors (mainly rainfall) will allow the creation of hazard or threat maps in the study area and other areas of the mountain range.

Author Contributions: Conceptualization, M.C.H.-C. and T.F.; methodology, M.C.H.-C., C.C. and T.F.; software, P.E.B.-L.; validation, M.C.H.-C., L.P.C., I.L.H.-P. and P.E.B.-L.; formal analysis, M.C.H.-C.; investigation, M.C.H.-C., L.P.C. and I.L.H.-P.; resources, M.C.H.-C., I.L.H.-P., P.E.B.-L., M.S.-G. and T.F.; data curation, M.C.H.-C. and P.E.B.-L.; writing—original draft preparation, M.C.H.-C., L.P.C., I.L.H.-P., P.E.B.-L. and T.F.; writing—review and editing, C.C., J.D., M.S.-G. and T.F.; supervision, C.C., M.S.-G. and J.D. All authors have read and agreed to the published version of the manuscript.

Funding: This research received no external funding.

Data Availability Statement: Not applicable.

Acknowledgments: The authors are grateful to Photogrammetric and Topometric Systems Research Group of the University of Jaen (TEP-213); Geologic Processes and Resources Group of the University of Jaen (RNM-325); Natural Hazards Lab of the Centre for Advanced Studies in Earth Sciences, Energy and Environment of the University of Jaen (CEACTEMA); Department of Earth and Marine Sciences (DiSTeM) of the University of Palermo (Italia); Department of Geographic and Environmental Engineering, UDCA and the Institute for Studies of Sectional Regime of Ecuador (IERSE), University of Azuay. We also acknowledge the help of Professor Sergio David Parra González, from the Faculty of Agricultural Sciences and Natural Resources of the Universidad de los Llanos (Colombia).

Conflicts of Interest: The authors declare no conflict of interest.

References

1. Schuster, R.L. Socioeconomic significance of landslides. In *Landslides: Investigation and Mitigation*; Turner, A.K., Schuster, R.L., Eds.; Transportation Research Board Special Report 247; National Academy of Sciences: Washington, DC, USA, 1996; pp. 12–35.
2. Petley, D.N. Global Patterns of Loss of Life from Landslides. *Geology* **2012**, *40*, 927–930. [CrossRef]
3. UNDRR. *Global Annual Report 2019*; UNDRR: Geneva, Switzerland, 2019; Available online: https://gar.undrr.org/report-2019 (accessed on 30 May 2023).
4. Guzzetti, F.; Carrara, A.; Cardinali, M.; Reichenbach, P. Landslide Hazard Evaluation: A Review of Current Techniques and Their Application in a Multi-Scale Study, Central Italy. *Geomorphology* **1999**, *31*, 181–216. [CrossRef]
5. Muñoz, E.; Poveda, G.; Ochoa, A.; Caballero, H. Multifractal Analysis of Spatial and Temporal Distributions of Landslides in Colombia. In *Advancing Culture of Living with Landslides*; Springer International Publishing: Cham, Switzerland, 2017; pp. 1073–1079. [CrossRef]
6. Aristizábal, E.; Sánchez, O. Spatial and Temporal Patterns and the Socioeconomic Impacts of Landslides in the Tropical and Mountainous Colombian Andes. *Disasters* **2020**, *44*, 596–618. [CrossRef]
7. Kühnl, M.; Sapena, M.; Wurm, M.; Geiß, C.; Taubenböck, H. Multitemporal Landslide Exposure and Vulnerability Assessment in Medellín, Colombia. *Nat. Hazards* **2022**, 1–24. [CrossRef]

8. García-Delgado, H.; Petley, D.N.; Bermúdez, M.A.; Sepúlveda, S.A. Fatal Landslides in Colombia (from Historical Times to 2020) and Their Socio-Economic Impacts. *Landslides* **2022**, *19*, 1689–1716. [CrossRef]
9. Gómez, D.; García, E.F.; Aristizábal, E. Spatial and Temporal Landslide Distributions Using Global and Open Landslide Databases. *Nat. Hazards* **2023**, *117*, 25–55. [CrossRef]
10. CRED: EM-DAT, The International Disaster DataBase, Centre for Research on the Epidemiology of Disasters (CRED). Available online: https://www.emdat.be/ (accessed on 30 May 2023).
11. UNDRR: DesInventar Sendai. Available online: https://db.desinventar.org/ (accessed on 30 May 2023).
12. NASA: Open Global Landslide Catalog. Available online: https://gpm.nasa.gov/landslides/index.html (accessed on 30 May 2023).
13. UN-SPIDER: Global Fatal Landslide Database (GFLD—University of Sheffield). United Nations Office for Outer Space Affairs UN-SPIDER Knowledge Portal. Available online: https://un-spider.org/links-and-resources/data-sources/global-fatal-landslide-database-gfld-university-sheffield (accessed on 30 May 2023).
14. Ojeda Moncayo, J.; Donnelly, L. Landslides in Colombia and Their Impact on Towns and Cities. *IAEG Pap.* **2006**, *112*, 1–13.
15. Servicio Geológico Colombiano: Sistema de Información de Movimientos en Masa—SIMMA. Available online: http://simma.sgc.gov.co/# (accessed on 30 May 2023).
16. CORPES. *Mapa de Amenazas Geológicas Por Remoción en Masa y Erosión del Departamento de Cundinamarca*; Ingeominas: Bogotá, Colombia, 1998.
17. Varnes, D.J. *Landslide Hazard Zonation: A Review of Principles and Practise*; UNDRR: Geneva, Switzerland, 1984.
18. Isaza-Restrepo, P.A.; Martínez Carvajal, H.E.; Hidalgo Montoya, C.A. Methodology for Quantitative Landslide Risk Analysis in Residential Projects. *Habitat Int.* **2016**, *53*, 403–412. [CrossRef]
19. Smith, H.; Coupé, F.; Garcia-Ferrari, S.; Rivera, H.; Castro Mera, W.E. Toward Negotiated Mitigation of Landslide Risks in Informal Settlements: Reflections from a Pilot Experience in Medellín, Colombia. *Ecol. Soc.* **2020**, *25*, art19. [CrossRef]
20. Ayala-García, J.; Dall'Erba, S. The Impact of Preemptive Investment on Natural Disasters. *Pap. Reg. Sci.* **2022**, *101*, 1087–1103. [CrossRef]
21. García-Delgado, H.; Machuca, S.; Medina, E. Dynamic and Geomorphic Characterizations of the Mocoa Debris Flow (March 31, 2017, Putumayo Department, Southern Colombia). *Landslides* **2019**, *16*, 597–609. [CrossRef]
22. Pertuz-Paz, A.; Monsalve, G.; Loaiza-Úsuga, J.C.; Caballero-Acosta, J.H.; Agudelo-Vélez, L.I.; Sidle, R.C. Linking Soil Hydrology and Creep: A Northern Andes Case. *Geosciences* **2020**, *10*, 472. [CrossRef]
23. Soeters, R.; Van Westen, C.J. Slope instability recognition, analysis and zonation. In *Landslides, Investigation and Mitigation*; Turner, A.K., Schuster, R.L., Eds.; Transportation Research Board, National Research Council, Special Report 247; National Academy Press: Washington, DC, USA, 1996; pp. 129–177.
24. Palacio Cordoba, J.; Mergili, M.; Aristizábal, E. Probabilistic Landslide Susceptibility Analysis in Tropical Mountainous Terrain Using the Physically Based r.Slope.Stability Model. *Nat. Hazards Earth Syst. Sci.* **2020**, *20*, 815–829. [CrossRef]
25. Hidalgo, C.A.; Vega, J.A. Probabilistic Landslide Risk Assessment in Water Supply Basins: La Liboriana River Basin (Salgar-Colombia). *Nat. Hazards* **2021**, *109*, 273–301. [CrossRef]
26. Marín, R.J.; Velásquez, M.F.; Sánchez, O. Applicability and Performance of Deterministic and Probabilistic Physically Based Landslide Modeling in a Data-Scarce Environment of the Colombian Andes. *J. S. Am. Earth Sci.* **2021**, *108*, 103175. [CrossRef]
27. Marín, R.J.; Mattos, Á.J.; Fernández-Escobar, C.J. Understanding the Sensitivity to the Soil Properties and Rainfall Conditions of Two Physically-Based Slope Stability Models. *Boletín Geol.* **2022**, *44*, 93–109. [CrossRef]
28. García-Delgado, H. The San Eduardo Landslide (Eastern Cordillera of Colombia): Reactivation of a Deep-Seated Gravitational Slope Deformation. *Landslides* **2020**, *17*, 1951–1964. [CrossRef]
29. Prada-Sarmiento, L.F.; Cabrera, M.A.; Camacho, R.; Estrada, N.; Ramos-Cañón, A.M. The Mocoa Event on March 31: Analysis of a Series of Mass Movements in a Tropical Environment of the Andean-Amazonian Piedmont. *Landslides* **2019**, *16*, 2459–2468. [CrossRef]
30. Cheng, D.; Cui, Y.; Su, F.; Jia, Y.; Choi, C.E. The Characteristics of the Mocoa Compound Disaster Event, Colombia. *Landslides* **2018**, *15*, 1223–1232. [CrossRef]
31. Vargas-Cuervo, G.; Rotigliano, E.; Conoscenti, C. Prediction of Debris-Avalanches and -Flows Triggered by a Tropical Storm by Using a Stochastic Approach: An Application to the Events Occurred in Mocoa (Colombia) on 1 April 2017. *Geomorphology* **2019**, *339*, 31–43. [CrossRef]
32. Braab, E.E. Innovative Approaches to Landslide Hazard and Risk Mapping. In Proceedings of the 4th International Symposium on Landslides, Toronto, ON, Canada, 16–21 September 1984; pp. 307–323.
33. Reichenbach, P.; Rossi, M.; Malamud, B.D.; Mihir, M.; Guzzetti, F. A Review of Statistically-Based Landslide Susceptibility Models. *Earth Sci. Rev.* **2018**, *180*, 60–91. [CrossRef]
34. Irigaray, C.; Fernández, T.; Chacón, J. Comparative Analysis of Methods for Landslide Susceptibility Mapping. In *Landslides, Proceedings of the 8th International Conference and Field Trip on Landslides, Granada, Spain, 27–28 September 1996*; CRC Press: Boca Raton, FL, USA; pp. 373–384.
35. Irigaray, C.; Fernández, T.; El Hamdouni, R.; Chacón, J. Verification of Landslide Susceptibility Mapping: A Case Study. *Earth Surf. Process. Landf.* **1999**, *24*, 537–544.
36. Fernández, T.; Irigaray, C.; El Hamdouni, R.; Chacón, J. Methodology for Landslide Susceptibility Mapping by Means of a GIS. Application to the Contraviesa Area (Granada, Spain). *Nat. Hazards* **2003**, *30*, 297–308. [CrossRef]

37. Chung, C.-J.F.; Fabbri, A.G. Probabilistic Prediction Models for Landslide Hazard Mapping. *Photogramm. Eng. Remote Sens.* **1999**, *65*, 1389–1399.
38. Guzzetti, F.; Reichenbach, P.; Cardinali, M.; Galli, M.; Ardizzone, F. Probabilistic Landslide Hazard Assessment at the Basin Scale. *Geomorphology* **2005**, *72*, 272–299. [CrossRef]
39. Baeza, C.; Corominas, J. Assessment of Shallow Landslide Susceptibility by Means of Multivariate Statistical Techniques. *Earth Surf. Process. Landf.* **2001**, *26*, 1251–1263. [CrossRef]
40. Merghadi, A.; Yunus, A.P.; Dou, J.; Whiteley, J.; ThaiPham, B.; Bui, D.T.; Avtar, R.; Abderrahmane, B. Machine Learning Methods for Landslide Susceptibility Studies: A Comparative Overview of Algorithm Performance. *Earth Sci. Rev.* **2020**, *207*, 103225. [CrossRef]
41. Pradhan, B.; Lee, S. Regional Landslide Susceptibility Analysis Using Back-Propagation Neural Network Model at Cameron Highland, Malaysia. *Landslides* **2010**, *7*, 13–30. [CrossRef]
42. Bui, D.T.; Tuan, T.A.; Klempe, H.; Pradhan, B.; Revhaug, I. Spatial Prediction Models for Shallow Landslide Hazards: A Comparative Assessment of the Efficacy of Support Vector Machines, Artificial Neural Networks, Kernel Logistic Regression, and Logistic Model Tree. *Landslides* **2016**, *13*, 361–378. [CrossRef]
43. Pourghasemi, H.R.; Rahmati, O. Prediction of the Landslide Susceptibility: Which Algorithm, Which Precision? *Catena* **2018**, *162*, 177–192. [CrossRef]
44. Salazar Gutiérrez, L.F.; Menjivar Flores, J.C.; Martínez Carvajal, H.E. Susceptibility Factors of Drainage Basins to Shallow Landslides in Coffee-Growing Areas in the Department of Caldas, Colombia. *Environ. Earth Sci.* **2021**, *80*, 145. [CrossRef]
45. Goyes-Peñafiel, P.; Hernandez-Rojas, A. Landslide Susceptibility Index Based on the Integration of Logistic Regression and Weights of Evidence: A Case Study in Popayan, Colombia. *Eng. Geol.* **2021**, *280*, 105958. [CrossRef]
46. Calderón-Guevara, W.; Sánchez-Silva, M.; Nitescu, B.; Villarraga, D.F. Comparative Review of Data-Driven Landslide Susceptibility Models: Case Study in the Eastern Andes Mountain Range of Colombia. *Nat. Hazards* **2022**, *113*, 1105–1132. [CrossRef]
47. Aristizábal, E.; López, S.; Sánchez, O.; Vásquez, M.; Rincón, F.; Ruiz-Vásquez, D.; Restrepo, S.; Valencia, J.S. Evaluación de La Amenaza Por Movimientos En Masa Detonados Por Lluvias Para Una Región de Los Andes Colombianos Estimando La Probabilidad Espacial, Temporal, y Magnitud. *Rev. Boletín Geol.* **2019**, *41*, 85–105. [CrossRef]
48. Aristizábal, E.; Morales-García, P.; Vásquez-Guarín, M.; Ruíz-Vásquez, D.; Palacio-Córdoba, J.; Ángel-Cárdenas, F.P.; Caballero-Acosta, H.; Ordóñez-Carmona, O. Metodologías Para La Evaluación de La Amenaza Por Movimientos En Masa Como Parte de Los Estudios Básico de Amenaza: Caso de Estudio Municipio de Andes, Antioquia, Colombia. *Boletín Geol.* **2022**, *43*, 199–217. [CrossRef]
49. Ruiz-Vásquez, D.; Aristizábal, E. Landslide Susceptibility Assessment in Mountainous and Tropical Scarce-Data Regions Using Remote Sensing Data: A Case Study in the Colombian Andes. In *EGU General Assembly*; EGU: Vienna, Austria; 8–13 April 2018; p. 3408.
50. Correa Muñoz, N.A.; Higidio Castro, J.F. Determination of Landslide Susceptibility in Linear Infrastructure. Case: Aqueduct Network in Palacé, Popayan (Colombia). *Ing. Investig.* **2017**, *37*, 17–24. [CrossRef]
51. Pradhan, A.M.S.; Lee, J.-M.; Kim, Y.-T. Semi-Quantitative Method to Identify the Vulnerable Areas in Terms of Building Aggregation for Probable Landslide Runout at the Regional Scale: A Case Study from Soacha Province, Colombia. *Bull. Eng. Geol. Environ.* **2019**, *78*, 5745–5762. [CrossRef]
52. Valencia Ortiz, J.A.; Martínez-Graña, A.M. A Neural Network Model Applied to Landslide Susceptibility Analysis (Capitanejo, Colombia). *Geomat. Nat. Hazards Risk* **2018**, *9*, 1106–1128. [CrossRef]
53. Renza, D.; Cárdenas, E.A.; Martinez, E.; Weber, S.S. CNN-Based Model for Landslide Susceptibility Assessment from Multispectral Data. *Appl. Sci.* **2022**, *12*, 8483. [CrossRef]
54. Cullen, C.A.; Al Suhili, R.; Aristizabal, E. A Landslide Numerical Factor Derived from CHIRPS for Shallow Rainfall Triggered Landslides in Colombia. *Remote Sens.* **2022**, *14*, 2239. [CrossRef]
55. Aristizábal Giraldo, E.V.; García Aristizábal, E.; Marín Sánchez, R.; Gómez Cardona, F.; Guzmán Martínez, J.C. Rainfall-Intensity Effect on Landslide Hazard Assessment Due to Climate Change in North-Western Colombian Andes. *Rev. Fac. Ing. Univ. Antioquia* **2022**, *103*, 51–66. [CrossRef]
56. Marín, R.J.; García, E.F.; Aristizábal, E. Effect of Basin Morphometric Parameters on Physically-Based Rainfall Thresholds for Shallow Landslides. *Eng. Geol.* **2020**, *278*, 105855. [CrossRef]
57. Marín, R.J.; Velásquez, M.F.; García, E.F.; Alvioli, M.; Aristizábal, E. Assessing Two Methods of Defining Rainfall Intensity and Duration Thresholds for Shallow Landslides in Data-Scarce Catchments of the Colombian Andean Mountains. *Catena* **2021**, *206*, 105563. [CrossRef]
58. Gutiérrez Alvis, D.E.; Bornachera Zarate, L.S.; Mosquera Palacios, D.J. Sistema de Alerta Temprana Por Movimiento En Masa Inducido Por Lluvia Para Ciudad Bolívar (Colombia). *Ing. Solidar.* **2018**, *14*, 26. [CrossRef]
59. Ramos-Cañón, A.M.; Prada-Sarmiento, L.F.; Trujillo-Vela, M.G.; Macías, J.P.; Santos-R, A.C. Linear Discriminant Analysis to Describe the Relationship between Rainfall and Landslides in Bogotá, Colombia. *Landslides* **2016**, *13*, 671–681. [CrossRef]
60. Velásquez, N. Assessment of Deep Convective Systems in the Colombian Andean Region. *Hydrology* **2022**, *9*, 119. [CrossRef]
61. Ortega, L.C.; Cañón, J.E. Correlative Analysis of Climate Impacts in an Andean Municipality of Colombia. *Rev. Cienc. Agrícolas* **2022**, *39*, 143–159. [CrossRef]

62. Parravano, V.; Teixell, A.; Mora, A. Influence of Salt in the Tectonic Development of the Frontal Thrust Belt of the Eastern Cordillera (Guatiquía Area, Colombian Andes). *Interpretation* **2015**, *3*, SAA17–SAA27. [CrossRef]
63. Chicangana, G.; Kammer, A. Evolución Tectónica de La Cordillera Oriental de Colombia. Desde La Apertura Del Océano Iapeto Hasta la Conformación de la Pangea: Una Visión Preliminar. Primera Parte: Aspectos Geológicos. *Geol. Colomb.* **2013**, *38*, 64–74.
64. Servicio Geológico Colombiano: Mapa geológico de Colombia. 2019. Available online: https://www2.sgc.gov.co/MGC/Paginas/mgc2M2019.aspx (accessed on 30 May 2023).
65. DANE: Censo Nacional de Población y Vivienda. 2018. Available online: https://www.dane.gov.co/index.php/estadisticas-por-tema/demografia-y-poblacion/censo-nacional-de-poblacion-y-vivenda-2018 (accessed on 30 May 2023).
66. Pulido, O.; Gómez, L.S. Geología Plancha 266 Villavicencio. In *Memoria Explicativa*; Ingeominas: Bogotá, Colombia, 2001.
67. JAXA-METI. ALOS Systematic Observation Strategy—PALSAR. Available online: https://www.eorc.jaxa.jp/ALOS/en/obs/palsar_strat.htm (accessed on 30 May 2023).
68. Alaska Satellite Facility. Available online: https://asf.alaska.edu/data-sets/sar-data-sets/alos-palsar/ (accessed on 30 May 2023).
69. Servicio Geológico Colombiano. Atlas Geológico de Colombia. 2015. Available online: https://www2.sgc.gov.co/MGC/Paginas/agc_500K2015.aspx (accessed on 30 May 2023).
70. Copernicus Open Access Hub. Available online: https://scihub.copernicus.eu/dhus/ (accessed on 30 May 2023).
71. Ideam. Atlas Climatológico de Colombia. Available online: http://atlas.ideam.gov.co/visorAtlasClimatologico.html (accessed on 30 May 2023).
72. Google Earth. Available online: https://www.google.es/intl/es/earth/index.html (accessed on 30 May 2023).
73. QGIS 3. A Free and Open Source Geographic Information System. 2023. Available online: https://www.qgis.org/en/site/ (accessed on 30 May 2023).
74. SAGA. System for Automated Geoscientific Analyses. Available online: https://saga-gis.sourceforge.io/en/index.html (accessed on 30 May 2023).
75. Rstudio. Available online: https://www.rstudio.com/categories/rstudio-ide/ (accessed on 30 May 2023).
76. Chacón, J.; Irigaray, C.; Fernández, T.; El Hamdouni, R. Engineering Geology Maps: Landslides and Geographical Information Systems. *Bull. Eng. Geol. Environ.* **2006**, *65*, 341–411. [CrossRef]
77. Guzzetti, F.; Mondini, A.C.; Cardinali, M.; Fiorucci, F.; Santangelo, M.; Chang, K. Landslide Inventory Maps: New Tools for an Old Problem. *Earth Sci. Rev.* **2012**, *112*, 42–66. [CrossRef]
78. Varnes, D.J. Slope Movements Types and Processes. In *Landslides Analysis and Control*; Schuster, R.L., Krizek, R.J., Eds.; National Academy of Sciences: Washington, DC, USA, 1978; pp. 11–33.
79. WP/WLI; Cruden, D.M. A Suggested Method for Describing the Activity of a Landslide. *Bull. Int. Assoc. Eng. Geol.* **1993**, *47*, 53–57.
80. Bravo-López, E.; Fernández Del Castillo, T.; Sellers, C.; Delgado-García, J. Analysis of Conditioning Factors in Cuenca, Ecuador, for Landslide Susceptibility Maps Generation Employing Machine Learning Methods. *Land* **2023**, *12*, 1135. [CrossRef]
81. Rouse, J.; Haas, R.; Schell, J.; Deering, D. Third Earth Resources Technology Satellite. In *Monitoring Vegetation Systems in the Great Plains with ERTS*; Technical Presentations; NASA: Washington, DC, USA, 1974; Volume I, pp. 309–317.
82. Pacheco-Quevedo, R.; Velastegui-Montoya, A.; Montalván-Burbano, N.; Morante-Carballo, F.; Korup, O.; Rennó, C.D. Land use and land cover as a conditioning factor in landslide susceptibility: A literature review. *Landslides* **2023**, *20*, 967–982. [CrossRef]
83. DeGraff, J.V.; Romesburg, H.C. Regional Landslide Susceptibility Assessment for Wildland Management: A Matrix Approach. In *Thresholds in Geomorphology*; Coates, D.R., Vitek, J.D., Eds.; Routledge: Boston, MA, USA, 1980; pp. 401–414.
84. Irigaray Fernández, C.; Fernández del Castillo, T.; El Hamdouni, R.; Chacón Montero, J. Evaluation and Validation of Landslide-Susceptibility Maps Obtained by a GIS Matrix Method: Examples from the Betic Cordillera (Southern Spain). *Nat. Hazards* **2007**, *41*, 61–79. [CrossRef]
85. Li, L.; Lan, H.; Wu, Y. How Sample Size Can Effect Landslide Size Distribution. *Geoenviron. Disasters* **2016**, *3*, 18. [CrossRef]
86. Shao, X.; Ma, S.; Xu, C.; Zhou, Q. Effects of Sampling Intensity and Non-Slide/Slide Sample Ratio on the Occurrence Probability of Coseismic Landslides. *Geomorphology* **2020**, *363*, 107222. [CrossRef]
87. Huang, F.; Cao, Z.; Jiang, S.-H.; Zhou, C.; Huang, J.; Guo, Z. Landslide Susceptibility Prediction Based on a Semi-Supervised Multiple-Layer Perceptron Model. *Landslides* **2020**, *17*, 2919–2930. [CrossRef]
88. Sameen, M.I.; Pradhan, B.; Bui, D.T.; Alamri, A.M. Systematic Sample Subdividing Strategy for Training Landslide Susceptibility Models. *Catena* **2020**, *187*, 104358. [CrossRef]
89. Dornik, A.; Drăguț, L.; Oguchi, T.; Hayakawa, Y.; Micu, M. Influence of Sampling Design on Landslide Susceptibility Modeling in Lithologically Heterogeneous Areas. *Sci. Rep.* **2022**, *12*, 2106. [CrossRef]
90. Carrara, A. Multivariate Models for Landslide Hazard Evaluation. *J. Int. Assoc. Math. Geol.* **1983**, *15*, 403–426. [CrossRef]
91. Breiman, L. Random Forests. *Mach. Learn.* **2001**, *45*, 5–32. [CrossRef]
92. Dou, J.; Yunus, A.P.; Tien Bui, D.; Merghadi, A.; Sahana, M.; Zhu, Z.; Chen, C.-W.; Khosravi, K.; Yang, Y.; Pham, B.T. Assessment of Advanced Random Forest and Decision Tree Algorithms for Modeling Rainfall-Induced Landslide Susceptibility in the Izu-Oshima Volcanic Island, Japan. *Sci. Total Environ.* **2019**, *662*, 332–346. [CrossRef] [PubMed]
93. Nefeslioglu, H.A.; Sezer, E.; Gokceoglu, C.; Bozkir, A.S.; Duman, T.Y. Assessment of Landslide Susceptibility by Decision Trees in the Metropolitan Area of Istanbul, Turkey. *Math. Probl. Eng.* **2010**, *2010*, 901095. [CrossRef]

94. Miner, A.; Vamplew, P.; Windle, D.J.; Flentje, P.; Warner, P. A Comparative Study of Various Data Mining Techniques as Applied to the Modeling of Landslide Susceptibility on the Bellarine Peninsula, Victoria, Australia. In *Geologically Active, Proceedings of the 11th IAEG Congress of the International Association of Engineering Geology and the Environment, Auckland, New Zealand, 5–10 September 2010*; CRC Press: Boca Raton, FL, USA, 2010.
95. Bravo-López, E.; Fernández Del Castillo, T.; Sellers, C.; Delgado-García, J. Landslide Susceptibility Mapping of Landslides with Artificial Neural Networks: Multi-Approach Analysis of Backpropagation Algorithm Applying the Neuralnet Package in Cuenca, Ecuador. *Remote Sens.* **2022**, *14*, 3495. [CrossRef]
96. Ciaburro, G.; Venkateswaran, B. *Neural Network with R: Smart Models Using CNN, RNN, Deep Learning, and Artificial Intelligence Principles*; Packt Publishing Ltd.: Birmingham, UK, 2017; Volume 91.
97. Günther, F.; Fritsch, S. Neuralnet: Training of Neural Networks. *R J.* **2010**, *2*, 30–38. [CrossRef]
98. Zare, M.; Pourghasemi, H.R.; Vafakhah, M.; Pradhan, B. Landslide Susceptibility Mapping at Vaz Watershed (Iran) Using an Artificial Neural Network Model: A Comparison between Multilayer Perceptron (MLP) and Radial Basic Function (RBF) Algorithms. *Arab. J. Geosci.* **2013**, *6*, 2873–2888. [CrossRef]
99. Pham, B.T.; Tien Bui, D.; Prakash, I.; Dholakia, M.B. Hybrid Integration of Multilayer Perceptron Neural Networks and Machine Learning Ensembles for Landslide Susceptibility Assessment at Himalayan Area (India) Using GIS. *Catena* **2017**, *149*, 52–63. [CrossRef]
100. Gholamy, A.; Kreinovich, V.; Kosheleva, O. Why 70/30 or 80/20 Relation between Training and Testing Sets: A Pedagogical Explanation. *Dep. Tech. Rep.* **2018**, *1209*, 1–6.
101. Vu, H.L.; Ng, K.T.W.; Richter, A.; An, C. Analysis of input set characteristics and variances on k-fold cross validation for a Recurrent Neural Network model on waste disposal rate estimation. *J. Environ. Manag.* **2022**, *311*, 114869. [CrossRef]
102. Zou, K.; O'Malley, A.; Mauri, L. Receiver-Operating Characteristic Analysis for Evaluating Diagnostic Tests and Predictive Models. *Circulation* **2007**, *115*, 654–657. [CrossRef]
103. Yilmaz, I. A Case Study from Koyulhisar (Sivas-Turkey) for Landslide Susceptibility Mapping by Artificial Neural Networks. *Bull. Eng. Geol. Environ.* **2009**, *68*, 297–306. [CrossRef]
104. Park, S.; Choi, C.; Kim, B.; Kim, J. Landslide Susceptibility Mapping Using Frequency Ratio, Analytic Hierarchy Process, Logistic Regression, and Artificial Neural Network Methods at the Inje Area, Korea. *Environ. Earth Sci.* **2013**, *68*, 1443–1464. [CrossRef]
105. Flórez-García, A.C.; Pérez Castillo, J.N. Técnicas Para La Predicción Espacial de Zonas Susceptibles a Deslizamientos. *Av. Investig. Ing.* **2019**, *16*, 20–48. [CrossRef]
106. Yi, Y.; Zhang, W.; Xu, X.; Zhang, Z.; Wu, X. Evaluation of Neural Network Models for Landslide Susceptibility Assessment. *Int. J. Digit. Earth* **2022**, *15*, 934–953. [CrossRef]
107. Aslam, B.; Zafar, A.; Khalil, U. Comparative Analysis of Multiple Conventional Neural Networks for Landslide Susceptibility Mapping. *Nat. Hazards* **2023**, *115*, 673–707. [CrossRef]
108. Chung, C.J.F.; Fabbri, A.G. Validation of Spatial Prediction Models for Landslide Hazard Mapping. *Nat. Hazards* **2003**, *30*, 451–472. [CrossRef]
109. Conforti, M.; Pascale, S.; Robustelli, G.; Sdao, F. Evaluation of Prediction Capability of the Artificial Neural Networks for Mapping Landslide Susceptibility in the Turbolo River Catchment (Northern Calabria, Italy). *Catena* **2014**, *113*, 236–250. [CrossRef]
110. Hungr, O.; Leroueil, S.; Picarelli, L. The Varnes Classification of Landslide Types, an Update. *Landslides* **2014**, *11*, 167–194. [CrossRef]
111. Fernández, T.; Pérez García, J.L.; Gómez López, J.M.; Cardenal, F.J.; Moya-Giménez, F.; Delgado, J. Multitemporal Landslide Inventory and Activity Analysis by Means of Aerial Photogrammetry and LiDAR Techniques in an Area of Southern Spain. *Remote Sens.* **2021**, *13*, 2110. [CrossRef]
112. Gariano, S.L.; Guzzetti, F. Landslides in a Changing Climate. *Earth Sci. Rev.* **2016**, *162*, 227–252. [CrossRef]
113. Correa, O.; García, F.; Bernal, G.; Cardona, O.D.; Rodriguez, C. Early Warning System for Rainfall-Triggered Landslides Based on Real-Time Probabilistic Hazard Assessment. *Nat. Hazards* **2020**, *100*, 345–361. [CrossRef]
114. García-Delgado, H.; Contreras, N.M. Historical Distribution for Landslides Triggered by Earthquakes in the Colombian Region. In Proceedings of the XIII International Symposium on Landslides, Cartagena, Colombia, 15–19 June 2020.
115. Bermúdez, M.A.; Velandia, F.; García-Delgado, H.; Jiménez, D.; Bernet, M. Exhumation of the Southern Transpressive Bucaramanga Fault, Eastern Cordillera of Colombia: Insights from Detrital, Quantitative Thermochronology and Geomorphology. *J. S. Am. Earth Sci.* **2021**, *106*, 103057. [CrossRef]
116. Grima, N.; Edwards, D.; Edwards, F.; Petley, D.; Fisher, B. Landslides in the Andes: Forests Can Provide Cost-Effective Landslide Regulation Services. *Sci. Total Environ.* **2020**, *745*, 141128. [CrossRef] [PubMed]
117. Vorpahl, P.; Elsenbeer, H.; Märker, M.; Schröder, B. How Can Statistical Models Help to Determine Driving Factors of Landslides? *Ecol. Modell.* **2012**, *239*, 27–39. [CrossRef]
118. Zhu, A.-X.; Miao, Y.; Yang, L.; Bai, S.; Liu, J.; Hong, H. Comparison of the Presence-Only Method and Presence-Absence Method in Landslide Susceptibility Mapping. *Catena* **2018**, *171*, 222–233. [CrossRef]
119. Costanzo, D.; Rotigliano, E.; Irigaray, C.; Jiménez-Perálvarez, J.D.; Chacón, J. Factors Selection in Landslide Susceptibility Modelling on Large Scale Following the Gis Matrix Method: Application to the River Beiro Basin (Spain). *Nat. Hazards Earth Syst. Sci.* **2012**, *12*, 327–340. [CrossRef]

120. Meena, S.R.; Puliero, S.; Bhuyan, K.; Floris, M.; Catani, F. Assessing the Importance of Conditioning Factor Selection in Landslide Susceptibility for the Province of Belluno (Region of Veneto, Northeastern Italy). *Nat. Hazards Earth Syst. Sci.* **2022**, *22*, 1395–1417. [CrossRef]
121. Liu, L.L.; Yang, C.; Wang, X.M. Landslide Susceptibility Assessment Using Feature Selection-Based Machine Learning Models. *Geomech. Eng.* **2021**, *25*, 1–16.
122. Chacón Montero, J.; Irigaray Fernández, C.; Fernández del Castillo, T. Large to Middle Scale Landslides Inventory, Analysis and Mapping with Modelling and Assessment of Derived Susceptibility, Hazards and Risks in a GIS. In *International Congress International Association of Engineering Geology*; A.A. Balkema: Rotterdam, The Netherlands, 1994; pp. 4669–4678.
123. van Westen, C.; van Asch, T.; Soeters, R. Landslide hazard and risk zonation—Why is it still so difficult? *Bull. Eng. Geol. Environ.* **2006**, *65*, 167–184. [CrossRef]
124. Huabin, W.; Gangjun, L.; Weiya, X.; Gonghui, W. GIS-based landslide hazard assessment: An overview. *Prog. Phys. Geogr. Earth Environ.* **2005**, *29*, 548–567. [CrossRef]
125. Korup, O.; Stolle, A. Landslide Prediction from Machine Learning. *Geol. Today* **2014**, *30*, 26–33. [CrossRef]
126. Sahin, E.K. Assessing the Predictive Capability of Ensemble Tree Methods for Landslide Susceptibility Mapping Using XGBoost, Gradient Boosting Machine, and Random Forest. *SN Appl. Sci.* **2020**, *2*, 1308. [CrossRef]
127. Deng, H.; Wu, X.; Zhang, W.; Liu, Y.; Li, W.; Li, X.; Zhou, P.; Zhuo, W. Slope-Unit Scale Landslide Susceptibility Mapping Based on the Random Forest Model in Deep Valley Areas. *Remote Sens.* **2022**, *14*, 4245. [CrossRef]
128. Wei, A.; Yu, K.; Dai, F.; Gu, F.; Zhang, W.; Liu, Y. Application of Tree-Based Ensemble Models to Landslide Susceptibility Mapping: A Comparative Study. *Sustainability* **2022**, *14*, 6330. [CrossRef]
129. Bilbao, I.; Bilbao, J. Overfitting problem and the over-training in the era of data: Particularly for Artificial Neural Networks. In Proceedings of the Eighth International Conference on Intelligent Computing and Information Systems (ICICIS), Cairo, Egypt, 5–7 December 2017; pp. 173–177. [CrossRef]
130. Lv, L.; Chen, T.; Dou, J.; Plaza, A. A hybrid ensemble-based deep-learning framework for landslide susceptibility mapping. *Int. J. Appl. Earth Obs. Geoinf.* **2022**, *108*, 102713. [CrossRef]

Disclaimer/Publisher's Note: The statements, opinions and data contained in all publications are solely those of the individual author(s) and contributor(s) and not of MDPI and/or the editor(s). MDPI and/or the editor(s) disclaim responsibility for any injury to people or property resulting from any ideas, methods, instructions or products referred to in the content.

Article

Deep Learning and Machine Learning Models for Landslide Susceptibility Mapping with Remote Sensing Data

Muhammad Afaq Hussain [1], Zhanlong Chen [1,*], Ying Zheng [2], Yulong Zhou [3] and Hamza Daud [4]

[1] School of Computer Science, China University of Geosciences, Wuhan 430074, China
[2] Ningbo Alatu Digital Science and Technology Corporation Limited, Ningbo 315000, China
[3] School of Geography and Information Engineering, China University of Geosciences, Wuhan 430074, China
[4] Badong National Observation and Research Station of Geohazards, China University of Geosciences, Wuhan 430074, China
* Correspondence: chenzl@cug.edu.cn

Abstract: Karakoram Highway (KKH) is an international route connecting South Asia with Central Asia and China that holds socio-economic and strategic significance. However, KKH has extreme geological conditions that make it prone and vulnerable to natural disasters, primarily landslides, posing a threat to its routine activities. In this context, the study provides an updated inventory of landslides in the area with precisely measured slope deformation (Vslope), utilizing the SBAS-InSAR (small baseline subset interferometric synthetic aperture radar) and PS-InSAR (persistent scatterer interferometric synthetic aperture radar) technology. By processing Sentinel-1 data from June 2021 to June 2023, utilizing the InSAR technique, a total of 571 landslides were identified and classified based on government reports and field investigations. A total of 24 new prospective landslides were identified, and some existing landslides were redefined. This updated landslide inventory was then utilized to create a landslide susceptibility model, which investigated the link between landslide occurrences and the causal variables. Deep learning (DL) and machine learning (ML) models, including convolutional neural networks (CNN 2D), recurrent neural networks (RNNs), random forest (RF), and extreme gradient boosting (XGBoost), are employed. The inventory was split into 70% for training and 30% for testing the models, and fifteen landslide causative factors were used for the susceptibility mapping. To compare the accuracy of the models, the area under the curve (AUC) of the receiver operating characteristic (ROC) was used. The CNN 2D technique demonstrated superior performance in creating the landslide susceptibility map (LSM) for KKH. The enhanced LSM provides a prospective modeling approach for hazard prevention and serves as a conceptual reference for routine management of the KKH for risk assessment and mitigation.

Keywords: convolutional neural network; recurrent neural networks; landslide susceptibility mapping; extreme gradient boosting; random forest

Citation: Hussain, M.A.; Chen, Z.; Zheng, Y.; Zhou, Y.; Daud, H. Deep Learning and Machine Learning Models for Landslide Susceptibility Mapping with Remote Sensing Data. *Remote Sens.* **2023**, *15*, 4703. https://doi.org/10.3390/rs15194703

Academic Editor: Rachid El Hamdouni

Received: 30 August 2023
Revised: 23 September 2023
Accepted: 23 September 2023
Published: 26 September 2023

Copyright: © 2023 by the authors. Licensee MDPI, Basel, Switzerland. This article is an open access article distributed under the terms and conditions of the Creative Commons Attribution (CC BY) license (https://creativecommons.org/licenses/by/4.0/).

1. Introduction

Landslides, one of the most common natural disasters, prevalent in mountainous regions worldwide, pose significant threats to the ecosystem [1]. Landslides are accounted as the downhill movement of debris, soil, and rocks under the force of gravity and can be classified based on the materials involved (mud, rock, soil, or debris) and their movement type (topple flow or slide) [2]. The factors leading to landslides are a combination of tectonics, geomorphology, and climate change, which culminate in a critical slope evolution [3,4]. Other triggering factors contribute to landslides depending on the specific features of the area. Natural variables such as rainfall, rapid snowmelt, earthquakes, and anthropogenic activities, e.g., habitation construction, irrigation, etc., can play a role in the occurrence of landslides [5]. While landslides are often regarded as a natural process, their occurrence mostly has been influenced by anthropogenic activity [6]. In recent years, exponentially growing populations, a surge in infrastructure development, and settlement growth in developing countries' mountainous regions have increased the probability of landslides, leading to an alarming increase in landslide-related fatalities [7].

The "China-Pakistan Economic Corridor" (CPEC), a significant project under the "One Belt and One Road" initiative, is centered around connecting Pakistan and China via the Karakoram Highway (KKH). The KKH was constructed from 1974 to 1978 and commenced operation in 1979. The highway encompasses most of the route of the CPEC. However, this vital route faces challenges due to the high mountainous terrain with overflowing loose debris and heavy rainfall, triggering frequent and severe geological catastrophes such as glacier debris flows, rock falls, landslides, debris and soil slippage, and avalanches [8]. Determining landslide probabilities along the KKH is a complex process influenced by limited data availability, technical limitations, and harsh environments. Since its completion, the reputation of the KKH has been marred by various geohazards [9]. Specifically, earth-induced landslides in 2005 caused considerable damage to the highway [10]. Enormous rockslides and rock avalanches have occurred, with over 115 incidents reported since 1987 [11]. Moreover, in 2010, a landslide blocked the Hunza River, inundating 19 km of the highway with a loss of 20 lives and damaging 350 houses [12]. The geological conditions along the KKH pose additional challenges, including fragile and weathered rock masses, varying climates, low and high terrains, diverse stratigraphy, and local variations in tectonic motion. Due to these factors, the study region has become a geohazard laboratory. Enhancing precise LSM along the KKH to mitigate the risks posed by these natural hazards is imperative.

Recently, remote sensing (RS) and geographic information systems (GIS) technology have made remarkable technological progress. The utilization of GIS spatial analysis tools and remote-sensing-derived data has enhanced the effectiveness of landslide susceptibility mapping for accurate assessment. Here, comprehensive landslide inventory data and knowledge of landslide conditioning factors are crucial for both data-driven spatial modeling and knowledge-based approaches [13]. Researchers have conducted numerous studies using bivariate analyses to quantify the spatial correlations between landslides and specific factors that influence their dispersion [14–17]. Several other studies have applied knowledge-based spatial approaches to produce natural risk vulnerability maps, fuzzy logic models [18,19], the analytical hierarchy process (AHP) [20], and the evidential belief function [21], as well as data-driven spatial approaches such as support vector machines [22–24], logistic regression methods [24,25], artificial neural network (ANN) models [26–28], alternating decision tree (ADTree) [29], principal component analysis (PCA) [30], deep belief network (DBN) [31], decision tree [25,32], superposable neural networks [33], and naïve Bayes [34]. Expertise-based models often encounter challenges due to their reliance on expert opinions, which can introduce biases [35,36].

The primary strengths of probabilistic and ML approaches lie in their objective statistical foundation, consistency, capacity for precisely analyzing the factors influencing landslide development, and capacity building for updates. In this perspective, researchers are continuously seeking new and relatively more robust algorithms that can generalize across different spatial scales [37,38]. Deep learning algorithms, which are specifically developed for large datasets but have seen limited application thus far, need to be implemented and evaluated in this context. Currently, deep learning models, particularly recurrent neural networks and convolutional neural networks, have demonstrated remarkable success across various applications, making them well suited for handling big data [39]. RNNs, like other DL models, comprise a loss function, learnable parameters, and layers [40]. On the other hand, CNNs differ from RNNs as they include convolutional and pooling layers and focus solely on the current input data, while RNNs consider both the earlier provided inputs and present input data [41]. CNNs have proven effective in tasks like semantic segmentation and object detection [7]. Conversely, RNNs show superior performance in tasks such as image recognition, characterization, and sequential data analysis, including time series spatial data [42]. Despite the acceptable results achieved by CNNs and RNNs in various domains, their true efficiency and capabilities in landslide modeling and large-scale landslide susceptibility mapping (LSM) on big data have not been thoroughly analyzed [13]. A few deep learning models have been utilized for natural hazard vulnerability mapping, containing landslide susceptibility mapping and flash floods [43–45]. However, these studies have separately employed different deep learning models, and their relative proficiency has not been evaluated yet.

In recent years, interferometric synthetic aperture radar (InSAR) methods have acquired universal approval and usage as tools for landslide monitoring and mapping. Over the past two decades, the RS technique, particularly In-SAR, has demonstrated substantial possibility across different fields, including the study of landslide deformation [46] and groundwater extraction [47]. PS-InSAR proves useful in automatic slow-moving landslide mapping using a spatial statistical technique, the detection of particular landslides and the delineation of extended unstable regions, redefining of the limits of historical landslides, the detection of landslides using a multitemporal analysis of SAR imagery, and the verification of the terrain elements causing slope deformation [48]. In areas prone to frequent and rapid large landslides, RS provides a solution through surveys and advanced detection methods [49]. These techniques can greatly aid in assessing and creating landslide inventory maps. Various methods of InSAR have been effectively used in mapping slope displacement, including that in [50], the assessment of land displacement places identified by using SBAS-InSAR [51], the D-InSAR technique for landslide observing and land deformation [51,52], the coherence pixel technique [53], the SqeeInSAR approach to measuring surface motion [51], interferometric point target analysis [54], the use of StaMPS to evaluate the displacement in a high-vegetation region [55,56], and the PSInSAR method to compute the movement of landslides. These approaches are related to detecting and mapping landslide events, as mentioned in [54,57,58].

In this study, a combination of optical RS analysis and the InSAR technique is utilized to identify landslides and create an updated landslide inventory. The main goals are as follows: (1) mapping all types of landslides along the KKH and estimating displacement maps to identify new landslides, identify unstable places, and redefine the boundaries of previously identified landslides based on the deformation model; (2) generating a landslide susceptibility map using state-of-the-art ML and deep learning (DL) models, including random forest, XGBoost, recurrent neural networks, and convolutional neural networks; (3) comparing the performance of these advanced ML and DL models in terms of landslide susceptibility; (4) assessing the significance and relationships of environmental and anthropogenic factors influencing landslides and their role in evaluating landslide susceptibility in the study area; and (5) determining the most accurate susceptible model reliant on precision and AUC value. Despite the fact that the KKH faces significant landslide threats every summer, previous research has not adequately addressed the issue. Therefore, the landslide susceptibility map produced in this study will aid urban planning and disaster reduction efforts in the area. Moreover, the final InSAR-based landslide inventory will assist in tracking risky areas to minimize future hazards and fatalities. It is imperative to highlight that no previous studies have applied RNNs and CNNs for LSM at KKH. As the first study to utilize and compare these ML and DL models for LSM in this region, it will substantially contribute to the scientific literature.

2. Materials and Methods

2.1. Study Area and Geological Settings

The KKH in northern Pakistan is significant as part of the CPEC but is prone to frequent disruptions caused by various geological and hydro-climatological hazards. The study area was focused along a 263 km section of the KKH, passing through different districts of Gilgit Baltistan. A 5 km buffer around this section, covering an area of 3320 km^2 (Figure 1), was examined for the study. The terrain in the study area is rugged, with elevations ranging from 822 to 5545 m above mean sea level. The area experiences mild summers and harsh winters, and the yearly rainfall varies from 120 to 130 mm. The minimum and maximum temperatures are −21 °C and 16 °C, respectively, according to data from the Meteorological Department of Pakistan (https://www.pmd.gov.pk, accessed on 15 March 2023).

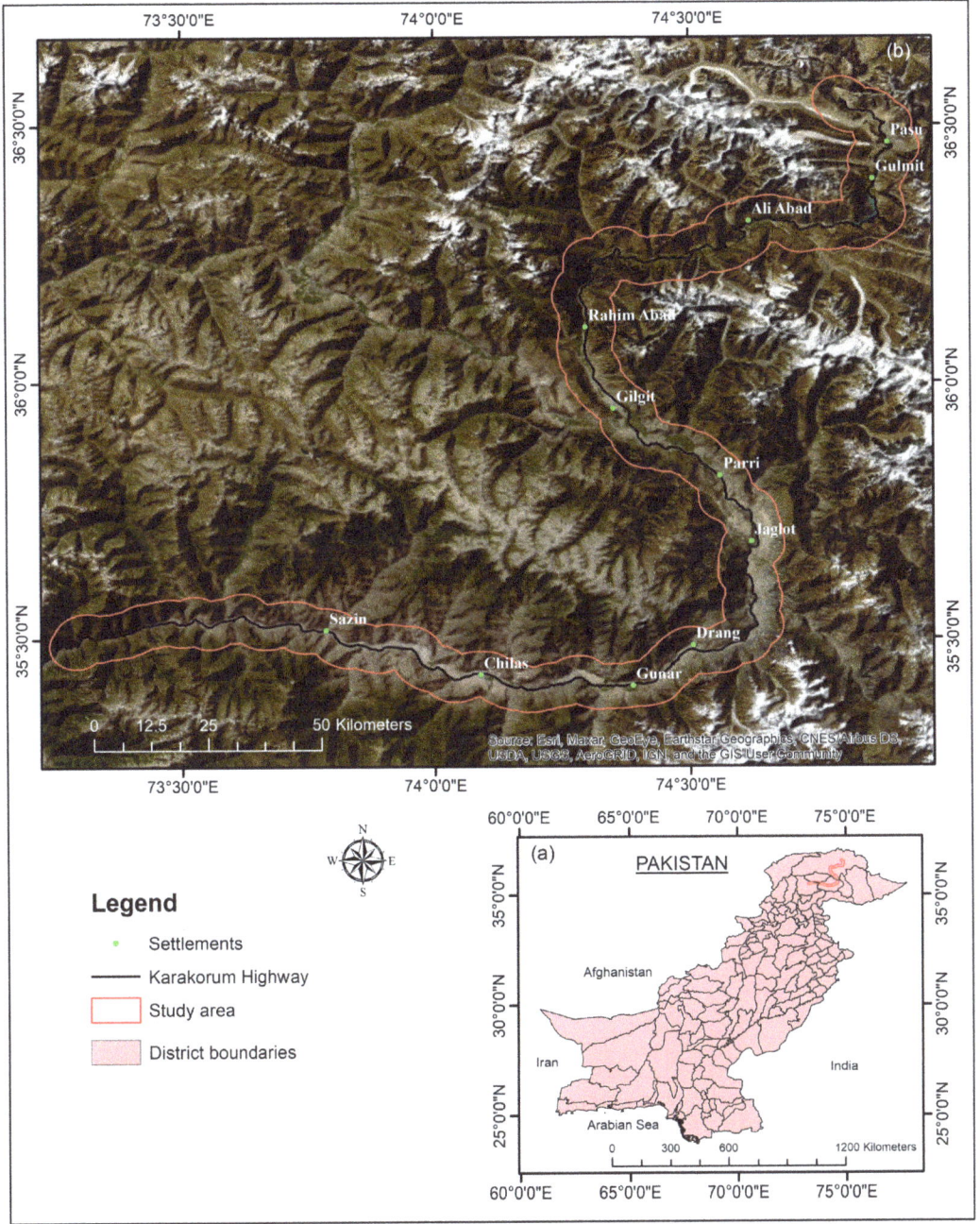

Figure 1. Location of the study area. (a) Pakistan, (b) Study area in red outline.

The lithology in the research area is significant in triggering landslides. The rocks of the area are primarily of Mesozoic and Paleozoic age. Based on the geological map produced by Searle et al. [59], the research area comprises a diverse range of rocks, including sedimentary, volcano-sedimentary, igneous, volcanic, and metamorphic rocks (Figure 2). These rocks are further stratified into various types, such as greenschist, siliciclastic, carbonates, basalt, andesite, granite, gabbro, and others. The Chalt schists, kilk formation, Quaternary sediments, deformed Misgar slates, and Gujhal dolomite are the most significant among the area's lithologic formations. All of these lithologies have been tectonically affected and have contributed to slope destabilization along the highway [9]. Over time, the lithologies exposed along the KKH have undergone weathering and weakening due to anthropogenic, hydro-climatic, and seismic events, resulting in significant landslides and surface distortion. Structurally, the area is sophisticated because it lies in a convergent boundary, specifically the Main Karakoram Thrust. This structural setting adds to the susceptibility of the area to geological hazards like landslides.

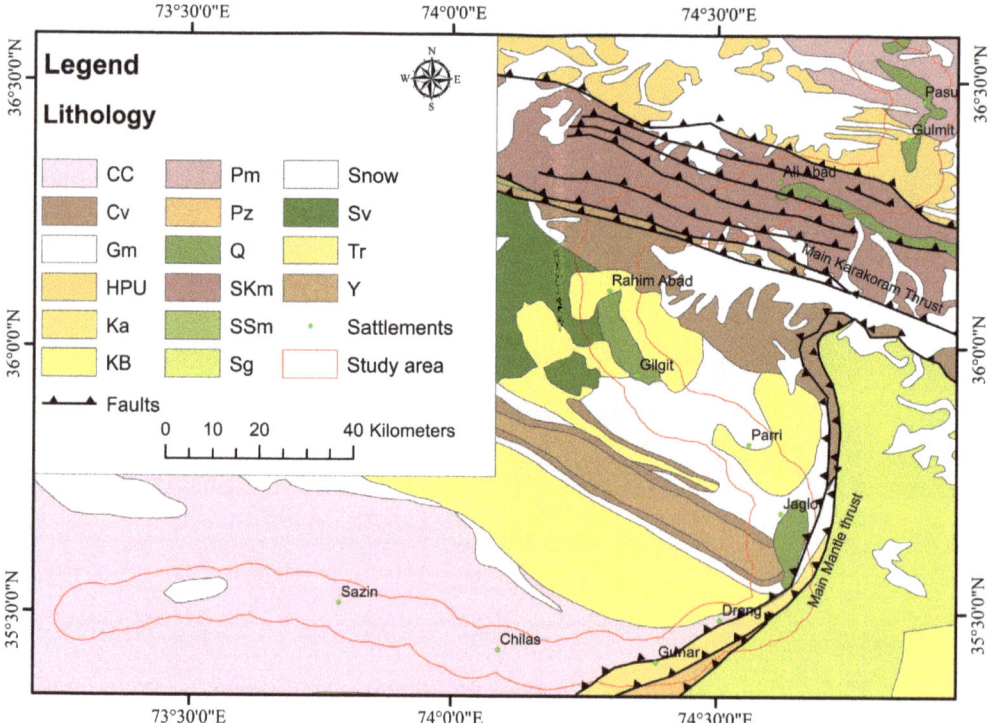

Figure 2. Regional geological map of the study region, which depicts the fault lines (MKT and MMT) and Geological units in the research area, where CC is Chilas Complex (Mafic and Ultra-mafic rocks), Gm is Gilgit complex metasedimentary rocks, Cv is Chalt group, Pm stands for Permian massive Limestone, HPU is a Hunza plutonic unit, Ka stands for Komila amphibolite Complex, Sg is Sumayar Leucogranite, KB is Kohistan Batholiths, SSm is Shyok Suture Melange, Q is Quaternary deposits, SKm stands for southern Karakoram complex, Pz is Palaeozoic Metasedimentary rocks, Sv is Kohistan Arc sequence, Tr is Triassic massive dolomite and limestone, and Y stands for Yasin group.

2.2. Datasets

The Alaska Satellite Facility (ASF) datasets provide an ALOS-PALSAR DEM (digital elevation model) with a resolution of 12.5 m, which was accessed from https://search.asf.alaska.edu/ (accessed on 10 February 2023). Additionally, Sentinel-2 images with a resolution of 10 m were derived from the Copernicus dataset (https://scihub.copernicus.eu, accessed on 10 February 2023) to produce a landcover map for the study area. Geological maps and fault lines for the research area were processed using the ArcGIS software 10.8 to understand the geological features [59,60]. To assess the relationship between rainfall and landslide events, annual precipitation data were obtained from the GIOVANNI online database system (https://giovanni.gsfc.nasa.gov/giovanni/, accessed on 15 March 2023), as rainfall and landslides are found to be directly proportionally related. For this study, two years (June 2021 to June 2023) of C-band Sentinel-1A SAR dataset imagery was obtained from the ASF (search.asf.alaska.edu) online system. The dataset contains scenes in descending and ascending tracks, as presented in Table 1. Figure 3 depicts the technical route of the research.

Table 1. Datasets used in PS-InSAR and SBAS-InSAR analysis.

Data Information	Ascending	Descending
Product type	Sentinel 1 SLC	
Acquisition mode	IW	
Polarization	VV	
Wavelength (m)	0.056	
No of images	63	60
Time period	June 2021–June 2023	
Frame	114	473
Track	100	107
Coverage (km^2)	250	
Incident angle	Horizontal (~45°) to vertical (~23°)	
Azimuth resolution and range (m)	5 × 20	

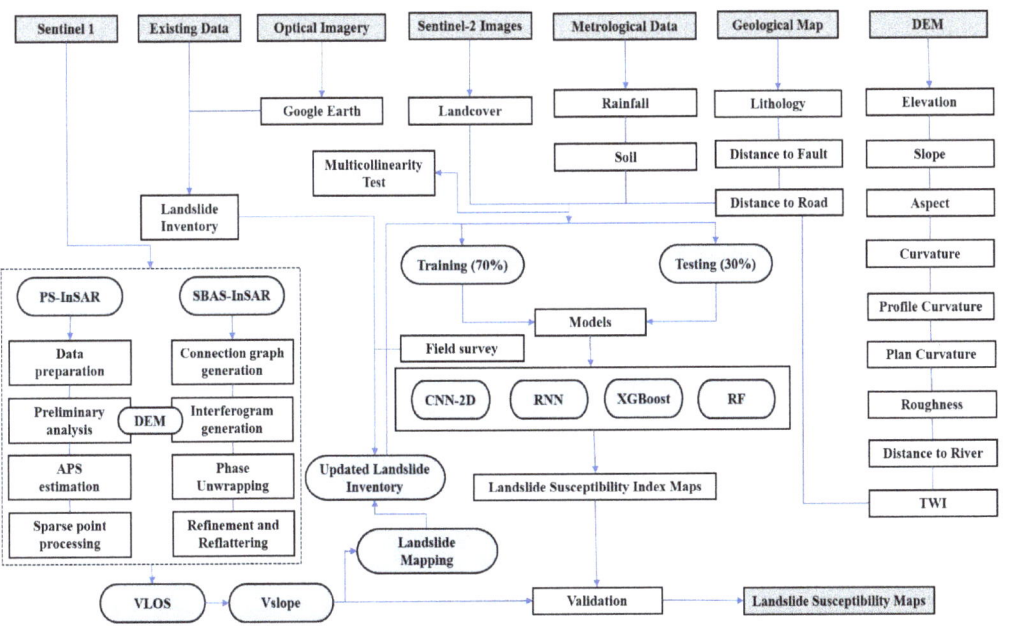

Figure 3. Technical route applied in the research.

2.3. Updated Landslide Inventory

Creating landslide inventory maps is a crucial step in LSM [61]. These maps provide essential information about the landscape's locations and types of landslides, serving as a foundation for predicting future landslides [1]. Landslide inventory maps are generated for a diversity of reasons, including identifying the type and location of landslides in a particular region; showing the impact of a single landslide-triggering incident, such as a rapid snowmelt incident, an intense rainfall event, or an earthquake; emphasizing the quantity of mass movements; calculating the frequency area statistics of slope failures; and providing relevant data to build landslide risk models or susceptibility models [62].

In this study, a total of 571 landslides were mapped using various techniques, including SBAS-InSAR and PS-InSAR, past studies [12,60,63], Frontier Works Organization road clearance logs, optical imagery analysis, Google Earth, and fieldwork in the study area. The landslide inventory, on the other hand, was developed through the visual interpretation of Sentinel-2 images with 10 m resolution (2022) and by using Google Earth, and it was checked using previous documents and a field evaluation of the study region. The inventory map was categorized into eight categories based on the material displacement, comprising 99 scree, 113 rockslides, 28 rock falls, 20 rock avalanches, 20 debris slides, 271 debris flows, 7 debris falls, and 13 complex slides (Figure 4).

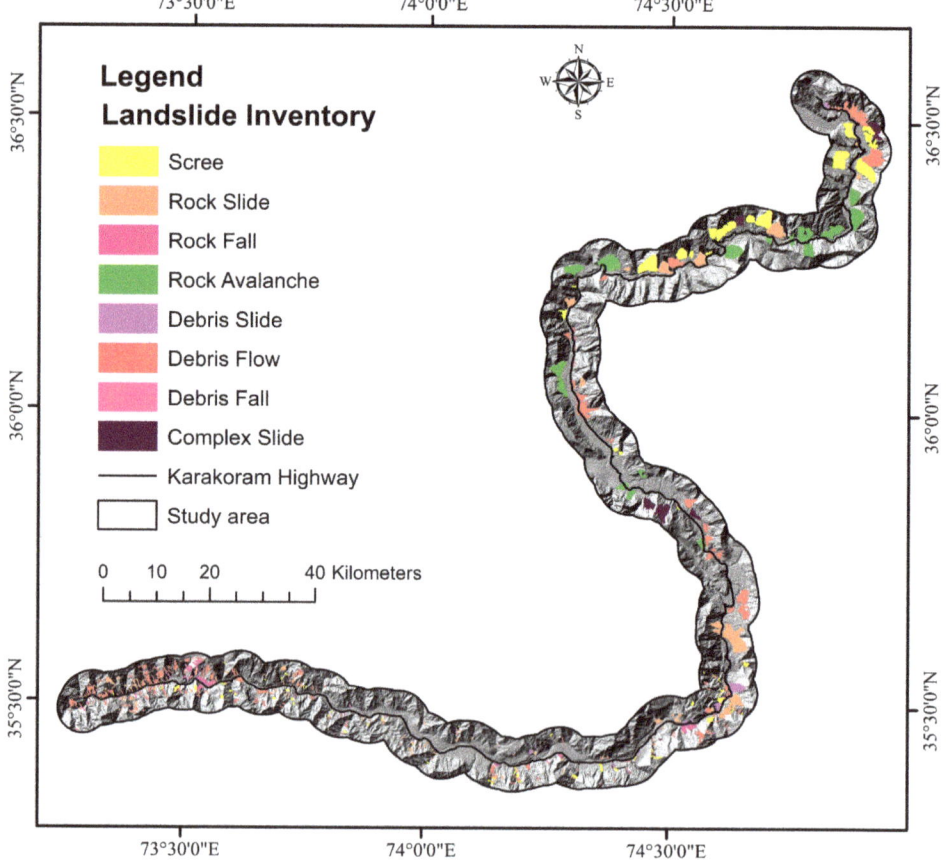

Figure 4. Landslide inventory categorization derived from movement along KKH (black line) with various colors.

The inventory map is applied to verify the identification of landslides through InSAR in the study area. Following the InSAR analysis, potential landslides are recognized based on their high displacement velocity, and these newly detected landslides are then incorporated into the updated version of the inventory map.

2.4. Landslide Conditioning Factors (LCFs)

The process of achieving high accuracy in the landslide susceptibility model and predicting vulnerable areas heavily relies on carefully selecting and preparing the Landslide Conditioning Factors (LCFs) database [64]. There are no universal standards for selecting independent variables for LSM [8,65,66]. The LCFs were chosen in this study based on information gathered from the relevant literature, data specific to the study area, and field investigations. Fifteen LCFs were selected for the current study, including slope, aspect, topographic wetness index (TWI), distance to roads, lithology, distance to rivers, roughness, distance to faults, curvature, precipitation, plan curvature, soil, profile curvature, elevation, and landcover (Table 2). Thematic layers with a spatial resolution of 12.5 × 12.5 m pixel size were prepared (Figures 5 and 6), all using the WGS84 Datum, UTM-Zone 43 coordinate system.

Table 2. List of landslide causative variables used in the research.

S.NO	Variables	Sources	Description/Extraction
1	Aspect, Elevation, Slope, Curvature, Plan Curvature, Profile Curvature, TWI, Distance to River, Roughness,	Digital Elevation Model	ALOS-PALSAR-DEM (https://search.asf.alaska.edu, accessed on 10 February 2023)
2	Lithology, Distance to Fault, Distance to road	Geological Map	Geological Survey of Pakistan
3	Landcover	Sentinel-2 images	Land use/Landcover (https://earthexplorer.usgs.gov/, accessed on 10 February 2023)
4	Soil	Soil map	Food and Agricultural Organization (FAO) website (http://www.fao.org/soils-portal/data-hub/soil-maps-and-databases/en/, accessed on 10 March 2023)
5	Precipitation	GIOVANNI	(https://giovanni.gsfc.nasa.gov/giovanni/, accessed on 15 March 2023)

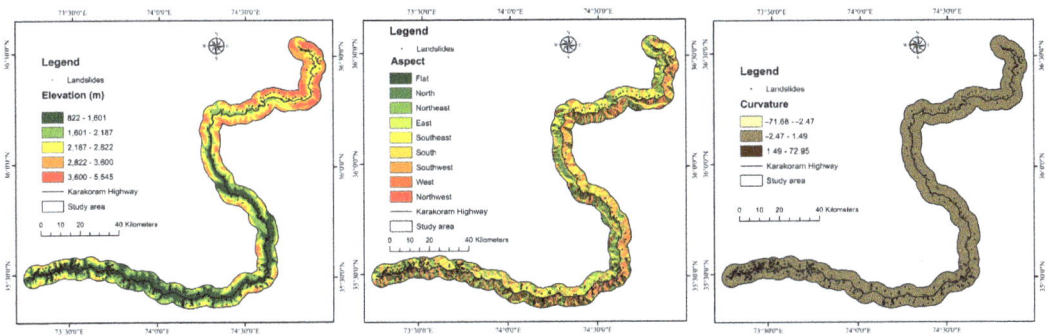

Figure 5. Cont.

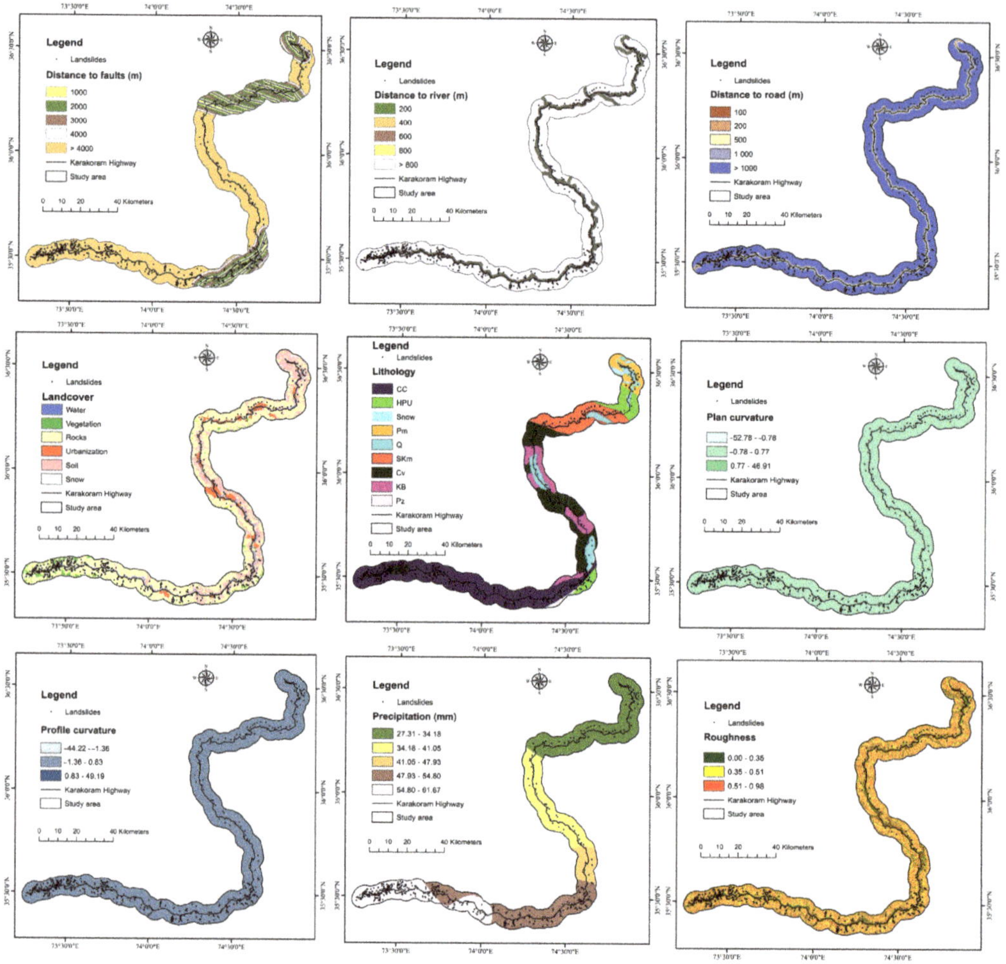

Figure 5. LCFs used in the study area.

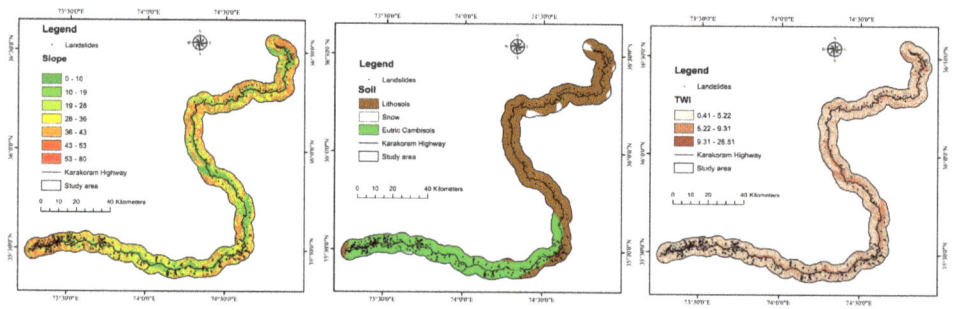

Figure 6. LCFs used in the study area.

2.5. Multicollinearity Assessment

In landslide susceptibility prediction, selecting the right variables is a critical step that significantly impacts the model's performance. To improve the accuracy of hazard mapping, it is essential to identify and choose appropriate variables for inclusion in the model [67]. One of the main challenges in variable selection is dealing with multicollinearity, which may arise due to the improper use of redundant or highly correlated factors [68]. To address the issue of multicollinearity, tolerance and variance inflation factors (VIF) are calculated. These measures help to identify closely and linearly related variables in a regression model, which could potentially lead to a reduction in the model's performance [69]. Through the standard equations for VIF and tolerance, a thorough evaluation of the degree of multicollinearity in the data is conducted. By identifying and removing variables with high levels of multicollinearity, the precision of landslide susceptibility models can be enhanced, and the most significant variables contributing to the prediction of landslide hazards can be identified more accurately.

$$\text{TOL} = 1 - R_j^2 \tag{1}$$

$$\text{VIF} = \frac{1}{1 - R_j^2} \tag{2}$$

where R_j^2 represents the regression value of j on various factors. Thus, multicollinearity problems usually happen if the TOL value is <0.10 and the variance inflation factors value is >10 [70].

2.6. Machine Learning Models

The modeling process included fitting, identifying, and developing an ML model.

The grid unit was used as the model unit, with a spatial resolution of 12.5 m for both the DEM and RS data, and all evaluation factors are recalculated at this level.

The model includes 15 conditioning factors and a landslide target variable (1 indicating landslide and 0 indicating non-landslide), with each row producing an object.

Each column illustrates an object's characteristic, and it is modified into a two-dimensional matrix for training (70%, 2138 samples) and testing (30%, 917 samples).

The models are constructed using training data, and predictions are made using test data. The landslide vulnerability index maps are produced by combining the prediction values for each model unit in each group. The results of the four models are transferred to a geographic information system. Landslide vulnerability is classified into five categories using Jenks natural breaks [71]: very low, low, moderate, high, and very high. The four models are tested using the ROC curve and the AUROC curve.

2.6.1. RF

Random forest (RF) is a well-known homogeneous bagging-based ensemble model developed by Breiman in 2001 [72]. It consists of multiple decision trees, making it an ensemble learning technique that aggregates the outputs of these trees to produce a classification [72–74]. The mechanism of the RF model can be outlined as follows:

I. Using the bootstrap approach, it creates numerous decision trees to randomly choose fresh sample sets from the initial training dataset with substitution.
II. At each resampling, a set of features is randomly selected, and the decision trees are built based on this subset of features.
III. The generated trees are combined into an RF, which is then used to categorize new data.

The RF method exhibits robustness against missing, unbalanced, and multicollinear data and is capable of handling high-dimensional data [75,76]. One of its primary advantages is its resistance to overfitting, even when a large number of random forest trees are grown. Additionally, there is no need to rescale, transform, or modify the data when applying the RF algorithm. In this research, the landslide susceptibility model was developed using the "randomForest" package in R 4.0.2 software [16]. After numerous attempts, the number of trees (ntree) was set to 500, and the mtry parameter was set to 6. To minimize the fluctuation of the model findings and to limit overfitting, RF was conducted using a

10-fold cross-validation technique. The hyperparameters used in the RF model are listed in Table 3.

Table 3. List of parameters used in random forest model.

Parameters	Values
Node size	14
mtry	06
ntree	500

2.6.2. XGBoost

The XGBoost supervised classification model is built on the gradient tree boosting algorithm [77,78], which is a powerful machine learning model designed by Chen and Guestrin in 2016 [79]. This model creates consecutive decision trees using the estimated residuals or errors from the preceding tree rather than integrating separate trees. This approach allows the algorithm to focus on samples with higher uncertainty, improving its performance. XGBoost offers several advantages, including scalability for various use cases with low computing resource essentials, fast processing speed, efficient handling of sparse data, and smooth integration [80]. The algorithm utilizes a loss function with an additional regularization term to smooth the final learned weights and prevent overfitting [79]. It also employs first- and second-order gradient statistics to optimize the loss function [81]. While XGBoost shares some parameters with other tree-based models, it involves additional hyperparameters to control the overfitting concern, enhance precision, and mitigate forecasting variance [82]. This study develops the landslide susceptibility model using the "XGBoost" package in R 4.0.2 software, which provides powerful capabilities for classification tasks [83]. In this research, three general parameters were chosen for us to alter in the XGBoost algorithm for LSM application: nrounds (the maximum number of boosting repetitions), subsample (the subsample ratio of the training instance), and colsample_bytree (the subsample ratio of columns while constructing each tree). The hyperparameters used in XGBoost model are listed in Table 4. The key points and usability of machine learning models are shown in Table 5.

Table 4. List of parameters used in extreme gradient boosting model.

Parameters	Values
nround	210
subsample	1
colsample_bytree	0.75
max_depth	6
gamma	0.01
eta	0.05

Table 5. The key points of RF and XGBoost models.

RF	XGBoost
Bagging ensemble method	Boosting ensemble method
Bagging-based algorithm where only a subset of features are selected at random to build a forest or collection of decision trees	Gradient boosting employs gradient descent algorithm to minimize errors in sequential models
Reduce risk overfitting	Regularization for avoiding overfitting
Maintain precision when a large proportion of data is missing.	Efficient handling of missing data
Time-consuming process	Less time-consuming process

2.7. Deep Learning Models

2.7.1. CNN-2D

As a supervised DL approach, the CNN excels in achieving high predictive performance in fields such as image and speech recognition. It accomplishes this by hierarchically composing simple local features into complex models. A typical CNN comprises one or more convolutional layers, fully associated layers, and max pooling layers, enabling it to classify and extract features from high-dimensional data [84]. In the context of landslide susceptibility mapping (LSM), the landslide occurrence potential in each grid cell is influenced by multiple factors. Each grid cell possesses a unique set of characteristic values that illustrate the likelihood of a landslide event. We must initialize the process to apply the CNN for LSM by transforming the 1D input grid cell containing various characteristic attributes into a 2D matrix.

We compared the number of landslide-impacting variables with the number of characteristic values for every variable in this study. The largest of these two integers was then chosen to define the size of the related 2D grid. For example, if the research region has 9 lithological classes and 15 landslide influencing factors, we create 9 × 9 matrices for each grid cell. In the matrix, each column vector represents an attribute value, and the element at the corresponding position is assigned the value of 1. In contrast, other elements in the vector are assigned the value of 0. Some of the predictive parameters utilized in the present research for CNN models are provided in Figure 7 and Table 6.

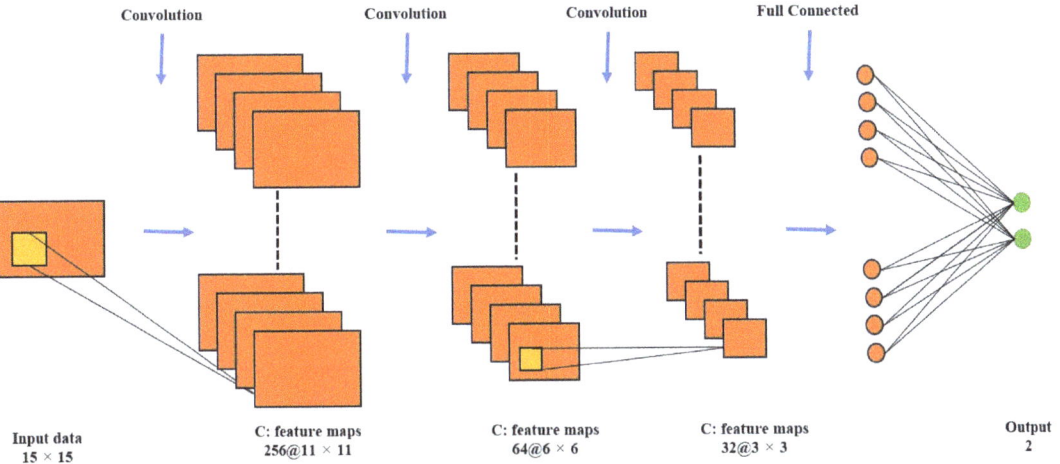

Figure 7. The structure of CNN-2D is illustrated in a schematic figure.

Table 6. List of parameters used in CNN-2D model.

Parameters	Values
Batch size	8
Epoch	250
Dropout	0.5
Learn rate	0.002
Activation function	ReLU
Optimizer	Adam

2.7.2. RNN

An RNN (recurrent neural network) has the ability to capture dynamic information in sequential data by creating connections between hidden layer nodes at different time steps. Unlike other neural networks, RNNs can effectively leverage sequential data. In traditional neural networks, all inputs are treated as self-reliant entities. However, in the RNN technique, each unit is linked to other units in the hidden layer at various time intervals, allowing data to be propagated from one layer to the next in the network [85]. This characteristic of RNNs is achieved through the concept of "loop feedback," where information is shared throughout the RNN. A simple RNN is typically executed using Jordan or Elman network architectures. At time step t, let x_t, y_t, and h_t represent the input vector, the output vector, and the hidden state vector, respectively. By utilizing these elements, we can acquire:

$$h_t = \sigma(W_h x_t + U_h h_{t-1} + b_h) \quad (3)$$

$$y_t = \sigma(W_y h_t + b_y) \quad (4)$$

where $\sigma(\cdot)$ is the training sample sequence's loss function, U and W are variable matrices, and b is the appropriate bias vector.

RNN is particularly adept at manipulating sequential inputs through its recurrent hidden states. Hence, the accurate visualization of data is crucial in realizing the forecasting capability of RNNs. This portion presents the data visualization method for landslide susceptibility mapping using RNNs, as illustrated in Figure 8. The parameters used in the RNN model are listed in Table 7.

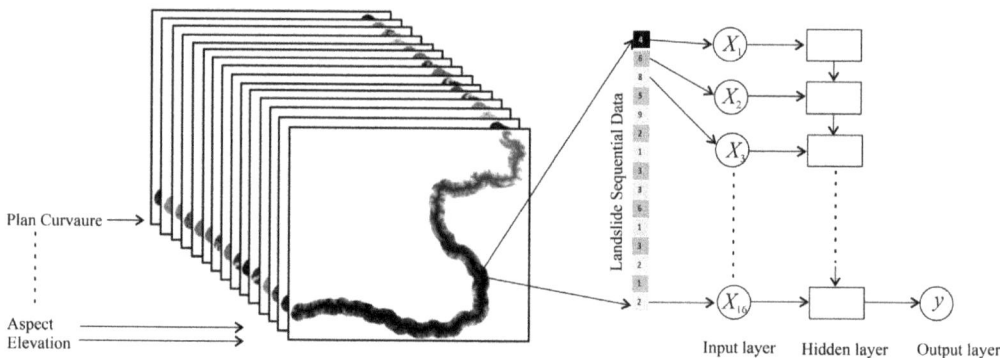

Figure 8. Data representation for RNN.

Table 7. List of parameters used in RNN model.

Parameters	Values
Batch size	64
Epoch	50
Dropout	0
Learn rate	0.001
Optimizer	Adam

First, each LCF is treated as a single-band image, and all variable categories are assembled. Subsequently, these variable categories are then ordered in a decreasing sequence of relevance. Hence, each pixel can be transformed into a sequential sample based on its importance level. This approach ensures that the most significant variables are fed to the RNN framework first, while the less crucial variables are sent to the model last. Because of the recurrent nature of the RNN, essential data contributing to landslide occurrences are reserved and passed to the next hidden state. This retention of important information is advantageous for the final LSM. By organizing the data representation in this manner, we can effectively leverage the sequential processing capabilities of the RNN to improve the precision of the LSM. The key points and usability of DL models are in Table 8.

Table 8. The key points of CNN-2D and RNN models.

	CNN 2D	RNN
Basics	Most popular type of neural networks	Most advanced and complex neural network
Structural layout	Structure is based on multiple layers of nodes including one or two conventional layers	Information flows in different direction, which gives it its self-learning feature and memory
Spatial recognition	Yes	No
Recurrent connection	No	Yes
Drawback	Large training data required	Slow and complex training and gradient concern

2.8. InSAR

The InSAR technique has proven to be highly valuable for the early identification of landslides due to its weather independence, wide monitoring coverage, and high accuracy. Among the various InSAR techniques, the small baseline subset (SBAS) is particularly useful for identifying slow-moving deformations with millimeter-level precision by utilizing a stack of SAR interferograms [86]. Additionally, the PS-InSAR and SBAS-InSAR techniques are utilized to evaluate the deformation in susceptible regions as generated by the models. This research collects and evaluates imagery from the Sentinel-1A IW sensor with a temporal resolution of 12 days.

2.8.1. SBAS-InSAR Processing

The SBAS-InSAR technology is used in this part to verify the LSM along the KKH. The basic data analysis chart, shown in Figure 3, incorporates the preprocessing of data, interferometric creation, phase unwrapping, refinement and re-flattening assessment, and displacement estimations.

The computation of time and spatial baselines between all Sentinel-1 picture pairs is part of data preparation. Following clipping and registration, the DEM data are utilized to finish image authorization, and the proportional conjunction that meets a particular threshold is chosen to generate a differential SAR interferogram set [87]. This investigation used a 30 m resolution SRTM-DEM to construct interferograms. The super primary image utilized comes from the images acquired on 15 July 2022, and 720 interferometric image pairings were created.

The key phase of SBAS-InSAR manipulation is an inversion, and the displacement computation is highly dependent on the investigation of inversion findings. The displacement rate and residual topography are estimated in the first inversion, and the input interferogram is optimized in the second unwrapping [88]. The second inversion expands upon the first by employing low-pass and high-pass filtering to calculate and eliminate the atmospheric phase, allowing for more accurate final displacement estimates and, ultimately, geocoding to determine the displacement rate dispersion in the research region. To avoid the effects of unwrapping inaccuracies, the line-of-sight displacement velocity was computed using a coherence threshold of 0.3 for SBAS-InSAR analysis [89].

2.8.2. PS-InSAR Processing

The PS-InSAR process analyzes the uniformity of the phase and amplitude using multitemporal SAR images wrapped around the same area, which determines the pixels that are not as caused by spatiotemporal decorrelation and then defines specific displacement information on each component of the phase, which must be conjuctly assessed and modeled to remove discrepancies [90,91].

The data preparation analysis steps (Figure 3) for importing SLC data with accurate paths comprise the following:

Acquiring imagery: Images with the same rotations are obtained, and both slave and master images are selected. The master images of the research area are obtained first, followed by the selection of slave images that overlap the same region.

Coregistration and examination: A specific area of interest is coregistered and evaluated. Various measures such as atmospheric phase screen (APS), track errors, and other factors are corrected and measured.

Phase constancy assessment: The phase constancy of the acquired data is evaluated. The pixels are projected to exhibit similar amplitudes and reduced phase distributions for these acquisitions. Absolute amplitude levels are not a significant concern in terms of manipulation disturbances.

Amplitude stability index (ASI): The ASI is used to choose persistent scatterers (PS) in the SARPROZ software (2023) procedure. PS points with ASI values greater than 0.7 are selected. This constraint parameter ensures that only a limited number of PS points are considered, which is necessary for accurate atmospheric phase screen computation.

Reference network and linear model: A reference network is built by linking PS points using Delaunay triangulation. The extracted linear model is removed, including residual height and linear deformation velocities. The APS is analyzed using an inverse network from the phase residual, and a single point of reference is defined to estimate the object's velocity.

Multi-image sparse point (MISP) processing: Second-order PS points are chosen using the criteria of ASI > 0.6 in this step. Thicker PS points are obtained at this phase. To eliminate APS, identical parameters and reference points used for APS estimates are applied.

Geocoding and visualization: Google Earth is used to geocode and map the PS points. The landslide susceptibility map only includes PS points with a coherence of 0.60, indicating their reliability [92].

3. Results

3.1. Multicollinearity Analysis

This research assumed that there are no significant linear associations among the landslide causative variables that can negatively impact the susceptibility models. A multicollinearity analysis is conducted on the 15 landslide conditioning variables, and the outcomes are presented in Table 9. The rainfall variable has the highest VIF score of 4.892, while the lowest VIF score of 1.017 is observed for Aspect. The TOL values ranged from 0.204 to 0.982.

Table 9. Multicollinearity assessment of the LCFs.

S.No.	Variables	Collinearity Statistics	
		TOL	VIF
1	Aspect	0.982	1.017
2	Landcover	0.826	1.209
3	Rainfall	0.204	4.892
4	Geology	0.549	1.821
5	TWI	0.655	1.524
6	Soil	0.284	3.509
7	Slope	0.531	1.881
8	Distance to Fault	0.979	1.021
9	Distance to Road	0.783	1.275
10	Distance to River	0.641	1.432
11	Roughness	0.637	1.542
12	Elevation	0.375	2.662
13	Profile Curvature	0.800	1.249
14	Plan Curvature	0.922	1.084
15	Curvature	0.779	1.282

3.2. Landslide Susceptibility Mapping

The landslide susceptibility maps are generated using deep learning and machine learning models: CNN 2D, RNN, XGBoost, and RF (Figure 9). The experiment showed that the CNN 2D model had the best performance among the four models. The LSMs were classified into five categories using the Jenks natural break [71] technique in ArcGIS.

The precision of the landslide susceptibility maps was assessed using the confusion matrix proposed by [32]. Table 10 shows the performance of the CNN 2D, RNN, XGBoost, and RF models during the training stage. The outcomes revealed that the CNN 2D model has a high precision rate of 0.836 in the study region. Further validation was performed using the ROC (receiver operating characteristic) method [93], which plots "sensitivity" vs. "specificity" for different cut-off estimates. However, to understand the model's performance, the ROC's AUC was used [94]. The AUC results for the CNN 2D, RNN, XGBoost, and RF models are 82.56%, 79.43%, 76.04%, and 75.37%, respectively (Figure 10).

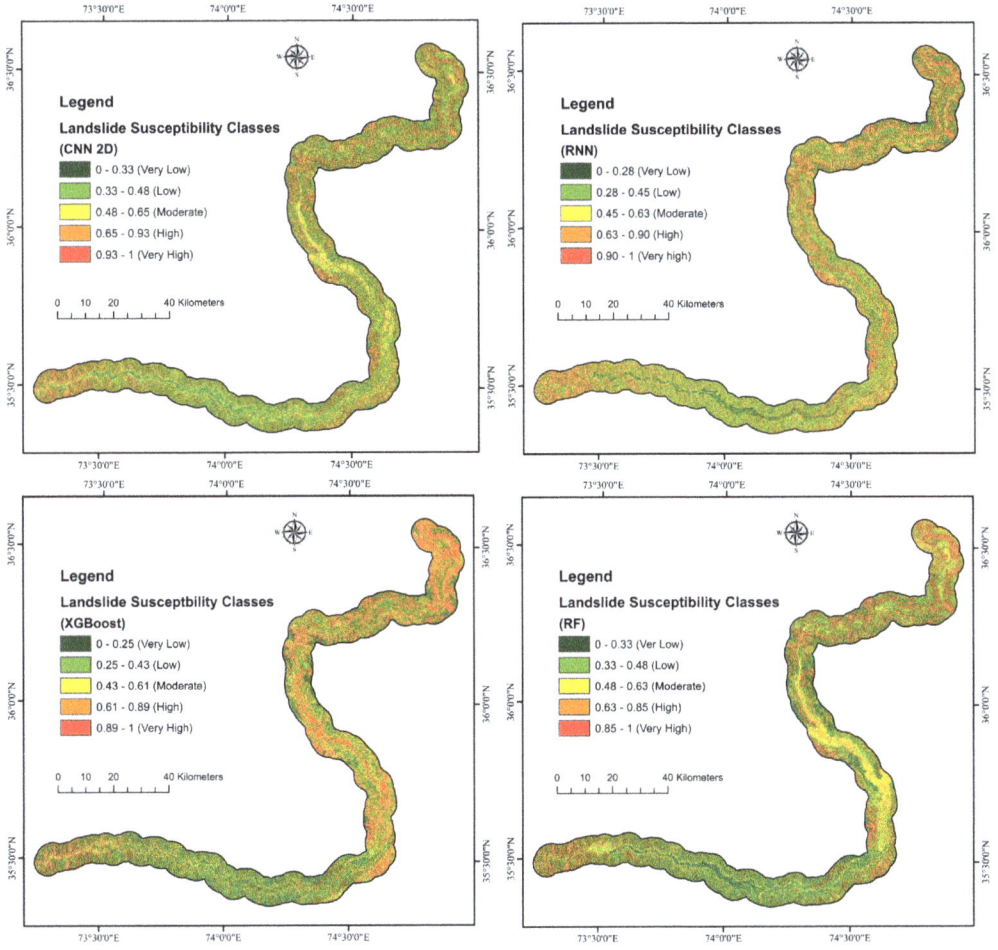

Figure 9. LSM using CNN 2D, RNN, XGBoost, and RF models.

Table 10. Confusion matrix of CNN 2D, RNN, XGBoost, and RF models.

Models	Observation	Predicted No	Predicted Yes	Precision
CNN 2D	No	46	53	83.61
	Yes	127	872	
RNN	No	38	49	83.24
	Yes	135	876	
Extreme Gradient Boosting	No	41	55	83.01
	Yes	132	870	
Random Forest	No	36	58	82.24
	Yes	137	876	

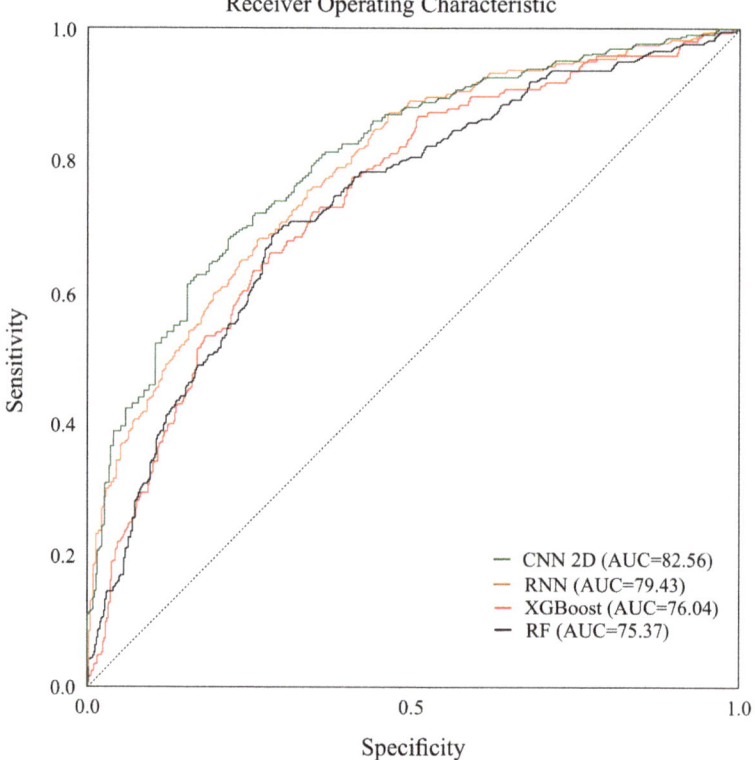

Figure 10. ROC plots of CNN 2D, RNN, XGBoost, and RF models.

3.3. Landslide Mapping Based on Deformation Velocity

A thorough landslide inventory was organized for the identification and analysis of landslides along KKH. This inventory was built by combining data from several sources, including displacement values acquired from both descending and ascending data gathered from PS-InSAR and SBAS-InSAR observation.

In this study, the InSAR analysis successfully detected the majority of previously mapped landslides. Moreover, based on the PS and SBAS data, several new landslides with significant deformation velocity were identified, along with the boundaries, which were calibrated through fieldwork. The PS and SBAS techniques were applied to derive displacement rates along the time series in a one-dimensional LOS direction using a set of highly coherent interferograms with small spatial and temporal baselines [95,96].

3.3.1. PS-InSAR Results

Surface deformation on the Earth's surface was calculated using a temporal coherence threshold of 0.7 for PS-InSAR analysis. A total of 324,747 PS/DS target points were obtained, representing the LOS deformation values ranging from −92.37 to 72.28 mm/year. These values were converted into Vslope using the transformation formula (5), resulting in a total of 212,373 points. The maximum slope displacement velocity was determined, with an appropriate threshold set between 0 and −20 mm/year (Figure 11). It was ascertained that barren terrain has a higher concentration of PS points than forested regions.

$$Vslope = \frac{VLOS}{\cos\emptyset} \quad (5)$$

where VLOS is displacement and $\emptyset$ is the incident angle.

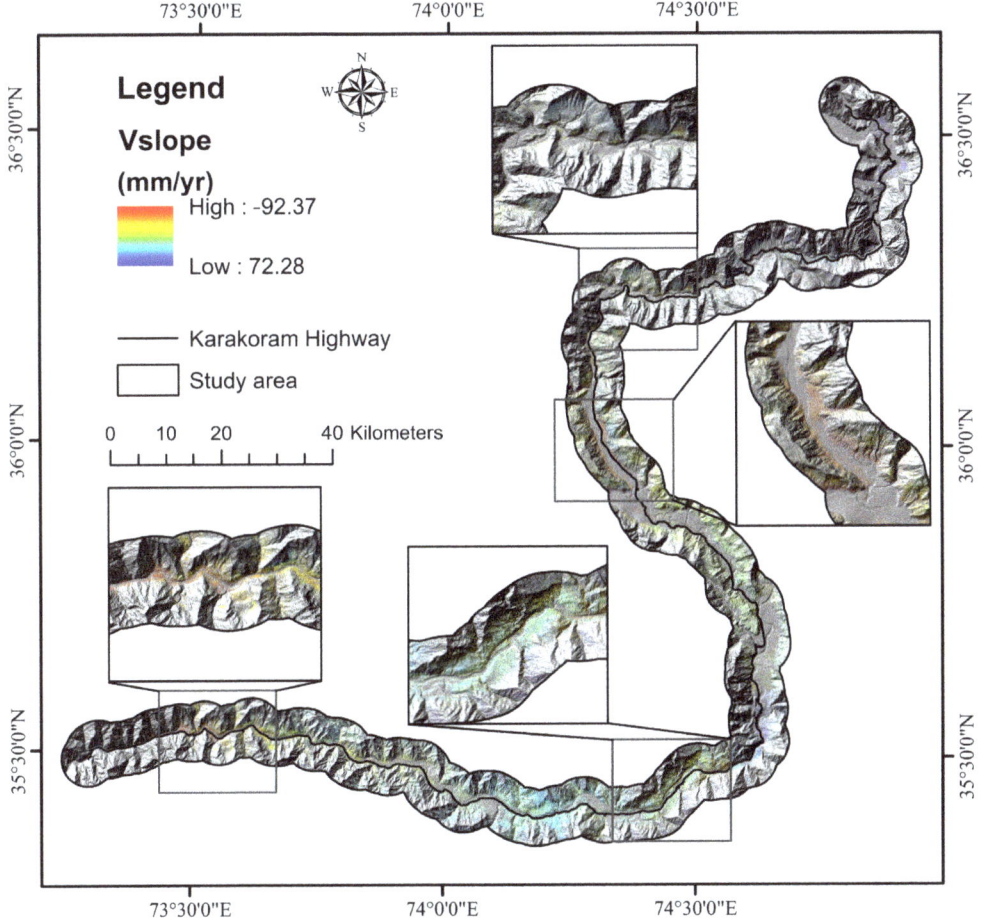

Figure 11. Displacement velocity along landslides and slope measured by using PS-InSAR applying the ascending and descending orbit data.

As a result, a total of 15 potential landslides were identified and detected based on the PS-InSAR-identified displacement velocity. Among the 15 landslides detected using PS-InSAR, 10 were exclusively identified using ascending Sentinel-1 datasets, while 5 were specifically detected using descending Sentinel-1 datasets (Figure 12). This observation demonstrates that the combination of descending and ascending datasets can overcome the constraints of acquiring data from a single scanning posture, improving the landslide detection process.

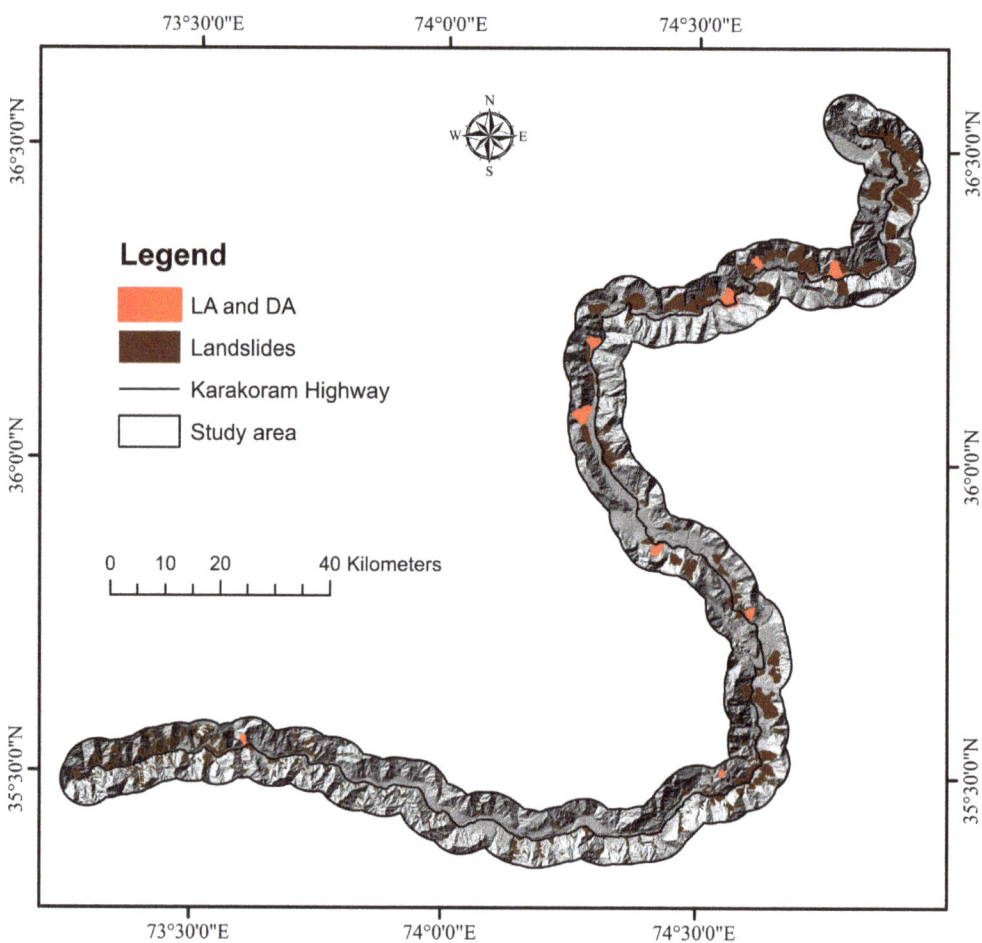

Figure 12. Landslide distribution identified through multi-track Sentinel-1 datasets based on PS-InSAR. LA and LD: the landslide identified from ascending and descending Sentinel-1 datasets.

In the identification and detection of landslides, the PS-InSAR identifies displacement data and characteristics from images and field photographs, which are found valuable when used in integration. Most of these landslides are concealed by PS-InSAR-detected coherent targets (Figure 13). However, some landslides may not strictly meet the criteria of higher deformation velocity to be identified as active landslides, as they might experience lower rates of deformation due to steeper slopes or being in an actual inactive state.

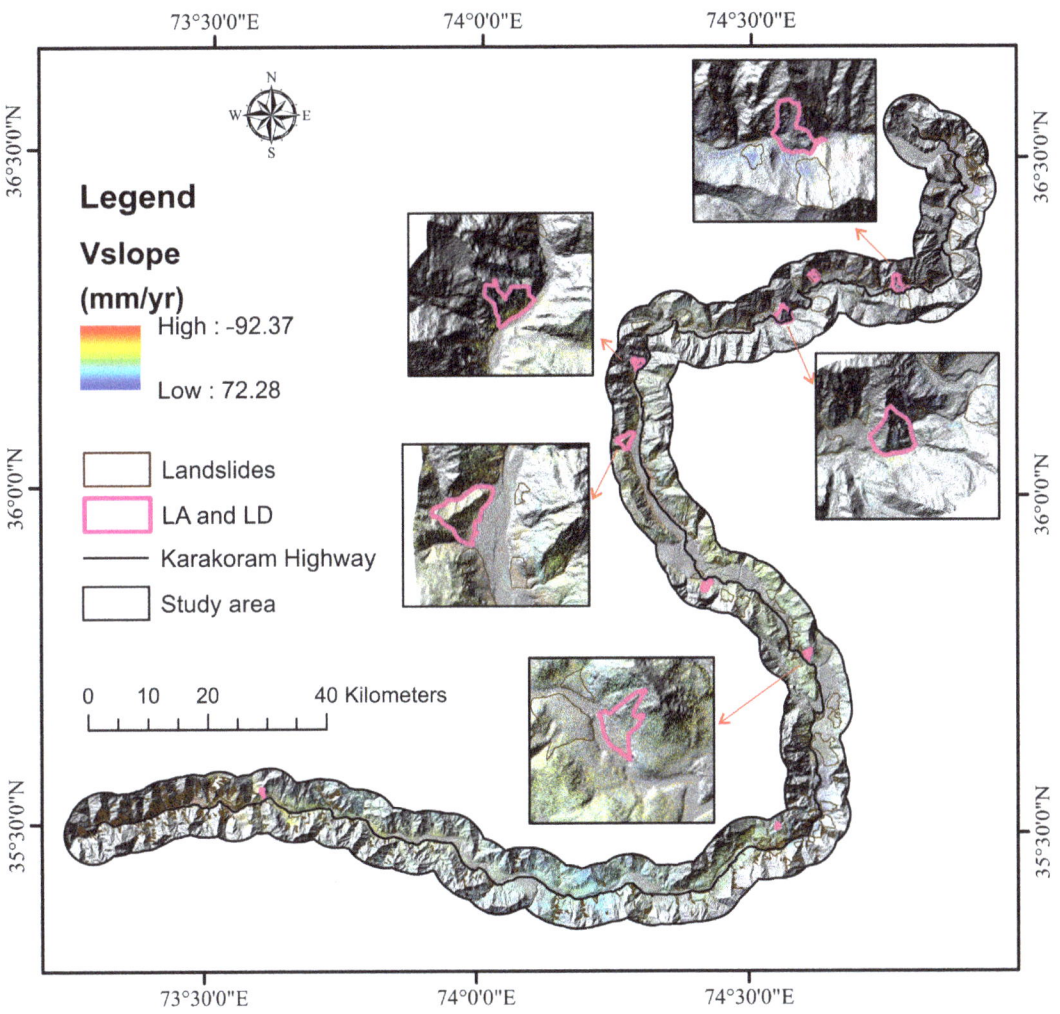

Figure 13. PS-InSAR displacement observation and optical image assessment of identified and interpreted landslides.

3.3.2. SBAS-InSAR Results

In the SBAS-InSAR processing, the LOS displacement velocity (VLos) was determined using a coherent threshold of 0.3. The slope orientation velocity (Vslope) was then derived from the satellite LOS data, showing only unidirectional displacement. Since landslides and Earth's surface deformations mostly happen over steep land, Vslope is an essential constituent used to forecast landslide evolution. The SBAS-InSAR results indicate that the displacement velocity along the LOS ranged from −81.89 to 75.40 mm/year (as depicted in Figure 14).

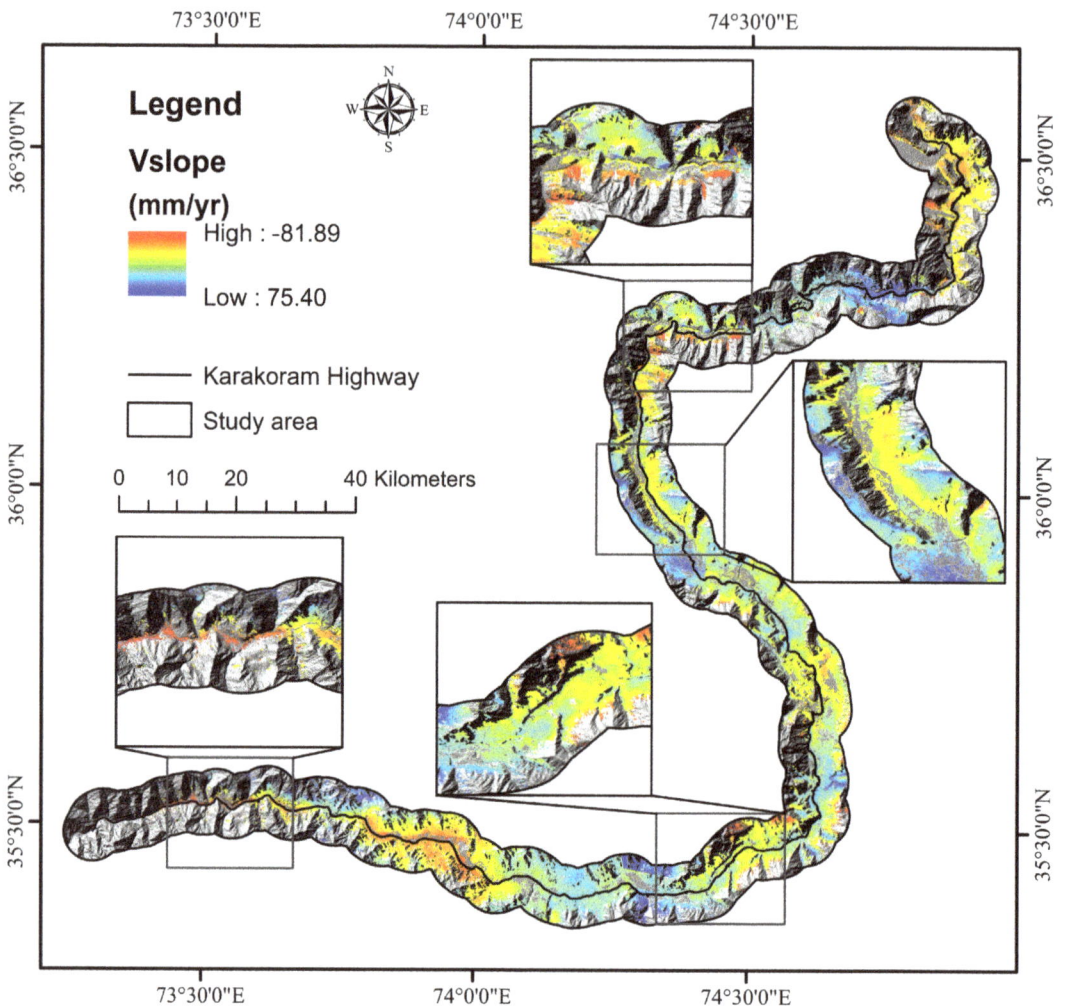

Figure 14. Displacement velocity along landslides and slope measured by using SBAS-InSAR applying the ascending and descending orbit data.

The preliminary delineation of landslide boundaries was conducted by combining the displacement velocity along the slope from both descending and ascending datasets with visual analysis of optical RS images and field assessments. This involves referring to areas with relatively high deformation velocity, topographic characteristics obtained from the digital elevation model (DEM), and features observed in the optical images.

As a result of the SBAS-InSAR analysis, a total of 9 potential landslides were detected and identified. Additionally, 547 landslides were represented through a combination of information from the literature [12,60,66] and field observations (Figure 15). Among the 9 SBAS-InSAR-identified landslides, the ascending Sentinel-1 dataset identified 6, and 3 were specifically identified using the descending Sentinel-1 dataset.

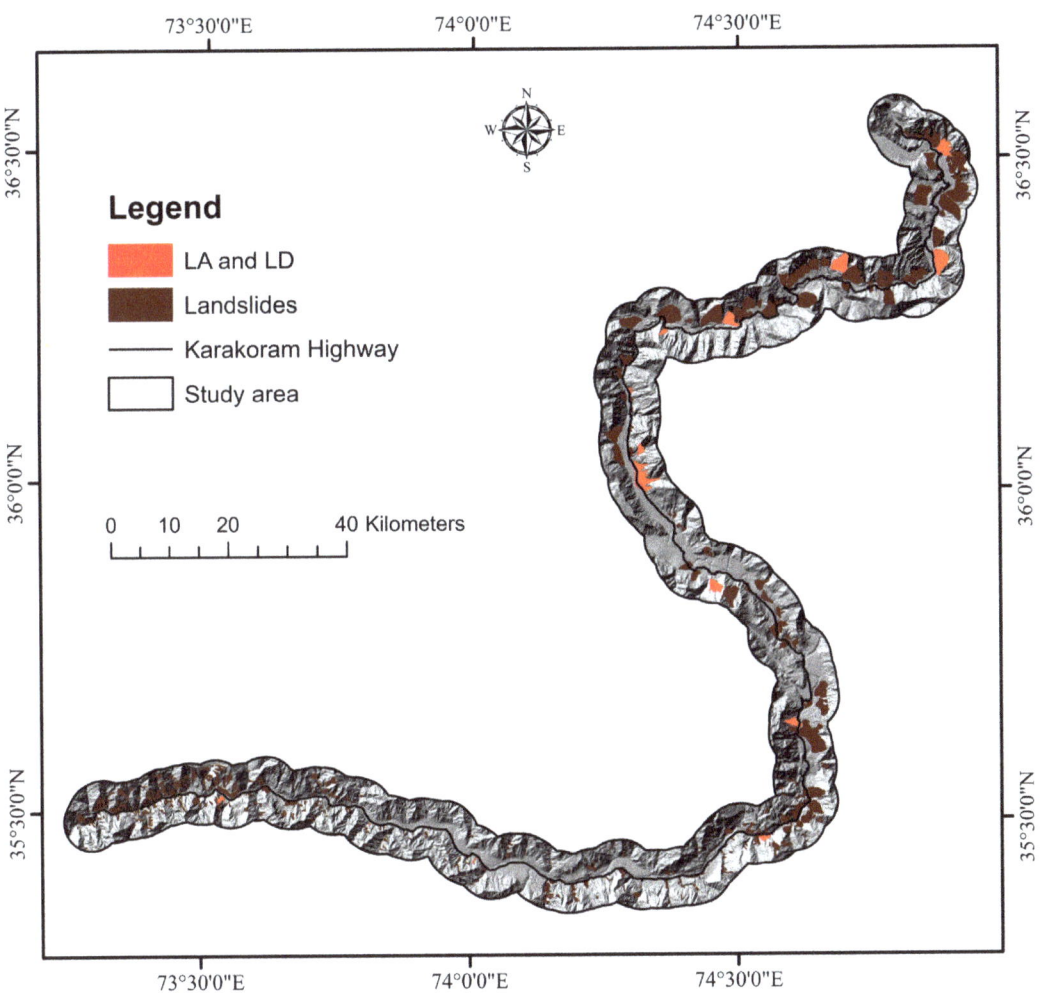

Figure 15. Landslide distribution identified through multi-track Sentinel-1 datasets based on SBAS-InSAR. LA and LD: the landslide detected by using ascending and descending Sentinel-1 dataset.

Figure 15 provides comparative examples of identified landslides, most of which are wrapped by SBAS-InSAR-identified coherent targets. The analysis of the landslides revealed that approximately 90% of them are associated with Quaternary deposits, the Hunza plutonic unit, southern Karakorum Metamorphic Complex, Permian massive limestone, and Chilas Complex formations. Their delineation was achieved through field investigations, analysis of optical RS images, and references to the existing literature (Figure 16).

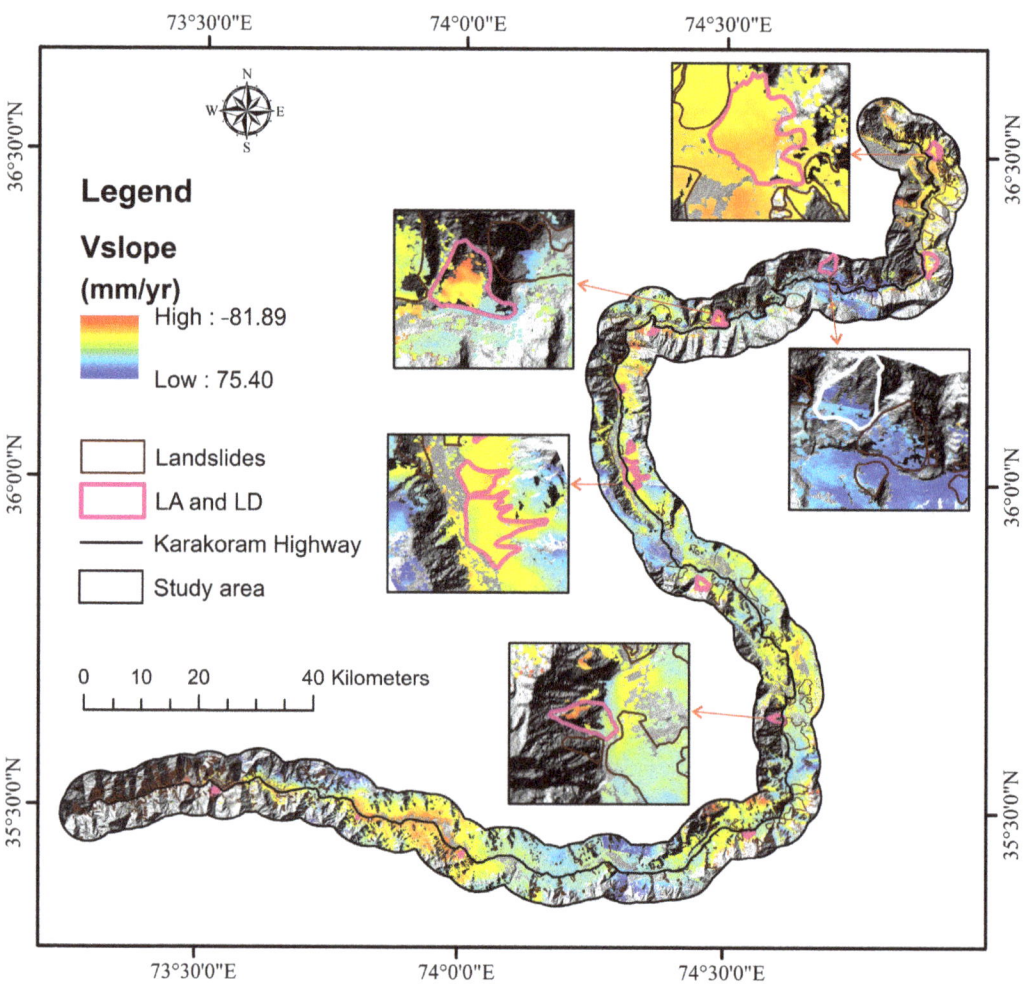

Figure 16. SBAS-InSAR displacement observation and the optical image assessment of identified and interpreted landslides.

The majority of KKH's landscape is barren, with nearly 90 percent of the landslides occurring in non-vegetated zones. While SBAS-InSAR and PS-InSAR methods may have limitations in vegetation-covered areas, they still apply to more than 60% of the land devoid of vegetation. This suggests that vegetation plays a critical role in controlling slope stability in the region, consistent with previous research findings [60,97,98].

4. Discussion

The current study utilized RS techniques, such as optical RS and InSAR, for risk assessment and landslide mapping along the KKH [63,99,100]. This study took benefit of the multi-azimuth interpretation provided by the descending and ascending Sentinel-1 dataset, allowing for more extensive monitoring of surface displacement. The PS-InSAR and SBAS-InSAR techniques effectively captured regions with high deformation rates in most areas along the KKH. Additionally, the comprehensive landslide inventory presented in this study includes the latest landslides, ensuring the database is up to date and its valuable information. By processing Sentinel-1 data from June 2021 to June 2023, utilizing the InSAR technique, 24 new prospective landslides were identified, and some existing landslides were redefined. This updated landslide inventory was then utilized to create a landslide susceptibility model, which investigated the link between landslide occurrences and the causal variables. By combining the findings from PS-InSAR, SBAS-InSAR, and field investigations, the inventory was updated with landslides that have the potential for future failure and pose risks for the region, contributing to improved landslide susceptibility mapping.

The selected landslide influencing factors were used to construct CNN 2D and RNN architectures for comparison with the XGBoost and RF methods. The LSMs were validated and compared based on the AUROC curve and accuracy. CNN is known for its ability to efficiently obtain spatial data using weight sharing and local connections, making it a promising method for landslide modeling [101]. Earlier research has shown that combining CNNs with additional statistical approaches can produce better accuracy in landslide susceptibility modeling than using CNNs alone [102]. According to [103], CNN models are an improved tool for landslide modeling due to their substantial outcomes and higher accuracy rate in spatial landslide forecasting. The outcomes also reveal that both DL and traditional ML algorithms give excellent precision in a variety of sectors, such as landslide assessment and earth science studies throughout the world [104], which is in line with our findings, which showed that the ROC for the four models varies from 82.56 to 75.37%.

In the subsequent experiments, the proposed CNN and RNN models demonstrated enhanced predictive capability compared to the popular XGBoost and RF classifiers. Specifically, CNN-2D attained the highest AUC value of 0.825 on the validation set, indicating its effectiveness in improving prediction performance and its potential as a potential approach for future research. Various statistical and machine learning approaches have been compared and applied for landslide spatial forecast in areas, including AHP and Scoops 3D [105], frequency ratio (FR) and weight of evidence [106], the weighted overlay technique and AHP [9], random forest [63], support vector classification (SVC) [107], and XGBoost [100], but DL techniques, such as CNNs, provide powerful improvements by automatically exploring representations from raw data, making them valuable in various fields, including landslide susceptibility assessments. The experimental findings highlighted that CNN-2D outperformed the traditional DL approach of RNNs and the classical XGBoost and RF ML techniques. Furthermore, the suggested data representation techniques offer an innovative approach to handling raw landslide data. By exploiting the power of DL methods and combining them with other approaches, there is enormous potential to advance landslide susceptibility analysis in the future. The proposed 2D CNN structure includes convolutional max pooling layers and a dropout layer. Overfitting is a common issue when utilizing a 2D CNN in LSM. To address overfitting, each convolution layer is subsequently followed by a dropout layer, which temporarily discards NN units during the training process of the CNN based on a certain probability. This helps to improve classification accuracies and enhance the model's generalization capability.

Different LCFs influence landslide triggers and relate to each other, making the selection of appropriate variables crucial for building an accurate landslide susceptibility model. The aim is to construct models with reduced noise and greater forecast ability. Before analyzing landslide susceptibility, it is essential to evaluate the forecasting potential of each contributing factor. To attain this, efforts are made to select the most relevant and impactful factors. Multicollinearity analysis is employed to evaluate correlations between the LCFs. In this study, 15 landslide conditioning variables were chosen as independent factors for evaluating landslide susceptibility, and the results are presented in Table 3. The variance inflation factor (VIF) was used to test the multicollinearity between these factors. Among the selected factors, rainfall had the highest VIF score of 4.892, while aspect had the lowest VIF score of 1.017. The tolerance (TOL) values ranged from 0.204 to 0.982. The outcomes revealed that there is no significant multicollinearity among the chosen variables, allowing all variables to be integrated into the models. It is worth noting that landslides can still occur in areas with significant vegetation due to rainfall and other external forces.

Despite the beneficial effect of the selected factors in evaluating landslide susceptibility, the current research could have been more effective if certain factors were adhered to. One major factor is data availability. This research relied on limited data diversity with a focus on historical landslide data, which limited the comprehensive analysis [60]. Another factor is that the input data resolution remains unpredictable during the data preparation phase, which has been a prevalent issue in past investigations [108,109]. Terrain condition factors were derived from a 12.5 m resolution DEM, while variables related to geological conditions were based on a 1:500,000 scale geological map. All factor layers were resampled at a 12.5 m resolution in ArcGIS 10.8 software to ensure data availability and computational convenience. The analysis of model performance in this study indicates that resampling processing was feasible. Secondly, because of restricted data availability, we examined several types of landslides with varying triggering conditions throughout a given time. While some investigators have previously explored this approach, a separate investigation of distinct types of landslides is more in line with the practical and current state factors [110,111].

5. Conclusions

The PS-InSAR and SBAS-InSAR techniques and multi-track ascending and descending Sentinel-1 SAR datasets were used to measure surface displacement velocity along the KKH. An updated and comprehensive landslide inventory was created by combining field surveys, image analysis, and a literature evaluation, identifying 571 landslides, including 24 newly detected active landslides and 547 landslides from previous records. To predict landslide susceptibility along KKH, two well-known deep learning (CNN-2D and RNN) and machine learning (XGBoost and RF) algorithms were utilized and compared. The CNN-2D algorithm demonstrated superior performance with an AUC of 82.56, outperforming RNN, XGBoost, and RF in terms of AUC, ROC, predictive power, and accuracy. The landslide susceptibility maps generated by these models can serve as valuable tools for decision-makers, land use planners, and various non-governmental and governmental organizations involved in resource and disaster management, infrastructure development, and human activity in the study area. In the future, studies can explore improved deep learning and machine learning architectures for landslide susceptibility mapping to improve accuracy and predictive capabilities utilizing this research as a baseline.

The current study is limited because of the absence of geotechnical and geophysical data. It is suggested that the developed dataset be used in future studies to improve algorithm prediction potency, create more precise LSM, and discover correlations between landslide incidence and these new geo-environmental variables. It is also advised to combine DL algorithms with metaheuristic techniques to optimize model parameters and boost algorithm prediction capabilities.

Author Contributions: Conceptualization, M.A.H.; Methodology, M.A.H.; Software, M.A.H.; Validation, Y.Z. (Yulong Zhou); Formal analysis, Y.Z. (Yulong Zhou); Investigation, Y.Z. (Ying Zheng); Resources, Y.Z. (Ying Zheng); Writing—original draft, M.A.H.; Writing—review & editing, H.D.; Visualization, H.D.; Supervision, Z.C.; Project administration, Z.C.; Funding acquisition, Z.C. All authors have read and agreed to the published version of the manuscript.

Funding: This research was funded by National Natural Science Foundation of China, grant number No. 41871305; National key R & D program of China, grant number No.2017YFC0602204; Fundamental Research Funds for the Central Universities, China University of Geosciences (Wuhan), grant number No. CUGQY1945; Opening Fund of Key Laboratory of Geological Survey and Evaluation of Ministry of Education; and Fundamental Research Funds for the Central Universities, grant number No. GLAB2019ZR02.

Data Availability Statement: The data presented in the study are available upon request from the first and corresponding authors. The data are not publicly available due to the thesis that is being prepared using these data.

Conflicts of Interest: The authors declare no conflict of interest.

References

1. Zhou, X.; Wen, H.; Zhang, Y.; Xu, J.; Zhang, W. Landslide susceptibility mapping using hybrid random forest with GeoDetector and RFE for factor optimization. *Geosci. Front.* **2021**, *12*, 101211. [CrossRef]
2. Ngo, P.T.T.; Panahi, M.; Khosravi, K.; Ghorbanzadeh, O.; Kariminejad, N.; Cerda, A.; Lee, S. Evaluation of deep learning algorithms for national scale landslide susceptibility mapping of Iran. *Geosci. Front.* **2021**, *12*, 505–519.
3. Chen, N.; Tian, S.; Wang, F.; Shi, P.; Liu, L.; Xiao, M.; Liu, E.; Tang, W.; Rahman, M.; Somos-Valenzuela, M. Multi-wing butterfly effects on catastrophic rockslides. *Geosci. Front.* **2023**, *14*, 101627. [CrossRef]
4. Berrocal, J.; Espinosa, A.; Galdos, J. Seismological and geological aspects of the Mantaro landslide in Peru. *Nature* **1978**, *275*, 533–536. [CrossRef]
5. Wilde, M.; Günther, A.; Reichenbach, P.; Malet, J.-P.; Hervás, J. Pan-European landslide susceptibility mapping: ELSUS Version 2. *J. Maps* **2018**, *14*, 97–104. [CrossRef]
6. Lima, P.; Steger, S.; Glade, T.; Tilch, N.; Schwarz, L.; Kociu, A. Landslide susceptibility mapping at national scale: A first attempt for Austria. In Proceedings of the Advancing Culture of Living with Landslides: Volume 2 Advances in Landslide Science; Springer: Ljubljana, Slovenia, 2017; pp. 943–951.
7. Ghorbanzadeh, O.; Blaschke, T.; Gholamnia, K.; Meena, S.R.; Tiede, D.; Aryal, J. Evaluation of different machine learning methods and deep-learning convolutional neural networks for landslide detection. *Remote Sens.* **2019**, *11*, 196. [CrossRef]
8. Hussain, M.A.; Chen, Z.; Zheng, Y.; Shoaib, M.; Shah, S.U.; Ali, N.; Afzal, Z. Landslide susceptibility mapping using machine learning algorithm validated by persistent scatterer In-SAR technique. *Sensors* **2022**, *22*, 3119. [CrossRef]
9. Ali, S.; Biermanns, P.; Haider, R.; Reicherter, K. Landslide susceptibility mapping by using a geographic information system (GIS) along the China–Pakistan Economic Corridor (Karakoram Highway), Pakistan. *Nat. Hazards Earth Syst. Sci.* **2019**, *19*, 999–1022. [CrossRef]
10. Sato, H.P.; Hasegawa, H.; Fujiwara, S.; Tobita, M.; Koarai, M.; Une, H.; Iwahashi, J. Interpretation of landslide distribution triggered by the 2005 Northern Pakistan earthquake using SPOT 5 imagery. *Landslides* **2007**, *4*, 113–122. [CrossRef]
11. Hewitt, K. Catastrophic landslides and their effects on the Upper Indus streams, Karakoram Himalaya, northern Pakistan. *Geomorphology* **1998**, *26*, 47–80. [CrossRef]
12. Abbas, H.; Hussain, D.; Khan, G.; ul Hassan, S.N.; Kulsoom, I.; Hussain, S. Landslide inventory and landslide susceptibility mapping for china pakistan economic corridor (CPEC)'s main route (Karakorum Highway). *J. Appl. Emerg. Sci.* **2021**, *11*, 18–30.
13. Ghorbanzadeh, O.; Blaschke, T.; Aryal, J.; Gholaminia, K. A new GIS-based technique using an adaptive neuro-fuzzy inference system for land subsidence susceptibility mapping. *J. Spat. Sci.* **2020**, *65*, 401–418. [CrossRef]
14. Sestras, P.; Bilașco, Ș.; Roșca, S.; Naș, S.; Bondrea, M.; Gâlgău, R.; Vereș, I.; Sălăgean, T.; Spalević, V.; Cîmpeanu, S. Landslides susceptibility assessment based on GIS statistical bivariate analysis in the hills surrounding a metropolitan area. *Sustainability* **2019**, *11*, 1362. [CrossRef]
15. Ding, Q.; Chen, W.; Hong, H. Application of frequency ratio, weights of evidence and evidential belief function models in landslide susceptibility mapping. *Geocarto Int.* **2017**, *32*, 619–639. [CrossRef]
16. Youssef, A.M.; Pourghasemi, H.R. Landslide susceptibility mapping using machine learning algorithms and comparison of their performance at Abha Basin, Asir Region, Saudi Arabia. *Geosci. Front.* **2021**, *12*, 639–655. [CrossRef]
17. Constantin, M.; Bednarik, M.; Jurchescu, M.C.; Vlaicu, M. Landslide susceptibility assessment using the bivariate statistical analysis and the index of entropy in the Sibiciu Basin (Romania). *Environ. Earth Sci.* **2011**, *63*, 397–406. [CrossRef]
18. Chen, W.; Panahi, M.; Tsangaratos, P.; Shahabi, H.; Ilia, I.; Panahi, S.; Li, S.; Jaafari, A.; Ahmad, B.B. Applying population-based evolutionary algorithms and a neuro-fuzzy system for modeling landslide susceptibility. *Catena* **2019**, *172*, 212–231. [CrossRef]
19. Ali, S.A.; Parvin, F.; Vojteková, J.; Costache, R.; Linh, N.T.T.; Pham, Q.B.; Vojtek, M.; Gigović, L.; Ahmad, A.; Ghorbani, M.A. GIS-based landslide susceptibility modeling: A comparison between fuzzy multi-criteria and machine learning algorithms. *Geosci. Front.* **2021**, *12*, 857–876. [CrossRef]
20. Jena, R.; Pradhan, B.; Beydoun, G.; Sofyan, H.; Affan, M. Integrated model for earthquake risk assessment using neural network and analytic hierarchy process: Aceh province, Indonesia. *Geosci. Front.* **2020**, *11*, 613–634. [CrossRef]
21. Bui, D.T.; Pradhan, B.; Lofman, O.; Revhaug, I.; Dick, O.B. Spatial prediction of landslide hazards in Hoa Binh province (Vietnam): A comparative assessment of the efficacy of evidential belief functions and fuzzy logic models. *Catena* **2012**, *96*, 28–40.
22. Balogun, A.-L.; Rezaie, F.; Pham, Q.B.; Gigović, L.; Drobnjak, S.; Aina, Y.A.; Panahi, M.; Yekeen, S.T.; Lee, S. Spatial prediction of landslide susceptibility in western Serbia using hybrid support vector regression (SVR) with GWO, BAT and COA algorithms. *Geosci. Front.* **2021**, *12*, 101104. [CrossRef]
23. Achour, Y.; Pourghasemi, H.R. How do machine learning techniques help in increasing accuracy of landslide susceptibility maps? *Geosci. Front.* **2020**, *11*, 871–883. [CrossRef]
24. Wang, H.; Zhang, L.; Yin, K.; Luo, H.; Li, J. Landslide identification using machine learning. *Geosci. Front.* **2021**, *12*, 351–364. [CrossRef]
25. Tien Bui, D.; Tuan, T.A.; Klempe, H.; Pradhan, B.; Revhaug, I. Spatial prediction models for shallow landslide hazards: A comparative assessment of the efficacy of support vector machines, artificial neural networks, kernel logistic regression, and logistic model tree. *Landslides* **2016**, *13*, 361–378. [CrossRef]
26. Pradhan, B.; Lee, S. Regional landslide susceptibility analysis using back-propagation neural network model at Cameron Highland, Malaysia. *Landslides* **2010**, *7*, 13–30. [CrossRef]

27. Elkadiri, R.; Sultan, M.; Youssef, A.M.; Elbayoumi, T.; Chase, R.; Bulkhi, A.B.; Al-Katheeri, M.M. A remote sensing-based approach for debris-flow susceptibility assessment using artificial neural networks and logistic regression modeling. *IEEE J. Sel. Top. Appl. Earth Obs. Remote Sens.* **2014**, *7*, 4818–4835. [CrossRef]
28. Gorsevski, P.V.; Brown, M.K.; Panter, K.; Onasch, C.M.; Simic, A.; Snyder, J. Landslide detection and susceptibility mapping using LiDAR and an artificial neural network approach: A case study in the Cuyahoga Valley National Park, Ohio. *Landslides* **2016**, *13*, 467–484. [CrossRef]
29. Arabameri, A.; Chen, W.; Loche, M.; Zhao, X.; Li, Y.; Lombardo, L.; Cerda, A.; Pradhan, B.; Bui, D.T. Comparison of machine learning models for gully erosion susceptibility mapping. *Geosci. Front.* **2020**, *11*, 1609–1620. [CrossRef]
30. Sabokbar, H.F.; Roodposhti, M.S.; Tazik, E. Landslide susceptibility mapping using geographically-weighted principal component analysis. *Geomorphology* **2014**, *226*, 15–24. [CrossRef]
31. Xiong, Y.; Zhou, Y.; Wang, F.; Wang, S.; Wang, J.; Ji, J.; Wang, Z. Landslide susceptibility mapping using ant colony optimization strategy and deep belief network in Jiuzhaigou Region. *IEEE J. Sel. Top. Appl. Earth Obs. Remote Sens.* **2021**, *14*, 11042–11057. [CrossRef]
32. Kavzoglu, T.; Sahin, E.K.; Colkesen, I. Landslide susceptibility mapping using GIS-based multi-criteria decision analysis, support vector machines, and logistic regression. *Landslides* **2014**, *11*, 425–439. [CrossRef]
33. Youssef, K.; Shao, K.; Moon, S.; Bouchard, L.-S. Landslide susceptibility modeling by interpretable neural network. *Commun. Earth Environ.* **2023**, *4*, 162. [CrossRef]
34. Tsangaratos, P.; Ilia, I. Comparison of a logistic regression and Naïve Bayes classifier in landslide susceptibility assessments: The influence of models complexity and training dataset size. *Catena* **2016**, *145*, 164–179. [CrossRef]
35. Khosravi, K.; Sartaj, M.; Tsai, F.T.-C.; Singh, V.P.; Kazakis, N.; Melesse, A.M.; Prakash, I.; Bui, D.T.; Pham, B.T. A comparison study of DRASTIC methods with various objective methods for groundwater vulnerability assessment. *Sci. Total Environ.* **2018**, *642*, 1032–1049. [CrossRef]
36. Hussain, M.A.; Chen, Z.; Wang, R.; Shoaib, M. PS-InSAR-based validated landslide susceptibility mapping along Karakorum Highway, Pakistan. *Remote Sens.* **2021**, *13*, 4129. [CrossRef]
37. Wang, Y.; Hong, H.; Chen, W.; Li, S.; Panahi, M.; Khosravi, K.; Shirzadi, A.; Shahabi, H.; Panahi, S.; Costache, R. Flood susceptibility mapping in Dingnan County (China) using adaptive neuro-fuzzy inference system with biogeography based optimization and imperialistic competitive algorithm. *J. Environ. Manag.* **2019**, *247*, 712–729. [CrossRef]
38. Chen, W.; Hong, H.; Panahi, M.; Shahabi, H.; Wang, Y.; Shirzadi, A.; Pirasteh, S.; Alesheikh, A.A.; Khosravi, K.; Panahi, S. Spatial prediction of landslide susceptibility using gis-based data mining techniques of anfis with whale optimization algorithm (woa) and grey wolf optimizer (gwo). *Appl. Sci.* **2019**, *9*, 3755. [CrossRef]
39. Wang, Y.; Fang, Z.; Wang, M.; Peng, L.; Hong, H. Comparative study of landslide susceptibility mapping with different recurrent neural networks. *Comput. Geosci.* **2020**, *138*, 104445. [CrossRef]
40. Zaremba, W.; Sutskever, I.; Vinyals, O. Recurrent neural network regularization. *arXiv* **2014**, arXiv:1409.2329.
41. Krizhevsky, A.; Sutskever, I.; Hinton, G.E. Imagenet classification with deep convolutional neural networks. *Adv. Neural Inf. Process. Syst.* **2012**, *25*, 1097–1105. [CrossRef]
42. Du, Y.; Wang, W.; Wang, L. Hierarchical recurrent neural network for skeleton based action recognition. In Proceedings of the IEEE Conference on Computer Vision and Pattern Recognition, Boston, MA, USA, 7–12 June 2015; pp. 1110–1118.
43. Wang, Y.; Fang, Z.; Hong, H.; Peng, L. Flood susceptibility mapping using convolutional neural network frameworks. *J. Hydrol.* **2020**, *582*, 124482. [CrossRef]
44. Bui, D.T.; Hoang, N.-D.; Martínez-Álvarez, F.; Ngo, P.-T.T.; Hoa, P.V.; Pham, T.D.; Samui, P.; Costache, R. A novel deep learning neural network approach for predicting flash flood susceptibility: A case study at a high frequency tropical storm area. *Sci. Total Environ.* **2020**, *701*, 134413.
45. Pradhan, B.; Lee, S.; Dikshit, A.; Kim, H. Spatial flood susceptibility mapping using an explainable artificial intelligence (XAI) model. *Geosci. Front.* **2023**, *14*, 101625. [CrossRef]
46. Shi, X.; Liao, M.; Li, M.; Zhang, L.; Cunningham, C. Wide-area landslide deformation mapping with multi-path ALOS PALSAR data stacks: A case study of three gorges area, China. *Remote Sens.* **2016**, *8*, 136. [CrossRef]
47. Hussain, M.A.; Chen, Z.; Shoaib, M.; Shah, S.U.; Khan, J.; Ying, Z. Sentinel-1A for monitoring land subsidence of coastal city of Pakistan using Persistent Scatterers In-SAR technique. *Sci. Rep.* **2022**, *12*, 5294. [CrossRef]
48. Scaioni, M.; Longoni, L.; Melillo, V.; Papini, M. Remote sensing for landslide investigations: An overview of recent achievements and perspectives. *Remote Sens.* **2014**, *6*, 9600–9652. [CrossRef]
49. Schlögel, R.; Doubre, C.; Malet, J.-P.; Masson, F. Landslide deformation monitoring with ALOS/PALSAR imagery: A D-InSAR geomorphological interpretation method. *Geomorphology* **2015**, *231*, 314–330. [CrossRef]
50. Dai, C.; Li, W.; Wang, D.; Lu, H.; Xu, Q.; Jian, J. Active landslide detection based on Sentinel-1 data and InSAR technology in Zhouqu county, Gansu province, Northwest China. *J. Earth Sci.* **2021**, *32*, 1092–1103. [CrossRef]
51. Ferretti, A.; Fumagalli, A.; Novali, F.; Prati, C.; Rocca, F.; Rucci, A. A new algorithm for processing interferometric data-stacks: SqueeSAR. *IEEE Trans. Geosci. Remote Sens.* **2011**, *49*, 3460–3470. [CrossRef]
52. Ali, M.; Shahzad, M.I.; Nazeer, M.; Kazmi, J.H. Estimation of surface deformation due to Pasni earthquake using SAR interferometry. *Int. Arch. Photogramm. Remote Sens. Spat. Inf. Sci.* **2018**, *42*, 23–29. [CrossRef]

53. Mora, O.; Mallorqui, J.J.; Broquetas, A. Linear and nonlinear terrain deformation maps from a reduced set of interferometric SAR images. *IEEE Trans. Geosci. Remote Sens.* **2003**, *41*, 2243–2253. [CrossRef]
54. Strozzi, T.; Wegmuller, U.; Keusen, H.R.; Graf, K.; Wiesmann, A. Analysis of the terrain displacement along a funicular by SAR interferometry. *IEEE Geosci. Remote Sens. Lett.* **2006**, *3*, 15–18. [CrossRef]
55. Jiaxuan, H.; Mowen, X.; Atkinson, P. Dynamic susceptibility mapping of slow-moving landslides using PSInSAR. *Int. J. Remote Sens.* **2020**, *41*, 7509–7529. [CrossRef]
56. Ciampalini, A.; Raspini, F.; Lagomarsino, D.; Catani, F.; Casagli, N. Landslide susceptibility map refinement using PSInSAR data. *Remote Sens. Environ.* **2016**, *184*, 302–315. [CrossRef]
57. Lu, P.; Stumpf, A.; Kerle, N.; Casagli, N. Object-oriented change detection for landslide rapid mapping. *IEEE Geosci. Remote Sens. Lett.* **2011**, *8*, 701–705. [CrossRef]
58. Righini, G.; Pancioli, V.; Casagli, N. Updating landslide inventory maps using Persistent Scatterer Interferometry (PSI). *Int. J. Remote Sens.* **2012**, *33*, 2068–2096. [CrossRef]
59. Searle, M.; Khan, M.A.; Fraser, J.; Gough, S.; Jan, M.Q. The tectonic evolution of the Kohistan-Karakoram collision belt along the Karakoram Highway transect, north Pakistan. *Tectonics* **1999**, *18*, 929–949. [CrossRef]
60. Su, X.-J.; Zhang, Y.; Meng, X.-M.; Yue, D.-X.; Ma, J.-H.; Guo, F.-Y.; Zhou, Z.-Q.; Rehman, M.U.; Khalid, Z.; Chen, G. Landslide mapping and analysis along the China-Pakistan Karakoram Highway based on SBAS-InSAR detection in 2017. *J. Mt. Sci.* **2021**, *18*, 2540–2564. [CrossRef]
61. Guzzetti, F.; Carrara, A.; Cardinali, M.; Reichenbach, P. Landslide hazard evaluation: A review of current techniques and their application in a multi-scale study, Central Italy. *Geomorphology* **1999**, *31*, 181–216. [CrossRef]
62. Galli, M.; Ardizzone, F.; Cardinali, M.; Guzzetti, F.; Reichenbach, P. Comparing landslide inventory maps. *Geomorphology* **2008**, *94*, 268–289. [CrossRef]
63. Hussain, M.A.; Chen, Z.; Kalsoom, I.; Asghar, A.; Shoaib, M. Landslide susceptibility mapping using machine learning algorithm: A case study along Karakoram Highway (KKH), Pakistan. *J. Indian Soc. Remote Sens.* **2022**, *50*, 849–866. [CrossRef]
64. Zeng, T.; Wu, L.; Peduto, D.; Glade, T.; Hayakawa, Y.S.; Yin, K. Ensemble learning framework for landslide susceptibility mapping: Different basic classifier and ensemble strategy. *Geosci. Front.* **2023**, 101645. [CrossRef]
65. Zhao, F.; Meng, X.; Zhang, Y.; Chen, G.; Su, X.; Yue, D. Landslide susceptibility mapping of karakorum highway combined with the application of SBAS-InSAR technology. *Sensors* **2019**, *19*, 2685. [CrossRef] [PubMed]
66. Hussain, M.A.; Chen, Z.; Wang, R.; Shah, S.U.; Shoaib, M.; Ali, N.; Xu, D.; Ma, C. Landslide susceptibility mapping using machine learning algorithm. *Civ. Eng. J* **2022**, *8*, 209–224. [CrossRef]
67. Pradhan, B.; Seeni, M.I.; Nampak, H. Integration of LiDAR and QuickBird data for automatic landslide detection using object-based analysis and random forests. *Laser Scanning Appl. Landslide Assess.* **2017**, 69–81. [CrossRef]
68. Pourghasemi, H.R.; Gokceoglu, C. *Spatial Modeling in GIS and R for Earth and Environmental Sciences*; Elsevier: Amsterdam, The Netherlands, 2019.
69. Hong, H.; Liu, J.; Zhu, A.-X. Modeling landslide susceptibility using LogitBoost alternating decision trees and forest by penalizing attributes with the bagging ensemble. *Sci. Total Environ.* **2020**, *718*, 137231. [CrossRef] [PubMed]
70. Arabameri, A.; Saha, S.; Mukherjee, K.; Blaschke, T.; Chen, W.; Ngo, P.T.T.; Band, S.S. Modeling spatial flood using novel ensemble artificial intelligence approaches in northern Iran. *Remote Sens.* **2020**, *12*, 3423. [CrossRef]
71. Jenks, G.F.; Caspall, F.C. Error on choroplethic maps: Definition, measurement, reduction. *Ann. Assoc. Am. Geogr.* **1971**, *61*, 217–244. [CrossRef]
72. Breiman, L. Random forests. *Mach. Learn.* **2001**, *45*, 5–32. [CrossRef]
73. Micheletti, N.; Foresti, L.; Robert, S.; Leuenberger, M.; Pedrazzini, A.; Jaboyedoff, M.; Kanevski, M. Machine learning feature selection methods for landslide susceptibility mapping. *Math. Geosci.* **2014**, *46*, 33–57. [CrossRef]
74. Calle, M.L.; Urrea, V. Stability of Random Forest importance measures. *Brief. Bioinform.* **2011**, *12*, 86–89. [CrossRef] [PubMed]
75. Park, S.; Kim, J. Landslide susceptibility mapping based on random forest and boosted regression tree models, and a comparison of their performance. *Appl. Sci.* **2019**, *9*, 942. [CrossRef]
76. Joshi, A.; Pradhan, B.; Chakraborty, S.; Behera, M.D. Winter wheat yield prediction in the conterminous United States using solar-induced chlorophyll fluorescence data and XGBoost and random forest algorithm. *Ecol. Inform.* **2023**, *77*, 102194. [CrossRef]
77. Bentéjac, C.; Csörgő, A.; Martínez-Muñoz, G. A comparative analysis of gradient boosting algorithms. *Artif. Intell. Rev.* **2021**, *54*, 1937–1967. [CrossRef]
78. Ma, B.; Meng, F.; Yan, G.; Yan, H.; Chai, B.; Song, F. Diagnostic classification of cancers using extreme gradient boosting algorithm and multi-omics data. *Comput. Biol. Med.* **2020**, *121*, 103761. [CrossRef]
79. Chen, T.; Guestrin, C. Xgboost: A scalable tree boosting system. In Proceedings of the 22nd ACM SIGKDD International Conference on Knowledge Discovery and Data Mining, San Francisco, CA, USA, 13–17 August 2016; pp. 785–794.
80. LeDell, E.; Poirier, S. H2o automl: Scalable automatic machine learning. In Proceedings of the AutoML Workshop at ICML, Vienna, Austria, 17 July 2020.
81. Zhang, W.; Wu, C.; Zhong, H.; Li, Y.; Wang, L. Prediction of undrained shear strength using extreme gradient boosting and random forest based on Bayesian optimization. *Geosci. Front.* **2021**, *12*, 469–477. [CrossRef]
82. Merghadi, A.; Yunus, A.P.; Dou, J.; Whiteley, J.; ThaiPham, B.; Bui, D.T.; Avtar, R.; Abderrahmane, B. Machine learning methods for landslide susceptibility studies: A comparative overview of algorithm performance. *Earth-Sci. Rev.* **2020**, *207*, 103225. [CrossRef]

83. Akinci, H.; Zeybek, M.; Dogan, S. Evaluation of landslide susceptibility of Şavşat District of Artvin Province (Turkey) using machine learning techniques. In *Landslides*; IntechOpen: London, UK, 2021.
84. Shin, H.-C.; Roth, H.R.; Gao, M.; Lu, L.; Xu, Z.; Nogues, I.; Yao, J.; Mollura, D.; Summers, R.M. Deep convolutional neural networks for computer-aided detection: CNN architectures, dataset characteristics and transfer learning. *IEEE Trans. Med. Imaging* 2016, *35*, 1285–1298. [CrossRef]
85. Xu, S.; Niu, R. Displacement prediction of Baijiabao landslide based on empirical mode decomposition and long short-term memory neural network in Three Gorges area, China. *Comput. Geosci.* 2018, *111*, 87–96. [CrossRef]
86. Hu, B.; Wang, H.-S.; Sun, Y.-L.; Hou, J.-G.; Liang, J. Long-term land subsidence monitoring of Beijing (China) using the small baseline subset (SBAS) technique. *Remote Sens.* 2014, *6*, 3648–3661. [CrossRef]
87. Oliver-Cabrera, T.; Jones, C.E.; Yunjun, Z.; Simard, M. InSAR phase unwrapping error correction for rapid repeat measurements of water level change in wetlands. *IEEE Trans. Geosci. Remote Sens.* 2021, *60*, 1–15. [CrossRef]
88. Xia, Z.; Motagh, M.; Li, T.; Roessner, S. The June 2020 Aniangzhai landslide in Sichuan Province, Southwest China: Slope instability analysis from radar and optical satellite remote sensing data. *Landslides* 2022, *19*, 313–329. [CrossRef]
89. Sataer, G.; Sultan, M.; Emil, M.K.; Yellich, J.A.; Palaseanu-Lovejoy, M.; Becker, R.; Gebremichael, E.; Abdelmohsen, K. Remote sensing application for landslide detection, monitoring along Eastern Lake Michigan (Miami Park, MI). *Remote Sens.* 2022, *14*, 3474. [CrossRef]
90. Zhou, C.; Cao, Y.; Yin, K.; Wang, Y.; Shi, X.; Catani, F.; Ahmed, B. Landslide characterization applying sentinel-1 images and InSAR technique: The muyubao landslide in the three Gorges Reservoir Area, China. *Remote Sens.* 2020, *12*, 3385. [CrossRef]
91. Hussain, M.A.; Chen, Z.; Zheng, Y.; Shoaib, M.; Ma, J.; Ahmad, I.; Asghar, A.; Khan, J. PS-InSAR Based Monitoring of Land Subsidence by Groundwater Extraction for Lahore Metropolitan City, Pakistan. *Remote Sens.* 2022, *14*, 3950. [CrossRef]
92. Khan, J.; Ren, X.; Hussain, M.A.; Jan, M.Q. Monitoring land subsidence using PS-InSAR technique in Rawalpindi and islamabad, Pakistan. *Remote Sens.* 2022, *14*, 3722. [CrossRef]
93. Bui, D.T.; Ngo, P.-T.T.; Pham, T.D.; Jaafari, A.; Minh, N.Q.; Hoa, P.V.; Samui, P. A novel hybrid approach based on a swarm intelligence optimized extreme learning machine for flash flood susceptibility mapping. *Catena* 2019, *179*, 184–196. [CrossRef]
94. Song, Y.; Niu, R.; Xu, S.; Ye, R.; Peng, L.; Guo, T.; Li, S.; Chen, T. Landslide susceptibility mapping based on weighted gradient boosting decision tree in Wanzhou section of the Three Gorges Reservoir Area (China). *ISPRS Int. J. Geo-Inf.* 2018, *8*, 4. [CrossRef]
95. Berardino, P.; Fornaro, G.; Lanari, R.; Sansosti, E. A new algorithm for surface deformation monitoring based on small baseline differential SAR interferograms. *IEEE Trans. Geosci. Remote Sens.* 2002, *40*, 2375–2383. [CrossRef]
96. Lanari, R.; Mora, O.; Manunta, M.; Mallorquí, J.J.; Berardino, P.; Sansosti, E. A small-baseline approach for investigating deformations on full-resolution differential SAR interferograms. *IEEE Trans. Geosci. Remote Sens.* 2004, *42*, 1377–1386. [CrossRef]
97. Bacha, A.S.; Shafique, M.; van der Werff, H. Landslide inventory and susceptibility modelling using geospatial tools, in Hunza-Nagar valley, northern Pakistan. *J. Mt. Sci.* 2018, *15*, 1354–1370. [CrossRef]
98. Rehman, M.U.; Zhang, Y.; Meng, X.; Su, X.; Catani, F.; Rehman, G.; Yue, D.; Khalid, Z.; Ahmad, S.; Ahmad, I. Analysis of landslide movements using interferometric synthetic aperture radar: A case study in Hunza-Nagar Valley, Pakistan. *Remote Sens.* 2020, *12*, 2054. [CrossRef]
99. Hussain, S.; Pan, B.; Afzal, Z.; Ali, M.; Zhang, X.; Shi, X. Landslide detection and inventory updating using the time-series InSAR approach along the Karakoram Highway, Northern Pakistan. *Sci. Rep.* 2023, *13*, 1–19. [CrossRef] [PubMed]
100. Kulsoom, I.; Hua, W.; Hussain, S.; Chen, Q.; Khan, G.; Shihao, D. SBAS-InSAR based validated landslide susceptibility mapping along the Karakoram Highway: A case study of Gilgit-Baltistan, Pakistan. *Sci. Rep.* 2023, *13*, 3344. [CrossRef]
101. Nhu, V.-H.; Shirzadi, A.; Shahabi, H.; Chen, W.; Clague, J.J.; Geertsema, M.; Jaafari, A.; Avand, M.; Miraki, S.; Talebpour Asl, D. Shallow landslide susceptibility mapping by random forest base classifier and its ensembles in a semi-arid region of Iran. *Forests* 2020, *11*, 421. [CrossRef]
102. Fang, Z.; Wang, Y.; Peng, L.; Hong, H. Integration of convolutional neural network and conventional machine learning classifiers for landslide susceptibility mapping. *Comput. Geosci.* 2020, *139*, 104470. [CrossRef]
103. Ding, A.; Zhang, Q.; Zhou, X.; Dai, B. Automatic recognition of landslide based on CNN and texture change detection. In Proceedings of the 2016 31st Youth Academic Annual Conference of Chinese Association of Automation (YAC), Wuhan, China, 11–13 November 2016; pp. 444–448.
104. Karantanellis, E.; Marinos, V.; Vassilakis, E.; Hölbling, D. Evaluation of machine learning algorithms for object-based mapping of landslide zones using UAV data. *Geosciences* 2021, *11*, 305. [CrossRef]
105. Rashid, B.; Iqbal, J.; Su, L.-J. Landslide susceptibility analysis of Karakoram highway using analytical hierarchy process and scoops 3D. *J. Mt. Sci.* 2020, *17*, 1596–1612. [CrossRef]
106. Hussain, M.L.; Shafique, M.; Bacha, A.S.; Chen, X.-Q.; Chen, H.-Y. Landslide inventory and susceptibility assessment using multiple statistical approaches along the Karakoram highway, northern Pakistan. *J. Mt. Sci.* 2021, *18*, 583–598. [CrossRef]
107. Qing, F.; Zhao, Y.; Meng, X.; Su, X.; Qi, T.; Yue, D. Application of machine learning to debris flow susceptibility mapping along the China–Pakistan Karakoram Highway. *Remote Sens.* 2020, *12*, 2933. [CrossRef]
108. Yi, Y.; Zhang, Z.; Zhang, W.; Jia, H.; Zhang, J. Landslide susceptibility mapping using multiscale sampling strategy and convolutional neural network: A case study in Jiuzhaigou region. *Catena* 2020, *195*, 104851. [CrossRef]
109. Sun, D.; Wen, H.; Wang, D.; Xu, J. A random forest model of landslide susceptibility mapping based on hyperparameter optimization using Bayes algorithm. *Geomorphology* 2020, *362*, 107201. [CrossRef]

110. Bui, D.T.; Tsangaratos, P.; Nguyen, V.-T.; Van Liem, N.; Trinh, P.T. Comparing the prediction performance of a Deep Learning Neural Network model with conventional machine learning models in landslide susceptibility assessment. *Catena* **2020**, *188*, 104426. [CrossRef]
111. Peethambaran, B.; Anbalagan, R.; Kanungo, D.; Goswami, A.; Shihabudheen, K. A comparative evaluation of supervised machine learning algorithms for township level landslide susceptibility zonation in parts of Indian Himalayas. *Catena* **2020**, *195*, 104751. [CrossRef]

Disclaimer/Publisher's Note: The statements, opinions and data contained in all publications are solely those of the individual author(s) and contributor(s) and not of MDPI and/or the editor(s). MDPI and/or the editor(s) disclaim responsibility for any injury to people or property resulting from any ideas, methods, instructions or products referred to in the content.

Article

Deformation Behavior and Reactivation Mechanism of the Dandu Ancient Landslide Triggered by Seasonal Rainfall: A Case Study from the East Tibetan Plateau, China

Sanshao Ren [1,2], Yongshuang Zhang [1,2,*], Jinqiu Li [1,2], Zhenkai Zhou [3], Xiaoyi Liu [4] and Changxu Tao [1]

[1] School of Engineering and Technology, China University of Geosciences, Beijing 100083, China; renss@cugb.edu.cn (S.R.); lijinqiu@mail.cgs.gov.cn (J.L.); 3002230018@cugb.edu.cn (C.T.)
[2] Institute of Hydrogeology and Environmental Geology, Chinese Academy of Geological Sciences, Shijiazhuang 050061, China
[3] Institute of Geomechanics, Chinese Academy of Geological Sciences, Beijing 100081, China; zhouzhenkai@mail.cgs.gov.cn
[4] China Aero Geophysical Survey and Remote Sensing Center for Natural Resources, Beijing 100083, China; 2101140091@cugb.edu.cn
* Correspondence: zhys100@cugb.edu.cn

Abstract: In recent years, numerous ancient landslides initially triggered by historic earthquakes on the eastern Tibetan Plateau have been reactivated by fault activity and heavy rainfall, causing severe human and economic losses. Previous studies have indicated that short-term heavy rainfall plays a crucial role in the reactivation of ancient landslides. However, the deformation behavior and reactivation mechanisms of seasonal rainfall-induced ancient landslides remain poorly understood. In this paper, taking the Dandu ancient landslide as an example, field investigations, ring shear experiments, and interferometric synthetic aperture radar (InSAR) deformation monitoring were performed. The cracks in the landslide, formed by fault creeping and seismic activity, provide pathways for rainwater infiltration, ultimately reducing the shear resistance of the slip zone and causing reactivation and deformation of the Dandu landslide. The deformation behavior of landslides is very responsive to seasonal rainfall, with sliding movements beginning to accelerate sharply during the rainy season and decelerating during the dry season. However, this response generally lags by several weeks, indicating that rainfall takes time to infiltrate into the slip zone. These research results could help us better understand the reactivation mechanism of ancient landslides triggered by seasonal rainfall. Furthermore, these findings explain why many slope failures take place in the dry season, which typically occurs approximately a month after the rainy season, rather than in the rainy season itself.

Keywords: ancient landslide; slip zone; reactivation mechanism; InSAR; seasonal rainfall

1. Introduction

There have been numerous giant ancient landslides triggered by paleo-earthquakes on the eastern Tibetan Plateau [1–3], and the platforms of these landslides are important living sites for people in high mountain canyon areas. However, in recent decades, many ancient landslides have been reactivated by fault activity and heavy rainfall (Figure 1), causing severe human and economic losses [4,5]. In 2018, the Jiangdingya ancient landslide in Zhouqu County, Gansu Province, was reactivated by heavy rainfall, blocking the Bailong River, flooding the upstream villages and towns, and destroying roads [6]. In 2018, influenced by the continuous cumulative rainfall in the previous 14 days, a giant ancient landslide in Boli Village, Yanyuan County, was reactivated, damaging 186 houses and causing significant economic losses [7]. In 2021, heavy rainfall led to the reactivation of the Moli landslide in Guoye Township, Zhouqu County, Gansu Province, causing the deformation of a large number of houses and threatening the lives of more than 1000 people [8]. In 2021, the

Citation: Ren, S.; Zhang, Y.; Li, J.; Zhou, Z.; Liu, X.; Tao, C. Deformation Behavior and Reactivation Mechanism of the Dandu Ancient Landslide Triggered by Seasonal Rainfall: A Case Study from the East Tibetan Plateau, China. *Remote Sens.* **2023**, *15*, 5538. https://doi.org/10.3390/rs15235538

Academic Editor: Rachid El Hamdouni

Received: 27 October 2023
Revised: 24 November 2023
Accepted: 26 November 2023
Published: 28 November 2023

Copyright: © 2023 by the authors. Licensee MDPI, Basel, Switzerland. This article is an open access article distributed under the terms and conditions of the Creative Commons Attribution (CC BY) license (https://creativecommons.org/licenses/by/4.0/).

Aniangzhai ancient landslide in Danba County, Sichuan Province, China, induced by heavy rainfall, was reactivated, and the Dadu River was dammed [9].

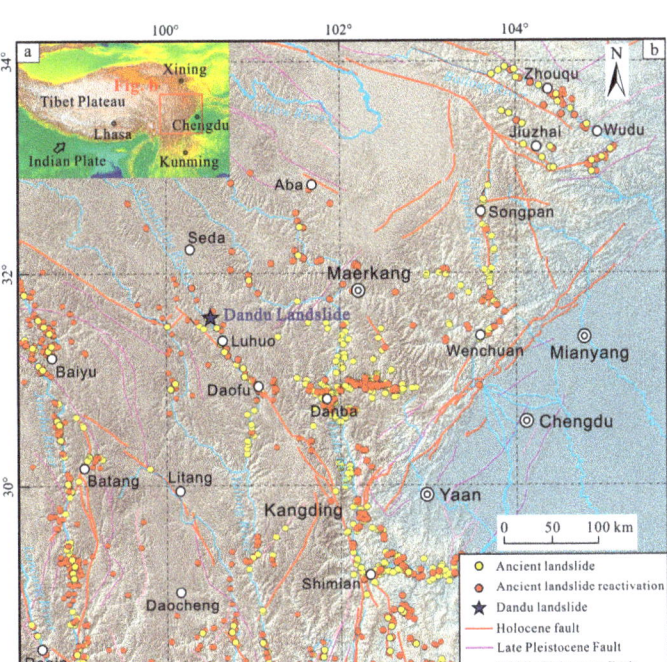

Figure 1. (**a**) Location of the study area on the Tibetan Plateau; (**b**) distribution of ancient landslides and reactivated landslides on the east Tibetan Plateau. The SRTM DEM was freely downloaded from USGS (https://earthexplorer.usgs.gov/, accessed on 10 October 2021), and the fault data were collected from the Geological Cloud, China Geological Survey (http://geocloud.cgs.gov.cn, accessed on 18 March 2019).

Currently, there are still numerous ancient landslides on the Tibetan Plateau that are slowly moving, with minimal displacement during the dry season but accelerated sliding during the rainy season [10,11]. These ancient landslides may also slide intermittently for several decades or even centuries [4]. Alternatively, they may experience rapid acceleration within a short period of time and catastrophic failure, leading to extensive destruction and fatalities [12–14]. Previous studies have indicated that short-term heavy rainfall is a key factor in the reactivation of ancient landslides [6,15], but the deformation behavior and reactivation mechanism of seasonal rainfall-induced ancient landslides, which are prerequisites for mitigating the hazards of landslide reactivation, are still not well understood [16–18].

Interferometric synthetic aperture radar (InSAR) is a measurement technique based on active microwave remote sensing that has been widely applied in the study of active landslides [19]. InSAR technology has the capability to capture ground deformations at the centimeter to millimeter level [20–23]. Time-series InSAR methods, such as the small baseline subset InSAR (SBAS-InSAR), can trace the historical deformation processes of landslides over time [23,24]. The SBAS-InSAR technique, with its excellent deformation-detection ability, has been used for studying surface deformation, especially landslide detection and reconstruction of the landslide evolution process [25], offering valuable information for analyzing the patterns and causes of active landslides. In recent years, small unmanned aerial vehicles (UAVs) have been increasingly utilized in the study of individual landslide deformations. They can acquire high-resolution optical orthophotos and precise digital surface model (DSM) data [26].

In general, ancient landslides are reactivated along pre-existing slip zones that have reached the residual state [27,28]. Therefore, understanding the mechanism of residual strength evolution in slip zones is a prerequisite for the stability assessment and engineering design of mitigation measures for ancient landslides [29–31]. The ring shear apparatus can reach almost unlimited shear displacements and is, therefore, widely used to assess the residual strength and resistance of soils [32,33].

In this paper, we utilized the Dandu ancient landslide, which is located upstream of the Xianshui River on the eastern Tibetan Plateau, as a typical case. Field surveys, unmanned aerial vehicle (UAV) mapping, ring shear experiments, and InSAR deformation monitoring were employed to study the deformation behavior and reactivation mechanism. The study results are important references for the disaster risk prevention of ancient landslide reactivation on the eastern Tibetan Plateau.

2. Study Area

2.1. Geological Setting

The Dandu landslide is situated in Luhuo County, Sichuan Province, China, on the eastern margin of the Tibet Plateau, upstream of the Xianshui River, at 100°27′14.00″E, 31°33′16.40″N. The G317 national road passes over the front edge of the landslide. The elevation of the surrounding mountain tops in the study area ranges from 3700 to 4000 m, and there is a relative elevation difference of 800–1000 m from the mountain peaks to the valley.

The Xianshui River active fault, the southwestern boundary fault of the Bayan Har block [34], passes through the middle of the landslide (Figure 2). The Xianshui River Fault, the most active fault on the Tibetan Plateau, is a left-trending strike-slip fault with a total length of approximately 350 km, an overall strike of 320~330°, and an average sliding rate of 10 mm/a [34–36]. Since 1725, the Xianshui River Fault has been involved in a total of 9 earthquakes with magnitudes higher than Ms 7.0, as well as 17 earthquakes with magnitudes of Ms 6–6.9 [35]. Notable examples include the Ms 7.5 earthquake in 1816 and the Ms 7.6 earthquake in Fuhuo County in 1973 [35,36]. These historical earthquakes have triggered numerous large-scale ancient landslides that are similar to the Dandu landslide [1].

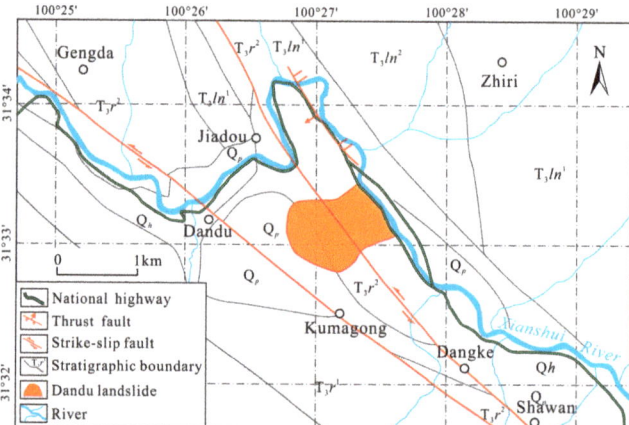

Figure 2. A geological background map of the Dandu landslide. Q_p: Holocene alluvium; Q_h: Pleistocene alluvium; T_3ln^1: metamorphic sandstone of the lower part Lianghekou Formation in the Upper Triassic; T_3ln^2: Metamorphic sandstone of the middle part of Lianghekou Formation in the Upper Triassic; T_3r^1: metamorphic sandstone interbedded with slate of the lower part of Ruganian Formation in the Upper Triassic; T_3r^2: basalt, sandstone, limestone, and slate of the upper part of Rugenian Formation in the Upper Triassic.

The climate type in the study area is sub-humid, which is typical in the Tibet Plateau, as the dry and rainy seasons are distinct. The average annual precipitation in the region is 672.8 mm. The majority of rainfall occurs in the rainy season, which spans from May to September and accounts for approximately 86.4% of the average annual precipitation [1,37].

2.2. Features of Ancient Landslide

The Dandu ancient landslide was a rockslide triggered by a historic earthquake. In plan view, the ancient landslide has a "long-tongued" shape, and its rear edge exhibits a "crown" geomorphology [38,39]. The longitudinal length of the landslide measures approximately 1180 m, while the average width is approximately 650 m. The total surface area of the landslide is about 70×10^4 m^2. The thickness of the landslide mass ranges from 20 to 30 m, for a total volume of approximately 1600×10^4 m^3. The profile of the landslide is characterized by stepped slopes, with an inclination ranging from 10° to 15°.

The lithology of the Dandu landslide area consists of basalt, sandstone, limestone, and slate of the Rugenian formation (T_3r^2). The origin of the strata is 300°∠75°, which was affected by the fault; the rock structure was broken, and the slickensides of the bedrock formed by extrusion and shearing are obvious. The lithology of the landslide mass is a mixture of soil and rock, with 30% to 60% gravel content and a gravel grain size of 5–20 cm. The lithology of the gravel is metamorphic sandstone and limestone, and most of the basalt in the landslide has been weathered into soil. Giant boulders, up to 6 m in diameter, are distributed in the channel at the leading edge of the landslide and in the middle of the landslide, and the boulder lithology is mainly weather-resistant limestone.

3. Materials and Methods

3.1. Field Investigation and UAV Photography

A field investigation was conducted to observe and study the deformation features, such as cracks and scarps, of the Dandu landslide. Additionally, the field characteristics of the slip zone soil were examined. The orthophotos were taken using a UAV (DJI Mavic 3E, manufactured by DJI Innovation Technology Co. in Shenzhen, China). The DJI Mavic 3E was equipped with a wide-angle camera with an effective pixel of 20 million, and it adopted network RTK for precise positioning, supporting high-precision, high-efficiency surveying and mapping operations (Figure 3a). The South Surveying and Mapping Company's reference station was used, and the UAV was 300–2000 m from the reference station during the flight. On 18 October 2022, four flights were conducted at a flight altitude of 300 m and a flight area of about 3.5 km^2, obtaining a total of 480 images (Figure 3b), with photo resolutions of 5472 × 3078. The planned routes have a 70 percent longitudinal overlap and a 50 percent transverse overlap. Furthermore, a digital surface model (DSM) was created using the DJI Smart Map (3.7.0) software (Figure 3c,d). The DSM, with a resolution of 0.5 m, synthesized the collected data and provided a comprehensive understanding of the morphological characteristics of the landslide and its recent deformation behavior.

3.2. Deformation Monitoring by InSAR

Taking into account the well-developed summer grasslands and shrubbery on the surface of the Dandu landslide, this study utilized the computation method of SBAS-InSAR. Compared to traditional persistent scatterer (PS-InSAR) algorithms, SBAS-InSAR technology utilizes short temporal and spatial baseline sets, improving the decorrelation issues caused by a single master image and enhancing the spatial coverage density of the measurement points [22]. Simultaneously, SBAS-InSAR technology holds advantages for measuring rapid deformations in landslides over short periods. The basic principle was as follows [40].

(1) N + 1 views of SAR images covering the study area were obtained, acquired at times: $t_0, t_1, \ldots, t_n$. Suitable spatial–temporal baseline thresholds were set to register the slave images with the master images. The interferometric pairs were obtained accordingly, at a number M.

(2) M pairs of interferometric pairs were used to generate time-series interferograms for multi-master images.
(3) The regression algorithm was applied to the deformation dataset to estimate and remove the elevation residuals; the residual phases, such as noise and atmospheric delays, were separated according to the selected combined filter methods.
(4) The deformation time series was reconstructed using the small baseline set time series deformation solution model. With t_0 as the reference moment, the differential interferometric phase was acquired during data processing with observed quantities. Time t_i was the relative time i to t_0 ($0 < i < N$) and obtained unknown quantities, and the interferometric phase value of the image element (r, c) was:

$$\delta\varphi_i(r,c) = \varphi(t_B,r,c) - \varphi(t_A,r,c) \approx \frac{4\pi}{\lambda}[d(t_B,r,c) - d(t_A,r,c)] \quad (1)$$

where λ is the radar wavelength and $d(t_B, r, c,)$ and $d(t_A, r, c,)$ are the deformation of pixels traveling at moments t_B and t_A along the radar line-of-sight direction, respectively.

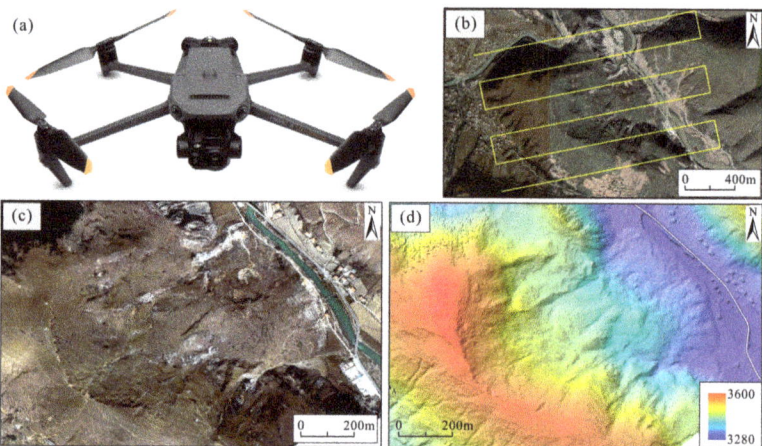

Figure 3. UAV image acquisition and processing: (**a**) unmanned aerial vehicle (DJI Mavic 3E); (**b**) planned flight routes; (**c**) orthophoto generated by DJI Smart Map software with a resolution of 0.1 m; (**d**) digital surface model (DSM) with a resolution of 0.5 m.

The deformation characteristics of the Dandu landslide were analyzed using a total of 149 Sentinel-1A SAR images acquired in ascending geometry. These images were collected from 7 January 2018 to 24 December 2022. The detailed parameters of the SAR dataset are shown in Table 1. The SAR image from 9 October 2020 was chosen as the master image. The remaining SAR images were registered with the master image to ensure azimuthal registration accuracy to within one-thousandth of a pixel. The spatial vertical baseline lengths between SAR data were mostly within 200 m, with the longest being 266.5 m. The time intervals for InSAR computations ranged from a maximum of 48 d to a minimum of 12 d, with a total of 441 interferometric data pairs (Figure 4).

Table 1. Parameters of SAR data.

Satellite	Track	Date Range	Number of Images	Revisit Cycle (Days)	Resolution (m)		Angle of Incidence (°)
					Azimuth	Range	
Sentinel-1	P26	7 January 2018–24 December 2022	149	12/24	13.98	2.33	34.71

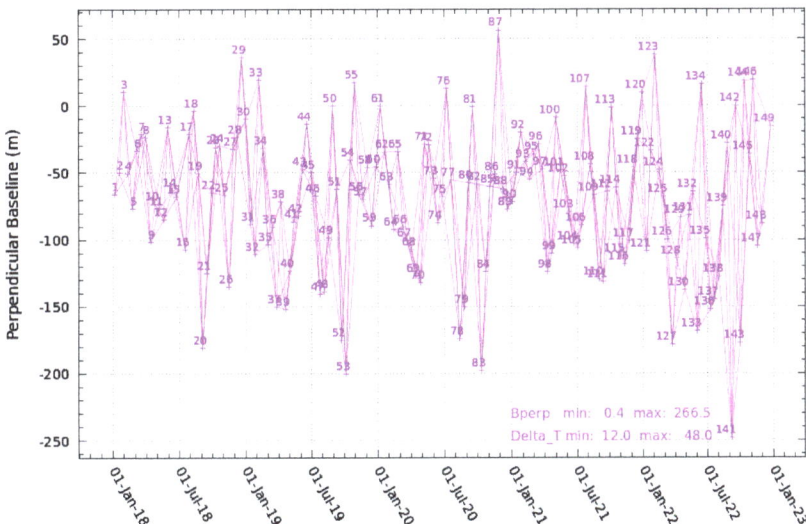

Figure 4. Spatial perpendicular baseline of the 149 SAR images in the SBAS-InSAR process.

3.3. Ring Shear Test of Slip Zone Soil

In this study, an ARS-3 ring shear machine from Wille-Geotechnik, Germany [41], was utilized. The machine featured a circular shear box with an inner diameter of 50 mm, an outer diameter of 100 mm, and a height of 25 mm. The normal stress and the torque were controlled by the servo-actuated loading piston and the servo hydraulic motor, respectively. The maximum axial pressure of the machine was 10 kN; the maximum shear stress was 1000 kPa; and the maximum shear rate was 100 mm/min. Transducers installed in the pressurized system measured the shear stress and normal stresses, and the test data was automatically collected by a data acquisition device and transferred to a computer (Figure 5). During shearing, the hanging wall was fixed, while the foot wall with the shear box led to a shear surface in the vicinity of the shearing gap.

Field investigations revealed that the Dandu landslide sheared out from the bed of the Xianshui River, where slip zone soil can be seen. The basic physical properties of the slip zone soil are presented in Table 2 (Figure 6). The slip zone is predominantly composed of gravelly, gray-green silt-clay, which is a result of the compression, kneading, and argillization of basalt and sericite slate. The thickness of the slip zone measures between 0.2 and 0.3 m and exhibits clear slickensides (Figure 7c,d). Additionally, there is a significant amount of groundwater outflow along the slip surface (Figure 7e). The dry slip zone is characterized by its dense and hard composition. However, the water-soaked slip zone has a mud-like soil with considerably low strength.

Table 2. Basic physical properties of the slip zone soil of the Dandu landslide.

Dry Density (g/cm^3)	Plastic Limit (%)	Liquid Limit (%)	Plasticity Index (I_P)	Particle Size Distribution (mm, %)			
				<0.005	0.005~0.075	0.075~2	>2
1.80~1.88	21.1~21.5	37.0~37.7	15.9~16.2	16	26	38	20

In this study, a total of eight groups of remolded samples were tested, each with different normal stresses. The particle size of the samples used in the tests was less than or equal to 2 mm. Four samples were tested at natural moisture content (10%), and the other four samples needed to be saturated. The water content of the saturated samples was approximately 21%. After saturation was complete, the sample was placed in the shear box. Prior to each shear test, the samples were consolidated for 24 h, then sheared to reach their residual states at a shear rate of 0.2 mm/min. The ring shear tests were conducted following the standard geotechnical test methods (GB/T50123-2019) [42]. The specific testing scheme for the ring shear tests is provided in Table 3.

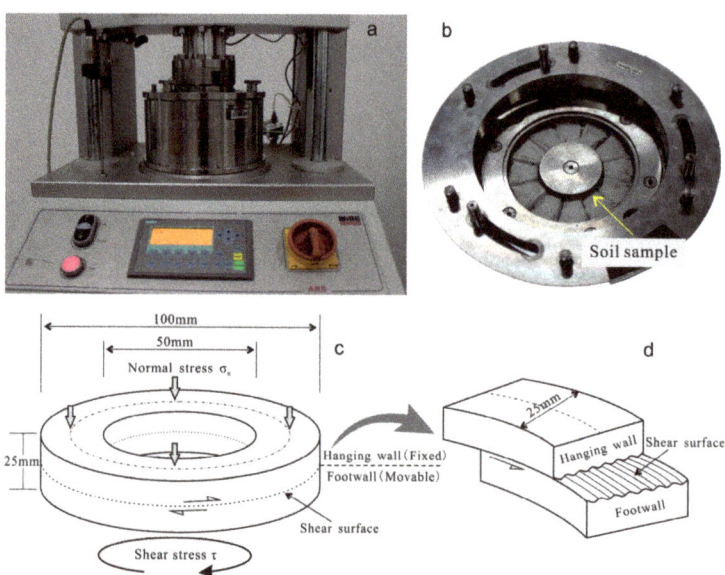

Figure 5. Schematic illustration of the ring shear test. (**a**) ARS-3 ring shear apparatus; (**b**) soil sample in the shear box; (**c**) size of the test specimen; and (**d**) shear surface.

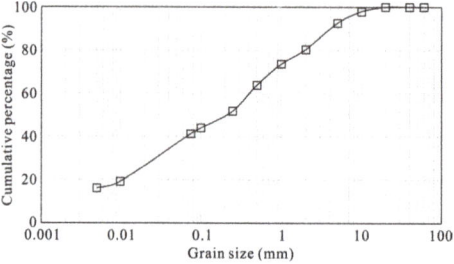

Figure 6. Particle size distribution of the slip zone.

Table 3. Ring shear test scheme for slip zone soil of the landslide.

Sample Number	Dry Density ρ (g/cm^3)	Normal Stress (σ_n/kPa)	Initial Water Content	Particle Size Distribution (mm, %)		
				<0.005	0.005~0.075	0.075~2
DD01	1.83	100	10%	20	33	47
DD02	1.84	200				
DD03	1.82	400				
DD04	1.84	800				
DD05	1.81	100	Saturated			
DD06	1.82	200				
DD07	1.80	400				
DD08	1.80	800				

Figure 7. Reactivation features of the Dandu landslide: (**a**) overall view of the Dandu landslide; (**b**) deformation zone at the front of the landslide; (**c**) slip zone at the landslide's toe; (**d**) slickensides on the slip surface; (**e**) groundwater outflow along the slip zone; and (**f**) cracks in the front of the landslide.

4. Results
4.1. Reactivation Features and Zonation of the Landslide

The Dandu ancient landslide can be divided into three zones based on deformation and cumulative displacement (Figures 7a and 8). Zone I, an ancient landslide, was initially triggered by a historic earthquake, and the main scarp is clear. Zone II is a secondary landslide that formed due to the reactivation of the ancient landslide. Zone II can be further divided into two subzones, namely, II-1 and II-2.

Zone II-1 is located on the left side of the front edge of the ancient landslide and is a rotational landslide. The landslide has a longitudinal length of 360 m and a transverse width of 300 m, with an area of about 9×10^4 m^2, a thickness of 15 to 25 m, and a volume of about 150×10^4 m^3 (Figure 7). Currently, significant deformation can be observed in Zone II-1, characterized by a steep scarp measuring 10–15 m in height. Additionally, numerous cracks have developed in this zone (Figures 7b, 8 and 9).

Zone II-2, located on the right side of the front edge of the ancient landslide, is relatively stable compared to Zone II-1. The longitudinal length of the landslide is 540 m, and the transverse width ranges from 250 to 320 m. The total area affected by the landslide is approximately 13×10^4 m^2. The thickness of the landslide ranges from 15 to 25 m, and the volume of the landslide is estimated to be around 200×10^4 m^3.

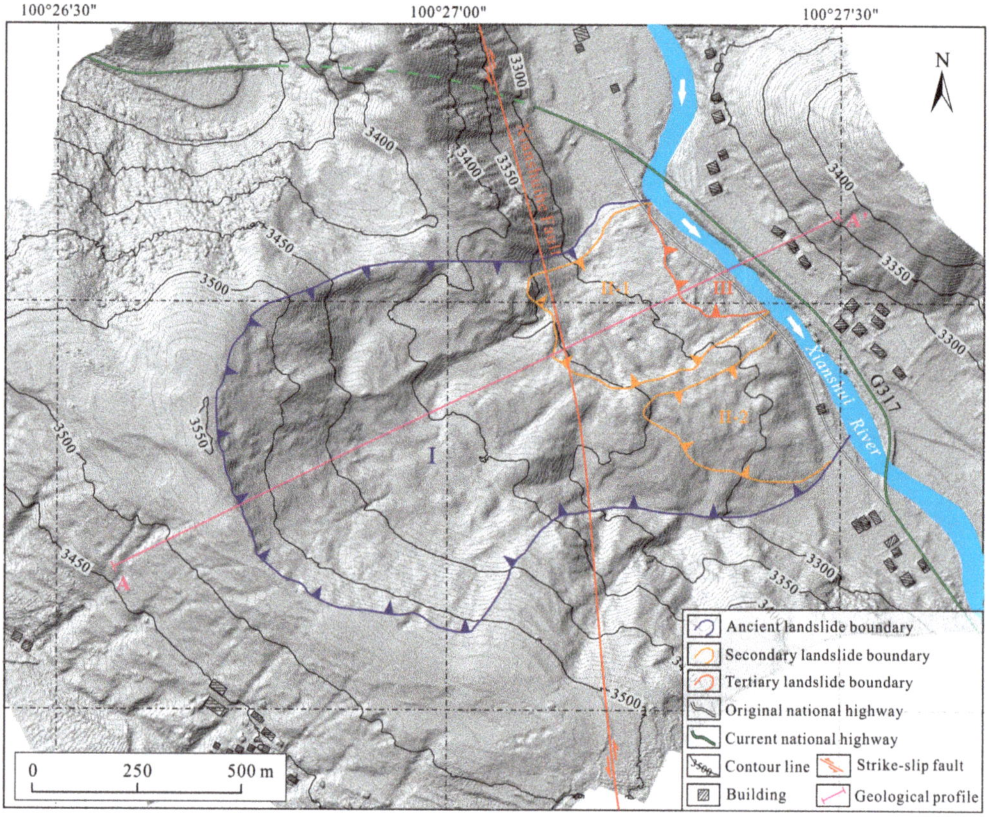

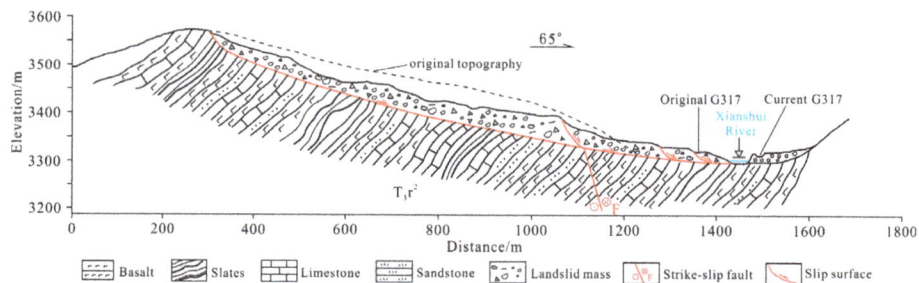

Figure 8. Engineering geological map of the Dandu landslide.

Figure 9. Engineering geological profile of the Dandu landslide (see A-A′ profile in Figure 8).

Zone III is a tertiary landslide formed by reactivation along the front edge of Zone II-1, with a scarp height of 5 m, a longitudinal length of 100 m, a transverse width of 100~250 m, an area of about 1.5×10^4 m², a thickness of about 20m, and a landslide volume of about 30×10^4 m³ (Figure 7a,b and Figure 8). The deformation was strongest in Zone III, where there are several tension cracks 10–30 cm wide and extending over 10 m in the middle and lower parts of the landslide, which has led to ground uplift and destruction of the retaining wall at the landslide toe. The original G317 road also suffered significant damage due to landslide deformation. As a result, a new G317 road had to be constructed on the opposite bank of the Xianshui River. Currently, zone III is still creeping, squeezing the channel of the

Xianshui River, and, as the foot of the landslide continues to be eroded by the river, sliding of the landslide could accelerate and pose a threat to the new G317 road (Figures 7–9).

4.2. Deformation Characteristics Monitored by InSAR

The SBAS-InSAR results in Figure 7 indicate deformation rates ranging from −45 to 19 mm/a from 2018 to 2022. The negative values represent deformation, with a moving trend against the sensors. It can be observed that most of the Dandu ancient landslide is in a stable state. However, significant deformations have been observed at the front edge of the landslide, specifically in zone II-1 and zone III, over the past five years. The highest deformation rate is observed in Zone III, with an approximate rate of 45 mm/y. Zone II-1 shows a deformation rate of approximately 25 mm/y. In contrast, the deformations in zone I and zone II-2 were relatively small, both less than 10 mm/y. These deformation characteristics, observed through InSAR monitoring, align well with the findings from field investigations (Figure 7). The reactivation of ancient landslides exhibits multiple periods and multiple zones of deformation, with lower-order sequences showing higher deformation rates and poorer stability (Figures 8 and 10).

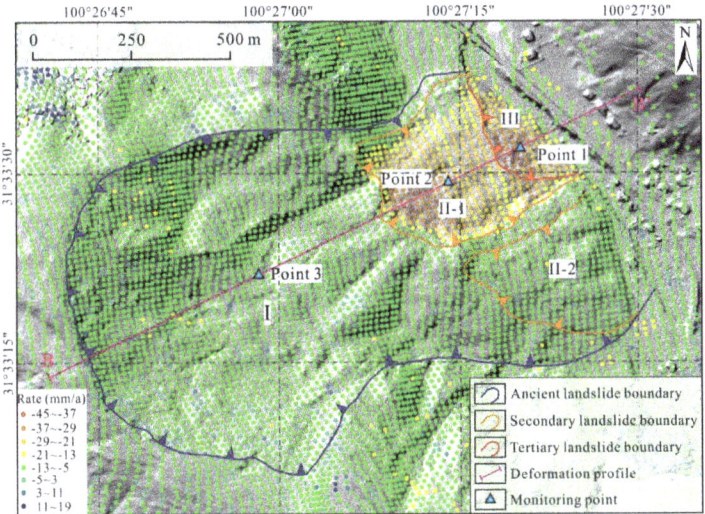

Figure 10. Deformation rate map of the Dandu landslide monitored by SBAS-InSAR from 2018 to 2022.

The analysis of the Dandu landslide involved selecting profile B-B' along the main sliding direction (Figure 10). A profile map of the deformation rate was then generated (Figure 11). The results indicate that the deformation rate varies significantly along the sliding direction. The high values, represented by negative values, are primarily concentrated at the front edge of the landslide. This observation suggests that the landslide predominantly exhibits overall tensile sliding.

To further investigate the dynamic behavior of the landslide in response to seasonal changes, we selected three points (labeled 1, 2, and 3) in Zones I, II-1, and III of the landslide (Figures 10 and 11). These points were then subjected to time series analysis using the finite-difference formula to determine the velocity of their movements.

$$vi = (di+1 - di)/(ti+1 - ti) \qquad (2)$$

In Equation (2), v_i represents the velocity of point P, at time t_i, and d_i and d_{i+1} represent the cumulative displacements of the points at times t_i and t_{i+1}, respectively.

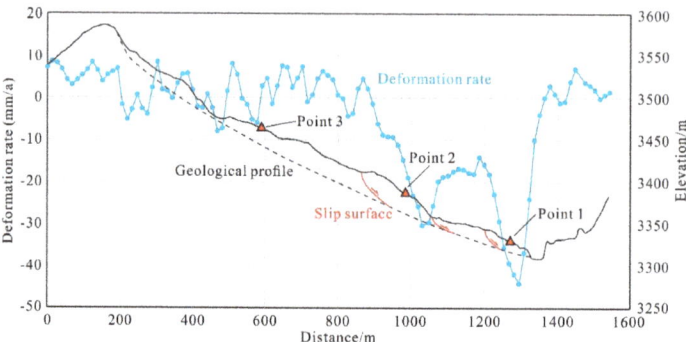

Figure 11. Deformation rate along the B-B′ profile (B-B′ profile in Figure 10).

The findings indicate that the maximum cumulative displacements observed at points 1, 2, and 3 between 2018 and 2022 were approximately 220 mm, 100 mm, and 20 mm, respectively. These points exhibited movement patterns that followed annual cycles consistently throughout the years.

The sliding motion initiated an acceleration phase with the onset of rainy seasons and significantly decelerated during dry seasons (Figure 12). A substantial amount of deformation was observed from mid-June to mid-December, while minimal displacements occurred during the dry season from mid-December to mid-June. These observations suggest that seasonal rainfall plays a significant role in triggering the reactivation and deformation of the Dandu landslide.

Based on the data presented in Figure 12b, it can be observed that even though the wet season consistently begins in mid-May each year, the acceleration of the landslide does not occur until a few weeks later, typically in early June. However, the specific time lag varies from year to year. The maximum recorded time lags are 24 days, 30 days, 34 days, and 26 days for the years 2018, 2019, 2020, and 2022, respectively (Figure 12b). It should be noted that no significant acceleration was observed in 2021. While it is possible that displacements may occur before the satellites capture the deformation, considering the minimum revisit period of Sentinel-1A datasets, which is 12 days, it can be inferred that the range of time lags fell between 12–24 days, 18–30 days, 22–34 days, and 14–26 days for the years 2018, 2019, 2020, and 2022, respectively.

4.3. Shear Strength of the Slip Zone

The ring shear tests showed that the stress–displacement curves exhibit significant strain-softening characteristics (Figure 13a,b). Shear stress can quickly reach its peak value (usually at 10 mm displacement) with relatively small displacements and then gradually decline to the residual state. Under saturated conditions, the shear strength decreased significantly compared to that under natural conditions (Figure 13c,d). Specifically, the peak internal friction angle decreased from 15.1° to 9.5°, a reduction of 37%. The peak cohesion also decreased from 33.4 kPa to 12.1 kPa, a reduction of 64%. Similarly, the residual internal friction angle decreased from 9.5° to 6.8°, a reduction of 28%, and the residual cohesion decreased from 14.2 kPa to 8.0 kPa, a reduction of 44%. In the saturated state, the shear surface was smoother (Figure 14), indicating that the presence of water not only softened the slip zone, thereby reducing the shear strength, but also promoted the directional arrangement of particles, leading to a reduction in the roughness of the shear surface. The smoother shear surface contributed to lowering the shear strength of the slip zone.

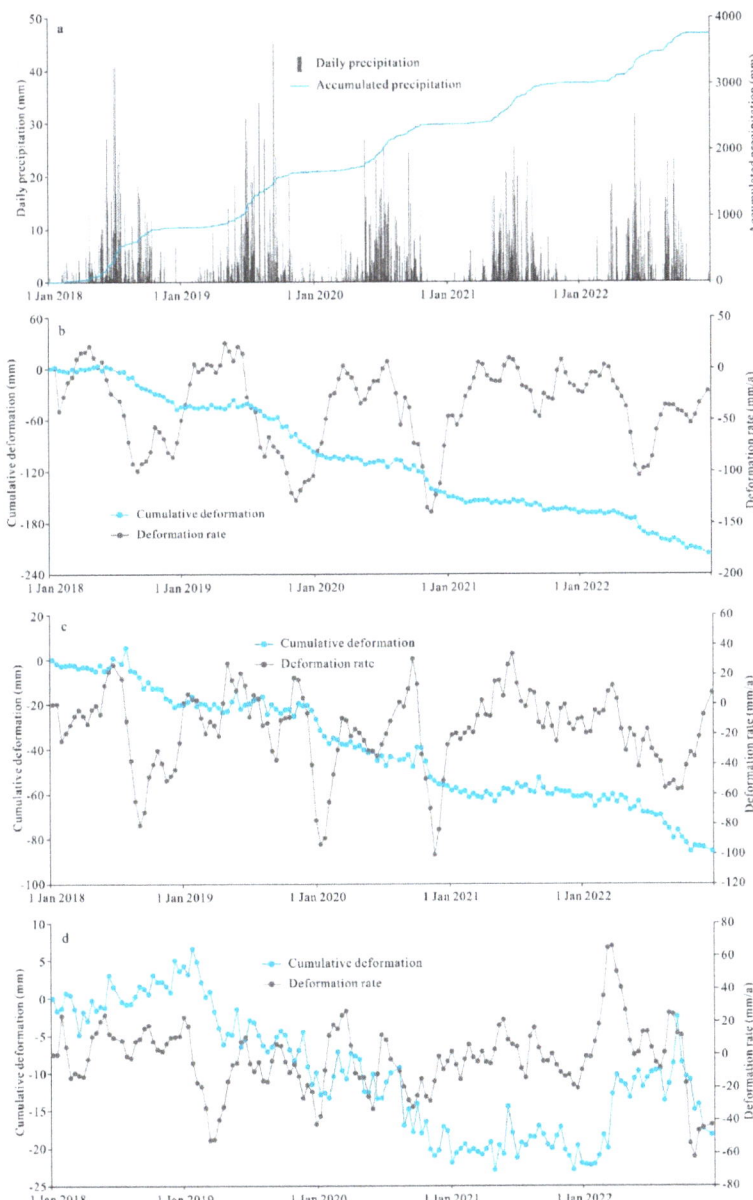

Figure 12. Relationship between landslide displacement and precipitation: (**a**) the gray line at the bottom represents daily precipitation, and the blue line represents accumulated precipitation; (**b**) relationship among landslide cumulative deformation, deformation rate, and time at point 1; (**c**) relationship among landslide cumulative deformation, deformation rate, and time at point 2; and (**d**) relationship among landslide cumulative deformation, deformation rate, and time at point 3. The positions of points 1, 2, and 3 are shown in Figure 8.

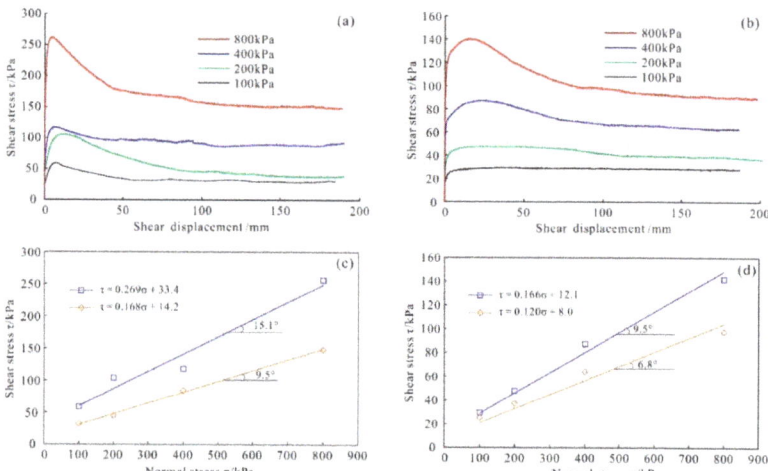

Figure 13. Shear stress against shear displacement curves and shear strength failure envelopes for the slip zone of the Dandu landslide: (**a**) shear stress against shear displacement curves; initial water content was 10%; (**b**) shear stress against shear displacement curves; saturated water content was approximately 21%; (**c**) shear strength failure envelopes; initial water content was 10%; and (**d**) shear strength failure envelopes; saturated water content was approximately 21%.

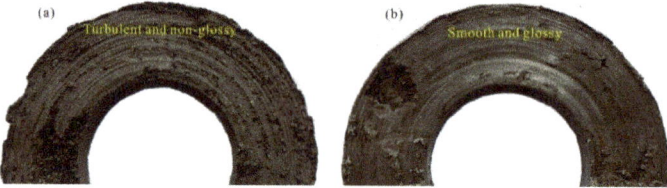

Figure 14. Shear surface characteristics: (**a**) the initial water content was 10%; (**b**) the saturated water content was approximately 21%.

5. Discussion

5.1. Active Faults Are the Driving Forces for the Formation of Landslide Cracks

Previous studies have claimed that creeping of active faults controls the local stress field and affects slope stability [43,44]. On the one hand, the field investigation found that the Xianshui River Fault passes through the middle of the Dandu landslide, and the horizontal slip rate of this fault has, since the Holocene, reached 10–20 mm/a [34–36]; its influence on the ancient landslide cannot be ignored. On the other hand, the Xianshui River Fault is prone to frequent earthquakes, with nine earthquakes of magnitudes of Ms ≥ 7.0 occurring along the fault since 1725. Notable examples include the Ms 7.5 earthquake in 1816 and the Ms 7.6 earthquake in Fuhuo County in 1973 [34–36]. These seismic events likely contributed to the formation of landslide cracks [45]. Therefore, it can be inferred that landslide cracks, formed by fault creeping and seismic activity, provide a preponderance of infiltration paths for rainwater and are key factors in landslide reactivation (Figure 15).

5.2. Pre-Existing Slip Zones Are the Essence of Landslide Reactivation

Ancient landslides tend to reactivate along pre-existing slip zones that have reached residual states [29,30]. Previous studies have confirmed, via experiments and back analysis, that when a landslide is reactivated, the initiation strength is essentially equal to the residual strength of the slip zone [46,47]. The Dandu landslide has been creeping for at least 10 years [48], and long-term creeping has induced a gradual directional alignment of the soil particles on the slip surface [31], which will lead to a gradual decrease in shear strength (Figure 15).

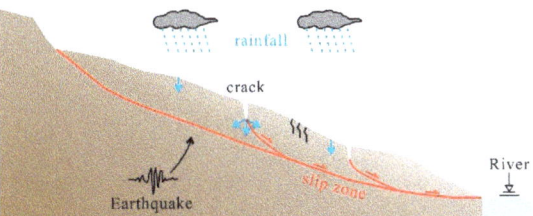

Figure 15. A model of ancient landslide reactivation triggered by seasonal rainfall.

5.3. Rainfall Is a Trigger Factor for Landslide Reactivation

Rainfall is a crucial trigger factor for landslide reactivation [49]. The InSAR monitoring results show that the deformation characteristics of the Dandu landslide respond very well to the rainy season, although there is some lag. Rainwater infiltrates into the landslide mass through cracks, which not only generates pore water pressure and reduces the effective stresses [50] but, more importantly, severely weakens the shear strength of the slip zone. Field investigations have also revealed the presence of multiple springs in the front portion of the landslide. Particularly, groundwater was observed flowing along the shear outlets of the landslide. The dry slip zone was dense and compact, while the water-soaked slip zone was muddy and exhibited low strength. Ring shear tests further confirmed that the strength of the slip zone in the Dandu landslide decreased significantly with the increasing water content.

It was also found that it takes time for rainwater to infiltrate into the slip zone [44,51], which explains why the response of landslide deformation to rainfall lags by several weeks. Furthermore, this may also explain why many landslides occur during the dry season [44,52], approximately a month after the rainy season, on the eastern margin of the Tibetan Plateau. In the past, it was commonly assumed that these landslides were not directly related to rainfall [53].

6. Conclusions

Taking the Dandu ancient landslide as a typical case, field surveys, ring shear experiments, and InSAR monitoring were performed to investigate the deformation behavior and reactivation mechanism. The following conclusions were drawn:

- The reactivation of the Dandu ancient landslides exhibits multiple periods and multiple zones of deformation, with lower-order sequences showing higher deformation rates and poorer stability. The deformation rates in zones III, II-1, and I were 40 mm/a, 20 mm/a, and less than 10 mm/a, respectively.
- The deformation characteristics of the Dandu landslide respond very well to seasonal rainfall. The sliding motion starts to accelerate after the rainy season arrives and decelerates substantially when the dry season arrives. However, this response generally lags by several weeks.
- The cracks in the landslide, formed by fault creeping and seismic activity, provide pathways for rainwater infiltration, ultimately reducing the shear resistance of the slip zone and causing the reactivation and deformation of the Dandu landslide. Meanwhile, rainfall infiltration takes time, which is why the response of landslide deformation to rainfall lags by several weeks.

Author Contributions: S.R. and Y.Z. framed the study plan and wrote the paper; S.R. and J.L. carried out field surveys; Z.Z. conducted the InSAR data treatment; X.L. processed rainfall data; and C.T. processed some figures. All authors have read and agreed to the published version of the manuscript.

Funding: This research was funded by the National Natural Science Foundation of China (Grant Nos. 41731287 and 42307229), the Key Laboratory of Airborne Geophysics and Remote Sensing Geology Foundation (No. 2023YFL22), and the Key Research and Development Program in Ningxia, China (No. 2021BEG03118).

Data Availability Statement: The Sentinel-1 datasets used in this study were provided by Copernicus and ESA, and the precipitation data were obtained from the National Meteorological Science Data Center of China free of charge. Other data presented in this paper are available on request from the corresponding author.

Acknowledgments: The authors would like to acknowledge the assistance of Dong Wenping and Guo Changbao in conducting the experiments and field surveys. The figures in this study were created using ArcGIS 10.7 and CorelDRAW X7 software. The authors also thank editors and reviewers for their insightful comments and suggestions.

Conflicts of Interest: The authors declare that there are no conflict of interest.

References

1. Guo, C.; Montgomery, D.R.; Zhang, Y.; Wang, K.; Yang, Z. Quantitative assessment of landslide susceptibility along the Xianshui fault zone, Tibetan Plateau, China. *Geomorphology* **2015**, *248*, 93–110. [CrossRef]
2. Chen, J.; Zhou, W.; Cui, Z.; Li, W.; Wu, S.; Ma, J. Formation process of a large paleolandslide-dammed lake at Xuelongnang in the upper Jinsha river, SE Tibetan Plateau: Constraints from OSL and 14C dating. *Landslides* **2018**, *15*, 2399–2412. [CrossRef]
3. Bao, Y.; Zhai, S.; Chen, J.; Xu, P.H.; Zhou, X.; Zhan, J.; Zhang, W.; Zhou, X. The evolution of the Samaoding paleo landslide river blocking event at the upper reaches of the Jinsha river, Tibetan Plateau. *Geomorphology* **2019**, *351*, 106970.
4. Lacroix, P.; Handwerger, A.L.; Bièvre, G. Life and death of slow-moving landslides. *Nat. Rev. Earth Environ.* **2020**, *1*, 404–419. [CrossRef]
5. García-Delgado, H. The San Eduardo Landslide (Eastern Cordillera of Colombia): Reactivation of a deep-seated gravitational slope deformation. *Landslides* **2020**, *17*, 1951–1964. [CrossRef]
6. Guo, C.; Zhang, Y.; Li, X.; Ren, S.; Yang, Z.; Wu, R.; Jin, J. Reactivation of giant Jiangdingya ancient landslide in Zhouqu County, Gansu Province, China. *Landslides* **2020**, *17*, 179–190. [CrossRef]
7. He, K.; Ma, G.; Hu, X.; Liu, B. Failure mechanism and stability analysis of a reactivated landslide occurrence in Yanyuan City, China. *Landslides* **2021**, *18*, 1097–1114. [CrossRef]
8. Yang, X.; Jiang, Y.; Zhu, J.; Ding, B.; Zhang, W. Deformation characteristics and failure mechanism of the Moli landslide in Guoye Town, Zhouqu County. *Landslides* **2023**, *20*, 789–800. [CrossRef]
9. Dai, K.; Li, Z.; Xu, Q.; Tomas, R.; Li, T.; Jiang, L.; Zhang, J.; Yin, T.; Wang, H. Identification and evaluation of the high mountain upper slope potential landslide based on multi-source remote sensing: The Aniangzhai landslide case study. *Landslides* **2023**, *20*, 1405–1417. [CrossRef]
10. Xie, M.; Zhao, W.; Ju, N.; He, C.; Huang, H.; Cui, Q. Landslide evolution assessment based on InSAR and real-time monitoring of a large reactivated landslide, Wenchuan, China. *Eng. Geol.* **2020**, *277*, 105781. [CrossRef]
11. Yan, Y.; Guo, C.; Li, C.; Yuan, H.; Qiu, Z. The Creep-Sliding Deformation Mechanism of the Jiaju Ancient Landslide in the Upstream of Dadu River, Tibetan Plateau, China. *Remote Sens.* **2023**, *15*, 592. [CrossRef]
12. Iverson, R.M.; George, D.L.; Allstadt, K.; Reid, M.E.; Collins, B.D.; Vallance, J.W.; Schilling, S.P.; Godt, J.W.; Cannon, C.M.; Magirl, C.S.; et al. Landslide mobility and hazards: Implications of the 2014 Oso disaster. *Earth Planet. Sci. Lett.* **2015**, *412*, 197–208. [CrossRef]
13. Palmer, J. Creeping earth could hold secret to deadly landslides. *Nature* **2017**, *548*, 384–386. [CrossRef] [PubMed]
14. Agliardi, F.; Scuderi, M.M.; Fusi, N.; Collettini, C. Slow-to-fast transition of giant creeping rockslides modulated by undrained loading in basal shear zones. *Nat. Commun.* **2020**, *11*, 1352. [CrossRef] [PubMed]
15. Zhang, Y.; Meng, X.; Novellino, A.; Dijkstra, T.; Chen, G.; Jordan, C.; Li, Y.; Su, X. Characterization of pre-failure deformation and evolution of a large earthflow using InSAR monitoring and optical image interpretation. *Landslides* **2022**, *19*, 35–50. [CrossRef]
16. Bayer, B.; Simoni, A.; Mulas, M.; Corsini, A.; Schmidt, D. Deformation responses of slow moving landslides to seasonal rainfall in the Northern Apennines, measured by InSAR. *Geomorphology* **2018**, *308*, 293–306. [CrossRef]
17. Luna, L.V.; Korup, O. Seasonal landslide activity lags annual precipitation pattern in the Pacific Northwest. *Geophys. Res. Lett.* **2022**, *49*, e2022GL098506. [CrossRef]
18. Liu, Y.; Qiu, H.; Yang, D.; Liu, Z.; Ma, S.; Pei, Y.; Zhang, J.; Tang, B. Deformation responses of landslides to seasonal rainfall based on InSAR and wavelet analysis. *Landslides* **2022**, *19*, 199–210. [CrossRef]
19. Zhao, C.; Kang, Y.; Zhang, Q.; Lu, Z.; Li, B. Landslide identification and monitoring along the Jinsha River catchment (Wudongde reservoir area), China, using the InSAR method. *Remote Sens.* **2018**, *10*, 993. [CrossRef]

20. Casu, F.; Manzo, M.; Lanari, R. A quantitative assessment of the SBAS algorithm performance for surface deformation retrieval from DInSAR data. *Remote Sens. Environ.* **2006**, *102*, 195–210. [CrossRef]
21. Ferretti, A.; Savio, G.; Barzaghi, R.; Borghi, A.; Musazzi, S.; Novali, F.; Prati, C.; Rocca, F. Submillimeter Accuracy of InSAR Time Series: Experimental Validation. *IEEE Trans. Geosci. Remote Sens.* **2007**, *45*, 1142–1153. [CrossRef]
22. Tizzani, P.; Berardino, P.; Casu, F.; Euillades, P.; Manzo, M.; Ricciardi, G.P.; Zeni, G.; Lanari, R. Surface deformation of Long Valley caldera and Mono Basin, California, investigated with the SBAS-InSAR approach. *Remote Sens. Environ.* **2007**, *108*, 277–289. [CrossRef]
23. Zhu, Y.; Yao, X.; Yao, L.; Zhou, Z.; Ren, K.; Li, L.; Yao, C.; Gu, Z. Identifying the Mechanism of Toppling Deformation by InSAR: A Case Study in Xiluodu Reservoir, Jinsha River. *Landslides* **2022**, *19*, 2311–2327. [CrossRef]
24. Su, X.; Zhang, Y.; Meng, X.; Rehman, M.U.; Khalid, Z.; Yue, D. Updating Inventory, Deformation, and Development Characteristics of Landslides in Hunza Valley, NW Karakoram, Pakistan by SBAS-InSAR. *Remote Sens.* **2022**, *14*, 4907. [CrossRef]
25. Sandwell, D.; Price, E. Phase gradient approach to stacking interferograms. *J. Geophys. Res. Solid Earth* **1998**, *103*, 30183–30204. [CrossRef]
26. Turner, D.; Lucieer, A.; De Jong, S.M. Time Series Analysis of Landslide Dynamics Using an Unmanned Aerial Vehicle (UAV). *Remote Sens.* **2015**, *7*, 1736–1757. [CrossRef]
27. Handwerger, A.L.; Fielding, E.J.; Huang, M.H.; Bennett, G.L.; Liang, C.; Schulz, W.H. Widespread initiation, reactivation and acceleration of landslides in the northern California Coast Ranges due to extreme rainfall. *J. Geophys. Res. Earth Surf.* **2019**, *124*, 1782–1797. [CrossRef]
28. Schulz, W.H.; Wang, G. Residual shear strength variability as a primary control on movement of landslides reactivated by earthquake-induced groundmotion: Implications for coastal Oregon, U.S. *J. Geophys. Res. Earth Surf.* **2014**, *119*, 1617–1635. [CrossRef]
29. Skempton, A.W. Residual strength of clays in landslides, folded strata and the laboratory. *Géotechnique* **1985**, *35*, 3–18. [CrossRef]
30. Di Maio, C.; Scaringi, G.; Vassallo, R. Residual strength and creep behaviour on the slip surface of specimens of a landslide in marine origin clay shales: Influence of pore fluid composition. *Landslides* **2015**, *12*, 657–667. [CrossRef]
31. Wen, B.P.; Jiang, X.Z. Effect of gravel content on creep behavior of clayey soil at residual state: Implication for its role in slow-moving landslides. *Landslides* **2017**, *14*, 559–576. [CrossRef]
32. Sassa, K.; Fukuoka, H.; Wang, G.; Ishikawa, N. Undrained dynamic-loading ring-shear apparatus and its application to landslide dynamics. *Landslides* **2004**, *1*, 1–5. [CrossRef]
33. Hu, W.; Li, Y.; Xu, Q.; Huang, R.; McSaveney, M.; Wang, G.; Fan, Y.; Wasowski, J.; Zheng, Y. Flowslide high fluidity induced by shear-thinning. *J. Geophys. Res. Solid Earth* **2022**, *127*, e2022JB024615. [CrossRef]
34. Bai, M.; Chevalier, M.L.; Pan, J.; Replumaz, A.; Philippe, H.L.; Marianne, M.; Li, H. Southeastward increase of the late Quaternary slip-rate of the Xianshui fault, eastern Tibet. Geodynamic and seismic hazard implications. *Earth Planet. Sci. Lett.* **2018**, *485*, 19–31. [CrossRef]
35. Zhang, Y.; Yao, X.; Yu, K.; Du, G.; Guo, C. Late-Quaternary slip rate and seismic activity of the Xianshui fault zone. *Acta Geol. Sin.* **2016**, *90*, 525–536. [CrossRef]
36. Chevalier, M.L.; Leloup, P.H.; Replumaz, A.; Pan, J.; Métois, M.; Li, H. Temporally constant slip rate along the Ganzi fault, NW Xianshui fault system, eastern Tibet. *GSA Bull.* **2017**, *130*, 396–410. [CrossRef]
37. Liu, X.; Zhang, Y.; Guo, C.; Wu, R.; Ren, S.; Shen, Y. Development characteristics and evolution process of the Garazong ancient rockslide along the Xianshuihe River in western Sichuan. *Acta Geol. Sin.* **2019**, *93*, 1767–1777. (In Chinese with English Abstract)
38. Hungr, O.; Leroueil, S.; Picarelli, L. The Varnes classification of landslide types, an update. *Landslides* **2014**, *11*, 167–194. [CrossRef]
39. Cruden, D.M.; Varnes, D.J. Landslide types and processes. In *Landslides Investigation and Mitigation*; Turner, A.K., Schuster, R.L., Eds.; Special Report 247; Transportation Research Board, US National Research Council: Washington, DC, USA, 1996; Chapter 3; pp. 36–75.
40. Berardino, P.; Fornaro, G.; Lanari, R.; Sansosti, E. A new algorithm for surface deformation monitoring based on small baseline differential SAR interferograms. *IEEE Trans. Geosci. Remote Sens.* **2002**, *40*, 2375–2383. [CrossRef]
41. Ren, S.; Zhang, Y.; Li, J.; Li, J.; Liu, X.; Wu, R. A new type of sliding zone soil and its severe effect on the formation of giant landslides in the Jinsha River tectonic suture zone, China. *Nat. Hazards* **2023**, *117*, 1847–1868. [CrossRef]
42. GB/T50123-2019; Standard for Geotechnical Testing Method. National Standard of the People's Republic of China: Beijing, China, 2019. (In Chinese)
43. Scheingross, J.S.; Minchew, B.M.; Mackey, B.H. Fault-zone controls on the spatial distribution of slow-moving landslides. *Geol. Soc. Am. Bull.* **2013**, *125*, 473–489. [CrossRef]
44. Zhang, Y.; Ren, S.; Liu, X.; Guo, C.; Li, J.; Bi, J.; Ran, L. Reactivation mechanism of old landslide triggered by coupling of fault creep and water infiltration: A case study from the east Tibetan Plateau. *Bull. Eng. Geol. Environ.* **2023**, *82*, 291. [CrossRef]
45. Yin, Y.; Wang, W.; Zhang, N.; Yan, J.; Wei, Y. The June 2017 Maoxian landslide: Geological disaster in an earthquake area after the Wenchuan Ms 8.0 earthquake. *Sci. China Technol. Sci.* **2017**, *60*, 1762–1766. [CrossRef]
46. Stark, T.D.; Hussain, M. Shear Strength in Preexisting Landslides. *J. Geotech. Geoenviron. Eng.* **2010**, *136*, 957–962. [CrossRef]
47. Ren, S.; Zhang, Y.; Xu, N.; Wu, R.; Liu, X.; Du, G. Mobilized strength of gravelly sliding zone soil in reactivated landslide: A case study of a giant landslide in the north-eastern margin of Tibet Plateau. *Environ. Earth Sci.* **2021**, *80*, 434. [CrossRef]

48. Wang, D.; Xiao, H.; Li, M. Discussion on the reconstruction method of landslide geological model before failure: A case study of Laohuzui landslide in Xianshui Fault Zone. *Hydrogeol. Eng. Geol.* **2013**, *40*, 111–116. (In Chinese with English Abstract)
49. Floris, M.; Bozzano, F. Evaluation of landslide reactivation: A modified rainfall threshold model based on historical records of rainfall and landslides. *Geomorphology* **2008**, *94*, 40–57. [CrossRef]
50. Xu, Y.; Kim, J.; George, D.L.; Lu, Z. Characterizing Seasonally Rainfall-Driven Movement of a Translational Landslide using SAR Imagery and SMAP Soil Moisture. *Remote Sens.* **2019**, *11*, 2347. [CrossRef]
51. Tu, G.; Huang, D.; Deng, H. Reactivation of a huge ancient landslide by surface water infiltration. *J. Mt. Sci.* **2019**, *16*, 806–820. [CrossRef]
52. Tian, S.; Chen, N.; Wu, H.; Yang, C.; Rahman, M. New insights into the occurrence of the baige landslide along the Jinsha river in Tibet. *Landslides* **2020**, *17*, 1207–1216. [CrossRef]
53. Notti, D.; Wrzesniak, A.; Dematteis, N.; Lollino, P.; Fazio, N.L.; Zucca, F.; Giordan, D. A multidisciplinary investigation of deep-seated landslide reactivation triggered by an extreme rainfall event: A case study of the Monesi di Mendatica landslide, Ligurian Alps. *Landslides* **2021**, *18*, 2341–2365. [CrossRef]

Disclaimer/Publisher's Note: The statements, opinions and data contained in all publications are solely those of the individual author(s) and contributor(s) and not of MDPI and/or the editor(s). MDPI and/or the editor(s) disclaim responsibility for any injury to people or property resulting from any ideas, methods, instructions or products referred to in the content.

Article

Rapid Mapping of Landslides Induced by Heavy Rainfall in the Emilia-Romagna (Italy) Region in May 2023

Maria Francesca Ferrario * and Franz Livio

Department of Science and High Technology, Università degli Studi dell'Insubria, Via Valleggio, 11, 22100 Como, Italy; franz.livio@uninsubria.it
* Correspondence: francesca.ferrario@uninsubria.it

Abstract: Heavy rainfall is a major factor for landslide triggering. Here, we present an inventory of 47,523 landslides triggered by two precipitation episodes that occurred in May 2023 in the Emilia-Romagna and conterminous regions (Italy). The landslides are manually mapped from a visual interpretation of satellite images and are mainly triggered by the second rainfall episode (16–17 May 2023); the inventory is entirely original, and the mapping is supplemented with field surveys at a few selected locations. The main goal of this paper is to present the dataset and to investigate the landslide distribution with respect to triggering (precipitation) and predisposing (land use, lithology, slope and distance from roads) factors using a statistical approach. The landslides occurred more frequently on steeper slopes and for the land use categories of "bare rocks and badlands" and woodlands. A weaker positive correlation is found for the lithological classes: silty and flysch-like units are more prone to host slope movements. The inventory presented here provides a comprehensive picture of the slope movements triggered in the study area and represents one of the most numerous rainfall-induced landslide inventories on a global scale. We claim that the inventory can support the validation of automatic products and that our results on triggering and predisposing factors can be used for modeling landslide susceptibility and more broadly for hazard purposes.

Keywords: landslide inventory; heavy rainfall; spatial distribution; Emilia-Romagna region

1. Introduction

Landslides are a movement of rock, earth or debris down a slope [1]. They can be triggered by several processes, including rainfall, earthquakes and human activities [2]. Heavy rainfall is among the most common triggering mechanisms and may result in thousands of landslides across a wide region. Landslides represent a major hazard source and have a relevant societal impact, because they cause heavy direct and indirect costs (e.g., [3]) on people, buildings, infrastructures and human activities. In Europe, the average economic loss during 1995–2014 is estimated to be 4.7 billion EUR per year [4]. Landslides are highly impacting in Italy: [5] describes a dataset covering the period 843-2008, reporting 1562 landslide events that have caused at least 7477 deaths. The human and economic costs of landslides are expected to increase in the future [6].

In order to assess landslide susceptibility and risk, it is essential to obtain information on landslide location. In this sense, landslide inventories represent the location either as a point or polygon and, if known, the date of occurrence and type of slope movement [7]; thus, landslide inventories are a basic prerequisite for hazard assessment [8,9]. Landslide inventories can be obtained using a variety of methods, including field reconnaissance and the interpretation of optical or Synthetic Aperture Radar (SAR) images (e.g., [7]); landslides can be drawn either manually or by semi-automatic and automatic methods [10–13]. The key goal of the current study is to realize a comprehensive landslide inventory, following a period of heavy precipitation that hit N Italy in May 2023.

Rainfall-triggered landslides have been a focus of scientific inquiry for decades, for instance, [14] compiled a list of more than 450 references dealing with this topic. The

availability of landslide inventories following heavy rainfall and storms is rapidly increasing, covering different geographic and climatic settings (e.g., [15–24]). A relatively high number of inventories are available for Italy, such as the seminal works by [25] on the Friuli seismic sequence and [26] on rapid snowmelt in Umbria. More recent efforts include the mapping of landslides triggered by heavy rainfall in the Liguria [15], Umbria [27] and Marche regions [28,29].

Contrarily to earthquake-induced landslides [30], a unified repository does not exist for rainfall-triggered landslides; moreover, some scholars claim that the number of inventories available in digital format is still limited [7,20,23]. One critical limitation in the quick identification of landslides triggered by precipitation is cloud coverage, which may persist for a long time in the area affected by storms.

Here, we present an inventory of landslides triggered by heavy rainfall that occurred in May 2023 in Northern Italy. The area was hit by prolonged rainfall, which caused extensive flooding and landslides, resulting in 17 deaths, tens of thousands displaced and at least 8.8 billion EUR in losses [31]. The inventory is entirely original and was produced by manual mapping, interpreting a set of pre- and post-event satellite images, and supplemented with a limited field survey. The purpose of our study is to (i) present the dataset of 47,523 mapped landslides, (ii) describe the amount and spatial pattern of precipitation and (iii) investigate the distribution of landslides with respect to predisposing and triggering factors. Our results may be compared to other mapping methods (e.g., automatic mapping); we envisage that our results can be used for hazard purposes, susceptibility modeling or the characterization of rainfall thresholds. Additionally, we stress that the availability of sound input data will result in more robust outcomes and provides the grounds for better informed decisions. In this sense, our dataset provides an additional case history that can be compared with extant knowledge and contributes to filling eventual data gaps.

This paper is organized as follows: in Section 2, we provide a description of the study area and of the May 2023 precipitation events; in Section 3, we introduce the materials and methods used to compile the inventory and investigate the predisposing and triggering factors; in Section 4, we present the obtained inventory and we analyze the spatial distribution of landslides with respect to conditioning factors; in Section 5, we compare this case study with other available data and we discuss the limitations of our approach; and finally, in Section 6, we draw some conclusions.

2. Regional Setting and the May 2023 Precipitation Events

2.1. The Study Area

Figure 1 shows the Digital Elevation Model (DEM) of the study area, in Northern Italy, mainly located in the Emilia-Romagna (ER) region. The area of interest (AOI), where we mapped the rainfall-triggered landslides, covers 5764 km^2, with elevations ranging between 10 and 1400 m asl. The DEM clearly shows a geomorphological boundary between the plain sector to the NE and the hilly sector of the Northern Apennines to the SW. The main rivers flow from the SW to the NE and reach either the Po River or the Adriatic Sea; in the plain, the rivers flow at elevations higher than the surrounding plain. Such a configuration implies that, following heavy rainfalls, the hilly sector is prone to slope movements, while the plain suffers from flooding.

In the hills, the slopes are generally gentle, with values in the range of 0–70°, with a broad modal peak centered at 10–20°. From the geological point of view, the chain sector of the AOI is the result of the tectonic superposition of two large sets that are different in lithology, structure and paleogeographic origin: an External Umbrian–Tuscan Domain, mainly outcropping to the SE and composed of turbiditic units, well-bedded marls and sandstones, and an Internal Ligurian–Emilian Domain, outcropping to the NW, composed of shales, chaotic *melange* and flysch-like units [32]. From the lithological point of view, the AOI is characterized by the outcropping of terrigenous formations with a high clay content, alternating sandstones marls and siltstones and turbiditic successions, resulting in a typical landscape characterized by badlands and diffuse slope movements. Lithological control on erosion results in aligned *cuestas* and hogbacks. Where the superimposition of flysch or sandstones occurs, sub-vertical slopes and mesa-like features are displayed (i.e., the so-called "Pietra di Bismantova"; [33]). The Italian inventory of slope movements [34] includes over 40,000 mapped landslides in the AOI (45% rotational/translational slides and 30% slow earth flows).

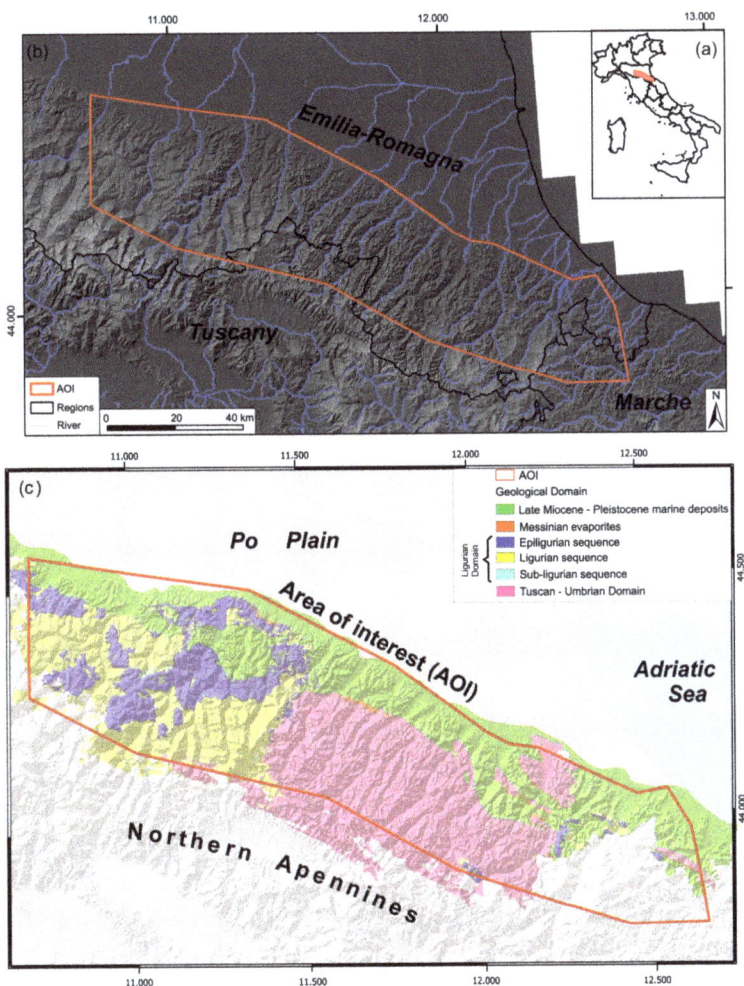

Figure 1. (**a**) Location of the area enlarged in panels b and c; (**b**) overview of the study area, the area of interest (AOI) is represented in red; and (**c**) simplified regional geological map.

The primary land covers, after the Emilia-Romagna region land use database (last updated in 2020, following the CORINE land cover classification scheme [35]), include agricultural fields and forested areas for the hillside sector of the AOI, followed by urbanized land use, mainly located at the foothills or in the plain area.

The mean annual rainfall is in the range 700–1400 mm (the climatic atlas of the Emilia-Romagna region, [36]), with the values increasing from the NE to SW; autumn represents the wettest season, followed by a secondary peak in precipitation in the spring.

2.2. The Rainfall Episodes of May 2023

During May 2023, the ER region experienced prolonged and intense rainfall, mainly distributed in two episodes, i.e., 1–2 May and 16–17 May. The gauge records (Figure 2, Table 1) show no significant rainfall in the 30 days preceding the first rainfall episode. At the beginning of May 2023, rainfall hit the entire AOI, with the cumulative values exceeding 200 mm (e.g., 254 mm at Le Taverne and 243 mm at Trebbio). The hourly intensities were generally lower than 3–5 mm/h, with few peaks exceeding 15–20 mm/h. This event was the most intense in a 2-day interval since 1997 and the most intense in the spring season since 1961 [37]. The geographic distribution of the rainfall is shown in Figure 2a. The highest values were recorded in the Reno, Lamone and Montone drainage basins; the precipitation generated an increase in river discharge, reaching hydrometric levels close to, and locally higher than, the levee levels. Sporadic slope movements were triggered as well [37].

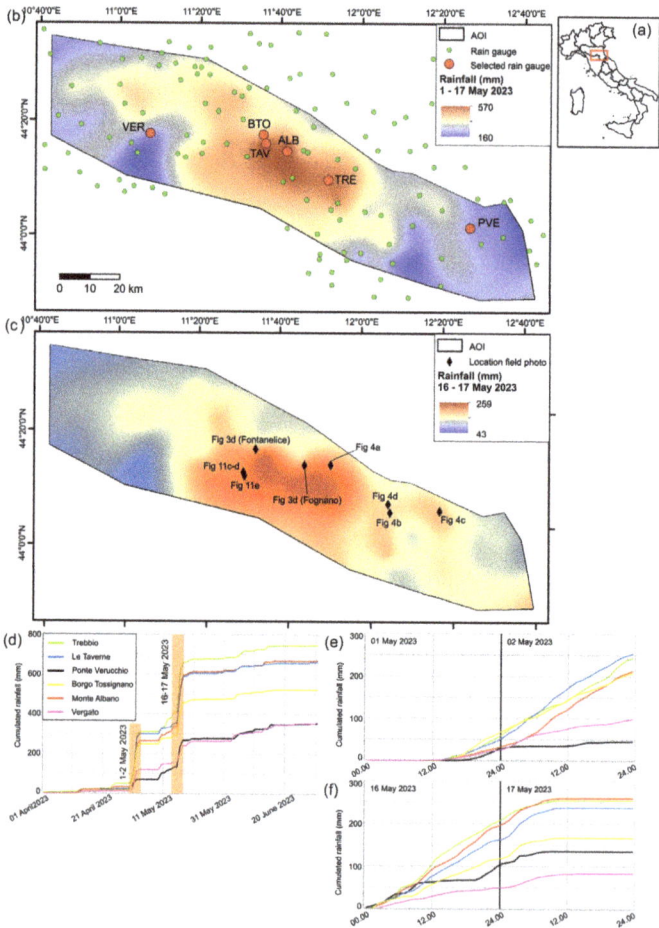

Figure 2. (**a**) Location of the area enlarged in panels (**b**,**c**); (**b**) rainfall amount during 1–17 May 2023 over the area of interest and location of the rain gauges; in red are labeled the stations depicted in (**d**–**f**); (**c**) rainfall amount during 16–17 May 2023 and location of the field photographs; (**d**) plot of cumulative rainfall for the period April–June 2023; (**e**) plot of cumulative rainfall for 1–2 May 2023; and (**f**) plot of cumulative rainfall for 16–17 May 2023.

Table 1. Details on the rainfall gauges plotted in Figure 2.

ID	Station	Municipality	Province	Elevation (m asl)	Longitude	Latitude	Hydrographic Basin
BTO	Borgo Tossignano	Borgo Tossignano	Bologna	98	11.578993	44.27467	Santerno
TAV	Le Taverne	Fontanelice	Bologna	486	11.587499	44.2492	Santerno
ALB	Monte Albano	Casola Valsenio	Ravenna	480	11.6734246	44.22432	Senio
PVE	Ponte Verucchio	Verucchio	Rimini	116	12.405109	43.9829	Marecchia
TRE	Trebbio	Modigliana	Forlì-Cesena	570	11.8371627	44.13697	Lamone
VER	Vergato	Vergato	Bologna	193	11.113128	44.2878	Reno

An even stronger precipitation event (Minerva storm) occurred on 16–17 May 2023 (Figure 2b); again, the most hit area was the hilly sector of the ER region. The cumulative values over the 2 days reached 260.8 mm at the Monte Albano gauge and 254.8 at Trebbio. The cumulative rainfall over the 17-day period (1–17 May 2023) is the highest historical record for about 65% of the gauge stations in central and eastern ER [38]; the maximum values reached 609.8 mm at Trebbio and 563.4 mm at Le Taverne (Figure 2c). The high discharge resulted in extensive floods of the plains and in thousands of landslides in the hills.

3. Materials and Methods

Figure 3 represents the methodological workflow adopted in this study. The approach consists of three subsequent phases, namely, (i) building the landslide inventory and performing the field reconnaissance, (ii) acquiring data on predisposing and triggering factors and (iii) conducting data analyses. At each step, input data are required, which may be obtained from external sources (e.g., satellite images and thematic maps) or generated in the previous steps (e.g., landslide polygons).

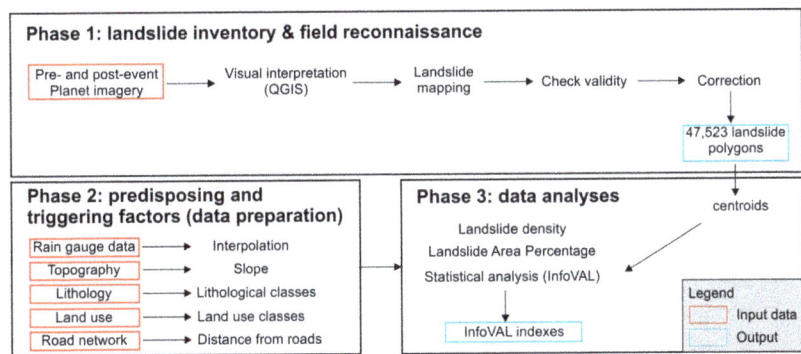

Figure 3. Methodological workflow adopted in this study.

3.1. Realization of the Landslide Inventory

The first step of this research is the realization of the landslide inventory, which is based on a visual inspection of pre- and post-event satellite images in an area of 5764 km². The satellite images were acquired by PlanetScope (https://www.planet.com/, last accessed on 3 November 2023) and were provided under an academic license as ortho-rectified products; 3 m resolution multiband tiles were used. A screening of the available imagery was initiated within hours of the 16–17 May precipitation episode, using the Planet website; for a few days the area was heavily clouded, and the first cloud-free images were available on 22 May. For the realization of the inventory, the images were accessed through the

Planet QGIS Plugin; the pre-event imagery refers to the Monthly Global Basemap products provided by Planet, and the April 2023 Basemap was used. The post-event images were acquired between 22 May and mid-June 2023. The cloud-free image closer to the event was used and multitemporal frames were checked in the selected areas (e.g., due to the presence of shadows or unclear images).

The landslides were manually mapped at a scale of 1:5000 by a single operator in a time interval of 5 weeks following the rainfall event; the inventory (version 1.0) was released online on 28 June 2023. The landslides were mapped as polygons encompassing both the source and deposit areas, because they are not easily discernible in satellite images; the landslides were mapped as individual polygons as much as possible, trying to avoid issues related to amalgamation [39].

We realized the field surveys in a limited subset of the AOI on 30 July–1 August 2023; we drove along the main and secondary roads, and we acquired photographic documentation of the landslides at selected spots along the roads or at scenery points for a broader view.

We computed the area–frequency distribution of the event inventory following [40]. The probability density for classes of a width equal in logarithmic coordinates is calculated as:

$$p = \frac{1}{N} \frac{\partial N_L}{\partial A_L}, \qquad (1)$$

where N is the total number of landslides, and N_L is the number of landslides with an area between A_L and $A_L + \delta A_L$.

3.2. Predisposing and Triggering Factors

The second step of the research includes the acquisition and preparation of the thematic data related to the predisposing and triggering factors. We compared the spatial distribution of the landslides with some descriptors of the local geological and geomorphological setting and with the spatial distribution of the precipitations during the May 2023 events. We selected five influencing factors for further analyses, namely, rainfall, slope, land use, lithology and distance from roads.

We interpolated the pluviometric data published in the ER report [38]. The cumulative pluviometric data refer to the period 1–17 May 2023 and for the two-day period of 16 and 17 May 2023. We interpolated the point data with a local polynomial interpolator algorithm (spline interpolation) to grant the representativeness of the data in terms of local trends and to avoid possible spikes in the data distribution.

For the topographical analyses, we used the 20 m DTM of Italy, released by ISPRA, and we derived the slope values (in degrees).

We obtained a lithological map of the area from the ER region WebGIS at a 1:50,000 scale. The geological map has been directly derived from the Italian national geological cartography program (CARG Project) and includes a lithological classification of each geological unit. For the sake of simplicity, we adopted the classification scheme summarized in Table 2.

Table 2. Lithological classes.

Lithology Class	Description
U1	Competent massive or well-bedded rocks (mainly limestones and dolostones)
U2	Siltstones, marls and limestones interbedded with marls
U3	Sandstones and sandstones interbedded with marls and siltstones
U4	Conglomerates and breccias
U5	Gypsum (massive or breccia facies)

Land use was downloaded from the ER region WebGIS. This database follows the CORINE land cover hierarchical classification scheme and was updated in 2020. We compared the occurrence of landslides with the main categories of land use as summarized in Table 3.

Table 3. Land use classes.

Land Use Class	Description
L1	Urban (residential and industrial)
L2	Arable land
L3	Agricultural (trees and vineyard)
L4	Meadows
L5	Woods (both evergreen and deciduous)
L6	Bare rocks and badlands
L7	Shrubs
L8	Water (lakes, rivers and swamps)

To estimate the possible influence of the vicinity of roads for landslide triggering, we used the Open Street map database of roads and streets, including all the road types except for paths in mountain areas.

3.3. Data Analysis

The last part of our work is dedicated to the statistical analysis of the inventory. As a first step, we carried out a topological check on the landslide inventory using the "check validity" QGIS tool to correct auto-intersecting polygons; this check resulted in 14 invalid polygons (0.03% of the mapped landslides), which were manually corrected. To investigate the spatial distribution of landslides, we computed the landslide number density (LND) and landslide area percentage (LAP), which later were statistically analyzed with respect to predisposing and triggering factors.

The landslide density was calculated by means of the kernel density estimation approach using a kernel density estimator (KDE) in a GIS environment (e.g., [41,42]). The KDE calculates the density of point features if a probability density function of event occurrence (i.e., the kernel shape) is centered at the observation point. The probability diminishes with increasing distance from the point, reaching zero at the search radius distance (h) from the point. Only a circular neighborhood is possible.

Following [43], we calculated h according to the following formula:

$$h = 0.9 \times \min\left(SD, \frac{IQR}{1.34}\right) \times n^{-0.2}, \qquad (2)$$

where IQR and SD are the interquartile range and the standard deviation, respectively, of the distances between each observation point and the centroid of the point population; n is the number of points.

The density at each output raster cell is calculated by adding the values of all the kernel surfaces where they overlay the raster cell center. We adopted an Epanechnikov kernel shape [44], namely, a kernel function built as the positive part of a parabola that minimizes the errors associated to the tails of the estimates [43]. We also produced a density raster of the calculated KDE density with 100 m cell spacing.

LAP represents the percentage of the territory covered by the rainfall-triggered landslides; it was computed on a regular grid of cells having an area of 1 km^2.

The quantitative relationship between landslides and predisposing factors affecting landslides are established by the data-driven Information Value (InfoVAL) method [45–47]. The method allows for the quantified prediction of susceptibility by means of a score (Wi), calculated according to landslide occurrence on each class and weighed according to the class distribution over the entire study area:

$$W_i = \ln \frac{\text{Densclass}_i}{\text{Densmap}_i}, \qquad (3)$$

where Wi is the score for the i-th class, Densclass is the landslide occurrence for the i-th class and Densmap is the i-th class occurrence on the whole area. This formula normalizes the event occurrences over the spatial distribution of each considered class. Positive values indicate a positive statistical correlation and negative values a negative one, while values close to zero indicate a random distribution of the data. We applied the InfoVAL method to the territory of the ER region, where consistent and homogenous information is available. Emilia-Romagna accounts for 97% of the mapped landslides, so we claim that the obtained results can be considered representative of the entire dataset.

4. Results
4.1. The Inventory

The dataset contains 47,523 landslides in an area of 5764 km^2, corresponding to an average density of about 8 landslides/km^2. Figure 4 shows two examples of multitemporal satellite images at the localities of Fontanelice and Fognano (for the location, see Figure 2c). The upper panels (Figure 4a) represent the images acquired before the rainfall episodes, i.e., on 27 April; the mid-panels (Figure 4b) were acquired after the first rain spell, i.e., on 5 May; and the lower panels (Figure 4c) were acquired a few days after the main rainfall episode, i.e., on 22–23 May. Finally, Figure 4d presents two field photographs of the sites.

The pre-event images provide an overview of the typical local setting, which includes a variable proportion of urban areas, agricultural fields, woodlands and outcropping rocks. From the comparison of the multitemporal images, it is evident that most of the landslides occurred after the 16–17 May rainfall episode; nevertheless, a few landslides are recognizable on the 5 May images as well (see red circles), mostly related to the reactivation of pre-existing slope movements. The landslides are not homogeneously distributed but tend to cluster at specific places, such as along the Santerno River (Fontanelice) or in the woodland and cultivated fields N of Fognano. The field photos show the occurrence of closely spaced slope movements with variable runouts, which scraped off the vegetative cover.

Figure 5 presents the selected examples of the landslides observed in the field and having variable characteristics in terms of the size, shape, geomorphological setting and involved materials. The location of the sites is presented in Figure 2c. Figure 5a includes two slides with long runouts coalescing into a single toe; the crown areas are at the top of a small hill, while the toe intersects an agricultural field. Figure 5b shows a single slope movement on the lower part of a steep slope in a vegetated region; in this case, the deposit area lies at the base of the slope. Figure 5c shows a panoramic view of the multiple landslides that occurred on the slope beneath the Sorrivoli Castle, a fortress built in the XI century. Figure 5d shows the trunks and sediments transported by a river course and deposited at the intersection with a road bridge. Figure 5c,d represent typical interactions between the landslide events or destabilized material and anthropic activities, infrastructures or cultural heritage.

The dataset is entirely original, and the shapefiles (coordinate system WGS84 UTM 32N) can be publicly accessed at https://zenodo.org/record/8102429 (last accessed 1 October 2023). We computed the LND and LAP values on a regular grid of cells having a dimension of 1 km^2. Figure 6 allows for appreciating the uneven spatial distribution of the landslides: the highest LND and LAP values are indeed located in a small portion in the central part of the AOI and decrease moving outwards. The maximum LND values reach 129 landslides/km^2, while the maximum LAP values reach around 30%.

The landslide polygons cover an area of 40.9 km^2, which represent 0.71% of the AOI; the largest mapped failure has an area of 98,000 m^2, while the average area is 860 m^2. The probability density of the dataset is presented in Figure 6c: the curve is characterized by the typical pattern of other landslide–area distributions, namely, a negative power law fit for medium to large landslides and a positive power law fit for small to medium landslides. The two limbs of the curve are separated by a rollover, which in our case is located at a size of 120–150 m^2. The exponent of the negative power law is -2.22, which is in broad agreement with other inventories worldwide [18,23,26,48].

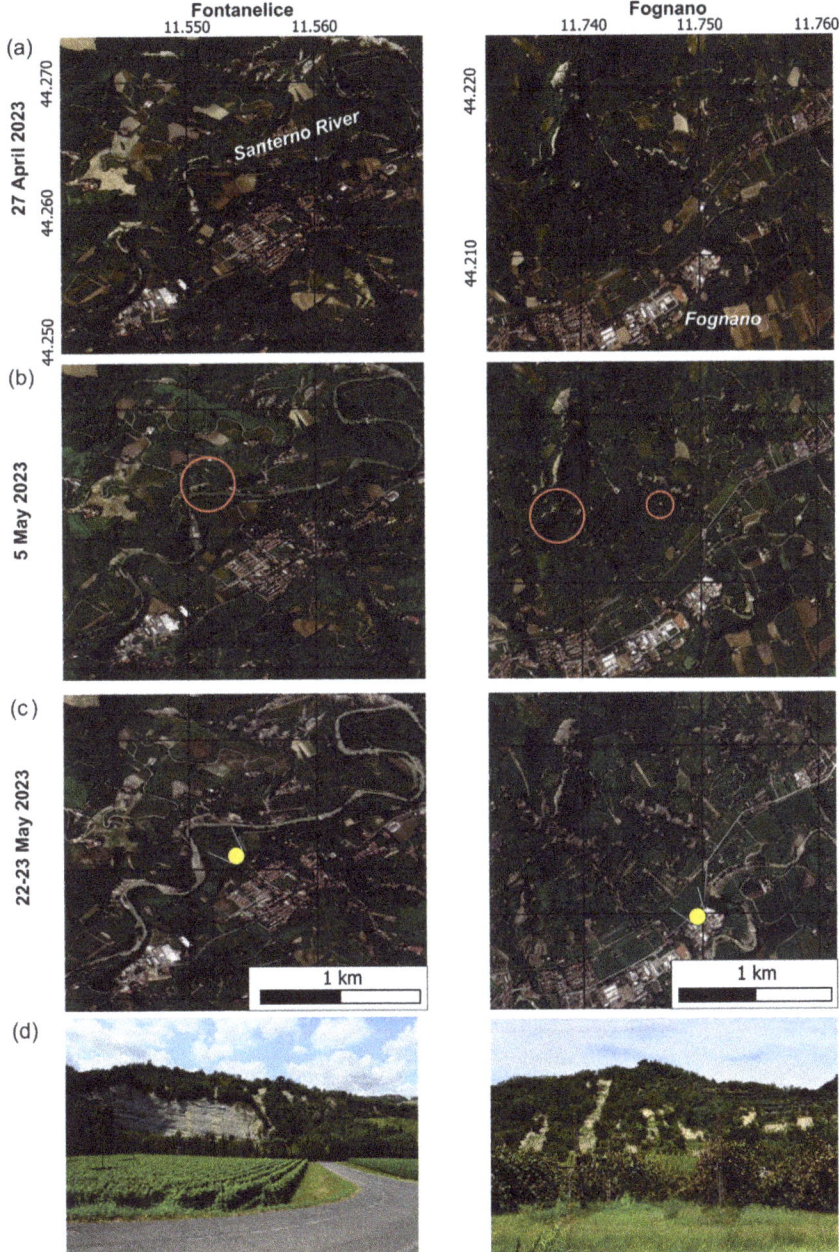

Figure 4. Selected examples of multitemporal Planet images at Fontanelice (left column) and Fognano (right column): (**a**) images acquired on 27 April 2023, before the rainfall events; (**b**) images acquired on 5 May 2023, after the first spell of heavy rain, red circles show the location of a few landslides triggered by the first rain episode; (**c**) images acquired on 22–23 May 2023, after the rainfall episodes, yellow dots are the location of field pictures; and (**d**) panoramic photographs of the two regions (photo shot by M. F. Ferrario on 30 July and 1 August 2023, respectively).

Figure 5. Field photos documenting some of the slope movements and related effects: (**a**) coalescing landslides in agricultural fields; (**b**) landslide on vegetated steep slopes, the material accumulates at the base of the slope; (**c**) interaction between landslides and cultural heritage (Sorrivoli Castle); and (**d**) material accumulated beneath a bridge pillar, enhancing the derived damage. All the photographs were shot by M. F. Ferrario between 30 July and 1 August 2023.

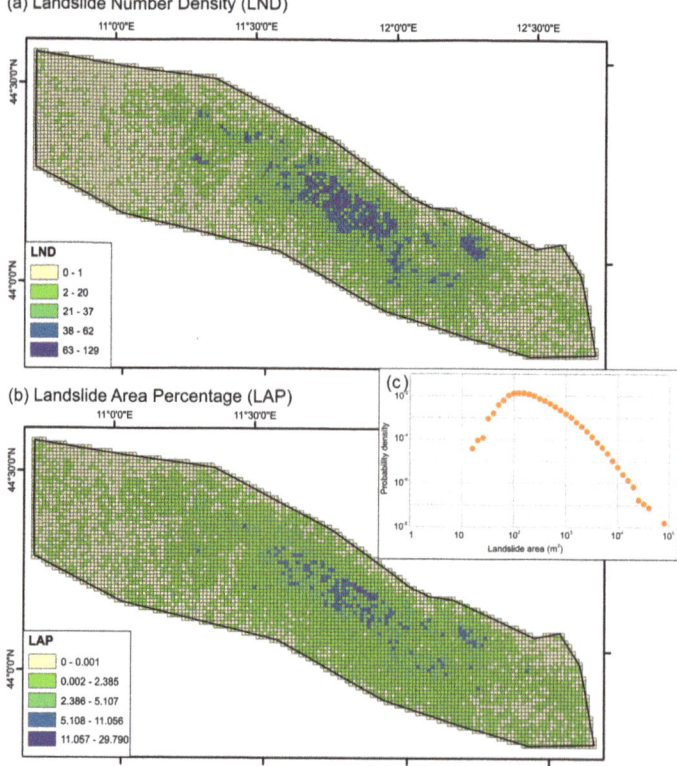

Figure 6. (**a**) Distribution of landslide number density (LND) values; (**b**) distribution of landslide area percentage (LAP) values; and (**c**) plot of the probability density as a function of landslide area.

4.2. Predisposing and Triggering Factors

We selected four predisposing factors for the subsequent analysis; one is related to the topography (i.e., slope) and one to the geologic setting (lithology), while the remaining two factors are the land use and distance from roads, which account for the anthropic territorial modifications. Figures 7 and 8 show the maps of the four factors, while the categories used to classify the lithology and land use are presented in Tables 2 and 3, respectively. The plots in the upper-right corner of Figures 6 and 7 show the repartition among the different units: the overall AOI is represented by the blue columns, while the orange dots represent the proportion of landslide centroids falling within each unit.

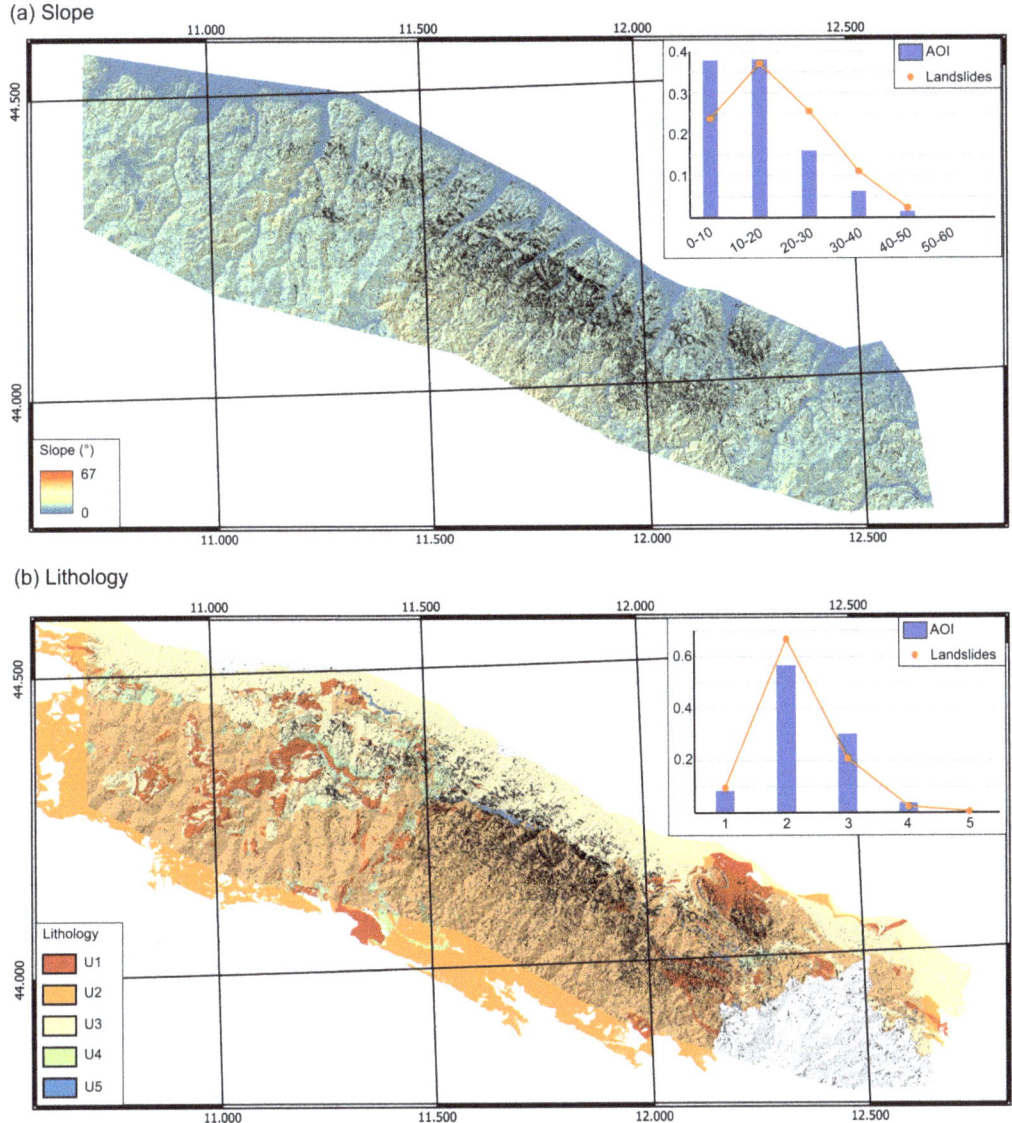

Figure 7. Maps of slope values (**a**) and lithology (**b**) in the AOI. Graphs in upper-right corner represent the repartition of the landscape (blue bars) and landslide centroids (dots) in the different classes.

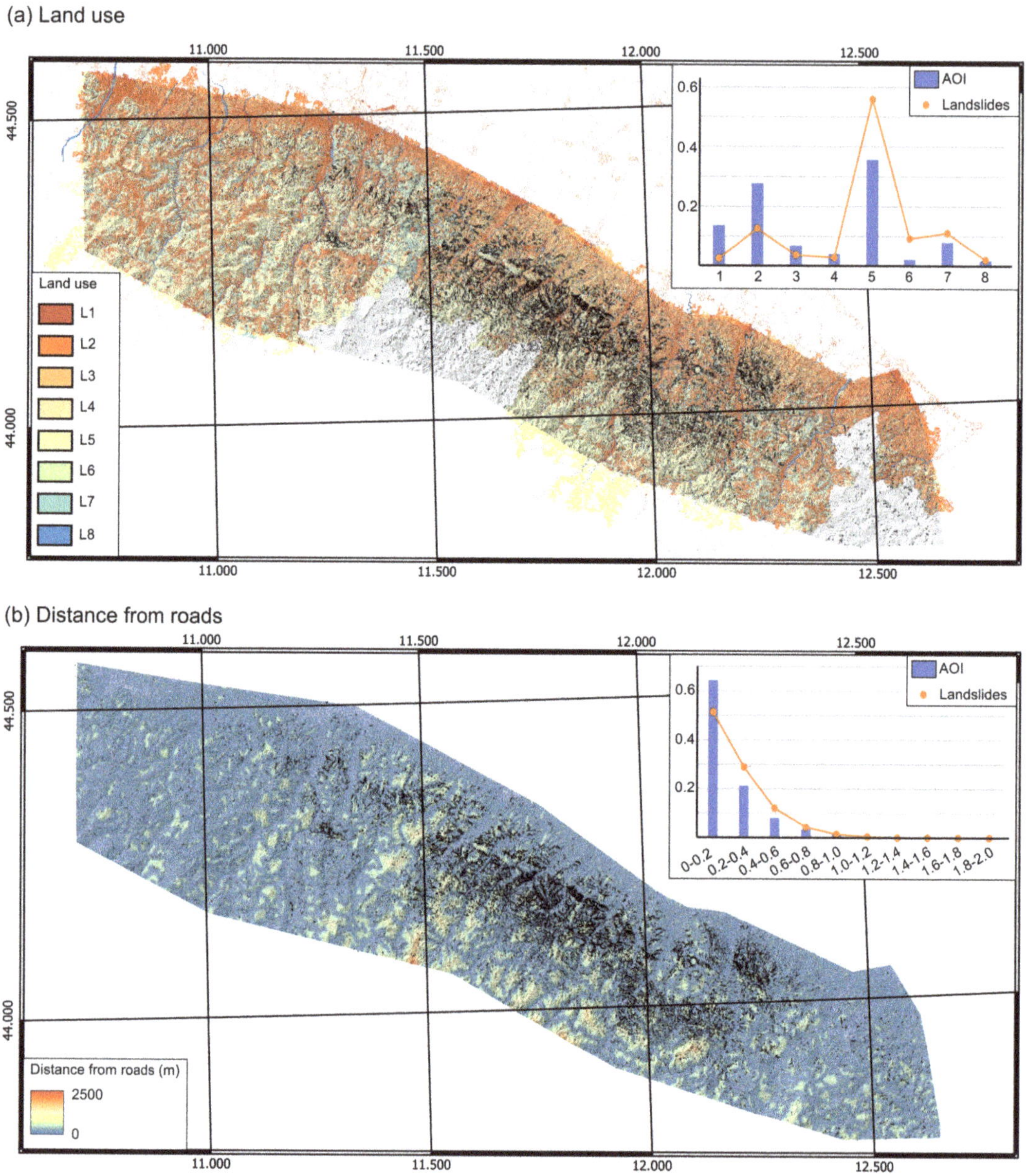

(a) Land use

(b) Distance from roads

Figure 8. Maps of land use (**a**) and distance from roads (**b**) in the AOI. Graphs in upper-right corner represent the repartition of the landscape (blue bars) and landslide centroids (dots) in the different classes.

Figure 7a presents the role of the slope, which has been categorized into units of 10° width. Categories 0–10° and 10–20° share an almost equal amount of the AOI (38% each), but the landslide centroids peak in the 10–20° class, with a value of 37%. The steeper slopes are less diffuse in the AOI, but the landslide centroids are over-represented, especially in the 20–30° and 30–40° classes.

Figure 7b presents the role of lithology, which has been categorized into five units (see Table 2). The most represented unit in the AOI comprises "siltstones, marls and limestones interbedded with marls"; it covers 57% of the AOI, and 67% of the landslide centroids lie within Unit 2. Overall, the abundance of landslides for each unit follows the lithological repartition in the AOI.

Figure 8a presents the role of land use, which has been categorized into eight units (see Table 3). The most represented unit in the AOI is L5 (i.e., woods, 36%), followed by L2 (arable land, 28%). When considering landslide centroids, more than half (56%) lie in Unit L5, pointing toward a strong influence of this land use category. Unit L6, bare rocks and badlands, accounts for only 2.2% of the AOI but includes 9% of the landslides, again suggesting that it may be a strong predisposing condition. On the contrary, Unit L1 (urban areas) covers 14% of the AOI, but only 2% of the landslides are located in Unit L1, probably because urban areas lie on flat regions and/or landslides are more difficult to identify in satellite imagery.

Figure 8b presents the role of the distance from roads, which has been categorized into units of 200 m width. The AOI shows an exponential decrease with distance: 86% of the AOI lies at less than 400 m from a road, testifying to the high density of infrastructures and thus the exposed assets. The distribution of landslide centroids mimics the AOI repartition, because the most represented class is within 200 m from a road (51%).

Precipitation is the triggering factor of this large amount of mass movements. A comparative analysis between the landslide KDE value and the corresponding precipitation, occurring at the same location, provides some additional insights. We considered both the distribution of the cumulative precipitations that occurred during the whole rainfall period (1–17 May 2023; Figure 9a) and the precipitations that occurred during the second rainfall event alone (16–17 May 2023; Figure 9b). Even considering a quite scattered distribution of the datapoints, it is apparent that there is a good positive correlation between the amount of rainfall and the spatial density of landslides. In particular, if we consider the 16–17 May 2023 event alone, the graph points to a threshold value for the inception of landslides of ca. 120 mm over the considered period.

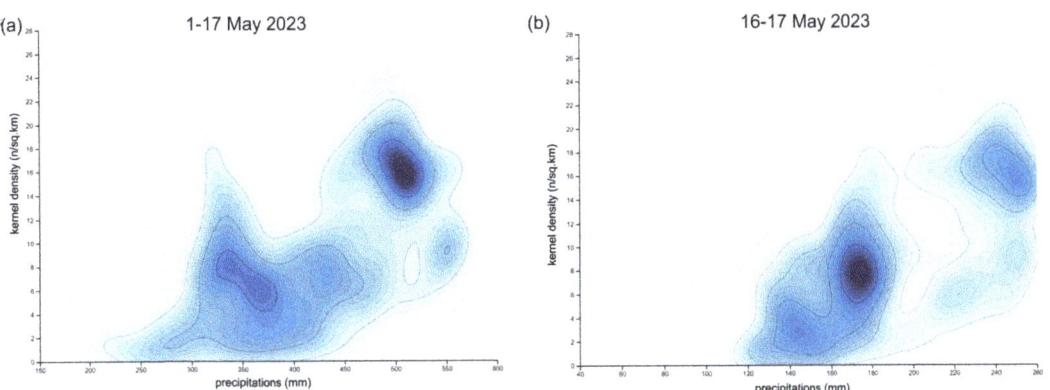

Figure 9. Density of datapoints contoured on a scatterplot showing the KDE value (number of events/km^2) compared with the corresponding value of precipitations recorded during 1–17 May (**a**) and during 16–17 May alone (**b**).

The analysis of the frequency distribution of the landslide KDE value for each land use and lithology class highlights how the landslides concentrate or not in certain zones (Figure 10). A lithological control on landslide occurrence is apparent only for conglomerate-like units (class U4), inhibiting the massive landsliding, and flysch-like units (class U2) that, on the contrary, look to host a higher landslide density.

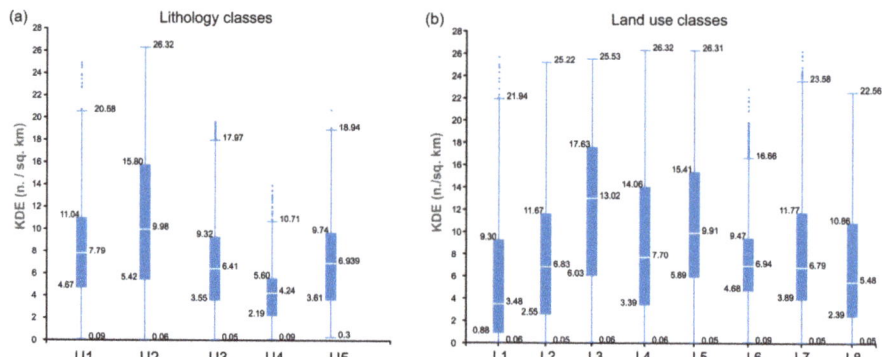

Figure 10. Statistics of the KDE value recorded for each of the classes of lithology (**a**) and land use (**b**), considered as possible predisposing factors (see Tables 2 and 3 for class codes); on the boxplots, the average (white line), 1 standard deviation (boxes), and 2 standard deviations (whiskers) are reported.

As for the land use, agricultural land (class L3) and woodland (class L5) are correlated to an average higher KDE value, also considering a quite broad distribution of values. On the other hand, urban areas (class L1), on average, are the only ones less affected by landsliding.

Finally, if we check the relative representation of the land use and lithology classes, and the reciprocal exchange of events among those classes (Figure 11), we can appreciate that, overall, most landslides belong to woodland (class L5) over flysch-like units (class U5), with most of the remaining landslides from the woodland distributed between the sandstones (class U3) and limestone (class U1) units. Flysch-like units are also associated with a considerable number of events that happened on arable land (class L2) and shrubs (class L7).

Figure 11. Stream plot summarizing the distribution of the datapoints among all the possible combinations of land use and lithology classes (see Tables 2 and 3 for class codes).

Another considerable number of events belong to bare lands (L6) on sandstones (U3) and to woodland (class L5) on limestones (class U1).

4.3. InfoVAL Analysis

From the calculated InfoVAL index (Table 4), it is apparent that lithology has a moderate effect on predisposing the slopes to landsliding: positive Wi values are associated with silty and flysch-like units, but only small positive correlations can be supposed. On the other hand, sandstones represent a more stable terrain over which landsliding is inhibited. The land use classes predisposing landsliding are woods and shrubs, with a striking high value associated with bare rocks and badlands. The slope classes are particularly indicative of a susceptibility between 20° and 40° and for extremely high values (60–80°).

Table 4. Summary of the considered classes as possible predisposing factors and calculated InfoVAL index (Wi).

Predictor	Class	InfoVAL Index (Wi)
Lithology	U1: Competent massive or well-bedded rocks (mainly limestones and dolostones)	0.09
	U2: Siltstones, marls and limestones interbedded with marls	0.14
	U3: Sandstones and sandstones interbedded with marls and siltstones	−0.41
	U4: Conglomerates and breccias	−0.48
	U5: Gypsum (massive or breccia facies)	0.08
Land use	L1: Urban (residential and industrial)	−1.60
	L2: Arable land	−0.67
	L3: Agricultural (trees and shrubs)	−0.53
	L4: Meadows	−0.23
	L5: Woods (both evergreen and deciduous)	0.56
	L6: Bare rock and badlands	1.57
	L7: Shrubs	0.45
	L8: Water (lakes, rivers and swamps)	0.22
Slope (°)	0–10	−0.31
	10–20	0.13
	20–30	0.62
	30–40	0.71
	40–50	0.60
	50–60	0.44
	60–70	0.79
	70–80	2.16
	80–90	-
Distance from roads (m)	0–200	−0.07
	200–400	0.46
	400–600	0.56
	600–800	0.40
	800–1000	0.19
	1000–1200	−0.24
	1200–1400	−0.81
	1400–1600	−0.46
	1600–1800	−0.70
	1800–2000	−0.14

Finally, the distance from roads is somehow indicating an augmented hazard in the range of 200–800 m. The majority of landslides mapped in our study lie within 200 m from a road. This result is consistent with those obtained by [28], who realized an inventory of 1687 landslides triggered by a rainfall episode that occurred in the Umbria and Marche regions on 15 September 2022. They found that 60% of the mapped landslides lie within 50 m from the roads, and the percentage grows to 89% within 200 m from the roads. It is important to note that the methods of [28] are different from the ones used here: indeed, they realized the extensive reconnaissance field surveys by driving along main and secondary roads and stopping at every landslide or scenery point.

5. Discussion

5.1. Comparison with Other Case Studies

The inventory presented in this study includes 47,523 landslides, thus representing one of the most numerous event inventories on a global scale. Here, we compare the figures of the Emilia-Romagna inventory with other case histories available in the literature to look for common patterns or, on the contrary, peculiarity in the case study presented here. Comparative studies have been realized on a dataset of 16 inventories pertaining to events that occurred between 2002 and 2019 [18,23]. The events cover a variety of precipitation patterns (cyclones, typhoons, local storms and prolonged intense rainfall) and geographic settings (Central and South America, East Africa, India and East Asia). The total event rainfall ranges between 45 and 2500 mm. A strong influence on the triggered landslides is played by the total storm rainfall, while topographic parameters (e.g., slope) have a lower impact [18]. Overall, a non-linear increase in total landsliding is observed with respect to total rainfall. The number of triggered landslides also scales with the total rainfall amount, although the scatter is higher; this latter point can be due to the fact that the inventories were delineated on images with a different resolution [18].

As a comparison, we recall that the rainfall in the Emilia-Romagna region reached 254 mm on 1–2 May 2023 and 260 mm on 16–17 May 2023, while the cumulated rainfall over the entire period (1–17 May 2023) was about 570 mm. These figures suggest that the case study presented here is consistent with the literature data.

Figure 12a presents the number of landslides versus the total rainfall amount for the dataset by [23], while Figure 12b presents the empirical relationship between the total rainfall and the landslide area presented by [18]. We supplemented the graph with three Italian case histories [15,27,28] and with the inventory presented in this study. It is quite evident that the Emilia-Romagna case history plots well above previous investigated events. Several factors may account for such a difference, including (i) the fact that the case study presented here was particularly effective in triggering landslides, or (ii) there is an issue of completeness in the landslide mapping, or (iii) there is some fundamental difference in the considered events or methodological assumptions. We prefer the former two hypotheses, because some of the inventories in the literature were derived from the same kind of satellite imagery and with comparable methods; the frequency–area distribution of the Emilia-Romagna inventory has a rollover at ca. 120–150 m^2, while the inventories have rollovers ranging between 10^2 and 10^3 m^2, suggesting that our inventory contains a larger number of small landslides.

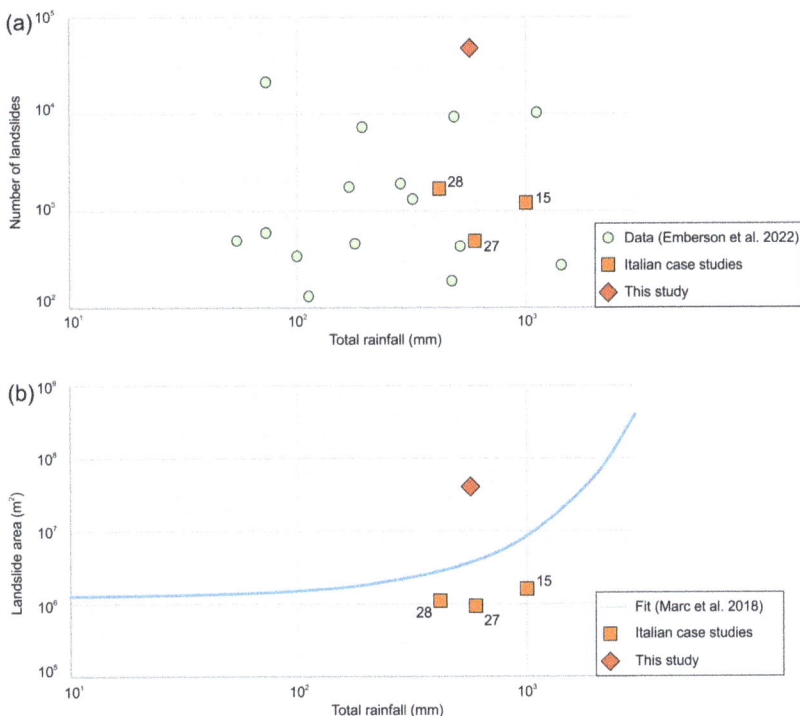

Figure 12. (**a**) Number of landslides versus total rainfall (mm), global data after [23], Italian cases after [15,27,28]; (**b**) total landslide area versus total rainfall, global data after [18], Italian cases after [15,27,28].

5.2. Limitations of the Inventory

Here, we highlight some critical aspects and limitations of our study, related either to the methodological mapping and analytical choices, or actual limitations in the input data.

In the first days after the 16–17 May rainfall episode, persistent cloud cover hampered the identification of landslides; the inventory is thus based on multitemporal images acquired on different dates. The high revisit time of the Planet images allows for closely bracketing the time of occurrence of the landslides. Figure 2d shows that no significant rainfall occurred after the 16–17 May episode; nevertheless, we cannot rule out that some slope movements may have been triggered a few days after the rainfall episode. The same reasoning applies to landslides that may have occurred in the first half of May (i.e., after the 1–2 May rainfall): Figure 4 shows that most of the movements have been triggered after the 16–17 May rainfall, but some landslides may have been mobilized before that episode. Overall, our inventory has to be intended as the cumulative effect of both the 1–2 and 16–17 May rainfall spells.

The inventory was released on 28 June 2023, i.e., 1.5 months after the main triggering event; in this sense, our inventory can be assimilated to rapid response products (e.g., [49]). Some operative choices represent a trade-off between the rapidity of execution and data quality/resolution, for instance, the mapping was realized at a 1:5000 scale and should be consulted at most at the same scale; a side effect is that our mapping is not recommended for higher resolution susceptibility studies, unless a field verification is undertaken. We underline that our inventory is one of the few that have been publicly released; to our knowledge only one other inventory has been published to date, realized through automatic mapping from Sentinel-1 images ([50]; publication date 1 June 2023).

The investigated AOI is 5764 km² wide, meaning that the inventory does not entirely cover the area hit by rainfall. However, we argue that the inventory effectively captures the worst hit area and represents a significant snapshot of the event.

We expect that our inventory may contain a certain number of false positives, especially in agricultural fields, built-up areas or close to riverbanks; indeed, in the post-event imagery, such areas show spectral characteristics similar to landslide areas (e.g., [10,11]). We also expect some completeness issues (false negatives) in the vicinity of roads and river networks: our limited field surveys allow us to ascertain the occurrence of a high number of slope movements affecting the road network. Figure 13 presents some representative cases of field photos and satellite images: the two areas shown in Figure 13a,b lie about 1 km apart on Province Road SP610 in the Castel del Rio Municipality. In the first case (Figure 13a,c,d), the slope movement is clearly visible in the satellite image and is included in the inventory; the landslide intercepts the road and flows down, reaching the Santerno River course. In the second case (Figure 13b,e), a small slope movement occurred along the road bank; this landslide is not visible in the 3 m resolution Planet image and thus is not included in our inventory. The latter example suggests that the abundance of slope movements within 200 m from roads (see Figure 8b) may be underestimated. The high influence of a dense road network as a predisposing factor for landslides has been already pointed out in previous studies (e.g., [51]).

Figure 13. Satellite images and field photographs of landslides impacting the road network along Province Road SP610 in the Castel del Rio Municipality: (**a,b**) Planet images acquired on 23 May, yellow dots are the locations where pictures were taken; (**c,d**) field photos of a landslide impacting the road, the deposit area lies in proximity of Santerno River; and (**e**) field photo of a smaller landslide, not recognizable on satellite imagery. All the photographs were shot by M. F. Ferrario on 30 July 2023.

Finally, we highlight that very limited field verification has been realized so far and no systematic evaluation of the reliability and accuracy of the inventory has been carried out. Indeed, the quality and reliability of the inventory control the overall quality of derivative products, such as the susceptibility and risk assessment [52,53]. Such aspects are beyond the scope of this paper but should be addressed whether the inventory presented here is used as input data for further studies.

6. Conclusions

In this paper, we present an inventory of 47,523 landslides triggered by heavy rainfall that occurred in May 2023 in the Emilia-Romagna and conterminous regions (Italy). A first precipitation episode occurred on 1–2 May 2023 and a second one on 16–17 May; the latter was responsible for most of the observed landslides.

The inventory was realized by a visual inspection of the pre- and post-event satellite imagery with a resolution of 3 m. The adopted methods are standard practice, and our effort is devoted to filling the data gap existing for the investigated triggering event. A limited field survey at selected locations was conducted to validate the inventory. We statistically investigate the relationship between landslide density and triggering (precipitation) and predisposing factors using the InfoVAL method. A strong influence is due to steep slopes and some land use categories (bare rocks and badlands, and woodlands), while a weaker positive correlation is found with respect to lithology because a higher landslide density is obtained for silty and flysch-like units.

The inventory presented here is one of the most numerous on a global scale and we argue that it can support the validation of other products, for instance, that are obtained through automatic mapping methods. The statistical results on triggering and predisposing factors may be useful for susceptibility assessment and land planning (e.g., landslide zoning) or the derivation of empirical rainfall thresholds for landslide triggering.

Author Contributions: Conceptualization, M.F.F. and F.L.; methodology, M.F.F. and F.L.; formal analysis, M.F.F. (realization of the inventory) and F.L. (analysis of influencing factors); investigation, M.F.F. and F.L.; writing—original draft preparation, M.F.F.; writing—review and editing, M.F.F. and F.L. All authors have read and agreed to the published version of the manuscript.

Funding: This research received no external funding.

Data Availability Statement: The landslide inventory realized in this study is publicly available at https://zenodo.org/record/8102429, last accessed 1 October 2023. Rainfall data retrieved from https://simc.arpae.it/dext3r/, last accessed 1 October 2023. Ancillary data for the Emilia-Romagna region can be downloaded at https://geoportale.regione.emilia-romagna.it/, last accessed 1 October 2023.

Acknowledgments: We wish to thank the Assistant Editor and three anonymous reviewers for their comments, which improved the quality of the manuscript.

Conflicts of Interest: The authors declare no conflicts of interest.

References

1. Cruden, D.M.; Varnes, D.J. Landslide types and processes. In *Landslides, Investigation and Mitigation, Special Report 247*; Turner, A.K., Schuster, R.L., Eds.; Transportation Research Board: Washington, DC, USA, 1996; pp. 36–75, ISBN 030906208X.
2. Hungr, O.; Leroueil, S.; Picarelli, L. The Varnes Classification of Landslide Types, an Update. *Landslides* **2014**, *11*, 167–194. [CrossRef]
3. Froude, M.J.; Petley, D.N. Global Fatal Landslide Occurrence from 2004 to 2016. *Nat. Hazards Earth Syst. Sci.* **2018**, *18*, 2161–2181. [CrossRef]
4. Haque, U.; Blum, P.; Da Silva, P.F.; Andersen, P.; Pilz, J.; Chalov, S.R.; Malet, J.-P.; Auflič, M.J.; Andres, N.; Poyiadji, E.; et al. Fatal Landslides in Europe. *Landslides* **2016**, *13*, 1545–1554. [CrossRef]
5. Salvati, P.; Bianchi, C.; Rossi, M.; Guzzetti, F. Societal Landslide and Flood Risk in Italy. *Nat. Hazards Earth Syst. Sci.* **2010**, *10*, 465–483. [CrossRef]

6. Haque, U.; Da Silva, P.F.; Devoli, G.; Pilz, J.; Zhao, B.; Khaloua, A.; Wilopo, W.; Andersen, P.; Lu, P.; Lee, J.; et al. The Human Cost of Global Warming: Deadly Landslides and Their Triggers (1995–2014). *Sci. Total Environ.* **2019**, *682*, 673–684. [CrossRef] [PubMed]
7. Guzzetti, F.; Mondini, A.C.; Cardinali, M.; Fiorucci, F.; Santangelo, M.; Chang, K.-T. Landslide Inventory Maps: New Tools for an Old Problem. *Earth-Sci. Rev.* **2012**, *112*, 42–66. [CrossRef]
8. Fell, R.; Corominas, J.; Bonnard, C.; Cascini, L.; Leroi, E.; Savage, W.Z. Guidelines for Landslide Susceptibility, Hazard and Risk Zoning for Land Use Planning. *Eng. Geol.* **2008**, *102*, 85–98. [CrossRef]
9. Corominas, J.; VanWesten, C.; Frattini, P.; Cascini, L.; Malet, J.-P.; Fotopoulou, S.; Smith, J. Recommendations for the quantitative analysis of landslide risk. *Bull. Eng. Geol. Environ.* **2014**, *73*, 209–263. [CrossRef]
10. Martha, T.R.; Kerle, N.; Jetten, V.; Van Westen, C.J.; Kumar, K.V. Characterising Spectral, Spatial and Morphometric Properties of Landslides for Semi-Automatic Detection Using Object-Oriented Methods. *Geomorphology* **2010**, *116*, 24–36. [CrossRef]
11. Mondini, A.C.; Chang, K.-T.; Yin, H.-Y. Combining Multiple Change Detection Indices for Mapping Landslides Triggered by Typhoons. *Geomorphology* **2011**, *134*, 440–451. [CrossRef]
12. Li, Z.; Shi, W.; Myint, S.W.; Lu, P.; Wang, Q. Semi-Automated Landslide Inventory Mapping from Bitemporal Aerial Photographs Using Change Detection and Level Set Method. *Remote Sens. Environ.* **2016**, *175*, 215–230. [CrossRef]
13. Scaioni, M.; Longoni, L.; Melillo, V.; Papini, M. Remote Sensing for Landslide Investigations: An Overview of Recent Achievements and Perspectives. *Remote Sens.* **2014**, *6*, 9600–9652. [CrossRef]
14. De Vita, P.; Reichenbach, P.; Bathurst, J.C.; Borga, M.; Crosta, G.; Crozier, M.; Glade, T.; Guzzetti, F.; Hansen, A.; Wasowski, J. Rainfall-Triggered Landslides: A Reference List. *Environ. Geol.* **1998**, *35*, 219–233. [CrossRef]
15. Guzzetti, F.; Cardinali, M.; Reichenbach, P.; Cipolla, F.; Sebastiani, C.; Galli, M.; Salvati, P. Landslides triggered by the 23 November 2000 rainfall event in the Imperia Province, Western Liguria, Italy. *Eng. Geol.* **2004**, *73*, 229–245. [CrossRef]
16. Kirschbaum, D.B.; Adler, R.; Hong, Y.; Hill, S.; Lerner-Lam, A. A Global Landslide Catalog for Hazard Applications: Method, Results, and Limitations. *Nat. Hazards* **2010**, *52*, 561–575. [CrossRef]
17. Tsai, F.; Hwang, J.-H.; Chen, L.-C.; Lin, T.-H. Post-Disaster Assessment of Landslides in Southern Taiwan after 2009 Typhoon Morakot Using Remote Sensing and Spatial Analysis. *Nat. Hazards Earth Syst. Sci.* **2010**, *10*, 2179–2190. [CrossRef]
18. Marc, O.; Stumpf, A.; Malet, J.-P.; Gosset, M.; Uchida, T.; Chiang, S.-H. Initial Insights from a Global Database of Rainfall-Induced Landslide Inventories: The Weak Influence of Slope and Strong Influence of Total Storm Rainfall. *Earth Surf. Dyn.* **2018**, *6*, 903–922. [CrossRef]
19. Liang, X.; Segoni, S.; Yin, K.; Du, J.; Chai, B.; Tofani, V.; Casagli, N. Characteristics of Landslides and Debris Flows Triggered by Extreme Rainfall in Daoshi Town during the 2019 Typhoon Lekima, Zhejiang Province, China. *Landslides* **2022**, *19*, 1735–1749. [CrossRef]
20. Ma, S.; Shao, X.; Xu, C. Characterizing the Distribution Pattern and a Physically Based Susceptibility Assessment of Shallow Landslides Triggered by the 2019 Heavy Rainfall Event in Longchuan County, Guangdong Province, China. *Remote Sens.* **2022**, *14*, 4257. [CrossRef]
21. Ma, S.; Shao, X.; Xu, C. Landslides Triggered by the 2016 Heavy Rainfall Event in Sanming, Fujian Province: Distribution Pattern Analysis and Spatio-Temporal Susceptibility Assessment. *Remote Sens.* **2023**, *15*, 2738. [CrossRef]
22. Burrows, K.; Marc, O.; Andermann, C. Retrieval of Monsoon Landslide Timings with Sentinel-1 Reveals the Effects of Earthquakes and Extreme Rainfall. *Geophys. Res. Lett.* **2023**, *50*, e2023GL104720. [CrossRef]
23. Emberson, R.; Kirschbaum, D.B.; Amatya, P.; Tanyas, H.; Marc, O. Insights from the Topographic Characteristics of a Large Global Catalog of Rainfall-Induced Landslide Event Inventories. *Nat. Hazards Earth Syst. Sci.* **2022**, *22*, 1129–1149. [CrossRef]
24. Roy, P.; Martha, T.R.; Vinod Kumar, K.; Chauhan, P.; Rao, V.V. Cluster Landslides and Associated Damage in the Dima Hasao District of Assam, India Due to Heavy Rainfall in May 2022. *Landslides* **2023**, *20*, 97–109. [CrossRef]
25. Govi, M. Carta delle frane prodotte dal terremoto (Map showing landslides triggered by the earthquake). *Riv. Ital. Paleontogia E Stratigr.* **1977**, *83*, Plate 1.
26. Guzzetti, F.; Malamud, B.D.; Turcotte, D.L.; Reichenbach, P. Power-Law Correlations of Landslide Areas in Central Italy. *Earth Planet. Sci. Lett.* **2002**, *195*, 169–183. [CrossRef]
27. Cardinali, M.; Galli, M.; Guzzetti, F.; Ardizzone, F.; Reichenbach, P.; Bartoccini, P. Rainfall Induced Landslides in December 2004 in South-Western Umbria, Central Italy: Types, Extent, Damage and Risk Assessment. *Nat. Hazards Earth Syst. Sci.* **2006**, *6*, 237–260. [CrossRef]
28. Donnini, M.; Santangelo, M.; Gariano, S.L.; Bucci, F.; Peruccacci, S.; Alvioli, M.; Althuwaynee, O.; Ardizzone, F.; Bianchi, C.; Bornaetxea, T.; et al. Landslides Triggered by an Extraordinary Rainfall Event in Central Italy on September 15, 2022. *Landslides* **2023**, *20*, 2199–2211. [CrossRef]
29. Santangelo, M.; Althuwaynee, O.; Alvioli, M.; Ardizzone, F.; Bianchi, C.; Bornaetxea, T.; Brunetti, M.T.; Bucci, F.; Cardinali, M.; Donnini, M.; et al. An Inventory of Landslides Triggered by an Extreme Rainfall Event in Marche-Umbria, Italy, on 15 September 2022. *Sci. Data* **2023**, *10*, 427. [CrossRef]
30. Schmitt, R.G.; Tanyas, H.; Jessee, M.A.N.; Zhu, J.; Biegel, K.M.; Allstadt, K.E.; Jibson, R.W.; Thompson, E.M.; van Westen, C.J.; Sato, H.P.; et al. *An Open Repository of Earthquake Triggered Ground-Failure Inventories*; US Geological Survey: Reston, VA, USA, 2017. [CrossRef]

31. Arrighi, C.; Domeneghetti, A. Brief communication: On the environmental impacts of 2023 flood in Emilia-Romagna. *Nat. Hazards Earth Syst. Sci.* **2023**. [CrossRef]
32. Conti, P.; Cornamusini, G.; Carmignani, L. An Outline of the Geology of the Northern Apennines (Italy), with Geological Map at 1:250,000 Scale. *IJG* **2020**, *139*, 149–194. [CrossRef]
33. Borgatti, L.; Tosatti, G. Slope Instability Processes Affecting the Pietra Di Bismantova Geosite (Northern Apennines, Italy). *Geoheritage* **2010**, *2*, 155–168. [CrossRef]
34. Iadanza, C.; Trigila, A.; Starace, P.; Dragoni, A.; Biondo, T.; Roccisano, M. IdroGEO: A Collaborative Web Mapping Application Based on REST API Services and Open Data on Landslides and Floods in Italy. *ISPRS Int. J. Geo-Inf.* **2021**, *10*, 89. [CrossRef]
35. Bossard, M.; Feranec, J.; Otahel, J. *CORINE Land Cover Technical Guide: Addendum 2000*; European Environment Agency: Copenhagen, Denmark, 2000; Volume 40.
36. ARPAE. *Atlante Climatico dell'Emilia-Romagna 1961–2015*; ARPAE: Washington, DC, USA, 2017; ISBN 978-88-87854-44-2. Available online: https://www.arpae.it/dettaglio_generale.asp?id=3811&idlivello=1591 (accessed on 1 September 2023).
37. ARPAE. *L'evento Meteo Idrogeologico del 1–4 Maggio*; ARPAE: Bologna, Italy, 2023. Available online: https://www.arpae.it/it/notizie/levento-meteo-idrogeologico-del-1-4-maggio (accessed on 25 August 2023).
38. ARPAE-SIMC. *L'evento Meteo Idrogeologico e Idraulico del 16–18 Maggio 2023*; ARPAE-SIMC: Bologna, Italy, 2023. Available online: https://www.arpae.it/it/notizie/levento-meteo-idrogeologico-del-16-18-maggio-2023 (accessed on 25 August 2023).
39. Marc, O.; Hovius, N. Amalgamation in Landslide Maps: Effects and Automatic Detection. *Nat. Hazards Earth Syst. Sci.* **2015**, *15*, 723–733. [CrossRef]
40. Malamud, B.D.; Turcotte, D.L.; Guzzetti, F.; Reichenbach, P. Landslides, earthquakes, and erosion. *Earth Planet. Sci. Lett.* **2004**, *229*, 45–59. [CrossRef]
41. Danese, M.; Lazzari, M. A kernel density estimation approach for landslide susceptibility assessment. In *Mountain Risks: Bringing Science to Society, Proceedings of the International Conference of Mountain Risks, Firenze, Italy, 24–26 November 2010*; CERG: Strasbourg, Italy, 2010; pp. 24–26.
42. Robinson, T.R.; Rosser, N.J.; Densmore, A.L.; Williams, J.G.; Kincey, M.E.; Benjamin, J.; Bell, H.J.A. Rapid post-earthquake modelling of coseismic landslide intensity and distribution for emergency response decision support. *Nat. Hazards Earth Syst. Sci.* **2017**, *17*, 1521–1540. [CrossRef]
43. Silverman, B.W. *Density Estimation for Statistics and Data Analysis*; Routledge: London, UK, 2018.
44. Epanechnikov, V.A. Non-parametric estimation of a multivariate probability density. *Theory Probab. Its Appl.* **1969**, *14*, 153–158. [CrossRef]
45. Van Westen, C.J. Statistical landslide hazard analysis. In *Application Guide, ILWIS 2.1 for Windows*; ITC: Enschede, The Netherlands, 1997; pp. 73–84.
46. Van Westen, C.J. The Modelling of Landslide Hazards Using GIS. *Surv. Geophys.* **2000**, *21*, 241–255. [CrossRef]
47. Yin, K.L.; Yan, T.Z. Statistical prediction model for slope instability of metamorphosed rocks. In *Landslides-Glissements de Terrain. Proceedings of the V International Symposium on Landslides, Lausanne, Switzerland, 10–15 July 1988*; A. A. Balkema: Rotterdam, The Netherlands, 1988; Volume 2, pp. 1269–1272.
48. Tanyaş, H.; Allstadt, K.E.; Van Westen, C.J. An Updated Method for Estimating Landslide-event Magnitude. *Earth Surf. Process. Landf.* **2018**, *43*, 1836–1847. [CrossRef]
49. Amatya, P.; Scheip, C.; Déprez, A.; Malet, J.-P.; Slaughter, S.L.; Handwerger, A.L.; Emberson, R.; Kirschbaum, D.; Jean-Baptiste, J.; Huang, M.-H.; et al. Learnings from Rapid Response Efforts to Remotely Detect Landslides Triggered by the August 2021 Nippes Earthquake and Tropical Storm Grace in Haiti. *Nat. Hazards* **2023**, *118*, 2337–2375. [CrossRef]
50. Notti, D.; Cignetti, M.; Cardone, D.; Godone, D.; Giordan, D. Rapid Mapping of Potential Ground Effects of the May 2023 Emilia-Romagna Rainstorms. Available online: https://www.researchgate.net/publication/371292106_Rapid_mapping_of_potential_ground_effects_of_the_May_2023_Emilia-Romagna_rainstorms (accessed on 25 August 2023).
51. Tanyaş, H.; Görüm, T.; Kirschbaum, D.; Lombardo, L. Could road constructions be more hazardous than an earthquake in terms of mass movement? *Nat. Hazards* **2022**, *112*, 639–663. [CrossRef]
52. Galli, M.; Ardizzone, F.; Cardinali, M.; Guzzetti, F.; Reichenbach, P. Comparing Landslide Inventory Maps. *Geomorphology* **2008**, *94*, 268–289. [CrossRef]
53. Pellicani, R.; Spilotro, G. Evaluating the Quality of Landslide Inventory Maps: Comparison between Archive and Surveyed Inventories for the Daunia Region (Apulia, Southern Italy). *Bull. Eng. Geol. Environ.* **2015**, *74*, 357–367. [CrossRef]

Disclaimer/Publisher's Note: The statements, opinions and data contained in all publications are solely those of the individual author(s) and contributor(s) and not of MDPI and/or the editor(s). MDPI and/or the editor(s) disclaim responsibility for any injury to people or property resulting from any ideas, methods, instructions or products referred to in the content.

MDPI
St. Alban-Anlage 66
4052 Basel
Switzerland
www.mdpi.com

Remote Sensing Editorial Office
E-mail: remotesensing@mdpi.com
www.mdpi.com/journal/remotesensing

Disclaimer/Publisher's Note: The statements, opinions and data contained in all publications are solely those of the individual author(s) and contributor(s) and not of MDPI and/or the editor(s). MDPI and/or the editor(s) disclaim responsibility for any injury to people or property resulting from any ideas, methods, instructions or products referred to in the content.

www.ingramcontent.com/pod-product-compliance
Lightning Source LLC
LaVergne TN
LVHW070220100526
838202LV00015B/2068